W9-BEJ-287

WARM AIR HEATING FOR CLIMATE CONTROL

WARM AIR HEATING FOR CLIMATE CONTROL

WILLIAM B. COOPER and **RAYMOND E. LEE**

Professors, Mechanical Technology
Macomb County Community College

RAYMOND A. QUINLAN

General Supervisor of Mechanical Facilities
General Motors Corporation Technical Center

PRENTICE-HALL, INC., *Englewood Cliffs, New Jersey 07632*

Library of Congress Cataloging in Publication Data

Cooper, William B.
Warm air heating for climate control.

Includes index.
1. Hot-air heating. 2. Dwellings—Heating and ventilation. I. Lee, Raymond E., joint author. II. Quinlan, Raymond A., joint author. II. Title.
TH7601.L38 697'.3 79-25156
ISBN 0-13-944231-6

Editorial production supervision
and interior design by: James M. Chege
Page Layout by: Gail Cocker
Cover design by: Edsal Enterprises
Manufacturing buyer: Anthony Caruso

Printed in the United States of America

10 9 8 7 6 5 4 3 2 1

PRENTICE-HALL INTERNATIONAL, INC., *London*
PRENTICE-HALL OF AUSTRALIA PTY. LIMITED, *Sydney*
PRENTICE-HALL OF CANADA, LTD., *Toronto*
PRENTICE-HALL OF INDIA PRIVATE LIMITED, *New Delhi*
PRENTICE-HALL OF JAPAN, INC., *Tokyo*
PRENTICE-HALL OF SOUTHEAST ASIA PTE. LTD., *Singapore*
WHITEHALL BOOKS LIMITED, *Wellington, New Zealand*

CONTENTS

PREFACE

The forced warm air heating system is the most popular. Millions of homes use this type of equipment. Like any other mechanism, an essential element for its satisfactory use is proper installation, maintenance, and service. Large numbers of trained people are required for this work.

Training requires good information. The authors of this book have attempted to place in one text as much data as possible to assist in training heating mechanics and technicians.

The first two chapters define comfort and explain heating technology. Information is supplied to indicate what conditions are desirable for warm air heating systems. Chapters three and four provide a quick method for checking the heating load and air distribution system. This is not for design purposes but rather to determine if the equipment has been properly selected and applied. Chapters seven, eight, and nine cover the parts of furnaces that are involved in installations, maintenance, and service. Some parts are common to all furnaces, while others are used only on specific types.

Combustion testing is covered in Chapter 10. This is extremely important, now that fossil fuels are becoming increasingly scarce and expensive. Chapters 11 through 19 deal with the electrical and control aspects of warm air furnaces. The major portion of service is concerned with electrical problems. Being able to understand wiring diagrams is important and the use of test instruments is absolutely essential. Electrical problems increase as manufacturers add accessories and energy conservation devices. In this text, information is supplied on a large number of manufactured products to better equip the service personnel to deal with field conditions.

Chapter 20 supplies information on air balancing and control adjustment, both of which are essential for a good operating system. Information specifically directed toward energy conservation and supplementary solar heating is supplied in Chapters 21 and 22.

No attempt has been made to adequately cover the service of residential cooling equipment. The authors believe that this requires treatment in a separate course of study.

WILLIAM B. COOPER
RAYMOND E. LEE
RAYMOND A. QUINLAN

Warren, Michigan

WARM AIR HEATING FOR CLIMATE CONTROL

1

CLIMATE

OBJECTIVES

After studying this chapter, the student will be able to:

- Define the common terms used in the text pertaining to warm air heating for climate control
- Determine the Btu added to a substance by the application of heat
- Convert Fahrenheit temperatures to Celsius temperatures
- Determine the relative humidity from dry and wet bulb temperatures

INDOOR CLIMATE

Climate is the condition of the weather. Weather refers to the character of the atmosphere. It may be hot or cold, wet or dry, calm or windy. This text is concerned with indoor climate. Broadly speaking indoor climate includes conditions of temperature, humidity, air motion, air cleanliness, noise or sound levels, odors, even conditions of some materials that occupy or surround the space.

The fundamental concepts that will be discussed in this unit are:

- Temperature
- Heat
- Humidity
- Air motion
- Laws of thermodynamics

Since people engaged in the heating business are responsible for controlling indoor climate conditions, this text discusses the means that are available to produce changes in the indoor climate. One of these is the addition or removal of heat to produce a temperature change.

TEMPERATURE

Temperature is a relative term that can be defined as something that is responsible for the sensations of hot and cold. Temperature can be accurately measured using a thermometer.

Thermometers

A *thermometer* is a device for measuring temperature. There are many shapes and sizes of thermometers available using a variety of scales. However, in general, two types of thermometers are most frequently used. One type is used in the British or English system and is known as the Fahrenheit thermometer. The other type is used in the metric system and is called the Celsius thermometer. Both of these thermometers measure temperature, but the scales are different. A comparison of the two thermometer scales is shown in Figure 1-1.

The scale selected for the Celsius thermometer is the easier of the two to understand. Zero degrees on the Celsius scale refers to the freezing point of water and 100 degrees refers to the boiling point of water. There are 100 equal divisions between the zero point and the 100-degree mark on the scale. On the Fahrenheit scale, the freezing point of water is 32 degrees and the boiling point of water is 212 degrees. The Fahrenheit thermometer has 180 equal divisions between the freezing point of water and its boiling point.

It is important to understand both of these thermometer scales. Some temperature information is given in Fahrenheit degrees and some in Celsius degrees. To prevent confusion about which scale is involved, a capital C is placed after the number of Celsius degrees and a capital F is placed after the

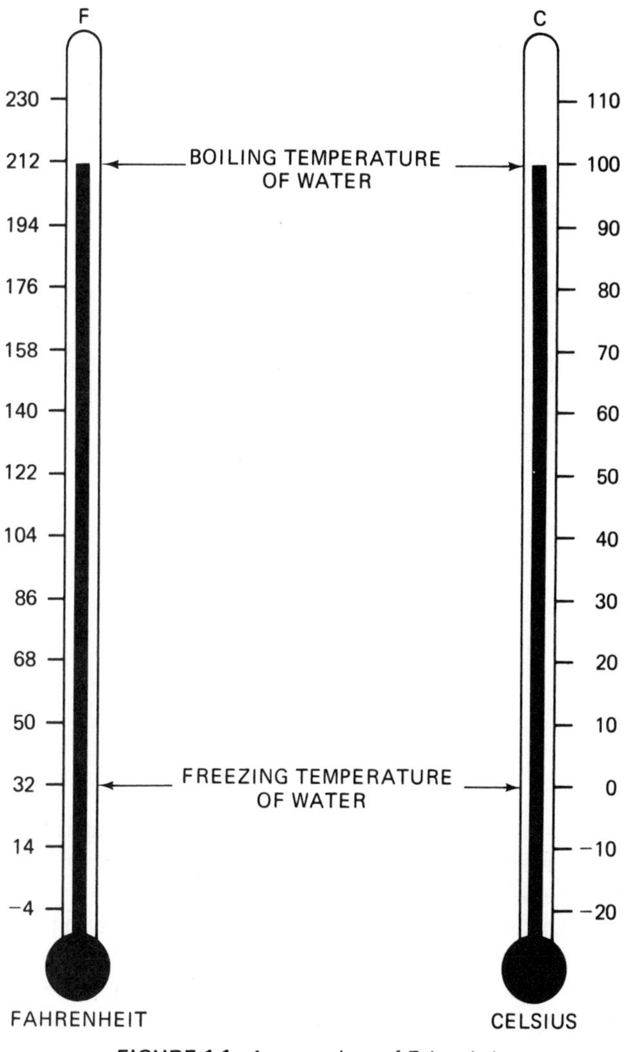

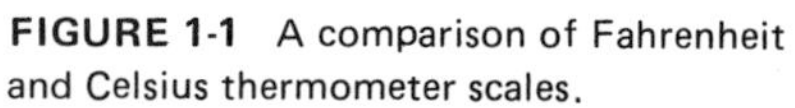

FIGURE 1-1 A comparison of Fahrenheit and Celsius thermometer scales.

number of Fahrenheit degrees. For example, 40° C means 40 degrees Celsius; 40° F means 40 degrees Fahrenheit.

When working continuously with one scale or the other, there is no problem in recording and using the information. However, there are times when it is necessary to change from one scale to the other. Such *conversions* can be done by using the following formulas:

$$\text{Degrees Fahrenheit} = (9/5 \times \text{degrees Celsius}) + 32 \text{ degrees}$$

$$\text{Degrees Celsius} = 5/9 \times (\text{degrees Fahrenheit} - 32 \text{ degrees})$$

EXAMPLE: 70°F is converted to Celsius degrees as follows:

$$
\begin{aligned}
°C &= 5/9 \times (°F - 32°) \\
&= 5/9 \times (70° - 32°) \\
&= 5/9 \times (38) \\
&= \frac{190}{9} \\
&= 21° \text{ Celsius (C)}
\end{aligned}
$$

EXAMPLE: 27°C is converted to Fahrenheit degrees as follows:

$$
\begin{aligned}
°F &= (9/5 \times C°) + 32° \\
&= (9/5 \times 27°) + 32° \\
&= \left(\frac{243}{5}\right) + 32° \\
&= 49 + 32 \\
&= 81° \text{ Fahrenheit (F)}
\end{aligned}
$$

HEAT

Heat is the quality that causes an increase in the temperature of a body when it is added, or a decrease in the temperature of a body when it is removed, provided that there is no change in state in the process.

The stipulation "no change in state" requires some explanation. A substance changes its state when it changes its physical form. For example, when a solid changes to a liquid, that is a change in state. When a liquid changes to a vapor, that is also a change in state. *Sublimation* is the change in state from a solid to a vapor without passing through the liquid state. A good example of this is the evaporation of mothballs or the evaporation of dry ice (solid carbon dioxide).

The unit of measurement of heat in the British system is the *British thermal unit* (*Btu*). One Btu is the amount of heat required to raise 1 pound of water 1 degree Fahrenheit, as shown in Figure 1-2.

In the original metric system the unit of heat was the *kilocalorie* (*kcal*). This is the amount of heat required to raise the temperature of 1 *kilogram* (*kg*), or 1000 *grams* (*g*), of water 1 degree Celsius (1°C). This unit of heat is still in use. However, the present metric unit or International System (SI) unit for heat, comparable to a Btu, is the joule (J). The joule is defined in terms of work, being equivalent to the movement of a force of one neuton

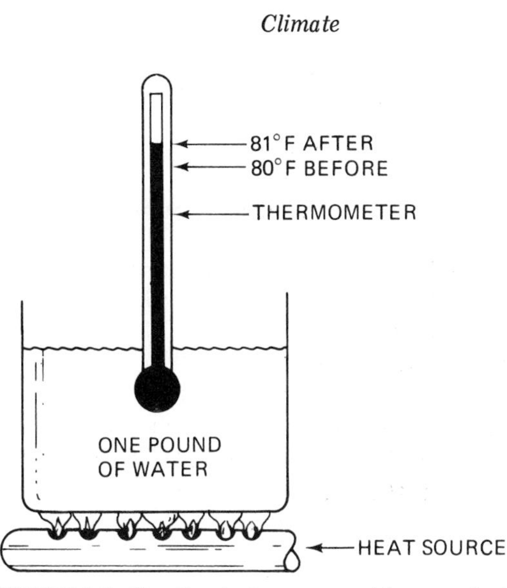

FIGURE 1-2 One Btu is the amount of heat required to raise one pound of water one degree Fahrenheit.

for a distance of 1 meter. When the unit of heat includes a time element comparable to Btuh, the metric unit is watts (W), which is equivalent to joules per second.

It is important in most cases to note the rate of heat change. How fast or how slow does a piece of heating equipment produce heat? Therefore, a more precise term is *Btu per hour* (*Btuh*). Sometimes figures get quite large, so, using *m* to represent 1000, the abbreviation *MBh* is used to represent *thousands of Btu per hour.* For example, a furnace may have an output of 100,000 Btu per hour, or 100 MBh.

Types of Heat

There are two types of heat: one type changes the temperature of a substance and the other changes its state. The type that changes the temperature is called *sensible heat.* The type that changes the state is called *latent heat.*

The formula used to determine the sensible heat added or removed from a substance is

$$Q = W \times \text{SH} \times \text{TD}$$

where: Q = quantity of heat, Btu

W = weight, lb

SH = specific heat, Btu/lb/°F

TD = temperature difference or change, °F

The formula used to determine the change in latent heat is:

$$Q = W \times L$$

where: L is the change in latent heat per pound.

When heat is added to a solid to change it to a liquid, the latent heat is called *heat of melting.* When a liquid is changed to a solid, the latent heat is called *heat of fusion.*

When a liquid is changed to a vapor, the latent heat is called *heat of vaporization.* When a vapor is changed to a liquid, the latent heat is called *heat of condensation.* Latent heat is always related to a change in state.

A good example to indicate the effects of heat is its reaction on water, as shown in Figure 1-3. In this figure, ice is taken at 0° F and heated to steam. Note that there are five parts to the diagram, and the following is a description of what takes place.

PART 1:

Ice at 0° F is heated to 32° F; 16 Btu is added. The amount of heat required is found by substituting values in the sensible heat formula.

$$\begin{aligned} Q &= 1 \text{ lb} \times 0.5 \text{ Btu/lb/°F} \times (32\text{°F} - 0\text{°F}) \\ &= 1 \times 0.5 \times 32 \\ &= 16 \text{ Btu} \end{aligned}$$

PART 2:

Ice at 32° F is changed to water at 32° F. Since the heat of melting of water is 144 Btu/lb, and using the latent heat formula, 144 Btu is added:

$$\begin{aligned} Q &= 1 \text{ lb} \times 144 \text{ Btu} \\ &= 144 \text{ Btu} \end{aligned}$$

PART 3:

Water at 32° F is heated to 212° F, by adding 180 Btu. Using the formula and substituting in the values yields

$$\begin{aligned} Q &= 1 \text{ lb} \times 1.0 \text{ Btu/lb/°F} \times (212\text{°F} - 32\text{°F}) \\ &= 1 \times 1 \times 180 \\ &= 180 \text{ Btu} \end{aligned}$$

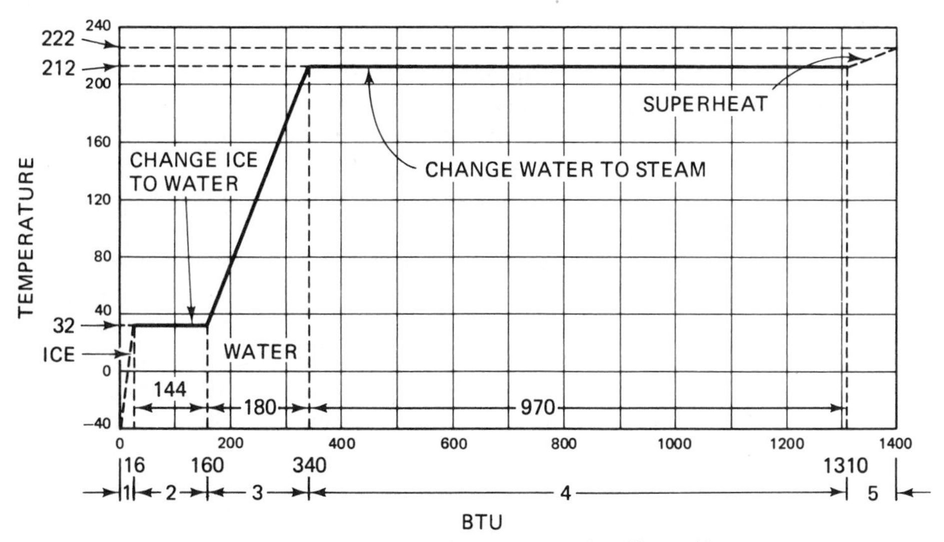

FIGURE 1-3 A temperature heat diagram, showing the effects of heat on water.

PART 4:

Water at 212° F is changed to steam by adding 970 Btu. Since the heat of vaporization of water is 970 Btu/lb, using this formula we obtain

$$Q = 1 \text{ lb} \times 970 \text{ Btu (latent heat)}$$
$$= 970 \text{ Btu}$$

PART 5:

The steam is superheated. Superheated means heating steam (vapor) above the boiling point. If the steam is superheated 10° F, the amount of heat added is 5 Btu. Since the specific heat of steam is 0.48 Btu/lb/° F, the amount of heat added is determined by the formula

$$Q = 1 \text{ lb} \times 0.48 \text{ Btu/lb/}^\circ\text{F} \times (222^\circ\text{F} - 212^\circ\text{F})$$
$$= 1 \times 0.48 \times 10$$
$$= 5 \text{ Btu (rounded off to the nearest whole number)}$$

Both heat of fusion and heat of vaporization are reversible processes. The change in state can occur in either direction. Water can be changed to ice, or ice to water. The amount of heat added or subtracted is the same in either case. Tables are available giving the heat of fusion and the heat of vaporization for various substances.

SPECIFIC HEAT:

When most substances are heated, they absorb heat, thereby raising their temperatures, except where a change in state occurs. The temperature of the substance increases rapidly or slowly, depending on the nature of the material. Since the standard heating unit is defined in terms of the temperature rise of water, it is fitting to use water as a basis for comparison for the heat-absorbing quality of other substances.

The definition of specific heat is much like the definition of Btu or kcal. The specific heat of a substance is the amount of heat required to raise 1 pound of a substance 1 degree Fahrenheit (1°F); or the amount of heat required to raise 1 kilogram of a substance 1 degree Celsius (1°C). Note that for both of these definitions the specific heat of water is 1.0. Thus, specific heat values are the same for both the British system and the metric system, as shown in Figure 1-4.

These specific heat values make it easy to calculate the amount of heat added to or removed from a substance when the temperature rise or drop is known. The amount of heat required is equal to the weight of the substance, times the specific heat, times the rise in temperature in degrees Fahrenheit.

FIGURE 1-4 A table of the specific heat values for some common substances.

SPECIFIC HEAT VALUES

MATERIAL	SPECIFIC HEAT BTU/LB./DEG F (KCAL/KG/DEG C)
WATER	1.00
ICE	0.50
AIR (DRY)	0.24
STEAM	0.48
ALUMINUM	0.22
BRICK	0.20
CONCRETE	0.16
COPPER	0.09
GLASS	0.20
IRON	0.10
WOOD (HARD)	0.45
WOOD (PINE)	0.67

THESE VALUES MAY BE USED FOR COMPUTATIONS WHICH INVOLVE NO CHANGE OF STATE.

EXAMPLE: How much heat is required to heat 10 lb of aluminum from 50°F to 60°F?

SOLUTION: Btu = pounds of aluminum × specific heat × °F temperature rise

$$= 10 \text{ lb} \times 0.22 \text{ Btu/lb/°F} \times (60°\text{F} - 50°\text{F})$$

(Note from Figure 1-4 that the specific heat of aluminum is 0.22 Btu/lb.)

$$\text{Btu} = 10 \times 0.22 \times 10$$

$$= 22$$

EXAMPLE: How much heat is required to raise 100 kg of copper from 5°C to 25°C?

SOLUTION: kcal = kg of copper × specific heat × °C temperature rise

$$= 100 \text{ kg} \times 0.09 \text{ kcal/kg/°C} \times (25°\text{C} - 5°\text{C})$$

(Note from table in Figure 1-4 that the specific heat of copper is 0.09 kcal/kg/°C.)

$$\text{kcal} = 100 \times 0.09 \times 20$$

$$= 180$$

TRANSFER OF HEAT:

Three different ways to transfer heat are by convection, conduction, and radiation, as shown in Figure 1-5. Modern heating plants make use of all three ways to transfer heat.

Convection is the circulatory motion in air due to the warmer portions rising, the denser cooler portions sinking. For convection to take place, there must be a difference in temperature between the source of heat and the surrounding air. The greater the difference in temperature, the greater the movement of air by convection. The greater the movement of air, the greater the transfer of heat.

Conduction is the flow of heat from one part of a material to another part in direct contact with it. The rate at which a material transmits heat is known as its conductivity. The amount of heat transmitted by conduction through a material is determined by the surface area of the material, the thickness, the temperature difference between two surfaces, and the conductivity.

Radiation is the transfer of heat through space by wave motion. Heat passes from one object to another without warming the space in between. The amount of heat transferred by radiation depends upon the area of the radiating body, the temperature difference, and the distance between the source of heat and the object being heated.

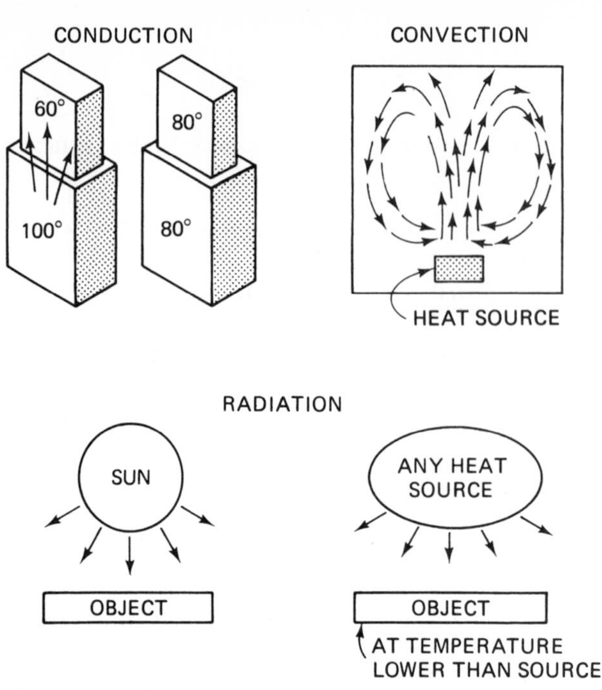

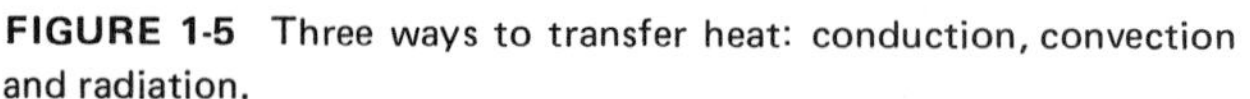

FIGURE 1-5 Three ways to transfer heat: conduction, convection and radiation.

HUMIDITY

Humidity is the amount of water vapor within a given space. There are two types of humidity: absolute and relative.

Absolute humidity is the weight of water vapor per unit of volume. *Relative humidity* is the ratio of the weight of water vapor in 1 pound (or in 1 kg) of dry air compared to the maximum amount of water vapor 1 pound (or 1 kg) of air will hold at a given temperature, expressed in percent.

In a heating system, during the winter months it is usually desirable to increase the percentage of humidity within the space being conditioned. For every pound of water that is evaporated, approximately 970 Btu is required.

Humidity is measured using a *sling psychrometer*, an instrument for determining the moisture content of air. The use of a sling psychrometer will be discussed in Chapter 2.

AIR MOTION AND ITS MEASUREMENT

Air motion means changing the position of air. Most heating systems transfer warm air to cold areas by the movement of air. As in other aspects of indoor climate, the control of air movement is a necessary factor in a satisfactory heating system.

The velocity (speed) of air can be measured by instruments such as an *anemometer.* The unit of measurement of indoor velocities is usually feet per minute (fpm) in the English system and meters per second (m/s) in the metric system.

LAWS OF THERMODYNAMICS

Thermodynamics is the science that deals with the relationship between heat and mechanical energy. Energy is the ability to do work. If a force is applied to an object moving it a given distance, work has been performed. Heat is a form of energy. Other forms of energy include light, chemical, mechanical, and electrical.

Since heat is a form of energy, it follows the natural laws that relate to energy. These laws are useful in the study of heating. From the laws of thermodynamics, two helpful facts are derived:

1. Energy can be neither created nor destroyed, but it can be converted from one form of energy to another
2. Heat flows from hot to cold

An example of the conversion of energy is the changing of electrical energy to heat in an electric heating unit. An example of heat flowing from hot to cold is illustrated in Figure 1-6.

FIGURE 1-6 Heat flows from hot to cold.

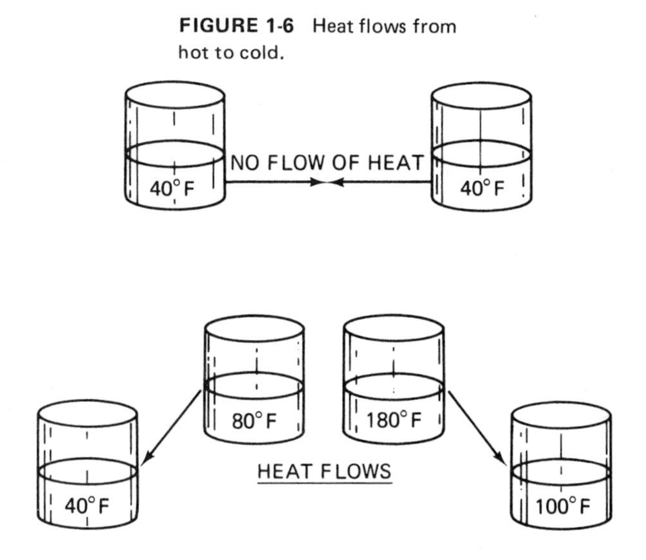

REVIEW QUESTIONS

Select the letter representing your choice of the correct answer.

1-1. Zero on the Celsius thermometer is the same as what reading on the Fahrenheit scale?

(a) 32°F

(b) 100°F

(c) 212°F

(d) 0°F

1-2. Using the conversion formula, -40°F would be equal to how many degrees on the Celsius scale?

(a) +60°C

(b) +10°C

(c) -40°C

(d) -60°C

1-3. A good example of change in state is:

(a) Heating water from 32°F to 212°F

(b) Changing water from 32° ice to 32° liquid

(c) Heating steam at 212°F above its boiling point at atmospheric pressure

(d) Changing ice at 0°F to ice at 32°F

1-4. The type of heat added in changing water from 32°F to water at 212°F is:

(a) Critical heat

(b) Latent heat

(c) Sensible heat

(d) Heat of fusion

1-5. The amount of heat required to raise 10 lb of water from 40°F to 100°F is:

(a) 40 Btu

(b) 400 Btu

(c) 60 Btu

(d) 600 Btu

1-6. A good example of sublimation is the evaporation of:

(a) Water

(b) Dry ice

(c) Milk

(d) Steam

1-7. Water has a specific heat of 1.0 Btu/lb/°F. The amount of heat required to raise the temperature of 5 lb of water from 10°F to 20°F is:

(a) 10 Btu

(b) 4.8 Btu

(c) 50 Btu

(d) 100 Btu

1-8. The specific heat of water vapor is:

(a) 0.50

(b) 1.00

(c) 1.15

(d) 2.00

1-9. The measurement of indoor velocities is usually in units of:

(a) Miles per minute

(b) Feet per minute

(c) Meters per hour

(d) Centimeters per second

1-10. When heat flows upward from a register, heat is transferred by:

(a) Vaporization

(b) Convection

(c) Conduction

(d) Sublimation

2

COMFORT

OBJECTIVES

After studying this chapter, the student will be able to:

- Describe the conditions, produced by the warm air heating system, that are necessary for human comfort

CONDITIONS THAT AFFECT COMFORT

Comfort is the absence of disturbing or distressing conditions—it is a feeling of contentment with the environment.

The study of human comfort concerns itself with two aspects:

1. How the body functions in respect to heat
2. How the area around a person affects the feeling of comfort

HUMAN REQUIREMENTS

The body can be compared to a heat engine. A heat engine has three characteristics:

1. It consumes fuel
2. It performs work
3. It dissipates heat

The body consumes fuel, in the form of food, which produces energy to perform work. The body also dissipates heat to the surrounding atmosphere.

The body has a unique characteristic in that it maintains a closely regulated internal body temperature, 97 to 100°F, (36 to 38°C). Any excess energy that is not used to produce work or to perform essential body functions is expelled to the atmosphere. It is, therefore, important that any loss of body heat be at the proper rate to maintain body temperature. If the body loses heat too fast, a person has the sensation of being cold. If the body loses heat too slowly, a person has the sensation of being too warm.

There are two additional factors that must be taken into consideration:

1. The amount of activity of a person. The greater the activity, the more heat that is produced by the body which must be dissipated to the atmosphere (Figure 2-1).
2. The amount of clothing worn by a person. Clothing is a form of in-

FIGURE 2-1 Heat from people.

ACTIVITY	TOTAL HEAT ADJUSTED* BTUH	SENSIBLE HEAT BTUH	LATENT HEAT BTUH
SCHOOL	420	230	190
OFFICE	510	255	255
LIGHT WORK	640	315	325
DANCING	1280	405	875
HEAVY WORK	1600	565	1035

*ADJUSTED TOTAL GAIN IS BASED ON NORMAL PERCENTAGE OF MEN, WOMEN AND CHILDREN FOR THE APPLICATION LISTED.

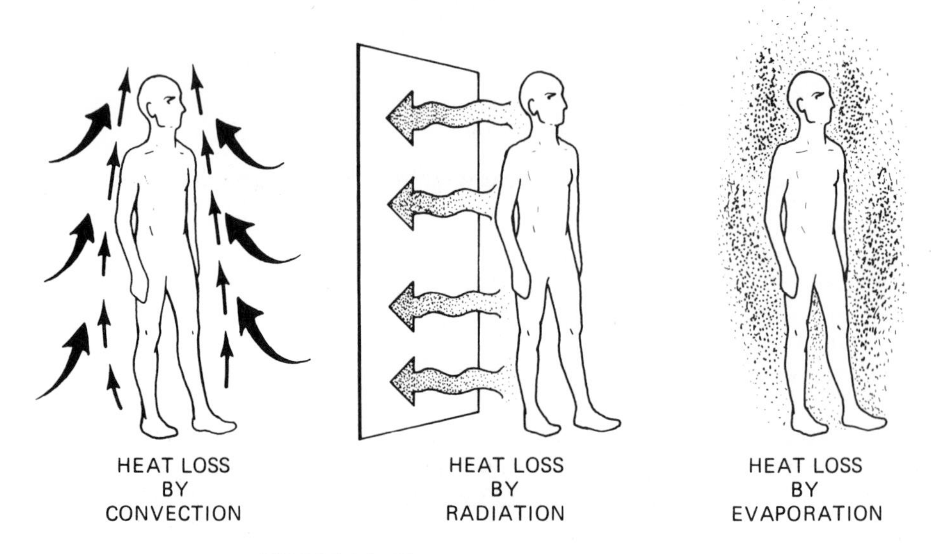

FIGURE 2-2 Three ways the body loses heat.

sulation and insulation slows down the transfer of heat. Therefore, the heavier the clothing, the greater the insulation value, and the less the amount of heat dissipated. Thus, heat that would normally be lost to the atmosphere is used to warm the surface of the body.

Excess heat must be transferred from the body to the area around it. This is accomplished in three ways (Figure 2-2):

1. Convection
2. Radiation
3. Evaporation

Convection

For convection to take place, the area around the body must be at a lower temperature. The greater the difference in temperature, the greater the movement of convection currents around the body, and the greater the heat transfer.

If the temperature difference is too great, a person will feel the sensation of being cold. If the temperature difference is too small, a person will feel the sensation of being hot.

Radiation

The body radiates heat to cooler surfaces just as the sun radiates heat to the earth. The greater the temperature difference between the body and the exposed surfaces, the greater the rate of heat transfer.

It has been found that the most comfortable conditions are maintained if the inside surfaces of the outside walls and windows are heated to the room temperature. If these surface temperatures are too low, the body radiates heat to them too fast and this produces a feeling of discomfort.

Evaporation

Water (or moisture) on the surface of the skin enters the surrounding air by means of evaporation. Evaporation is the process of changing water to vapor by the addition of heat. As the water absorbs heat from the body it evaporates, thus transferring heat to the air.

If the rate of evaporation is too great, the skin has a dry, uncomfortable feeling. The membranes of the nose and throat require adequate humidity to maintain their flexible condition. If the rate of evaporation is too slow, the skin has a clammy, sticky feeling.

Evaporation of moisture requires approximately 970 Btu of heat per pound of water evaporated. Heat required for evaporation is absorbed from the body, cooling it.

Space

Space is defined as the enclosed area in which one lives or works. Conditions for comfort are maintained within the space by mechanical heating. The requirement for comfort is that the space supply a means for dissipating the heat from the body at the proper rate to maintain proper body temperature.

The conditions for comfort in the space may be divided into two groups: thermal and environmental. The thermal conditions include temperature, relative humidity, and air motion. The environmental conditions include clean air, freedom from disturbing noise, and freedom from disagreeable odors. Because of the importance of the thermal factors, these will be discussed first.

Temperature

The ability to produce the correct space temperature is probably the most important single factor in providing comfortable conditions. Two types of temperature are considered:

1. The temperature of the air that surrounds the body
2. The temperature of exposed surfaces enclosing the room

For the maximum degree of comfort, both the air temperature and the surface temperature should be the same.

Air temperature is measured with an ordinary thermometer. Surface temperature is more difficult to measure, and is usually read by using either an electronic thermometer or a special surface temperature thermometer. The best comfort temperature level for both air and surfaces within the space is 76°F (24.5°C), according to the American Society of Heating, Refrigerating and Air Conditioning Engineers (ASHRAE) Comfort Standard 55-66.

However, in view of the need for conserving energy, interior temperature levels for heating have been lowered. This text uses 70°F (21.1°C) as a design inside temperature. In some cases lower temperatures are recommended when energy must be conserved. It is acknowledged that some individuals will have to put on additional clothing to remain comfortable at this temperature.

Relative Humidity

Relative humidity (rh) is measured with a sling psychrometer. A psychrometer has two identical thermometers. One has a wetted wick covering the bulb. Readings are taken on the dry bulb (db) and wet bulb (wb) temperatures after the psychrometer has been rotated in the air for approximately 1 min. (See Figure 2-3.)

Unless the air is completely saturated with moisture (contains all the moisture it can hold), the wet bulb thermometer will register a lower temperature than the dry bulb thermometer. This is due to the cooling effect of the moisture evaporating from the wick.

The difference in readings between the dry bulb thermometer and the wet bulb thermometer is an indication of the relative humidity of the air. To determine the relative humidity, the wet and dry bulb temperatures are plotted on a *psychrometric chart.* A psychrometric chart is a graph on which various properties of air are plotted. These properties include dry bulb temperature, wet bulb temperature, pressure, volume, moisture content, and relative humidity. Figure 2-4 is a simplified psychrometric chart that can be used to determine the relative humidity of a sample of air. Note that the wet bulb lines slope downward to the right, the dry bulb lines are vertical, and the relative humidity lines curve upward to the right. Any point on the chart, therefore, represents some wet bulb, dry bulb, and relative humidity condition.

EXAMPLE: What is the relative humidity for the condition of 75°F db (dry bulb) and 60°F wb (wet bulb)?

SOLUTION: Locate 75° on the base line and trace vertically upward until it meets the 60° line. The relative humidity at this point is 40%.

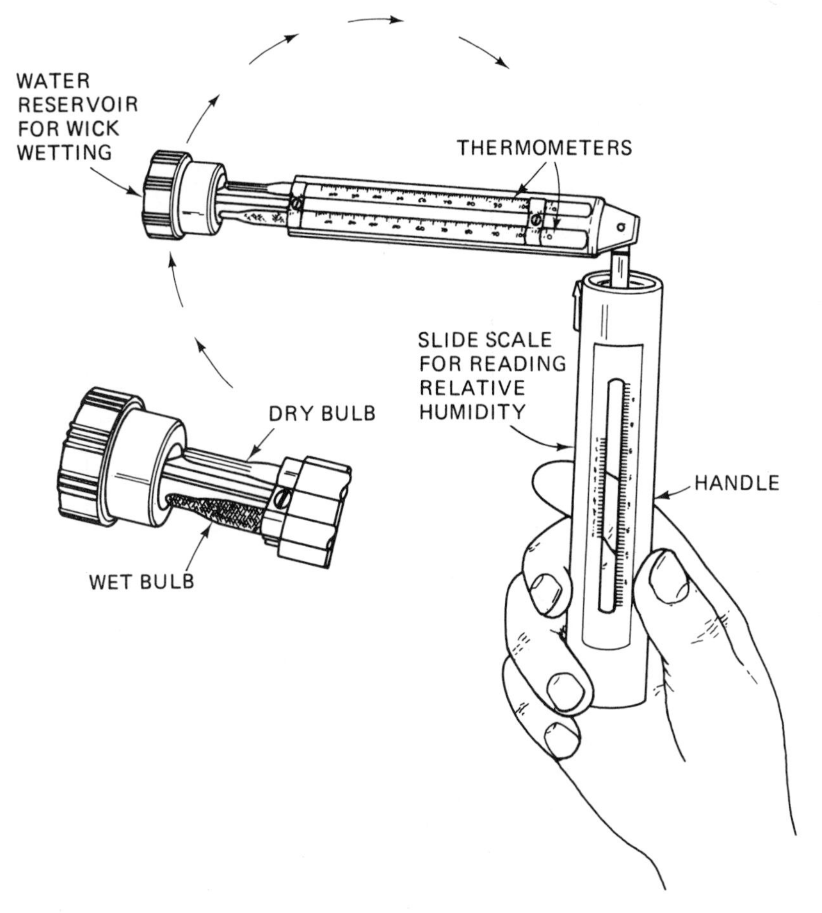

FIGURE 2-3 The sling psychrometer.

The ASHRAE Comfort Standard indicates that the preferred relative humidity is 40% and that any condition within the range 20 to 60% rh is satisfactory.

Comfort Chart

The text has mentioned a comfort standard, recommended by ASHRAE, which indicates the temperature and humidity at which most people are comfortable. These conditions assume sedentary activity (any activity requiring little movement) and medium-weight clothing. Studies by ASHRAE have indicated, however, that there is some range of comfort level as a result of individual preferences. Therefore, in the psychrometric chart shown in Figure 2-4, the shaded area gives a range of conditions that the heating system should be capable of meeting to provide maximum flexibility to suit individual tastes.

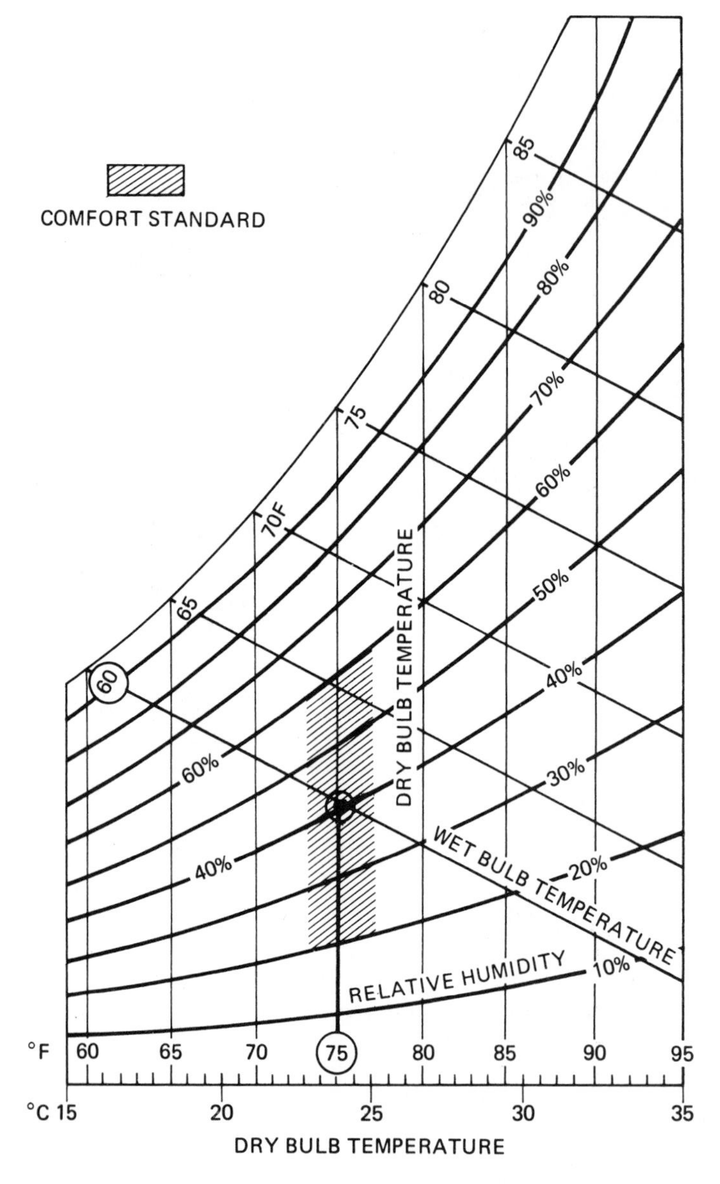

FIGURE 2-4 The psychrometric chart.

The greater the amount of warm clothing worn by the person, the lower the acceptable comfort level. The greater the amount of physical activity of a person, the lower the acceptable comfort level. The greater the air movement within the space, the higher the acceptable comfort level. The lower the temperature of exterior walls and windows enclosing the space, the higher the acceptable air temperature.

Air Motion

Most heating systems require some air movement to distribute heat within the space. In a forced air heating system, air is supplied through registers and returned to the heating unit through grilles. High velocity air, measured in feet per minute (fpm), enters the room through the supply registers. Air at these velocities may be required to properly heat cold exterior walls. The velocity is greatly reduced by natural means before the air reaches the occupants of the room.

According to the ASHRAE Comfort Standard, provided that other recommended conditions are maintained (76°F db, 40% rh), the air that comes in contact with people should not exceed a velocity of 45 fpm or 0.23 meters per second (m/s).

Clean Air

One of the desirable qualities of a forced warm air heating system is its ability, through air filters, to remove certain portions of the dust and dirt that are carried by the air. There are three types of residential air filters: disposable (throwaway), permanent (cleanable), and electronic, as shown in Figure 2-5.

Disposable filters are made of oil-impregnated fibers. When dirty, they are noncleanable and are replaced with new ones.

Permanent filters are made of metal or specially constructed fibrous material. The dirt is collected on the surface of the filtering material as the air passes through. When properly cleaned, these filters can be reused.

Electronic filters remove dust particles by first placing an electrical charge on the dust particle and then attracting the dust to collector plates having the opposite electrical charge. They are usually used with cleanable-type pre-filters because the electronic filter removes only extremely small particles.

Figure 2-6 shows the size of various dust particles. The size is indicated in *micronmeters* (μm). A micronmeter is 1/25,400 of an inch [one thousandth of a millimeter (mm)]. The ordinary throwaway or cleanable-type home air filter will remove particles down to 10 μm. The electronic air filter will remove particles from 0.1 to 10 μm.

Freedom From Disturbing Noise

Noise or objectionable sound can be airborne or travel through the structure of a building.

Airborne noise can be caused by mechanical equipment being located too near a building's occupants or by the mechanical noise being transmitted through connecting air ducts.

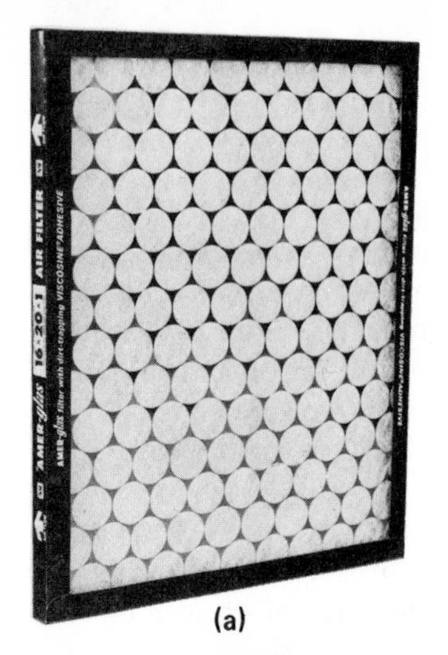

(a)

(b)

(c)

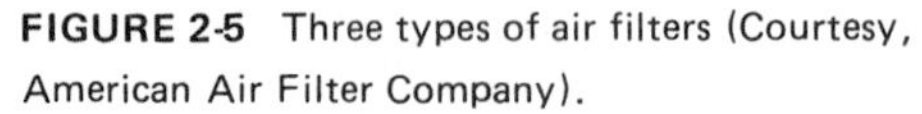

FIGURE 2-5 Three types of air filters (Courtesy, American Air Filter Company).

FIGURE 2-6 The size of dust particles (Courtesy, Heat Controller, Inc.).

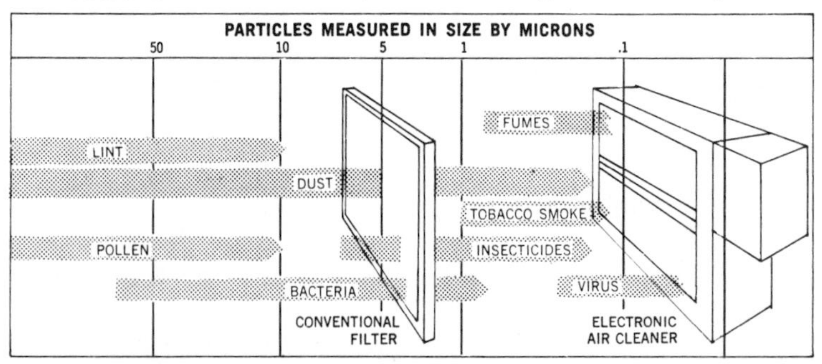

A micron measures 1/25,400 of an inch. The above illustration shows sizes of various common particles found in air we breathe everyday.

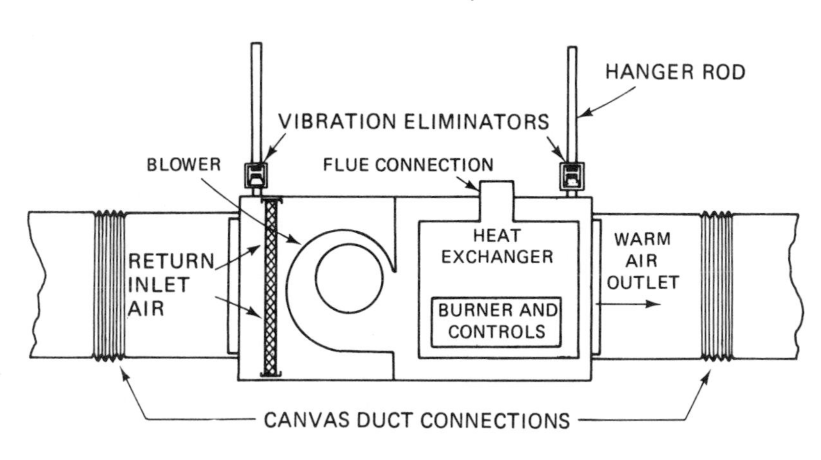

FIGURE 2-7 Noise and vibration eliminators (Reprinted with permission from the 1975 Equipment Volume, ASHRAE Handbook & Product Directory).

Noise that travels through the structure of a building is caused by the vibration of equipment mounted in direct contact with the building.

To remove noise problems, equipment should be properly selected from a noise-level standpoint. Where noise or vibration does exist, equipment should be properly isolated (separated from sound-transmitting materials) (Figure 2-7.)

Sound can be measured by accoustical instruments. The unit of sound-level measurement is the decibel (dB).

Freedom from Disagreeable Odors

Any closed space occupied by people will develop odors. Ventilation is the most effective means of reducing odors.

Some outside air leaks into a building through the cracks around windows and through the building construction. In modern construction this is occasionally not enough to provide adequate ventilation.

A provision can be made in the air supply to the heating unit to provide an additional quantity of outside air. Some designers use 10 cubic feet per minute (cfm) per occupant of the house as a reasonable amount of ventilation air. A damper arrangement can be used to regulate or close off the outside air if desired.

Ventilation air should not be confused with combustion air, since combustion air must be provided to properly burn the furnace fuel. This must not in any way subtract from the air (oxygen) available for the occupants of the house.

REVIEW QUESTIONS

Select the letter representing your choice of the correct answer.

2-1. Comfort is:

(a) The feeling of contentment with the environment

(b) An atmosphere of 86° db, 40% rh

(c) Air conditioning, like an ocean breeze

(d) Warm, moist, clean air

2-2. The body temperature is normally maintained at:

(a) 97 to 100°F

(b) 35°C

(c) 97 to 100°C

(d) 40 to 45°C

2-3. To be comfortable, one must:

(a) Wear warm clothing

(b) Lose heat at the proper rate to the surrounding air

(c) Be fairly active

(d) Use only forced warm air heat

2-4. The body transfers heat:

(a) By air that blows past it

(b) Mainly by conduction

(c) By the normal body functions

(d) By radiation, convection, and evaporation

2-5. The amount of heat required to evaporate 1 lb of water at 212°F is approximately:

(a) 970 Btu

(b) 144 Btu

(c) 300 kcal

(d) 500 kcal

2-6. Thermal comfort conditions include:

(a) Freedom from noise

(b) Freedom from disagreeable odors

(c) Clean air

(d) Temperature of air

2-7. ASHRAE Comfort Standard recommends the following temperature and relative humidity conditions:

(a) 75°F, 75% rh

(b) 76°F, 40% rh

(c) 80°F, 40% rh

(d) 27°C, 50% rh

2-8. A psychrometer is:

(a) A chart for measuring relative humidity

(b) A metric gauge for wire

(c) A device with two thermometers for measuring wet and dry bulb temperature

(d) The name of a heating unit

2-9. A psychrometric chart is:

(a) A device for measuring humidity

(b) A graph plotting air motion and temperature

(c) A chart where the various properties of air are plotted

(d) A chart showing how to assemble a heating unit

2-10. An electronic filter will remove dust particles in the size range:

(a) 0.1 to 10 μm

(b) 10 to 100 μm

(c) 0.001 to 0.001 μm

(d) 100 μm or more

3

ESTIMATING THE HEATING LOAD

OBJECTIVES

After studying this chapter the student will be able to:

- Estimate the heating load of a residence

EVALUATING THE HEATING SYSTEM

Many heating service technicians struggle to correct problems in the mechanical operation of heating equipment which are caused originally by improper selection, application, and installation of the heating system. This chapter and Chapters 4 and 5, will supply the service technician with information needed to evaluate the heating system. These chapters discuss estimating the heat load, equipment selection and air distribution, and installation practice. Thus, a technician can make appropriate recommendations to improve comfort conditions which are outside the realm of the normal repair or maintenance procedures.

Complete data for designing and laying out the system is not provided in this section, because this work is the function of a design engineer. The material presented in this chapter will supply a service technician with a quick, simplified method of determining the heating requirements.

MAINTAINING CORRECT TEMPERATURE

One of the most important functions of the heating system is to maintain the correct temperature. This chapter will explain how to determine the amount of heat needed to produce the correct temperature in the occupied space during extreme winter conditions. The outside design temperature selected is not the coldest temperature ever recorded in the area. It is the coldest sustained temperature normally experienced. Records show that colder temperatures persist only for a short time and a building holds heat during these periods without being materially affected by extreme outside conditions. Recommended outside design temperatures for various localities are shown in Figure 3-1.

SOURCES OF HEATING LOAD

There are two main sources of heating load (or loss) in a structure:

1. Transmission losses
2. Infiltration or ventilation losses

Heat moves from higher temperatures to lower temperatures. *Transmission losses* are the heat units that travel through the exposed areas of a building where there is a difference between the inside and outside temperatures. These exposed areas include walls, ceilings, floors, windows, and doors. They do not include inside partitions of the building between heated rooms.

Infiltration is the outside air that enters the building through cracks around windows and doors. Normally it is good practice, for ventilation purposes, to constantly add some outside air to the return air entering the furnace. This replaces air expelled by exhaust fans and dilutes any disagreeable odors which may be produced by cooking or smoking.

Heat must be supplied in order to warm outside air that enters a building by infiltration or ventilation.

OUTSIDE DESIGN CONDITIONS - UNITED STATES*

STATE AND CITY	WINTER DRY BULB (db) °F	WINTER DRY BULB (db) °C
ALABAMA		
ANNISTON	20	−6.7
BIRMINGHAM	15	−9.4
DECATUR	10	−12.2
MOBILE	25	−3.9
MONTGOMERY	20	−6.7
TUSCALOOSA	20	−6.7
ALASKA		
ANCHORAGE	−25	−31.7
FAIRBANKS	−50	−45.6
JUNEAU	− 5	−20.6
NOME	−30	−34.4
ARIZONA		
FLAGSTAFF	0	−17.8
KINGMAN	20	− 6.7
PHOENIX	30	− 1.1
TUCSON	30	− 1.1
WINSLOW	5	−15.0
YUMA	35	1.7
ARKANSAS		
FORT SMITH	10	−12.2
HOT SPRINGS	15	− 9.4
LITTLE ROCK	15	− 9.4
PINE BLUFF	15	− 9.4
TEXARKANA	20	− 6.7
CALIFORNIA		
BAKERSFIELD	30	− 1.1
BURBANK	35	1.7
EL CENTRO	35	1.7
EUREKA	30	− 1.1
FRESNO	30	− 1.1
LONG BEACH	40	4.4
LOS ANGELES	40	4.4
MONTEREY	35	1.7
NEEDLES	30	− 1.1
OAKLAND	35	1.7
SACRAMENTO	30	− 1.1
SAN BERNARDINO	30	− 1.1
SAN DIEGO	40	4.4
SAN FRANCISCO	35	1.7
SAN JOSE	35	1.7
YREKA	15	− 9.4

STATE AND CITY	WINTER DRY BULB (db) °F	WINTER DRY BULB (db) °C
COLORADO		
BOULDER	−5	−20.6
COLORADO SPRINGS	−5	−20.6
DENVER	−5	−20.6
GRAND JUNCTION	0	−17.8
PUEBLO	−5	−20.6
CONNECTICUT		
HARTFORD	5	−15.0
NEW HAVEN	5	−15.0
NEW LONDON	5	−15.0
DELAWARE		
DOVER	10	−12.2
WILMINGTON	10	−12.2
DIST. OF COLUMBIA		
WASHINGTON	15	− 9.4
FLORIDA		
FORT MYERS	40	4.4
JACKSONVILLE	30	− 1.1
KEY WEST	55	12.8
MIAMI	45	7.2
ORLANDO	45	1.7
PENSACOLA	25	− 3.9
SARASOTA	40	4.4
TALLAHASSEE	25	− 3.9
TAMPA	35	1.7
GEORGIA		
ATHENS	20	− 6.7
ATLANTA	15	− 9.4
AUGUSTA	20	− 6.7
COLUMBUS	20	− 6.7
MACON	20	− 6.7
ROME	15	− 9.4
SAVANNAH	25	− 3.9
HAWAII		
HONOLULU	60	15.6

*TABLE 1 CLIMATIC CONDITIONS FOR THE UNITED STATES ASHRAE, 1977 FUNDAMENTALS HANDBOOK, CHAPTER 23

FIGURE 3-1 Outside design conditions — United States.

STATE AND CITY	WINTER DRY BULB (db) °F	°C
IDAHO		
BOISE	5	−15.0
IDAHO FALLS	−10	−23.3
LEWISTON	0	−17.8
POCATELLO	−10	−23.3
TWIN FALLS	− 5	−20.6
ILLINOIS		
BLOOMINGTON	+ 5	−15.0
CHICAGO	−10	−23.3
DECATUR	− 5	−20.6
JOLIET	− 5	−20.6
MOLINE	−10	−23.3
PEORIA	−10	−23.3
SPRINGFIELD	− 5	−20.6
INDIANA		
EVANSVILLE	5	−15.0
FORT WAYNE	−5	−20.6
INDIANAPOLIS	0	−17.8
MUNCIE	−5	−20.6
SOUTH BEND	−5	−20.6
TERRE HAUTE	0	−17.8
IOWA		
AMES	−10	−23.3
BURLINGTON	− 5	−20.6
CEDAR RAPIDS	−10	−23.3
DES MOINES	−10	−23.3
SIOUX CITY	−10	−23.3
KANSAS		
DODGE CITY	0	−17.8
SALINA	0	−17.8
TOPEKA	0	−17.8
WICHITA	5	−15.0
KENTUCKY		
ASHLAND	5	−15.0
BOWLING GREEN	5	−15.0
LEXINGTON	5	−15.0
LOUISVILLE	5	−15.0

STATE AND CITY	WINTER DRY BULB (db) °F	°C
LOUISIANA		
ALEXANDRIA	25	−3.9
BATON ROUGE	25	−3.9
NEW ORLEANS	30	−1.1
SHREVEPORT	20	−6.7
MAINE		
AUGUSTA	− 5	−20.6
BANGOR	−10	−23.3
PORTLAND	− 5	−20.6
MARYLAND		
BALTIMORE	10	−12.2
CUMBERLAND	5	−15.0
MASSACHUSETTS		
BOSTON	5	−15.0
SPRINGFIELD	−5	−20.6
WORCESTER	0	−17.8
MICHIGAN		
BATTLE CREEK	0	−17.8
DETROIT	5	−15.0
FLINT	− 5	−20.6
GRAND RAPIDS	0	−17.8
LANSING	− 5	−20.6
MUSKEGON	0	−17.8
SAGINAW	0	−17.8
SAULT STE. MARIE	−10	−23.3
TRAVERSE CITY	− 5	−20.6
MINNESOTA		
ALEXANDRIA	−20	−28.9
DULUTH	−20	−28.9
INTERNATIONAL FALLS	−30	−34.4
MINNEAPOLIS	−15	−26.1
ST. CLOUD	−15	−26.1
ST. PAUL	−15	−26.1

FIGURE 3-1 Continued (Reprinted with permission from the 1977 Fundamental Volume, ASHRAE Handbook & Product Directory).

STATE AND CITY	WINTER DRY BULB (db) °F	°C
MISSISSIPPI		
BILOXI	30	−1.1
COLUMBUS	15	−9.4
JACKSON	20	−6.7
MERIDIAN	20	−6.7
VICKSBURG	20	−6.7
MISSOURI		
COLUMBIA	0	−17.8
JOPLIN	5	−15.0
KANSAS CITY	0	−17.8
KIRKSVILLE	−5	−20.6
ST. JOSEPH	−5	−20.6
ST. LOUIS	0	−17.8
SPRINGFIELD	5	−15.0
MONTANA		
BILLINGS	−15	−26.1
BUTTE	−25	−31.7
GREAT FALLS	−20	−28.9
HELENA	−20	−28.9
KALISPELL	−15	−26.1
MILES CITY	−20	−28.9
MISSOULA	−15	−16.1
NEBRASKA		
GRAND ISLAND	−10	−23.3
LINCOLN	− 5	−20.6
NORTH PLATTE	−10	−23.3
OMAHA	−10	−23.3
NEVADA		
CARSON CITY	5	−15.0
ELKO	−10	−23.3
LAS VEGAS	25	− 3.9
RENO	5	−15.0
TONOPAH	5	−15.0
WINNEMUCCA	0	−17.8
NEW HAMPSHIRE		
BERLIN	−15	−26.1
CONCORD	−10	−23.3
MANCHESTER	−10	−23.3

STATE AND CITY	WINTER DRY BULB (db) °F	°C
NEW JERSEY		
ATLANTIC CITY	10	−12.2
NEWARK	10	−12.2
PATERSON	5	−15.0
TRENTON	10	−12.2
NEW MEXICO		
ALBUQUERQUE	10	−12.2
LAS CRUCES	15	− 9.4
ROSWELL	15	− 9.4
SANTA FE	5	−15.0
TUCUMCARI	10	−12.2
NEW YORK		
ALBANY	5	−15.0
BINGHAMTON	0	−17.8
BUFFALO	0	−17.8
ELMIRA	− 5	−20.6
GENEVA	− 5	−15.0
NEW YORK	10	−12.2
ROCHESTER	0	−17.8
SYRACUSE	− 5	−20.6
UTICA	−10	−23.3
NORTH CAROLINA		
ASHEVILLE	10	−12.2
CHARLOTTE	20	− 6.7
DURHAM	15	− 9.4
GREENSBORO	15	− 9.4
NEW BERN	20	− 6.7
RALEIGH	15	− 9.4
WILMINGTON	25	− 3.9
WINSTON-SALEM	15	− 9.4
NORTH DAKOTA		
BISMARK	−25	−31.7
DICKINSON	−20	−28.9
FARGO	−20	−28.9
GRAND FORKS	−25	−31.7
JAMESTOWN	−20	−28.9
MINOT	−25	−31.7

FIGURE 3-1 Continued

STATE & CITY	WINTER DRY BULB (db)	
	°F	°C
OHIO		
AKRON	0	−17.8
CINCINNATI	0	−17.8
CLEVELAND	0	−17.8
COLUMBUS	0	−17.8
DAYTON	0	−17.8
TOLEDO	−5	−20.6
YOUNGSTOWN	0	−17.8
ZANESVILLE	0	−17.8
OKLAHOMA		
ARDMORE	15	− 9.4
MUSKOQEE	10	−12.2
NORMAN	10	−12.2
OKLAHOMA CITY	10	−12.2
TULSA	10	−12.2
OREGON		
BAKER	0	−17.8
EUGENE	15	− 9.4
MEDFORD	20	− 6.7
PENDLETON	0	−17.8
PORTLAND	15	− 9.4
SALEM	20	− 6.7
PENNSYLVANIA		
ALTOONA	0	−17.8
ERIE	5	−15.0
HARRISBURG	5	−15.0
PHILADELPHIA	10	−12.2
PITTSBURGH	0	−17.8
SCRANTON	0	−17.8
WILLIAMSPORT	0	−17.8
YORK	10	−12.2
RHODE ISLAND		
NEWPORT	5	−15.0
PROVIDENCE	5	−15.0
SOUTH CAROLINA		
CHARLESTON	25	−3.9
COLUMBIA	20	−6.7
FLORENCE	20	−6.7
GREENVILLE	20	−6.7
SPARTANBURG	20	−6.7

STATE AND CITY	WINTER DRY BULB (db)	
	°F	°C
SOUTH DAKOTA		
ABERDEEN	−20	−28.9
HURON	−20	−28.9
PIERRE	−15	−26.1
RAPID CITY	−10	−23.3
SIOUX FALLS	−15	−26.1
WATERTOWN	−20	−28.9
TENNESSEE		
CHATTANOOGA	15	− 9.4
JACKSON	10	−12.2
KNOXVILLE	15	− 9.4
MEMPHIS	15	− 9.4
NASHVILLE	10	−12.2
TEXAS		
ABILENE	15	− 9.4
AMARILLO	5	−15.0
AUSTIN	25	− 3.9
BROWNSVILLE	35	+ 1.7
CORPUS CHRISTI	30	− 1.1
DALLAS	20	− 6.7
DEL RIO	25	− 3.9
EL PASO	20	− 6.7
FORT WORTH	15	− 9.4
GALVESTON	30	− 1.1
HOUSTON	25	− 3.9
PORT ARTHUR	25	− 3.9
SAN ANTONIO	25	− 3.9
WACO	20	− 6.7
UTAH		
OGDEN	0	−17.8
PROVO	0	−17.8
SALT LAKE CITY	5	−15.0
VERMONT		
BARRE	−15	−26.1
BURLINGTON	−10	−23.3
VIRGINIA		
DANVILLE	15	− 9.4
LYNCHBURG	10	−12.2
NORFOLK	20	− 6.7
RICHMOND	15	− 9.4
ROANOKE	10	−12.2

FIGURE 3-1 Continued

STATE AND CITY	WINTER DRY BULB (db)	
	°F	°C
WASHINGTON		
BELLINGHAM	10	−12.2
EVERETT	20	− 6.7
OLYMPIA	15	− 9.4
SEATTLE	20	− 6.7
SPOKANE	−5	−20.6
TACOMA	20	− 6.7
WALLA WALLA	0	−17.8
YAKIMA	0	−17.8
WEST VIRGINIA		
CHARLESTON	5	−15.0
ELKINS	0	−17.8
HUNTINGTON	5	−15.0
MARTINSBURG	5	−15.0
PARKERSBURG	5	−15.0
WHEELING	0	−17.8
WISCONSIN		
EAU CLAIRE	−15	−26.1
GREEN BAY	−15	−26.1
LACROSSE	−15	−26.1
MADISON	−10	−23.3
MILWAUKEE	−10	−23.3
RACINE	− 5	−20.6
WYOMING		
CASPER	−10	−23.3
CHEYENNE	−10	−23.3
LARAMIE	−15	−26.1
SHERIDAN	−15	−26.1

FIGURE 3-1 Continued

Transmission Heating Losses

Heat moves through exposed parts of a building at various rates, depending on the following conditions:

1. The construction or material composition of the exposures
2. The area of the exposures
3. The difference between the inside and outside temperatures

The tables in Figures 3-2 through 3-7 give the factors for estimating the heat loss through various types of construction. At the top of each table is the outside design temperature. The inside design temperature on which these tables are based is 70° F (21.1° C) (Figure 3-8).

FIGURE 3-3 Heat loss factors for windows.

CONSTRUCTION	OUTSIDE DESIGN TEMPERATURE °F												
	−30	−25	−20	−15	−10	−5	0	+5	+10	+15	+20	+25	+30
A	155	145	140	130	125	115	110	100	90	85	75	70	60
B	90	85	80	75	70	65	60	55	50	45	45	40	35
C	70	65	65	60	55	55	50	45	40	40	35	30	30

A – SINGLE GLASS
B – DOUBLE GLASS OR INSULATING GLASS
C – SINGLE GLASS WITH STORM WINDOW

1. FACTORS ARE BASED ON SQUARE FEET OF WINDOW OPENING
2. THIS TABLE INCLUDES AN ALLOWANCE FOR AVERAGE INFILTRATION

FIGURE 3-3 Heat loss factors for doors.

TYPE OF CONSTRUCTION	OUTSIDE DESIGN TEMPERATURE °F												
	−30	−25	−20	−15	−10	−5	0	+5	+10	+15	+20	+25	+30
A	220	210	200	190	180	165	155	145	130	120	110	100	90
B	175	165	160	150	140	130	125	115	105	95	90	80	70
C	115	110	105	95	90	85	80	75	70	65	55	50	45
D	210	200	190	180	170	160	145	135	125	115	105	95	85
E	420	400	380	355	335	315	295	275	250	230	210	190	170

A – SINGLE GLASS
B – DOUBLE GLASS OR INSULATING GLASS
C – WITH WEATHERSTRIPPING AND WITH STORM DOORS
D – WITH WEATHERSTRIPPTING OR STORM DOOR
E – WITHOUT WEATHERSTRIPPING OR STORM DOOR

1. FACTORS ARE BASED ON SQUARE FEET OF DOOR OPENING.

FIGURE 3-4 Heat loss factors for exterior walls and partitions.

TYPE OF CONSTRUCTION	OUTSIDE DESIGN TEMPERATURE °F												
	−30	−25	−20	−15	−10	−5	0	+5	+10	+15	+20	+25	+30
A	25	24	23	22	21	19	18	17	16	14	13	12	10
B	13	12	12	11	10	10	9	9	8	7	7	6	5
C	7	7	6	6	6	5	5	5	4	4	4	3	3
D	55	52	49	47	44	41	39	36	33	30	28	25	22
E	31	29	28	26	25	23	21	20	18	17	15	14	12
F	8	7	7	6	6	6	5	5	5	4	4	3	3
G	44	42	40	37	35	33	31	29	26	24	22	20	18
H	26	24	23	22	20	20	18	17	15	14	13	12	10
I	27	26	24	23	22	20	19	18	16	15	14	12	11
J	58	55	52	50	46	44	41	38	35	32	30	26	23
K	14	13	12	12	11	10	10	9	8	7	7	6	5
L	4	4	4	4	4	3	3	3	3	2	2	2	2
M	2	2	2	2	2	2	2	2	2	1	1	1	1

WALL CONSTRUCTION

WOOD FRAME WITH SHEATHING AND SIDING OR BRICK VENEER.

A – NO INSULATION

B – INSULATION, R-5 RIGID POLYSTYRENE BOARD

C – INSULATION, R-11, 3-$3\frac{1}{2}$".

PARTITIONS – WOOD FRAME

D – FINISHED ONE SIDE ONLY, NO INSULATION

E – FINISHED BOTH SIDES, NO INSULATION

F – FINISHED BOTH SIDES, R-11, 3-$3\frac{1}{2}$" INSULATION.

SOLID MASONRY – BLOCK OR BRICK

G – PLASTERED OR PLAIN

H – FURRED, NO INSULATION

I – FURRED, R-5, RIGID POLYSTYRENE BOARD

BASEMENT OR CRAWL SPACE

J – ABOVE GRADE, NO INSULATION

K – ABOVE GRADE INSULATION, R-5, RIGID POLYSTYRENE BOARD

L – BELOW GRADE, NO INSULATION

M – BELOW GRADE INSULATION, R-5, RIGID POLYSTYRENE BOARD

1. FACTORS ARE BASED ON SQUARE FEET OF EXPOSED WALL.

FIGURE 3-5 Heat loss factors for ceilings and roofs.

CONSTRUCTION	OUTSIDE DESIGN TEMPERATURE °F												
	−30	−25	−20	−15	−10	−5	0	+5	+10	+15	+20	+25	+30
A	65	62	59	55	52	49	46	42	39	36	33	29	26
B	8	8	7	7	6	6	6	5	5	4	4	4	3
C	4	4	4	4	3	3	3	3	3	2	2	2	2
D	31	30	28	26	25	23	22	20	19	17	16	14	12
E	8	8	7	7	6	6	6	5	5	4	4	4	3
F	4	4	4	4	3	3	3	3	3	2	2	2	2

CEILING UNDER UNCONDITIONED SPACE OR VENTED ROOF

A – NO INSULATION

B – INSULATION, R-11, 3-3½" BATT OR FILL TYPE

C – INSULATION, R-22, 6-7" BATT OR FILL TYPE

ROOF – CEILING COMBINATION

D – NO INSULATION

E – INSULATION, R-11, 3-3½" BETWEEN CEILING AND ROOFING

F – INSULATION, R-22, 6-7" BETWEEN CEILING AND ROOFING

1. FACTORS ARE BASED ON SQUARE FEET OF EXPOSED CEILING OR ROOF.

FIGURE 3-6 Heat loss factors for floors (Btuh per square foot).

CONSTRUCTION	OUTSIDE DESIGN TEMPERATURE °F												
	−30	−25	−20	−15	−10	−5	0	+5	+10	+15	+20	+25	+30
A	22	21	20	19	18	17	15	14	13	12	11	10	9
B	44	42	40	38	36	34	30	28	26	24	22	20	18
C	8	8	7	7	6	6	6	5	5	4	4	4	3
D	0	0	0	0	0	0	0	0	0	0	0	0	0
E	5	4	4	4	4	4	3	3	3	3	2	2	2

FLOORS OVER UNCONDITIONED SPACE

A – OVER UNCONDITIONED ROOM

B – OVER OPEN SPACE, GARAGE, OR VENTED SPACE – NO INSULATION

C – OVER OPEN SPACE, GARAGE, OR VENTED SPACE, R-11, 3-3½" INSULATION

FLOORS OVER CONDITIONED SPACE

D – OVER HEATED BASEMENT OR CRAWL SPACE

BASEMENT FLOOR

E – CONCRETE ON GROUND

1. FACTORS ARE BASED ON SQUARE FEET OF EXPOSED FLOOR AREA.

FIGURE 3-7 Heat loss factors for slab floors (Btuh per lineal foot of perimeter).

TYPE OF CONSTRUCTION	OUTSIDE DESIGN TEMPERATURE °F												
	−30	−25	−20	−15	−10	−5	0	+5	+10	+15	+20	+25	+30
A	65	60	60	55	50	50	45	40	40	35	30	30	25
B	60	55	50	50	45	45	40	35	35	30	30	25	20
C	130	125	115	110	105	95	90	85	80	70	65	60	50
D	115	110	105	95	90	85	80	75	70	65	60	50	45
E	70	70	65	60	55	55	50	45	45	40	35	30	28
F	60	55	50	50	45	45	40	35	35	30	30	25	20

CONCRETE SLAB FLOOR, UNHEATED

A – INSULATION, 1" AT EDGE

B – INSULATION, 2" AT EDGE

CONCRETE SLAB FLOOR WITH PERIMETER HEATING

C – INSULATION, 1" AT EDGE

D – INSULATION, 2" AT EDGE

FLOOR OF HEATED CRAWL SPACE

E – BELOW GRADE, LESS THAN 18".

F – BELOW GRADE, MORE THAN 18".

1. FACTORS ARE BASED ON BTUH PER LINEAL FOOT OF EXPOSED PERIMETER.

FIGURE 3-8 Estimating the heat loss based on inside design temperature of 70°F (21.1°C).

TRANSMISSION OF HEAT THROUGH BUILDING EXPOSURES

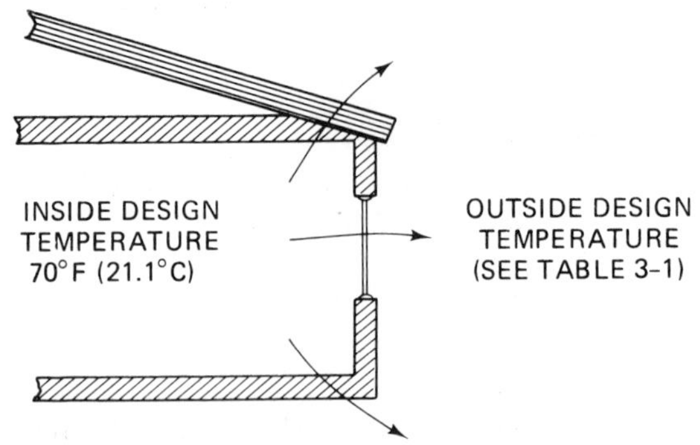

To simplify calculations, some of these factors are based on square-foot measurements of the surface, while others are based on using linear or perimeter (length along the outside wall) measurements (Figure 3-9). In measuring each exposure it is necessary to use a square-foot or a linear-foot measurement as required for each factor.

The table in Figure 3-10 gives transmission-loss factors through ducts located in unheated basements or garages. These factors increase the load by the percentage indicated.

FIGURE 3-9 Dimensions used in estimating heat loss.

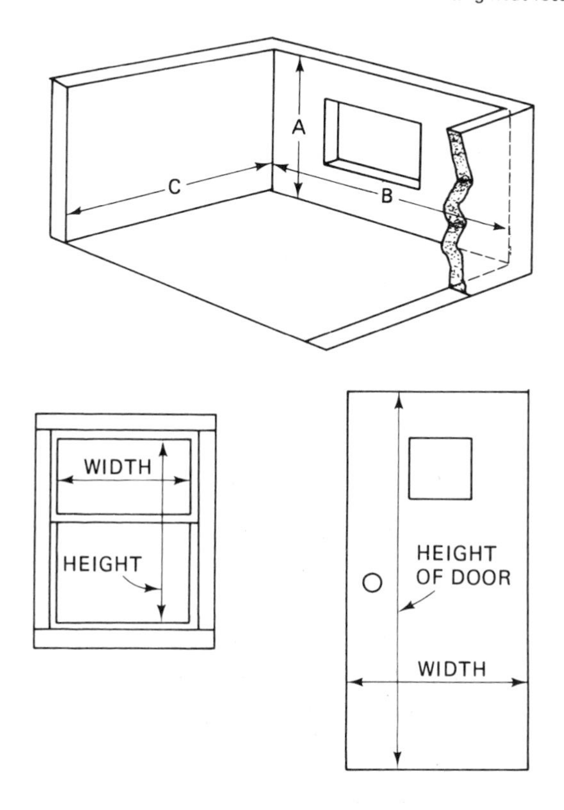

A x B = GROSS AREA OF WALL IN SQUARE FEET

B x C = AREA OF FLOOR OR CEILING IN SQUARE FEET

B + C = LINEAR FEET OF OUTSIDE WALL

DIMENSIONS TO USE WHEN WORKING UP LOAD CALCULATIONS:

- LENGTH OF THE EXPOSED WALL IS RECORDED TO THE NEAREST FOOT.
- HEIGHT OF THE EXPOSED WALL IS RECORDED TO THE NEAREST ONE-HALF FOOT.
- GROSS EXPOSED AREA OF A WALL IS EQUAL TO THE INSIDE LENGTH MULTIPLIED BY ITS HEIGHT. TO SAVE TIME IT IS RECORDED TO THE NEAREST SQUARE FOOT.
- NET EXPOSED AREA OF A WALL IS THE GROSS AREA LESS THE AREA OF THE WINDOWS AND DOORS IN THAT WALL.

FIGURE 3-10 Heat loss through ducts in unheated areas.

		OUTSIDE DESIGN TEMPERATURE °F												
		−30	−25	−20	−15	−10	−5	0	+5	+10	+15	+20	+25	+30
CONSTRUCTION	A	30%	25%	22%	20%	18%	15%	15%	15%	10%	10%	8%	5%	5%
	B	20%	18%	15%	12%	10%	10%	10%	8%	8%	5%	3%	2%	2%
	C	20%	18%	15%	12%	10%	10%	10%	8%	8%	5%	3%	2%	2%

DUCT IN UNHEATED ATTIC, CRAWL SPACE OR GARAGE

A – 1" INSULATION ON DUCT

B – 2" INSULATION ON DUCT

DUCT IN UNCONDITIONED BASEMENT

C – NO INSULATION ON DUCT

Infiltration and Ventilation Heating Losses

Heating losses due to infiltration and ventilation depend upon the following conditions:

1. The amount or volume of air that enters the space from the outside, usually measured in cubic feet per hour
2. The amount of heat required to raise the temperature of a cubic foot of air from outside to inside temperature

Since the infiltration rate is based on the crackage (the length of cracks) around windows and doors, to simplify the calculations the infiltration losses are included with the transmission losses in Figures 3–2 and 3–3. Ventilation requirements in residences are normally supplied by the infiltration. However, when residences are tightly vapor-sealed, some provision must be made for the introduction of outside air, and the heating load must be adjusted accordingly.

Heated garages require special consideration. Air cannot be returned to the furnace from a garage. A volume of outside air equal to that supplied to the garage must be supplied to the furnace and included with the ventilation air. Back-draft louvers and a fire damper, as shown in Figure 3-11, must be installed in the supply outlet to the garage. Usually, a garage can be maintained at a lower temperature than the rest of the house, normally 50°F. Therefore, in calculating garage heating losses, factors selected can be reduced in proportion to the temperature difference.

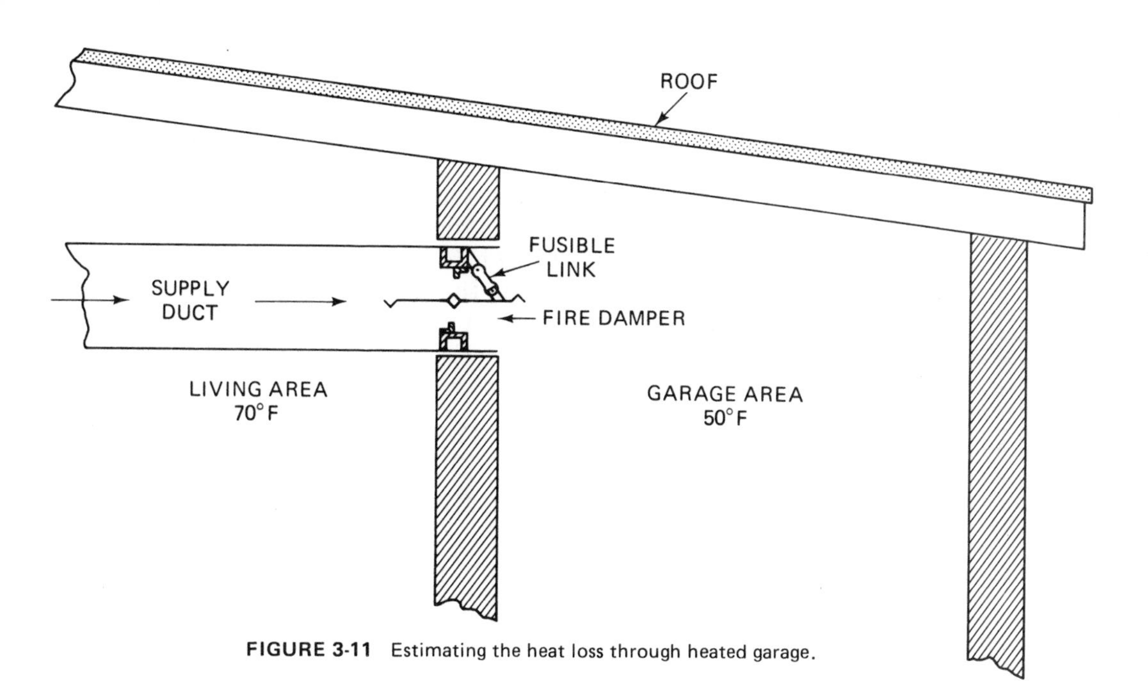

FIGURE 3-11 Estimating the heat loss through heated garage.

MATERIALS AND INFORMATION

The materials and information needed for estimating the heating load are:

1. Dimensioned house plan
2. Description of construction
3. Tables for heating-load factors
4. Data sheet

The preferred house plans are those drawn by an architect, giving dimensioned floor plans and elevations of a building. If these are not available, floor plans can be drawn with a ruler and $^1/_4$ in. cross-section paper (four squares to the inch). Measure the dimensions of the building selected and draw the floor plans to a scale of $^1/_4$ in. equaling 1 ft.

The details of construction include the materials used for windows, doors, walls, ceilings, and floors. It is very important to determine the thickness and type of insulation used in the walls, ceilings, and floors; also whether the windows are single or double glass and if there are storm windows and doors. The heat-loss factors shown in Figures 3-2 through 3-7, and Figure 3-10 cover the most commonly used types of construction.

The data sheet provides a means of recording information and calculating the heating load by rooms and for the entire house.

Sample Calculation

Based on construction information given in Figures 3-12, 3-13, 3-14, a sample calculation is provided on the data sheet, Figure 3-15. The procedure for using this form is shown in the Table below in the left-hand column. The explanation for each procedure appears in the right-hand column.

In the preceding example the basement was heated, so there was no heat loss through the ducts.

Procedure	*Explanation*
1) Record room names at top of the sheet and their overall dimensions to the nearest foot.	The sample starts with the kitchen and proceeds clockwise around the house, ending with the hall. The basement is not partitioned, so is treated as one room.
2) Record heat loss factors for each exposure.	The first section of the data sheet deals with walls. In figure 3-4, for the construction described in figure 3-12 and in the 0°F column, the factor on line B is 9. Note that both the B and 9 are recorded. The basement walls are type L and the factor is 3. Both are recorded. The windows, including the basement are found in figure 3-2. Doors are in figure 3-3. The net wall factors are same as those recorded above. The ceiling factor is from figure 3-5 and the floor factor is from figure 3-6. If this were a slab floor, the factor would come from figure 3-7. Partition walls are from 3-4.
3) Measure and record the dimensions for each exposure.	The kitchen is 11 $\times$ 9 ft and has two outside walls. One is 11 $\times$ 8 ft = 88 ft^2 and the other is 9 $\times$ 8 ft = 72 ft^2. The total gross wall is 88 + 72 ft^2 = 160 ft^2. This figure is recorded above the diagonal line in the kitchen column. The kitchen has two windows, each 3 $\times$ 4 ft, so the total glass area is 24 ft^2. This is recorded above the diagonal line, in figure 3-15. There is also a 3 $\times$ 7 ft door, 21 ft^2, which must be recorded. The net area of the exposed wall is the gross area 160 ft^2 - (the area of the windows and doors). This then would be 160 - (24 + 21) = 160 - 45 = 115 ft^2. The kitchen ceiling is 11$'$ $\times$ 9$'$ = 99 ft^2 and the floor is the same.

Procedure	Explanation
4) Calculate the heat and loss for	In each case the area in ft^2 is multiplied by the recorded factor. The kitchen windows would be 24 ft^2 X 60 = 1440 Btuh. The doors would be 21 ft^2 X 80 = 1680 Btuh. The net exposed wall would be 115 ft^2 X 9 = 1035 Btuh. The ceiling would be 99 ft X 6 = 594 Btuh.
5) Add the subtotal heat loss for each room.	In the example for the kitchen, only four exposures need to be added. (1440 + 1680 + 1035 + 594 = 4749 Btuh)
6) The basement is handled in the last column on the data sheet.	To calculate the gross area of the basement wall add up all of the inside lengths of the wall. Starting at the upper left corner and proceeding clockwise around the house. 23 + 40 + 29 + 15 + 6 + 25 = 138 linear ft X 7 ft high = 966 ft^2 The basement windows would be 18 ft^2 X 110 = 1980 Btuh. The net area of the exposed wall is the gross area 966 ft^2 - 18 ft^2 = 948 ft^2 The wall loss is 948 ft^2 X 3 = 2844 Btuh. The floor area is calculated as 1010 ft^2 with the loss = 1010 ft^2 X 3 = 3030 Btuh. The total basement loss = 1980 + 2844 + 3030 = 7854 Btuh.
7) Add the room heat load and record their total at the bottom of the data sheet.	This figure is the total of all 10 rooms and is 34,881 Btuh.
8) Add the duct loss to the subtotal to obtain the total room heat load for each room.	Since no ducts are placed in unheated areas, no addition for duct loss needs to be made.

FIGURE 3-12 Data for sample calculations.

ONE STORY HOUSE WITH BASEMENT, LOCATED IN CHICAGO, ILLINOIS
OUTDOOR DESIGN 0°F

WINDOWS – DOUBLE GLASS (FIRST FLOOR)

DOORS – WOOD, BUT WITH WEATHER STRIPPING AND STORM DOORS

WALLS – WOOD FRAME WITH SHEATHING AND BRICK VENEER, R-5 INSULATION. CEILING HEIGHT 8′

CEILING – VENTED ATTIC, R-11, 3″ INSULATION

FLOOR – OVER CONDITIONED BASEMENT

BASEMENT – ALL BELOW GRADE, CEILING HEIGHT 7′
WINDOWS – SINGLE GLASS
WALLS – SOLID MASONRY, NO INSULATION
FLOOR – CONCRETE ON GROUND

NOTE: FURNACE IS IN BASEMENT AND BASEMENT IS HEATED

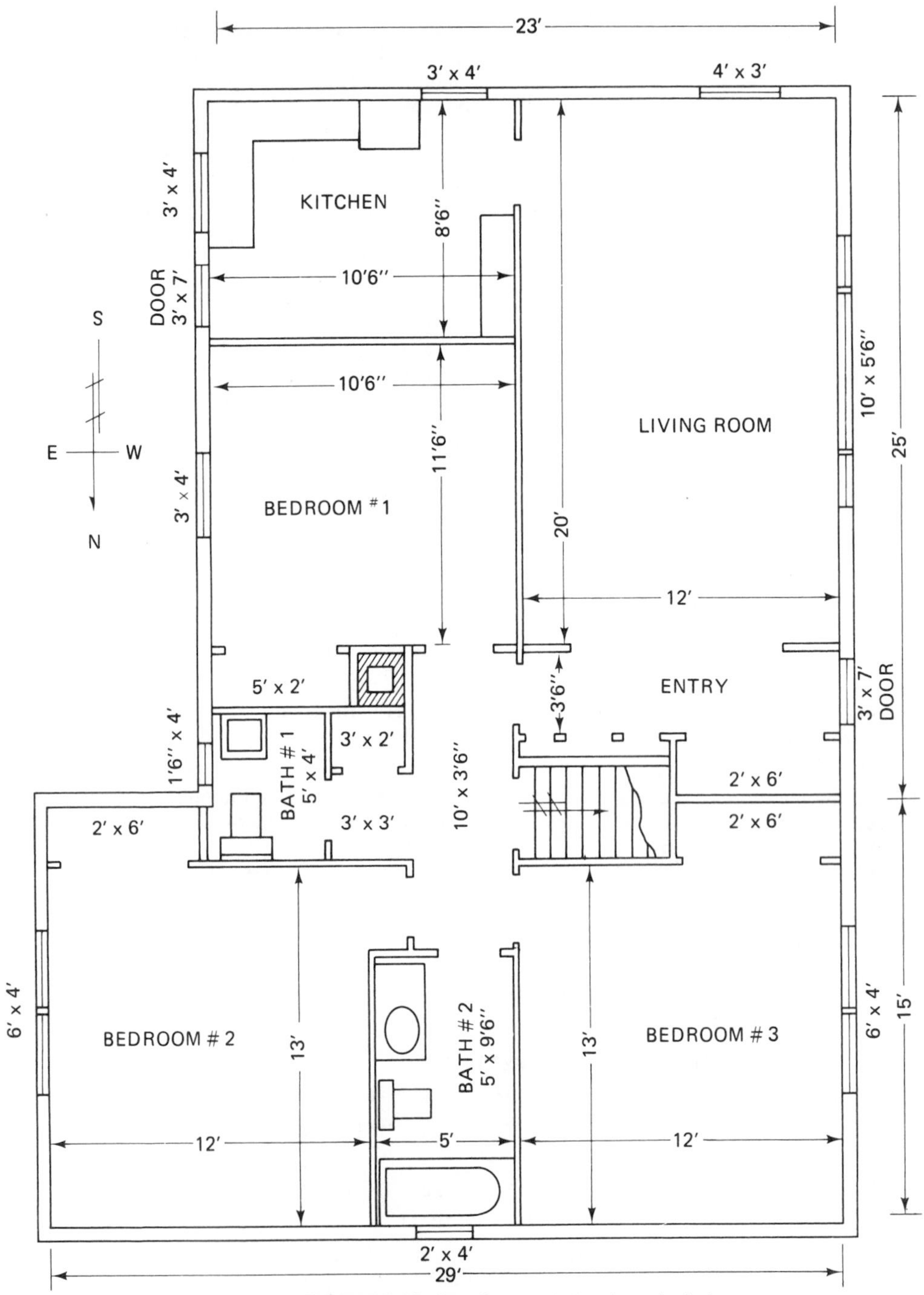

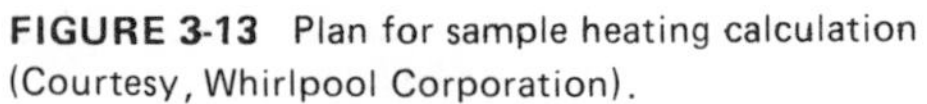
FIGURE 3-13 Plan for sample heating calculation (Courtesy, Whirlpool Corporation).

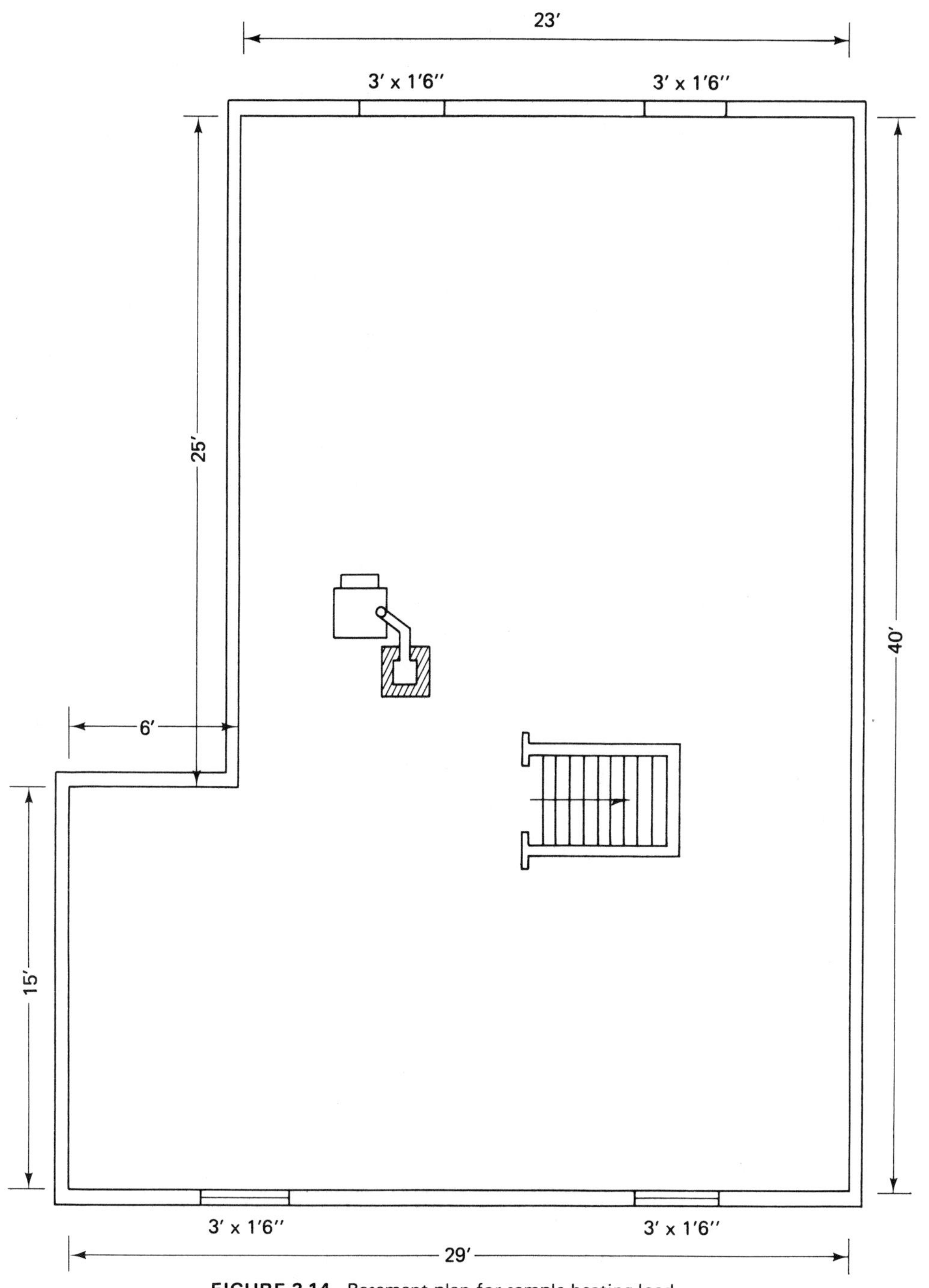

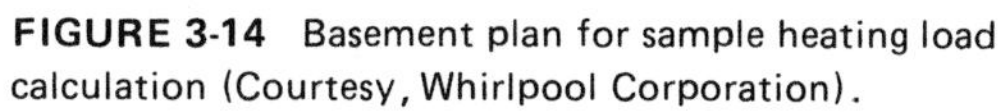
FIGURE 3-14 Basement plan for sample heating load calculation (Courtesy, Whirlpool Corporation).

FIGURE	ROOM NAME				KITCHEN	LIV. RM.	ENTRY	B.R. #3	BATH #2	B.R. #2	BATH #1	B.R. #1	HALL	BASEMENT
	DIMENSIONS, FEET				11′ x 9′	20′ x 12′	6′ x 12′	12′ x 15′	5′ x 10′	12′ x 15′	5′ x 7′	11′ x 14′	10′ x 4′	SEE PLAN
			FACTOR											
3-4	GROSS EXPOSED WALL SQ. FT. INCLUDING BASEMENT	1			160	256	48	200	40	200	24	112		
		2												966
		3												
3-2	WINDOWS SQ. FT. INCL. BASEMENT	1	B	60	24 / 1440	67 / 4020		24 / 1440	8 / 480	24 / 1440	6 / 360	12 / 720		
		2	A	110										18 / 1980
3-3	WINDOWS SQ. FT. INCL. BASEMENT	1	C	80	21 / 1680		21 / 1680							
		2												
3-4	NET EXPOSED WALL SQ. FT. INCLUDING BASEMENT	1	B	9	115 / 1035	189 / 1701	27 / 243	176 / 1584	32 / 288	176 / 1584	18 / 162	100 / 900		
		2	L	3										948 / 2844
		3												
3-5	CEILING SQ. FT.	1	B	6	99 / 594	240 / 1440	72 / 432	180 / 1080	50 / 300	180 / 1080	35 / 210	154 / 924	35 / 210	
		2												
3-6	FLOORS SQ. FT.	1	D	0										
		2	E	3										1010 / 3030
3-7	SLAB FLOOR LINEAR FT.	1												
3-4	PARTITIONS SQ. FT.	1												
	ROOM SUB-TOTAL				4749	7161	2355	4104	1068	4104	732	2544	210	7854
3-10	DUCT LOSS	1												
	ROOM TOTAL BTUH													
	BUILDING TOTAL (TOTAL OF ALL ROOMS) = 34,881 BTUH													
	FURNACE MODEL													

FIGURE 3-15 Sample calculation data sheet.

Assuming that the same basement is unconditioned (unheated) and that there is no insulation on the ducts, and the outdoor design is still 0° F.

Enter the 0° F column at the top of figure 3-10 and go down to line C; the factor is 10%. Since the room sub-total for the kitchen is 4749 Btuh; this figure would be increased by 10% or 475 Btuh. The new room total would be 4749 + 475 = 5224 Btuh. All of the other room sub-totals would be adjusted in the same manner.

REVIEW QUESTION

Using the step described above and the details supplied in Figure 3-16, the floor plans (Figures 3-17 and 3-18); and the blank data sheet (Figure 3-19), calculate the total heat loss.

FIGURE 3-16 Data for review calculation.

ONE STORY HOUSE WITH BASEMENT, LOCATED IN MINNEAPOLIS, MINN,; OUTDOOR DESIGN – 10°F

WINDOWS – SINGLE GLASS (FIRST FLOOR) WITH STORMS

DOORS – WOOD, WITHOUT WEATHER STRIPPING AND STORM DOORS

WALLS – WOOD FRAME WITH SHEATHING AND SIDING, R-11 $3\frac{1}{2}''$ INSULATION, CEILING HEIGHT 8′

CEILING – VENTED ATTIC, R-19, 6″ FILL TYPE INSULATION.

FLOOR – OVER UNCONDITIONED BASEMENT

BASEMENT – ALL BELOW GRADE, CEILING HEIGHT 7′
- WINDOWS – 3′ x 1′6″
- WALLS – SOLID MASONRY, R-5, RIGID POLYSTYRENE BOARD INSULATION
- FLOOR – CONCRETE ON GROUND

NOTE: THE FURNACE IS IN THE BASEMENT, BUT THE BASEMENT IS NOT HEATED. SEE FIGURE 3-10 FOR DUCT LOSSES.

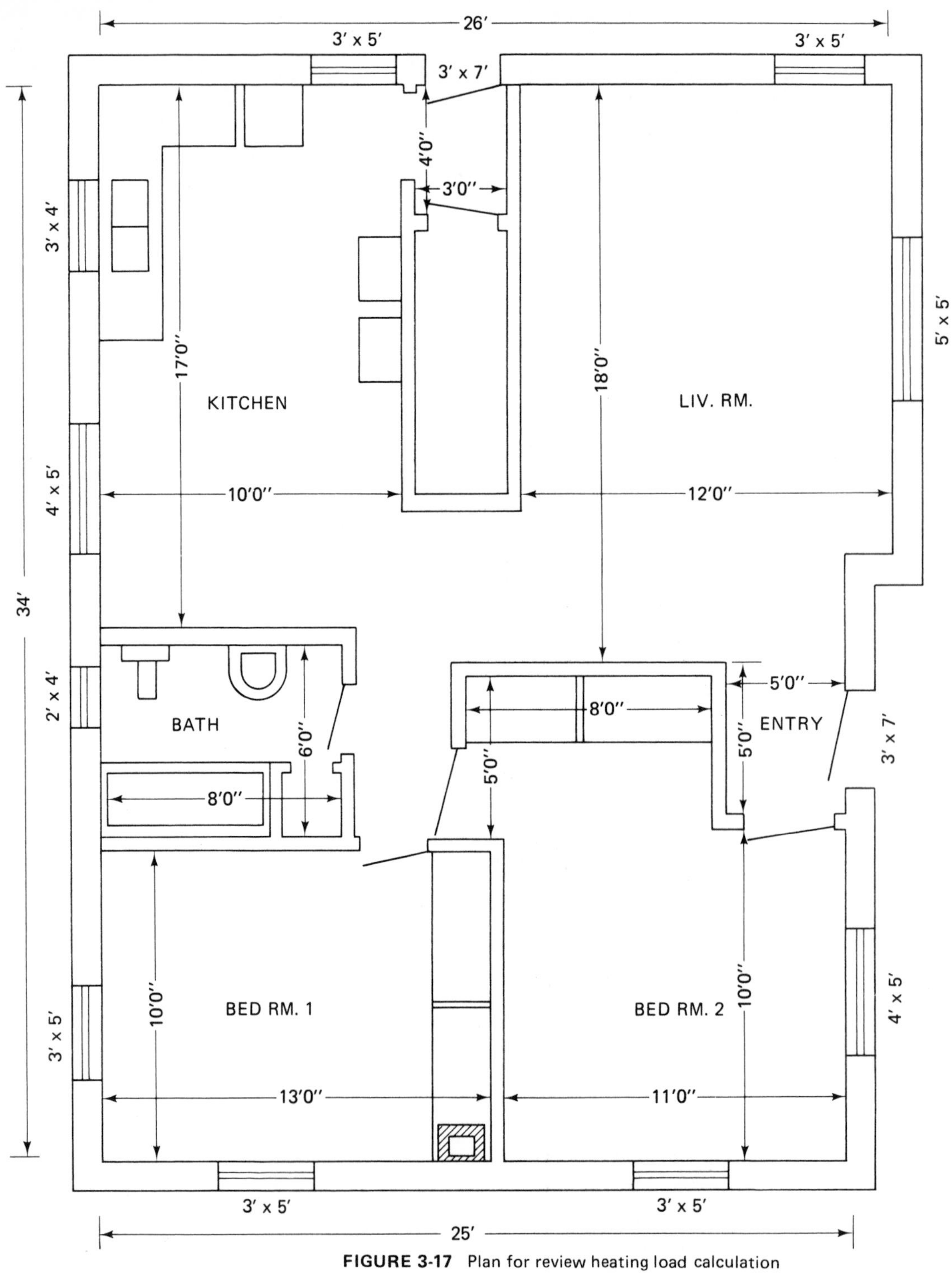

FIGURE 3-17 Plan for review heating load calculation (Courtesy, ITT Fluid Handling Division).

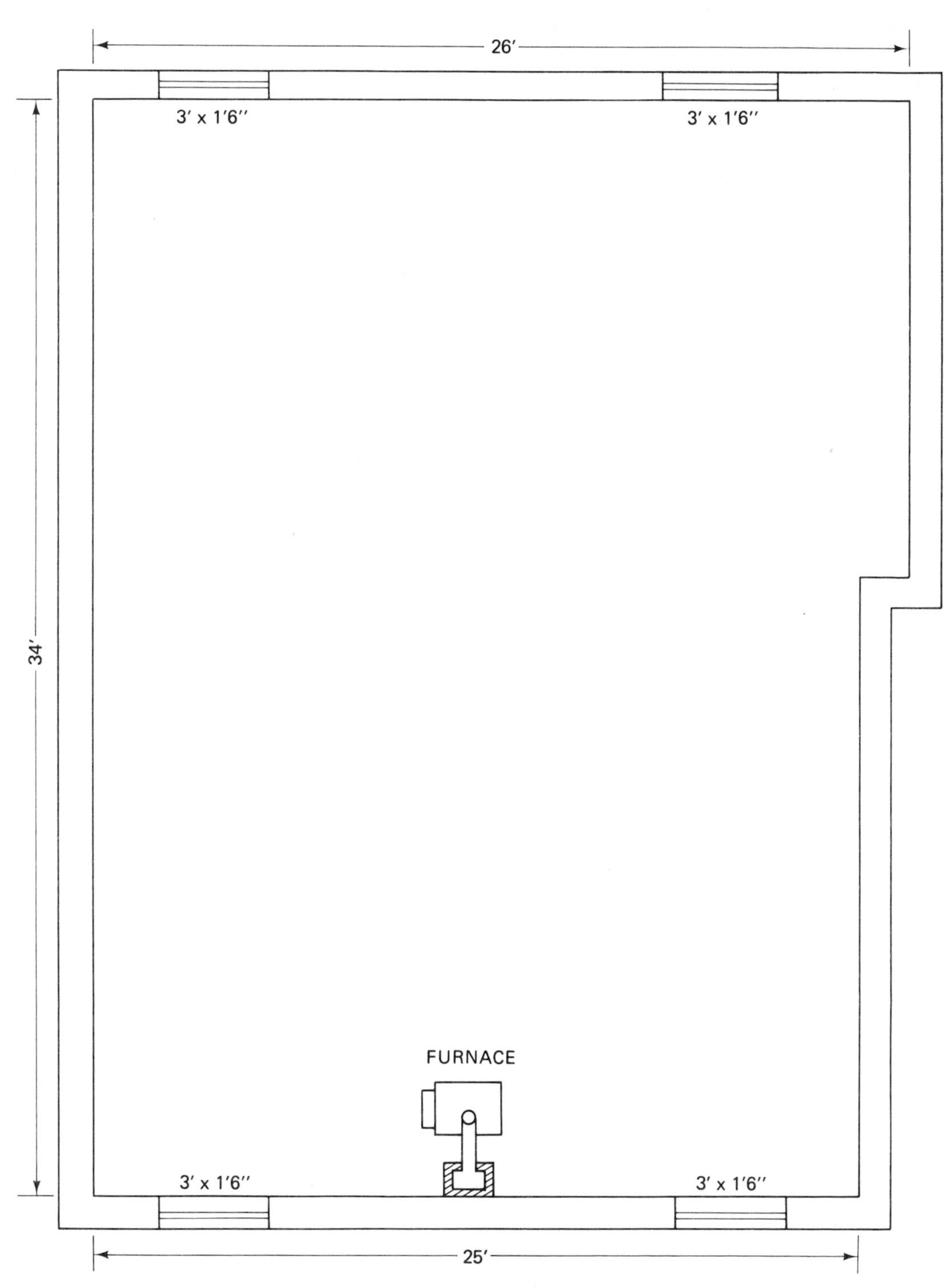

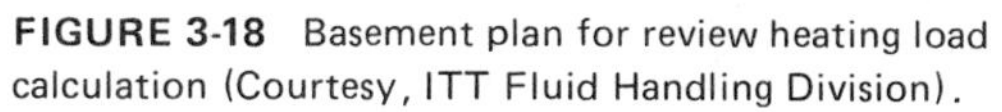

FIGURE 3-18 Basement plan for review heating load calculation (Courtesy, ITT Fluid Handling Division).

FIGURE	ROOM NAME												
	DIMENSIONS, FEET												
			FACTOR										
3-4	GROSS EXPOSED SQ. FT. INCLUDING BASEMENT	1											
		2											
		3											
3-2	WINDOWS SQ. FT. INCLUDING BASEMENT	1											
		2											
3-3	DOORS SQ. FT.	1											
		2											
3-4	NET EXPOSED WALL SQ. FT. INCLUDING BASEMENT	1											
		2											
		3											
3-5	CEILING SQ. FT.	1											
		2											
3-6	FLOORS SQ. FT.	1											
		2											
3-7	SLAB FLOOR LINEAR FT.	1											
3-4	PARTITIONS SQ. FT.	1											
	ROOM SUB-TOTAL												
3-10	DUCT LOSS	1											
	ROOM TOTAL BTUH												
	BUILDING TOTAL (TOTAL OF ALL ROOMS) = BTUH												
	FURNACE MODEL												

FIGURE 3-19 Data sheet for review calculation.

4

EQUIPMENT SELECTION AND AIR DISTRIBUTION

OBJECTIVES

After studying this chapter, the student will be able to

- Evaluate the furnace selection and the air distribution system for an existing residence
- Determine the input value of fuel used for a heating system
- Measure the air quantity actually being circulated in a forced warm air heating system

DIAGNOSING A PROBLEM

To diagnose a problem and then decide what is needed to correct it, a service technician must first determine whether the difficulty is a system problem or a mechanical problem. Some typical system problems are:

- Drafts
- Uneven temperature
- Not enough heat

Drafts are currents of relatively cold air that cause discomfort when they come in contact with the body. Drafts may be caused by a downflow of cold air from an outside wall or window.

Uneven temperatures can cause discomfort and are generally due to the following:

1. Different temperatures in a room near the floor and near the ceiling (sometimes called *stratification*) (Figure 4-1)
2. Different temperature in one room compared to another
3. Different temperature on one floor level as compared to another

Uneven room temperatures may be caused by incorrect supply diffuser locations. Cold outside surfaces of a room should be warmed by properly located supply air outlets.

FIGURE 4-1 Stratification of room air.

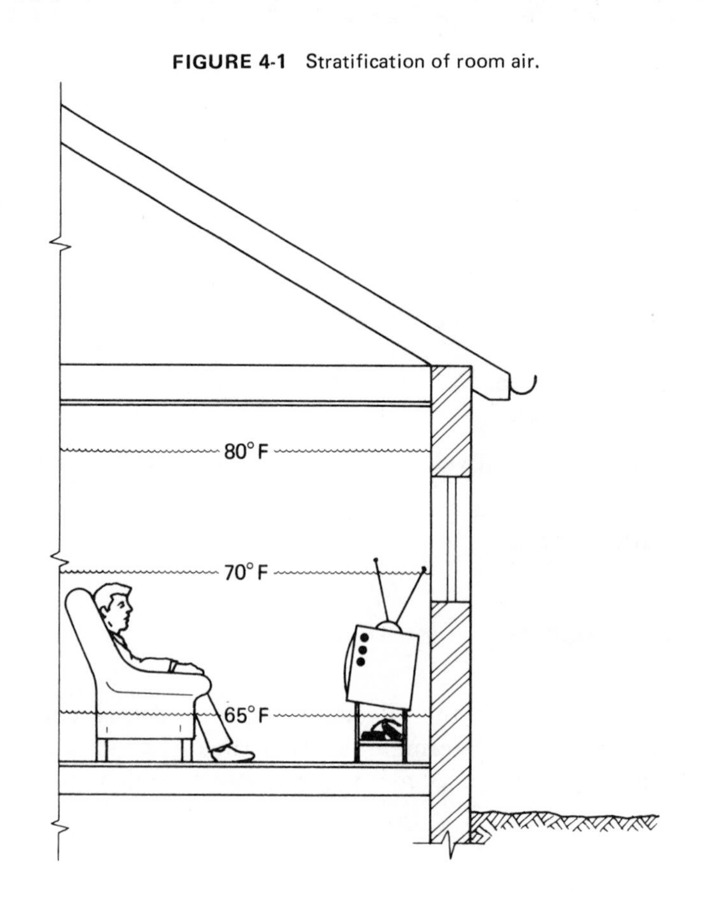

Temperature difference between rooms may be caused by improper balancing of the system due to lack of duct dampers in the branch run to each outlet. Balancing is the process of regulating the flow of air into each room to produce even temperatures.

Not enough heat may be caused by too small a furnace or improper fuel input to the unit to match the heating load of the house.

Some system problems can be solved, or the condition improved, by the service technician. However, some problems are built into the design of the system and can only be corrected by redesign or replacement of major components.

EVALUATING THE SYSTEM

The three items that should be considered are the

- Furnace
- Fan
- Air distribution system

Furnace

Two questions relate to the furnace:

1. Has the proper size been selected
2. Is the furnace adjusted to produce its rated output

In Chapter 3 a method was given for determining the heating load of a building. Any losses that occurred, such as duct loss or ventilation air, are added to the total room loss to determine the total required heating load. The furnace output rating should be equal to, but not greater than, 15% higher than the total required heating load. If the furnace is too small, it will not heat properly in extreme weather. If it is too large, the "off" cycles will be too long and the result will be uneven heating.

To determine if a furnace is producing its rated capacity, it is necessary to check the

- Fuel input
- Combustion efficiency

FUEL INPUT GAS:

Many manufacturers void their warranty if the gas input is not adjusted within 2% of the rated input of the furnace. This can be easily done by:

1. Obtaining the Btu/ft^3 rating of the gas from the local utility*
2. Turning off all other appliances and operating the gas furnace continuously during the test
3. Measuring the length of time it takes for the furnace to consume 1 ft^3 of gas

For example, if it takes 37 seconds to use 1 ft^3 of gas (using the 1-ft^3 dial on the meter) and the gas rating is 1025 Btu/ft^3, then the input is 1025 Btu/ft^3 $\times$ 97 ft^3 = 99,425 Btuh. The 97 is found on the 37-s line in the 1-ft^3 dial column in Figure 4-2. If the furnace is rated at 100,000 Btuh input, the usage would be within requirements.†

Some meters have low input dials other than 1 ft^3. Using the table shown in Figure 4-3 and the method described, a similar check on the gas input can be made. If the input is not correct, it should be changed to the furnace input rating by adjusting the gas-pressure regulating valve or by changing the size of the burner orifices (see the manufacturer's instructions for details).

FUEL INPUT OIL:

The input rating to an oil furnace can be checked by determining the oil burner nozzle size and measuring the oil pressure.

FIGURE 4-2 Typical gas furnace nameplate; face of typical gas meter.

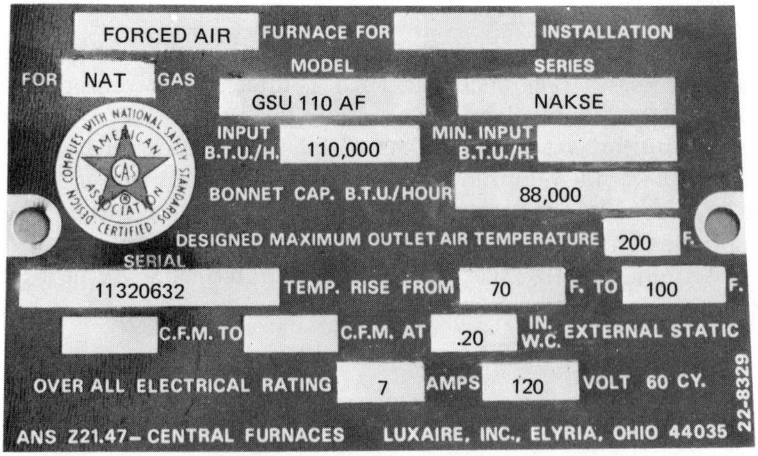

* Consult the local gas company for this information.

† Input rating for a gas furnace is given on the furnace nameplate.

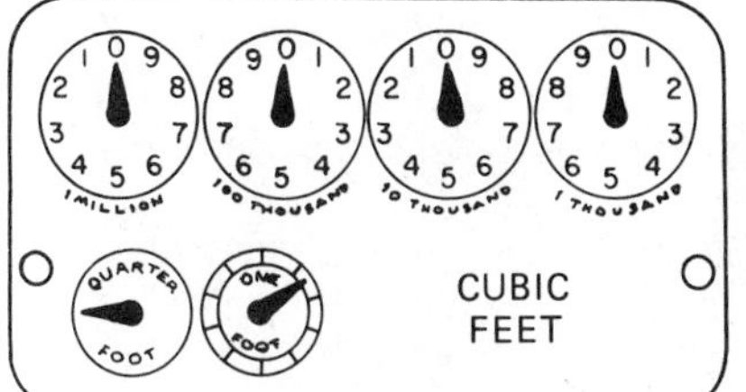

THE DIAL IS MARKED AS TO HOW MUCH GAS IS MEASURED FOR EACH REVOLUTION. USING THE NUMBER OF SECONDS FOR ONE REVOLUTION, AND THE SIZE OF THE TEST DIAL, FIND THE CUBIC FEET OF GAS CONSUMED PER HOUR FROM THE TABLE

SECONDS FOR ONE REV.	SIZE OF TEST DIAL ¼ CU.FT.	½ CU.FT.	1 CU.FT.	2 CU.FT.	5 CU.FT.	SECONDS FOR ONE REV.	SIZE OF TEST DIAL ¼ CU.FT.	½ CU.FT.	1 CU.FT.	2 CU.FT.	5 CU.FT.
10	90	180	360	720	1800						
11	82	164	327	655	1636	36	25	50	100	200	500
12	75	150	300	600	1500	→37	–	–	→97	195	486
13	69	138	277	555	1385	38	23	47	95	189	474
14	64	129	257	514	1286	39	–	–	92	185	462
15	60	120	240	480	1200	40	22	45	90	180	450
16	56	113	225	450	1125	41	–	–	–	176	439
17	53	106	212	424	1059	42	21	43	86	172	429
18	50	100	200	400	1000	43	–	–	–	167	419
19	47	95	189	379	947	44	–	41	82	164	409
20	45	90	180	360	900	45	20	40	80	160	400
21	43	86	171	343	857	46	–	–	78	157	391
22	41	82	164	327	818	47	19	38	76	153	383
23	39	78	157	313	783	48	–	–	75	150	375
24	37	75	150	300	750	49	–	–	–	147	367
25	36	72	144	288	720	50	18	36	72	144	360
26	34	69	138	277	692	51	–	–	–	141	355
27	33	67	133	267	667	52	–	–	69	138	346
28	32	64	129	257	643	53	17	34	–	136	240
29	31	62	124	248	621	54	–	–	67	133	333
30	30	60	120	240	600	55	–	–	–	131	327
31	–	–	116	232	581	56	16	32	64	129	321
32	28	56	113	225	563	57	–	–	–	126	316
33	–	–	109	218	545	58	–	31	62	124	310
34	26	53	106	212	529	59	–	–	–	122	305
35	–	–	103	206	514	60	15	30	60	120	300

FIGURE 4-3 Gas meter measuring dials and table for determining gas input based on dial readings (Courtesy, Bard Manufacturing Company).

Nozzles are rated for a given amount of oil flow at 100 pounds per square inch gauge (psig) oil pressure. One gallon of Grade No. 2 oil contains 140,000 Btu. The oil burner nozzle is examined and the input flow is read on the nozzle in gallons per hour (gal/h). Thus, a nozzle rated at 0.75 gal/h operating with an oil pressure of 100 pounds per square inch (psi) would produce an input rating of 105,000 Btu (0.75 gal/h × 140,000 Btu). The oil pressure is measured and adjusted, if necessary, to 100 psig within 3%. Figure 4-4 illustrates a typical oil furnace nameplate and oil burner nozzle.

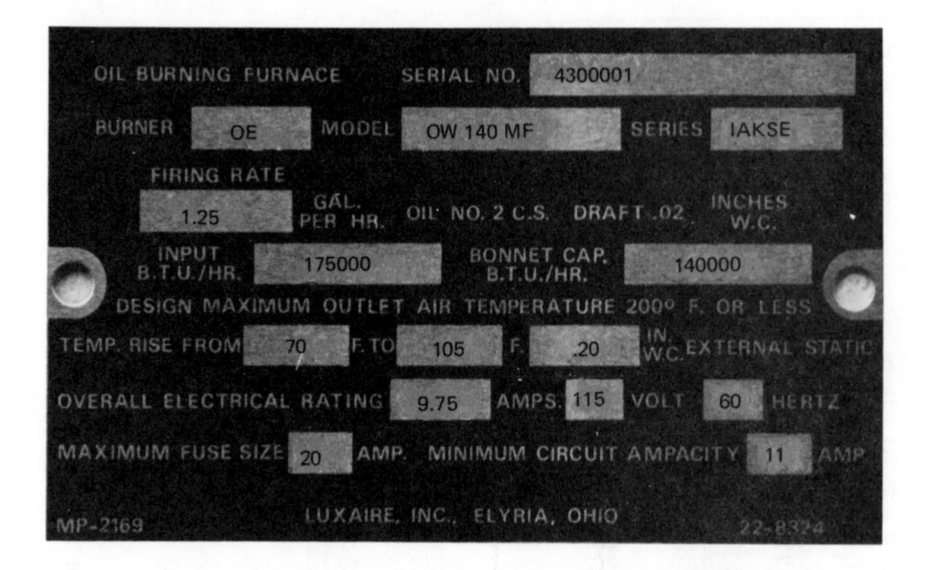

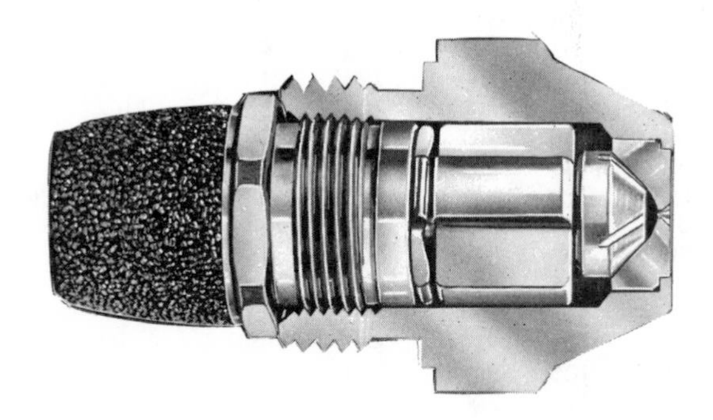

FIGURE 4-4 Typical oil furnace nameplate and oil burner nozzle (Courtesy, Delavan Corporation).

FUEL INPUT ELECTRIC:

For an electric furnace the input is usually rated in watt-hours (Wh) or kilowatt-hours (kWh): (1000 Wh = 1 kWh of electricity consumed). Watts (W) can be converted to Btu by multiplying by 3.4131 (1 W = 3.4131 Btu). For example, if an electric furnace is rated at 30 kWh, the input is 102,390 Btuh (30 kWh × 1000 h/kWh × 3.4131 Btu). Figure 4-5 illustrates a typical electric furnace nameplate.

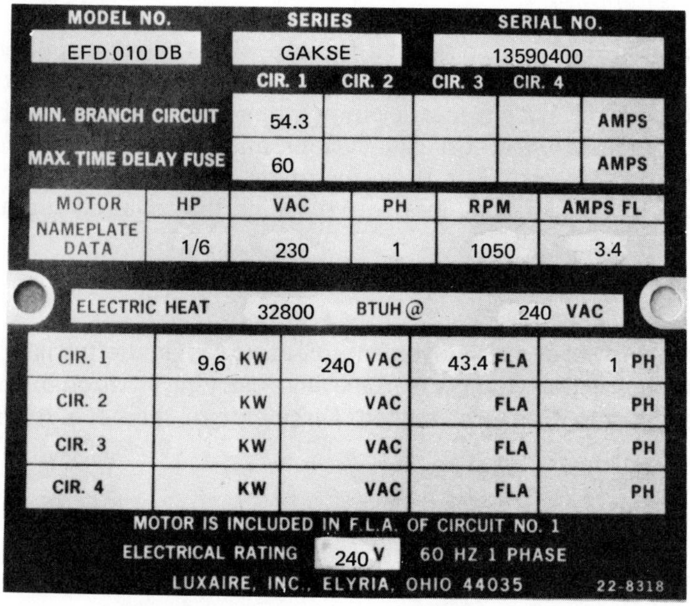

MODEL NO.	SERIES	SERIAL NO.
EFD 010 DB	GAKSE	13590400

	CIR. 1	CIR. 2	CIR. 3	CIR. 4	
MIN. BRANCH CIRCUIT	54.3				AMPS
MAX. TIME DELAY FUSE	60				AMPS

MOTOR NAMEPLATE DATA	HP	VAC	PH	RPM	AMPS FL
	1/6	230	1	1050	3.4

ELECTRIC HEAT 32800 BTUH @ 240 VAC

CIR. 1	9.6 KW	240 VAC	43.4 FLA	1 PH
CIR. 2	KW	VAC	FLA	PH
CIR. 3	KW	VAC	FLA	PH
CIR. 4	KW	VAC	FLA	PH

MOTOR IS INCLUDED IN F.L.A. OF CIRCUIT NO. 1
ELECTRICAL RATING 240 V 60 HZ 1 PHASE
LUXAIRE, INC., ELYRIA, OHIO 44035 22-8318

FIGURE 4-5 Typical electrical furnace nameplate.

EFFICIENCY:

For both gas and oil furnaces, the output rating is reduced from the input rating by an efficiency factor. However, for an electric furnace, the input is equal to the output since there is no heat loss up the chimney. The general formula is

$$\text{Btuh output} = \text{Btuh input} \times \text{efficiency factor}$$

The output rating of both gas and oil furnaces is shown on the nameplate as 80% of input (efficiency factor = 0.80). However, the efficiency of the installed furnace is usually lower. A service technician should perform an efficiency test to determine if the rated efficiency is actually being maintained. The method of determining and adjusting combustion efficiency for both gas and oil furnaces is covered in Chapter 10. In most cases where the combustion efficiency is low, the technician can adjust the fuel burner so that the furnace might again approach the 80% figure. Having measured the efficiency of the furnace, the actual output can be calculated from the formula given. For example, if a gas furnace is operating at 80% efficiency* and has an input of 100,000 Btuh, the output (bonnet capacity) is 80,000 Btuh (100,000 Btuh × 0.80 efficiency). An electric furnace with 30 (kWh) (102,390 Btuh) input, operates at 100% efficiency and has an output of 102,390 Btuh (102,390 Btuh × 1.00 efficiency).

* Manufacturer's rating.

SYSTEM EFFICIENCY:

It is seldom desirable to select a furnace with a rated output equal to the heat loss of the house. If this were done the furnace could be too small due to system losses or system inefficiency. Other losses can occur that are not accounted for in the heat-loss calculations, and combustion efficiencies may not always be achieved. For these reasons it is good practice to add 10 or 15% to the calculated heat loss to determine the required furnace output.

Fan

The fan produces the movement of air through the furnace, absorbing heat from the heat-exchanger surface and carrying it through the distribution system to the areas to be heated. Discussion of the fan involves

- Air volume
- Static pressure
- Causes of poor air distribution

AIR VOLUME:

For the system to operate satisfactorily, the fan must deliver the proper air volume. Most furnaces permit some flexibility (variation) in the air volume capacity of the furnace. A furnace having an output of 80,000 Btuh may be able to produce 800, 1200 and 1600 cubic feet per minute (cfm) of air volume, depending on the requirements. Systems designed for heating only usually require less air than do systems designed for both heating and cooling. The proper air volume for heating is usually determined by the required temperature rise. The temperature rise is the supply air temperature minus the return air temperature at the furnace. Systems used for heating only should be capable of heating the air 85° F. Systems designed for heating and cooling should be capable of heating the air a minimum of 70° F.

To find the quantity of air the furnace is actually circulating, a service technician measures the temperature rise. One thermometer is placed in the return air plenum and the other is placed in the supply air duct. While the furnace is operating continuously the readings are taken and the temperature rise (difference) computed. Thermometer locations for checking temperature rise are shown in Figure 4-6.

For a typical heating-only application, the return air temperature is 65° F and supply air temperature 150° F, indicating an 85° F temperature rise (150° F - 65° F = 85° F). For a typical heating/cooling system, the return air temperature is 65° F and the supply air temperature is 125° F, indicating a temperature rise of 70° F (135° F - 65° F = 70° F). The air volume circulated is proportionately different for the two types of systems. To determine air

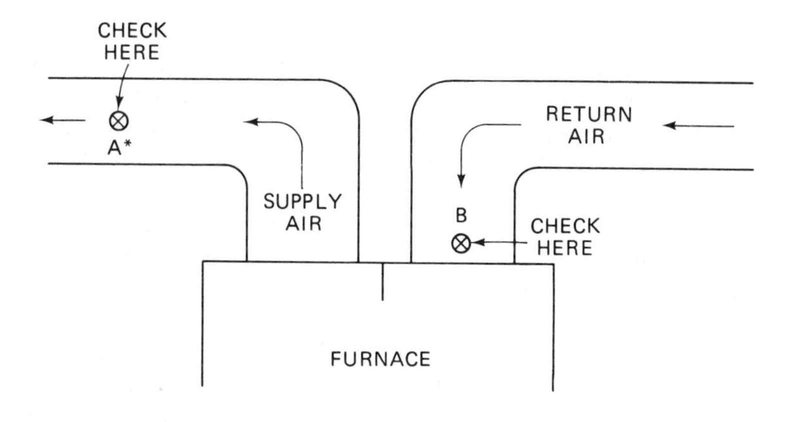

* CHECK POINT "A" MUST BE FAR ENOUGH DOWNSTREAM THAT THE THERMOMETER IS NOT EXPOSED TO RADIANT HEAT FROM THE HEAT EXCHANGER.

FIGURE 4-6 Thermometer locations for checking temperature rise.

volume the following formula is used:

$$\text{cfm} = \frac{\text{Btuh output of furnace}}{\text{temperature rise } (^\circ\text{F}) \times 1.08}$$

Thus, if the temperature rise on an 80,000-Btu output furnace is 85° F, the air volume is

$$\text{cfm} = \frac{80{,}000 \text{ Btuh}}{85^\circ \text{F} \times 1.08} = 870 \text{ cfm}$$

If the temperature rise on an 80,000-Btuh output furnace is 70° F, the air volume is

$$\text{cfm} = \frac{80{,}000 \text{ Btuh}}{70^\circ \text{F} \times 1.08} = 1058 \text{ cfm}$$

Many furnaces have two-speed fans, with the low speed used for heating and the high speed for cooling. Adjustments in the fan speed can be made on most furnaces to regulate the air volume to meet the requirements of the heating system. This will be discussed further in Chapter 7.

EXTERNAL STATIC PRESSURE OF THE FAN:

A manufacturer rates the fan air volume of a furnace to produce each quantity of air at a certain external static pressure. *Static pressure* is the resistance to air flow offered by any component through which the air passes.

External static pressure is the resistance of all components outside the furnace itself. Thus, if a unit is rated to supply 600 cfm at an external static pressure of 0.20 in., it means that the total resistance offered by supply ducts + return ducts + supply diffusers + return grilles must not exceed 0.20 in. of static pressure. A system external and internal static pressure diagram is shown in Figure 4-7.

Static pressure is measured in inches of water column (W.C.) that the pressure of the fan is capable of raising on a water gauge *manometer* (Figure 4-8). A manometer is an instrument for measuring pressure of gases and vapors.

The pressures in residential systems are small, so an inclined tube manometer is used to increase the accuracy of the readings taken. A manometer indicates the air pressure delivered by the fan above atmospheric pressure.

To illustrate how small these fan readings are, 1 atmosphere 14.7 psi is equal to 408 in. of water column (W.C.). Two tenths of an inch (0.20 in. W.C.) of pressure is 1/2040 of an atmosphere.

A great deal of care must be taken in designing an air distribution system to stay within the rated external static pressure of the furnace. If the system resistances are too high, the amount of air flow of the furnace is reduced.

FIGURE 4-7 System external and internal static pressure diagram.

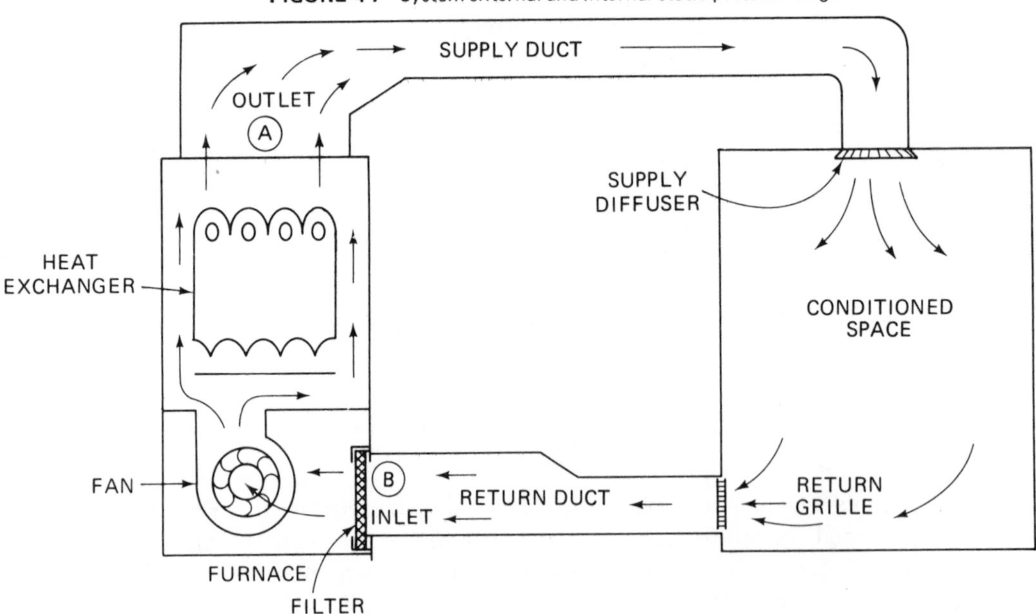

THE EXTERNAL STATIC –
FROM THE OUTLET AT "A" THROUGH THE SUPPLY DUCT, SUPPLY DIFFUSER, RETURN GRILLE, AND BACK THROUGH THE RETURN DUCT TO "B"

THE INTERNAL STATIC –
FROM THE INLET OF THE FURNACE AT "B" THROUGH THE FILTER, HEAT EXCHANGER AND FAN TO THE OUTLET AT "A".

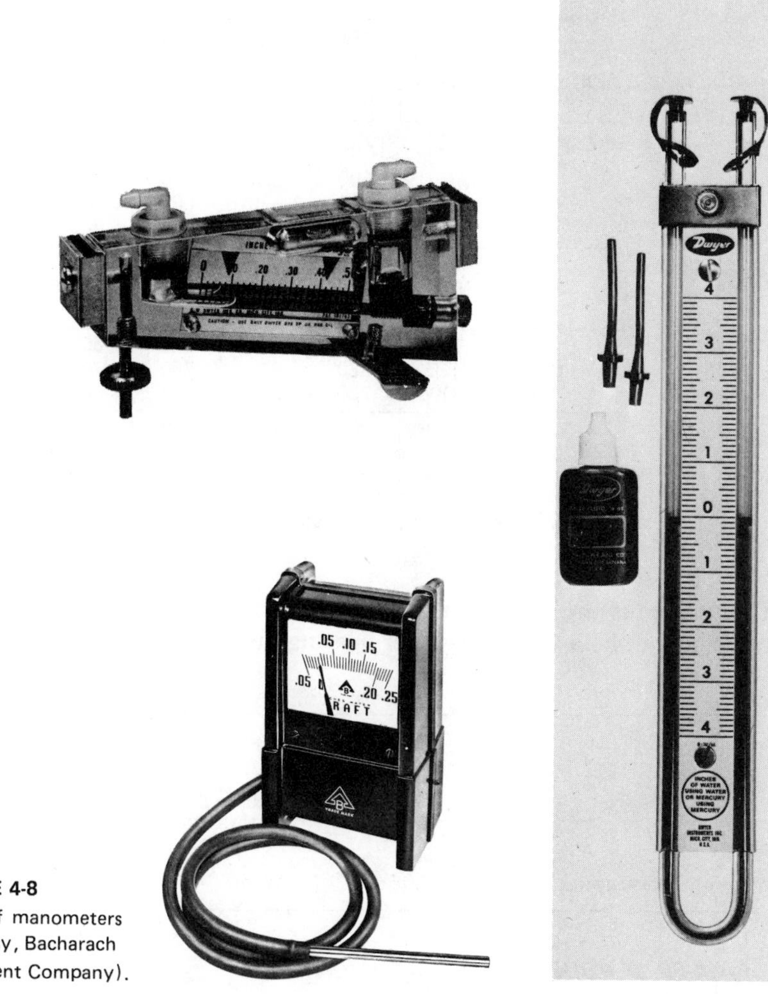

FIGURE 4-8
Types of manometers (Courtesy, Bacharach Instrument Company).

All manufacturers' ratings of external static pressure are based on clean filters. Dirty filters can cause reduced air volume and poor heating. A table showing the relationship between increased external static pressure and decreased air volume (cfm) is illustrated in Figure 4-9.

The greater the air volume and static pressure of a furnace, the larger the horsepower of the motor required to deliver the air. Therefore, units used for cooling, operating at higher air volumes and higher static pressures, require higher-horsepower motors than do units used for heating only. The speed or revolutions per minute (rpm) of the fan wheel must often be increased for cooling. The wet cooling coil installed external to the basic furnace has a resistance to the air flow in addition to duct work and grilles. The total external static pressure of a cooling system may be as high as 0.50 in.

UPFLOW FURNACES

FURNACE MODEL NO.	SPEED RANGE	CFM AIR FLOW AT EXTERNAL STATIC PRESSURE (IN. H_2O)			
		0.20	0.30	0.40	0.50
UPFLOW – (DIRECT DRIVE)					
A	HIGH MED. LOW	600 550 500	580 520 470	560 500 440	540 470 400
B	HIGH MED. LOW	800 720 665	750 680 630	700 640 590	650 600 550
C	HIGH MED. LOW	1000 850 700	940 800 670	870 750 630	800 700 600
D	HIGH MED. LOW	800 — 710	750 — 760	700 — 690	650 — 540

FIGURE 4-9 Manufacturers' data showing static pressures.

W.C. static pressure. Following are given two typical performance ratings for a unit that can be applied to either heating only or heating and cooling.

Use	*Air volume (cfm)*	*External static (in. W.C.)*	*Motor hp*	*Temperature rise (°F)*
Heating only	870	0.20	$\frac{1}{8}$	85
Heating and cooling	1058	0.50	$\frac{1}{4}$	70

CAUSES OF POOR AIR DISTRIBUTION:

One of the first jobs of a service technician in evaluating the distribution system is to determine whether it has been designed for heating only or for heating and cooling. If the air volume is found to be too low, it can be caused by one or more of the following:

1. Incorrect fan speed
2. Closed or partially closed dampers
3. Dirty filters
4. Incorrectly sized ducts
5. High-pressure-drop duct fittings

If improper fan speed is the problem, the technician can check on the possibility of speeding it up. This may require a larger motor. The increased speed must be kept within permissible noise levels.

If closed dampers are the problem, adjustments can be made. Dampers are installed for balancing the system (regulating the air flow to each room) (Figure 4-10). Misadjustments should be corrected and, if necessary, the system rebalanced.

If dirty filters are the problem, the owner should be advised of proper preventive maintenance measures.

If duct sizes are the problem, some improvement may be made by speeding up the fan. However, there are limitations on improving this condition without a major redesign.

High-pressure-drop duct fittings are a problem that can often be corrected by minor changes in the duct work. The pressure drop in duct fittings is usually given in terms of the equivalent length of straight duct of the same size having an equal pressure drop.

Figure 4-11 shows pressure drops in terms of equivalent lengths for various duct fittings. Abrupt turns and restrictions should be avoided whenever possible. As indicated in Figure 3-10, where duct work is placed in an unheated attic, crawl space, or garage, insulated duct work should be used. Various types of flexible ducts and methods of insulating ducts are illustrated in Figures 4-12 and 4-13.

Supply Air Distribution System

In general, there are three types of supply air distribution systems:

1. Radial
2. Perimeter loop
3. Trunk duct and branch

FIGURE 4-10 Balancing dampers.

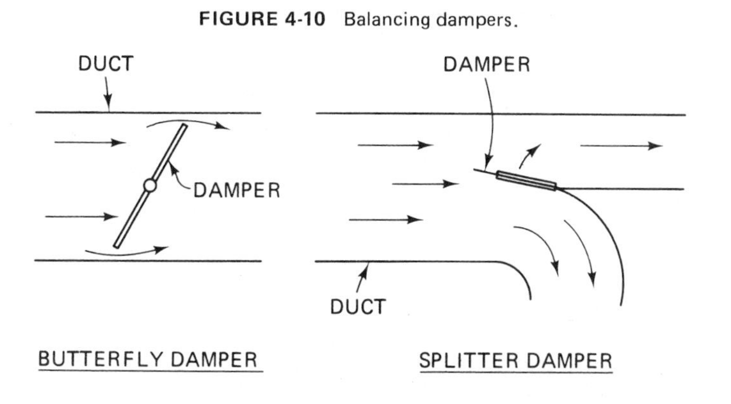

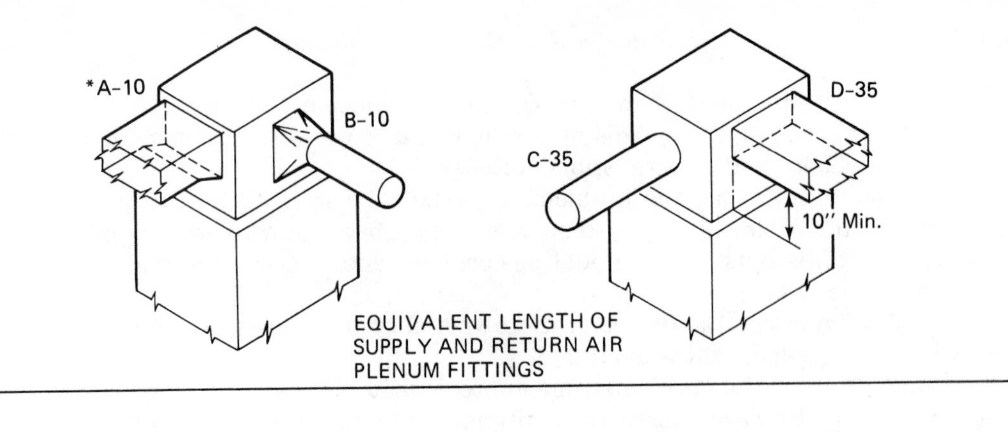

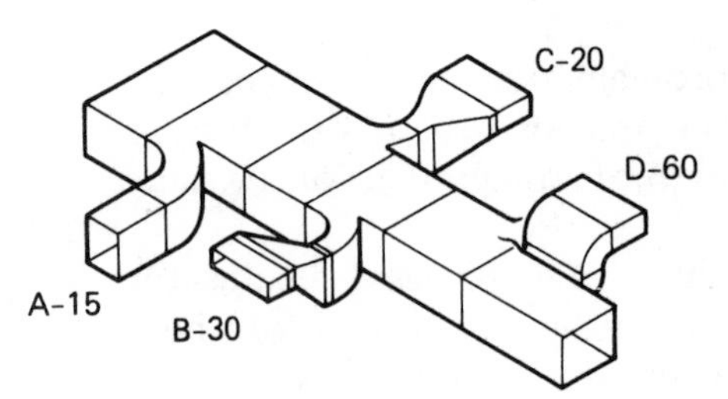

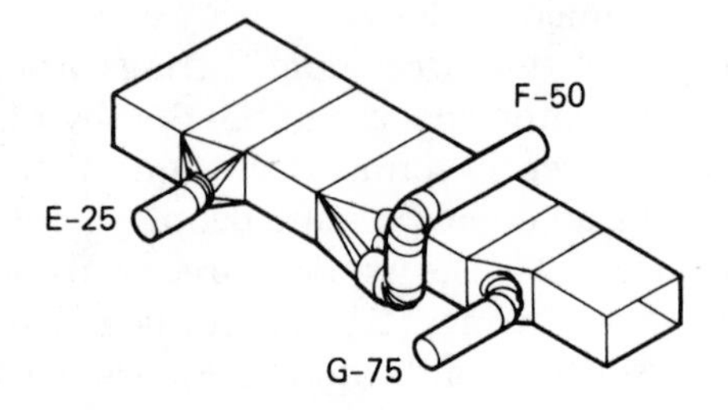

EQUIVALENT LENGTH OF REDUCING TRUNK FITTINGS

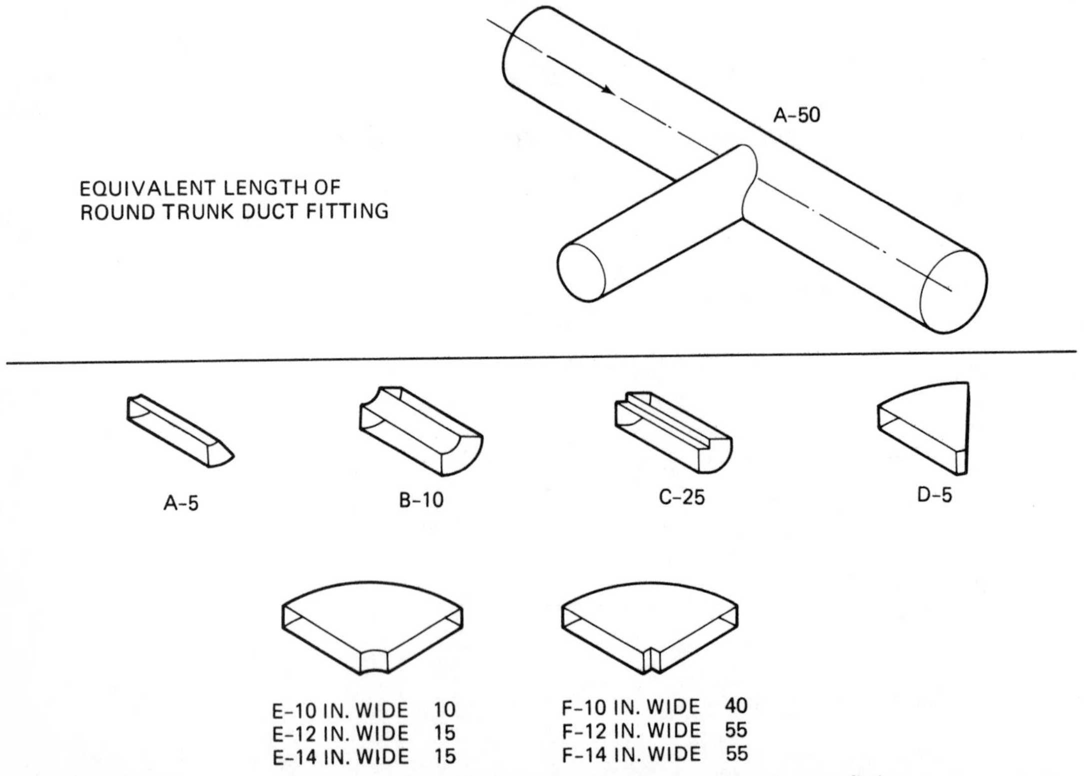

FIGURE 4-11 Equivalent length of supply and return fittings.

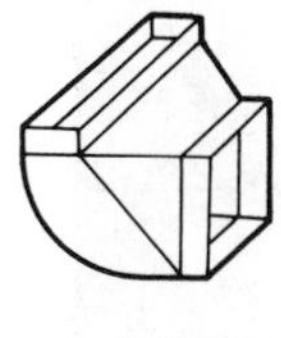

G-30

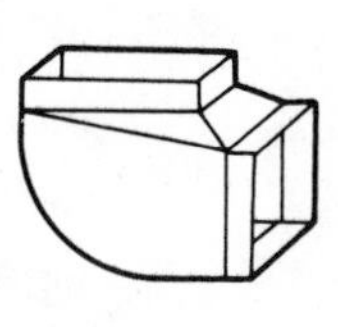

H-35

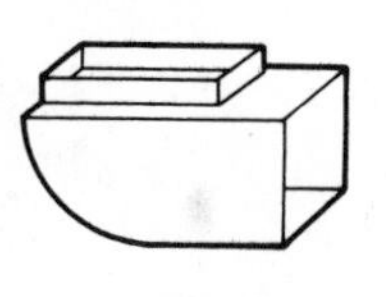

J-60

K-5

L-10

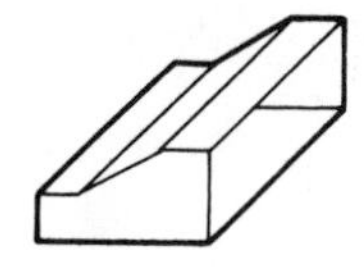

M-5

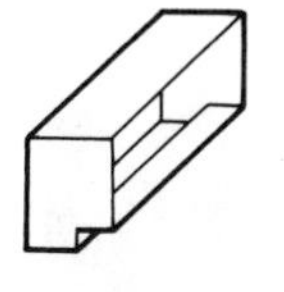

N-15

O-15

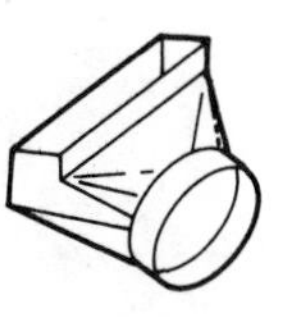

P-30

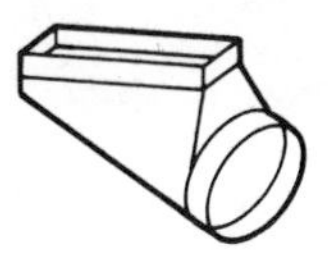

Q-50

EQUIVALENT LENGTH OF ANGLES AND ELBOWS FOR INDIVIDUAL AND BRANCH DUCT AND BOOT FITTINGS

*NOTE: THE LETTER NEXT TO EACH FITTING IDENTIFIES IT'S STYLE; THE FIGURE GIVEN IS THE EQUIVALENT LENGTH OF STRAIGHT DUCT HAVING THE SAME AIR RESISTANCE.

FIGURE 4-11 Continued (Reprinted with permission from the 1976 Systems Volume, ASHRAE Handbook & Product Directory).

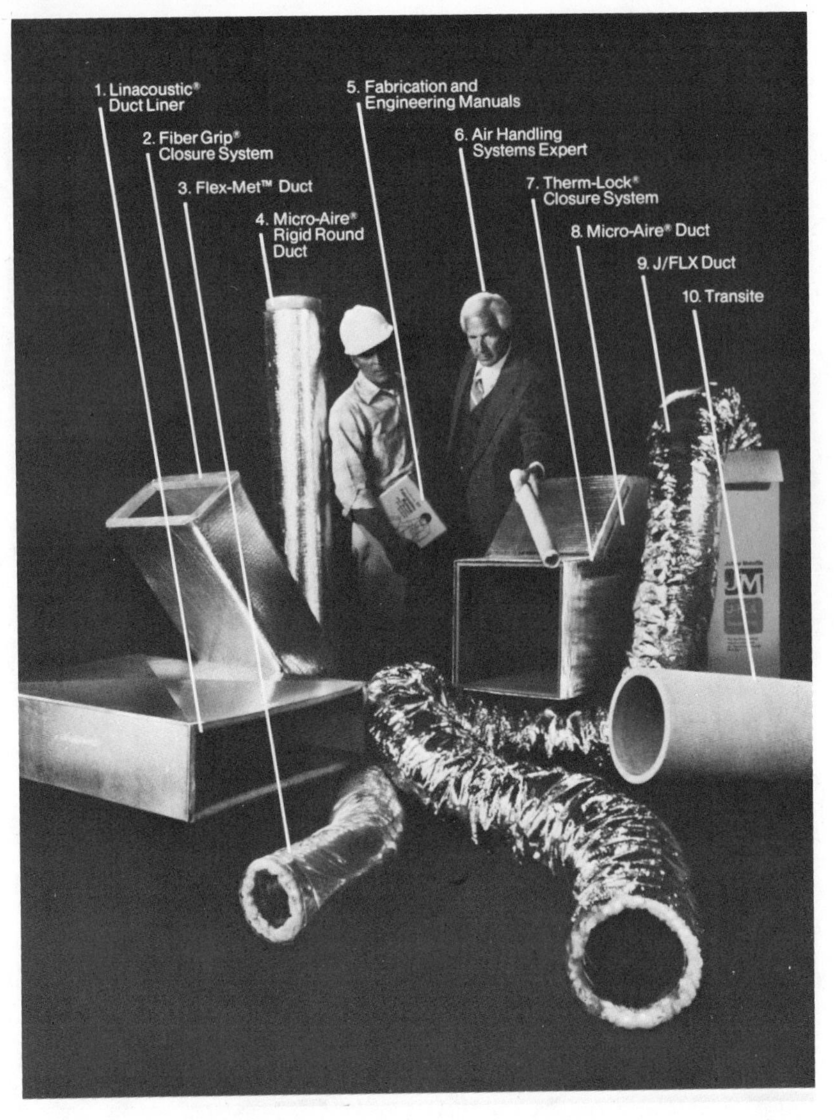

FIGURE 4-12 Example of flexible ducts and duct insulation (Courtesy, Johns-Manville Sales Corporation).

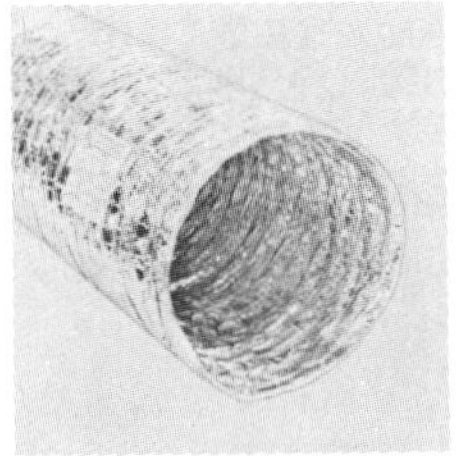

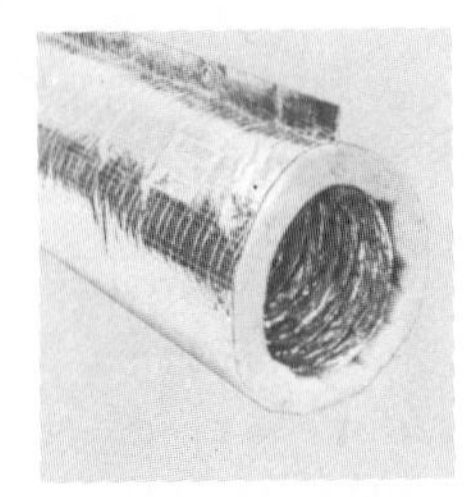

FIGURE 4-13 Various types of flexible ducts (Courtesy, Anco Products, Inc.).

Selection of a system depends upon the house construction and room arrangement. Each system has its advantages for certain types of applications.

RADIAL:

A radial system consists of a number of single pipes running from the furnace to the supply air outlets (Figure 4-14). In these systems the furnace is located near the center of the house so that the various runouts (supply air ducts) will be as nearly equal in length as possible. This system helps to provide warm floors and is used with both crawl-space construction and concrete slab floors.

PERIMETER LOOP:

The perimeter loop system (Figure 4-15) is a modification of the radial system. A single duct running along the perimeter (outer edge) of the house supplies air to each supply air diffuser located on the floor above. Radial ducts connect the perimeter loop to the furnace. The runout ducts from the furnace to the loop are larger and fewer than those in the radial system. This system lends itself for use with slab floor construction. The perimeter loop supplies extra heat along the perimeter of the house, warming the floor and heating the exterior walls of the house. The greatest heat loss in a slab floor is near its perimeter.

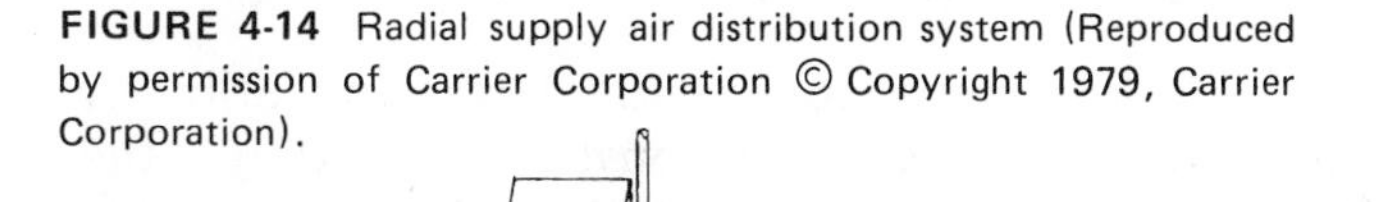

FIGURE 4-14 Radial supply air distribution system (Reproduced by permission of Carrier Corporation © Copyright 1979, Carrier Corporation).

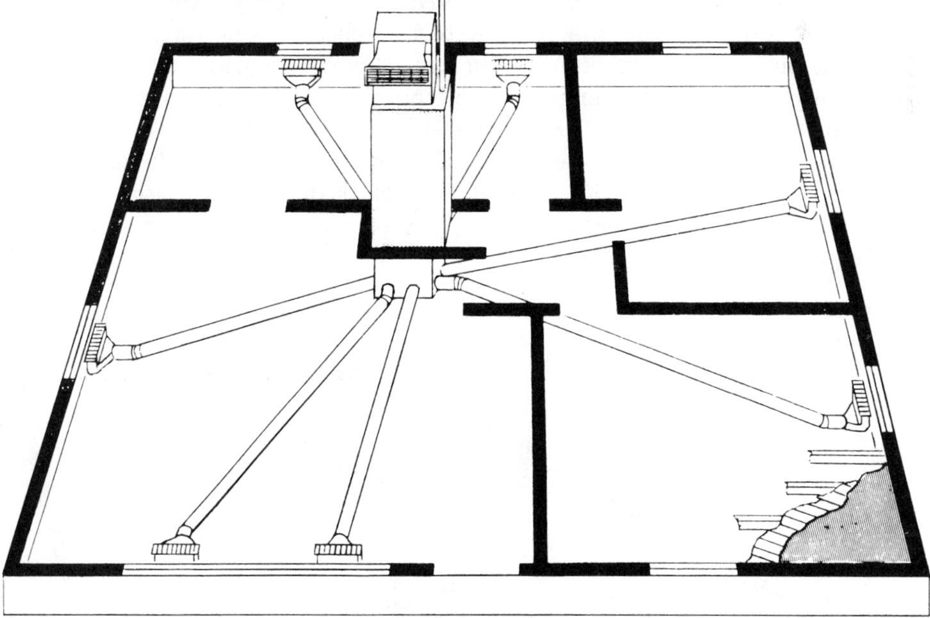

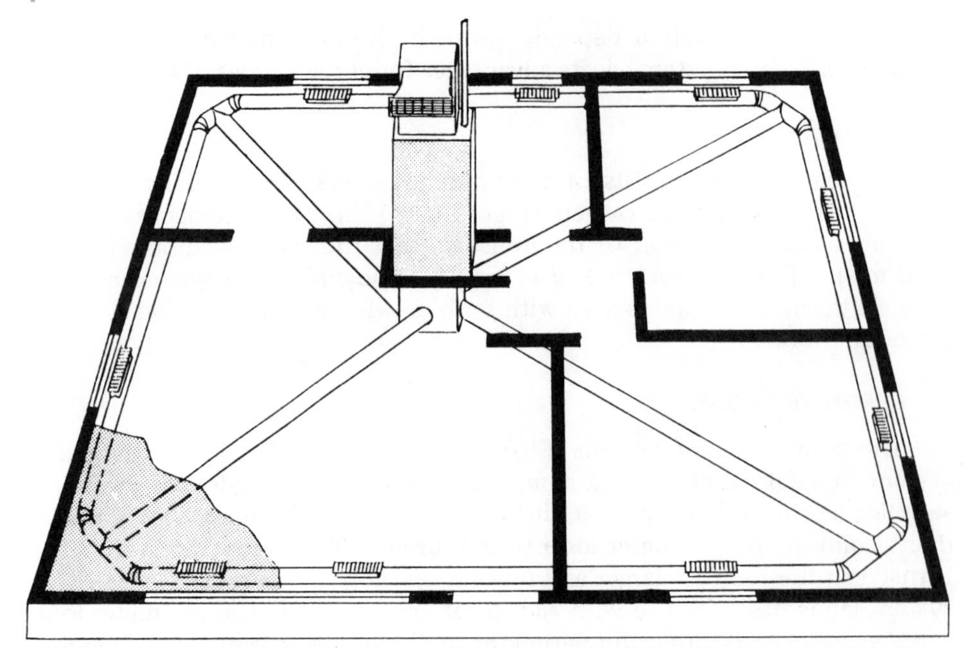

FIGURE 4-15 Perimeter loop supply air distribution system (Reproduced by permission of Carrier Corporation © Copyright 1979, Carrier Corporation).

TRUNK DUCT AND BRANCH:

The trunk-and-branch system is the most versatile of the systems. It permits the furnace to be located in any convenient location in the house. Large ducts carry the air from the furnace to branches or runouts going to individual air outlets. The trunk duct (large duct from furnace) may be the same size throughout its entire length if the length does not exceed 25 ft. This is called an extended plenum (Figure 4-16). On large systems the trunk duct should be reduced in size, after branches remove a portion of the air, so that it tapers toward the end (Figure 4-17). A trunk duct can be used for basement installations or overhead installations of duct work. It can be used for both supply and return systems.

FIGURE 4-16 Extended plenum air supply distribution system (Reproduced with permission of Carrier Corporation © Copyright 1979, Carrier Corporation).

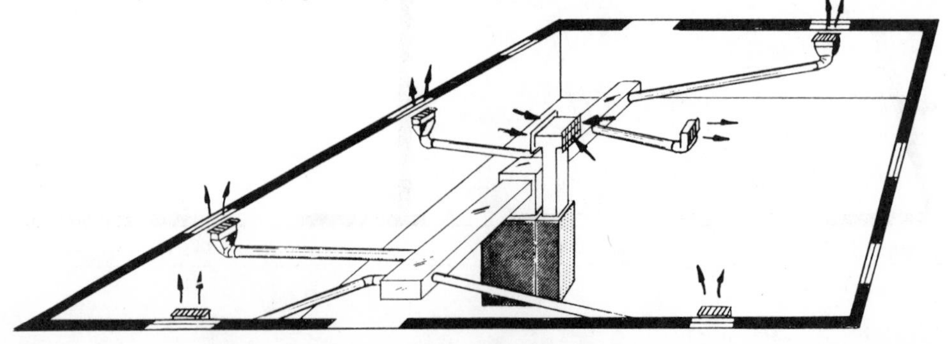

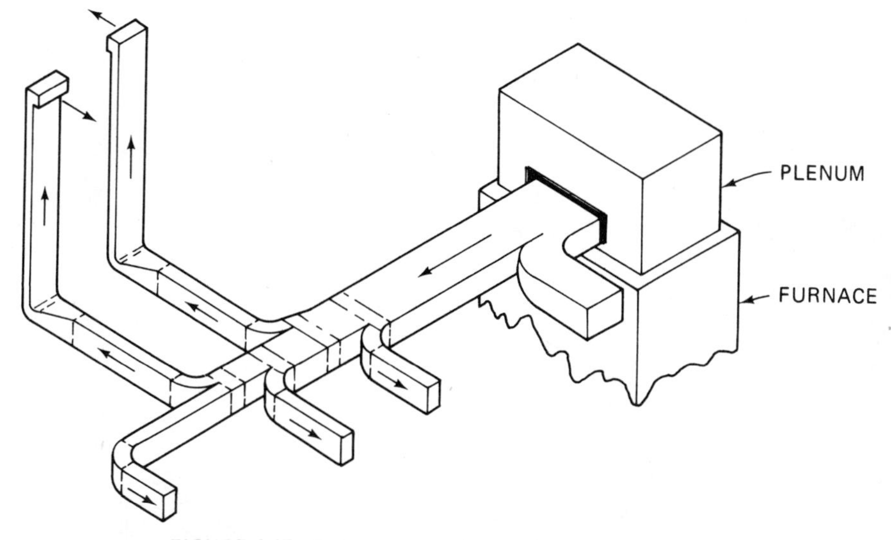

FIGURE 4-17 Reducing trunk duct used in supply or return air distribution systems.

SUPPLY AIR DIFFUSER PLACEMENT:

The placement of supply air diffusers, or registers, depends to some extent on the climate of the area in which the house is located. In northern climates, where cold floors can be a problem and where outside exposures must be thoroughly heated, a perimeter location of the supply air diffusers is best. The register can be located in the floor adjacent to each exposed wall or in the baseboard. In many southern climates, where cooling air distribution is more important than heating air distribution, diffusers can be located on an inside wall as high as 6 ft above the floor. In extremely warm climates, the supply air outlet can be placed in the ceiling. Figure 4-18 shows various locations for the placement of air supply distribution outlets. Figure 4-19 shows various types of diffusers, registers, and grilles, some of which are used for supply outlets and some for return inlets.

Return Air Distribution System

The return air distribution system is usually of simple construction, having less pressure drop than the supply system. The radial system or the trunk-and-branch system can be used. Where possible, return air is carried in boxed-in (enclosed) joist spaces (Figure 4-20).

RETURN AIR GRILLE PLACEMENT:

In northern climates, where supply air diffusers are located along the perimeter, return air grilles (returns) are usually placed near the baseboard

[A] REGISTERS SET TO DIRECT AIR UPWARD ALONG THE WALL AT AS WIDE AN ANGLE AS POSSIBLE

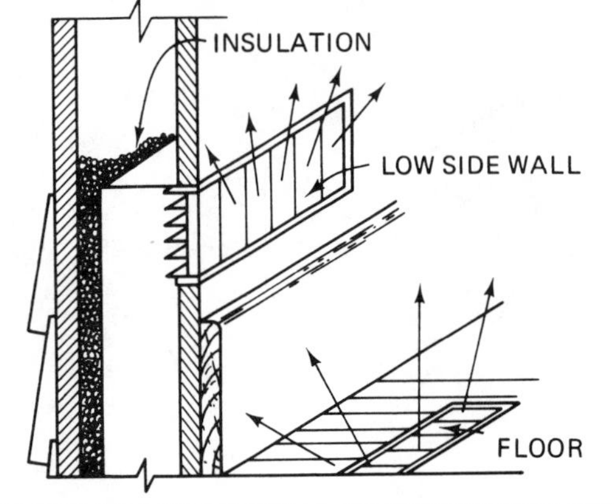

FLOOR OR LOW SIDEWALL PERIMETER OUTLETS[A]

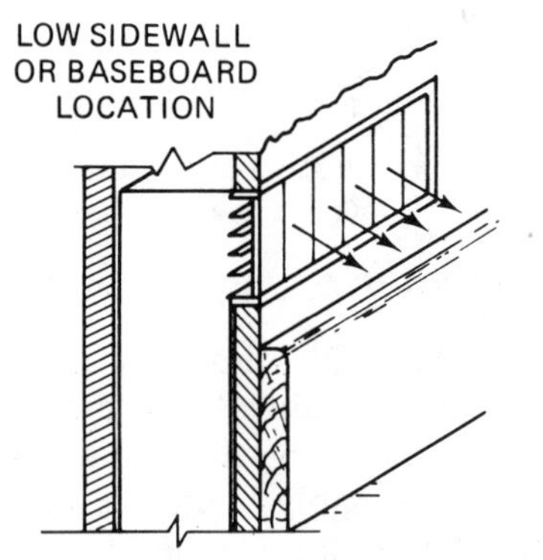

[A] VERTICAL BARS WITH ADJUSTABLE DEFLECTION, OR FIXED VERTICAL BARS WITH DEFLECTION TO RIGHT AND LEFT NOT EXCEEDING ABOUT 22 DEG. FOR LOW SIDEWALL LOCATION, THE DEFLECTION FOR HORIZONTAL, MULTIPLE VANE REGISTERS SHOULD NOT EXCEED 22 DEG. FOR BASEBOARD LOCATIONS, THE DEFLECTION FOR HORIZONTAL, MULTIPLE VANE REGISTERS SHOULD NOT EXCEED ABOUT 10 DEG.

RECOMMENDED TYPE OF BASEBOARD AND LOW SIDEWALL INSTALLATION ON WARM WALL[A]

CEILING

HIGH SIDEWALL LOCATION

[A] HORIZONTAL VANES, IN BACK OR FRONT, TO GIVE DOWNWARD DEFLECTIONS NOT TO EXCEED 15 TO 22 DEG.

RECOMMENDED TYPE OF HIGH SIDEWALL INSTALLATION ON WARM WALL[A]

FIGURE 4-18 Placement of air supply distribution outlets (Reprinted with permission from the 1976 Systems Volume, ASHRAE Handbook & Product Directory).

GENERAL CHARACTERISTICS OF OUTLETS

GROUP	OUTLET TYPE	OUTLET FLOW PATTERN	MOST EFFECTIVE APPLICATION	PREFERRED LOCATION	SIZE DETERMINED BY
1	CEILING AND HIGH SIDEWALL	HORIZONTAL	COOLING	NOT CRITICAL	MAJOR APPLICATION—HEATING OR COOLING
2	FLOOR REGISTERS, BASEBOARD AND LOW SIDEWALL	VERTICAL, NONSPREADING	COOLING AND HEATING	NOT CRITICAL	MAXIMUM ACCEPTABLE HEATING TEMPERATURE DIFFERENTIAL
3	FLOOR REGISTERS, BASEBOARD AND LOW SIDEWALL	VERTICAL, SPREADING	HEATING AND COOLING	ALONG EXPOSED PERIMETER	MINIMUM SUPPLY VELOCITY DIFFERS WITH TYPE AND ACCEPTABLE TEMPERATURE DIFFERENTIAL
4	BASEBOARD AND LOW SIDEWALL	HORIZONTAL	HEATING ONLY	LONG OUTLET—PERIMETER, SHORT OUTLET—NOT CTITICAL	MAXIMUM SUPPLY VELOCITY SHOULD BE LESS THAN 300 FPM

FIGURE 4-18 Continued

FIGURE 4-19 Various types of diffusers, grilles and registers (Courtesy, Hart & Cooley Manufacturing Company).

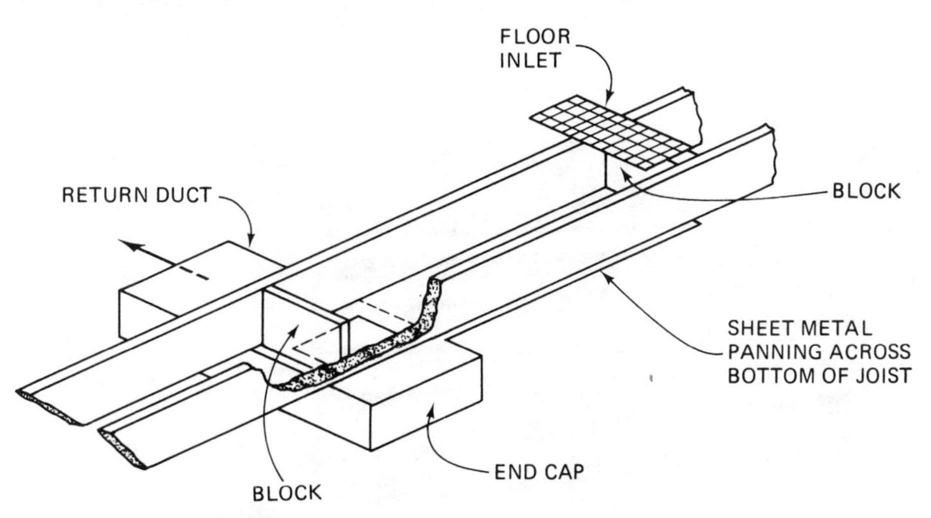

FIGURE 4-20 Boxed-in joist spaces in return air distribution system.

on inside walls. However, since the room temperature is not affected by the location of returns, they may be placed high on the inside walls to prevent drafts. In southern climates, where high inside wall outlets or ceiling diffusers are used, returns can be placed in any convenient location on inside walls. However, short-circuiting of the supply air directly into the return must be avoided. Whereas supply air outlets (one or more) are placed in every room to be heated, connecting rooms with open doorways can share a common return. Returns are not placed in bathrooms, kitchens, or garages. Air for these rooms must be taken from some other area. The general rule is that the total duct area for the return air system must be equal to the total duct area for the supply air system. Figure 4-21 shows possible locations for the placement of air return grilles in the return air system.

Provisions for Cooling

Where cooling as well as heating is supplied from the same outlets, it is necessary to direct the supply air upward for cooling. Cold air is heavier than warm air and tends to puddle (collect near the floor). To raise cold air, baseboard diffusers should have adjustable vanes or baffles (Figure 4-22).

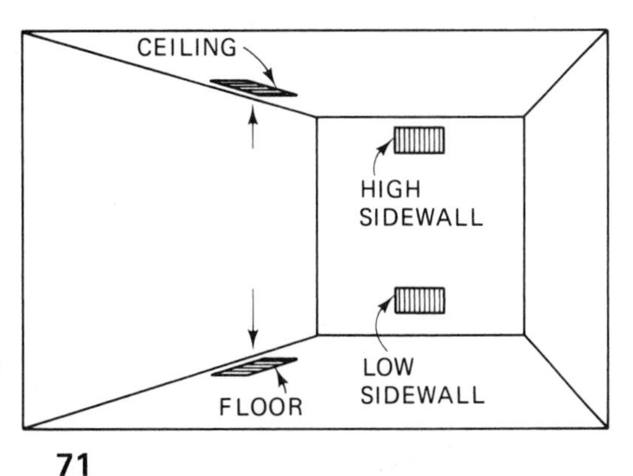

FIGURE 4-21 Placement of air return grilles in return air distribution system.

FIGURE 4-22 Adjustable air diffuser used in cooling/heating supply air distribution system (Courtesy, Environmental Elements Corporation).

Other Provisions

- Dampers must be provided in each branch supply run from the furnace for balancing the system.
- All ducts running in unconditioned areas such as attics or garages must be insulated.

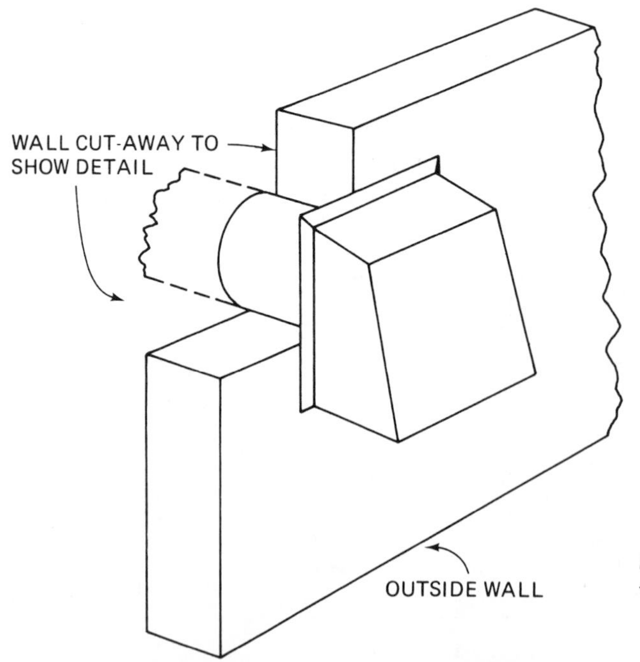

FIGURE 4-23 Outside air inlet to furnace for combustion.

- Ventilation air should be connected directly to the furnace or return air plenum, not to the intermediate return air duct work.
- Outside air can be made available to the furnace for combustion (Figure 4-23).

REVIEW QUESTIONS

Select the letter representing your choice of the correct answer(s).

4-1. Drafts are currents of relatively cold air that cause:

(a) Even heating

(b) Discomfort

(c) Not enough heat

(d) Recirculation

4-2. Cold outside surfaces of a room should be:

(a) Disregarded

(b) Painted

(c) Cooled

(d) Warmed

4-3. Not enough heat may be caused by:

(a) No moisture barrier provided in walls

(b) Chimney being too large

(c) Furnace which is too small

(d) Stratification

4-4. If the furnace is not producing its rated capacity, two factors should be considered (select two):

(a) Fuel input

(b) Amount of insulation in the house walls

(c) Size of chimney

(d) Combustion efficiency

4-5. The amount of input to a gas furnace can be determined by timing the gas meter during continuous operation:

(a) For a minute

(b) For a whole day

(c) For an hour

(d) In the middle of winter

4-6. The input to an oil furnace can be determined by (select two):

(a) Measuring the size of the combustion chamber

(b) Referring to the furnace nameplate

(c) Measuring oil pressure

(d) Oil burner nozzle size

4-7. One watt is equal to how many Btu?

(a) 4.1313

(b) 3.1413

(c) 1.4331

(d) 3.4131

4-8. To find the quantity of air the furnace is actually circulating, the service technician measures the:

(a) Temperature rise

(b) Size of the fan

(c) Size of outlet ducts

(d) Pressure drop

4-9. External static pressure is the total resistance of all components:

(a) In the return air system

(b) In the supply air system

(c) Throughout the whole system

(d) Outside the furnace

4-10. The air volume for cooling is usually:

(a) Greater than

(b) Less than

(c) The same as the air volume for heating

4-11. A perimeter loop system is a modification of:

(a) The single-pipe system

(b) The trunk duct system

(c) The radial system

(d) The extended plenum system

4-12. In northern climates where supply air diffusers are located along the perimeter, returns are placed:

(a) In all rooms

(b) In a central location

(c) On outside walls

(d) On inside walls

5

INSTALLATION PRACTICE

OBJECTIVES

After studying this chapter, the student will be able to:

- Evaluate the quality of a heating installation in a residence
- Determine the proper wire size and fuse size for electrical circuits
- Determine the proper gas pipe size for a residential gas furnace installation

EVALUATING CONSTRUCTION AND INSTALLATION

When a service technician is called upon to correct a system complaint, there are two areas to be evaluated.

1. **Building Construction:** Does the heating system conform to the building construction?
2. **Furnace Installation:** Is the equipment properly installed?

System complaints involve such conditions as drafts, uneven temperatures, cold floors, and other conditions that cause discomfort, even when the equipment may be operating well mechanically. Some of these complaints are due to the design of the system, which cannot be modified without considerable expense. Other system problems can be improved, if not fully corrected, when a service technician understands the cause.

Building Construction

As illustrated in Figure 5-1, the types of building construction that require special treatment for best heating results include:

- Structures with basements
- Structures over crawl space
- Concrete slab construction
- Split-level structures

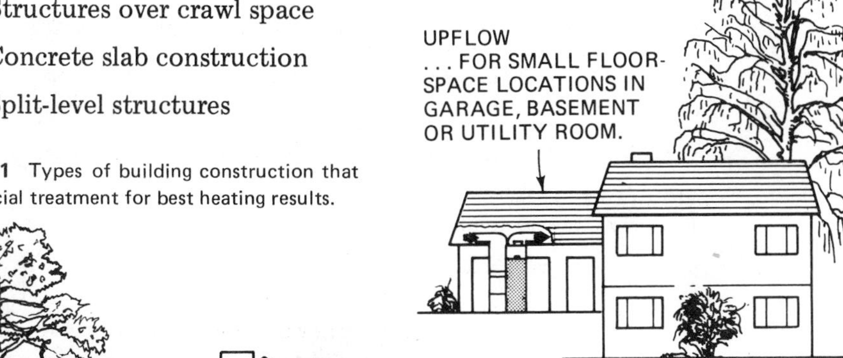

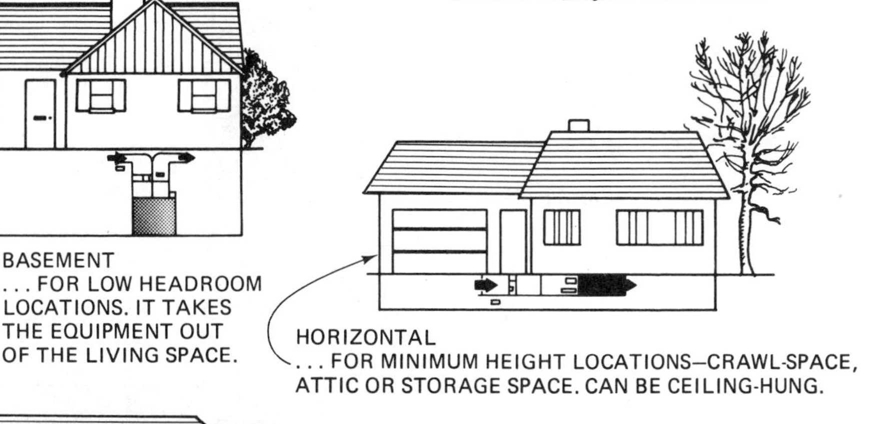

FIGURE 5-1 Types of building construction that require special treatment for best heating results.

STRUCTURES WITH BASEMENTS:

These structures should have some heat in the basement to produce warm floors on the level above. This practice should be followed whether or not the basement is finished for a recreation room. The structure should also be designed so that the basement can be properly heated at reasonable cost. The recommended construction for a properly heated basement is shown in Figures 5-2 and 5-3.

The following are desirable construction features of structures with basements:

- Subsoil below floor is well drained
- A moisture barrier is placed below the floor, and between the outside walls and the ground
- Insulation is placed in the joist around the perimeter of the structure
- Vaporproof insulating board is applied to the inside basement walls of existing structures and on the outside of new structures
- Any piping that pierces the vapor barrier should be properly sealed

FIGURE 5-2 Application of insulation in basement of existing structure (Courtesy, The Detroit Edison Company).

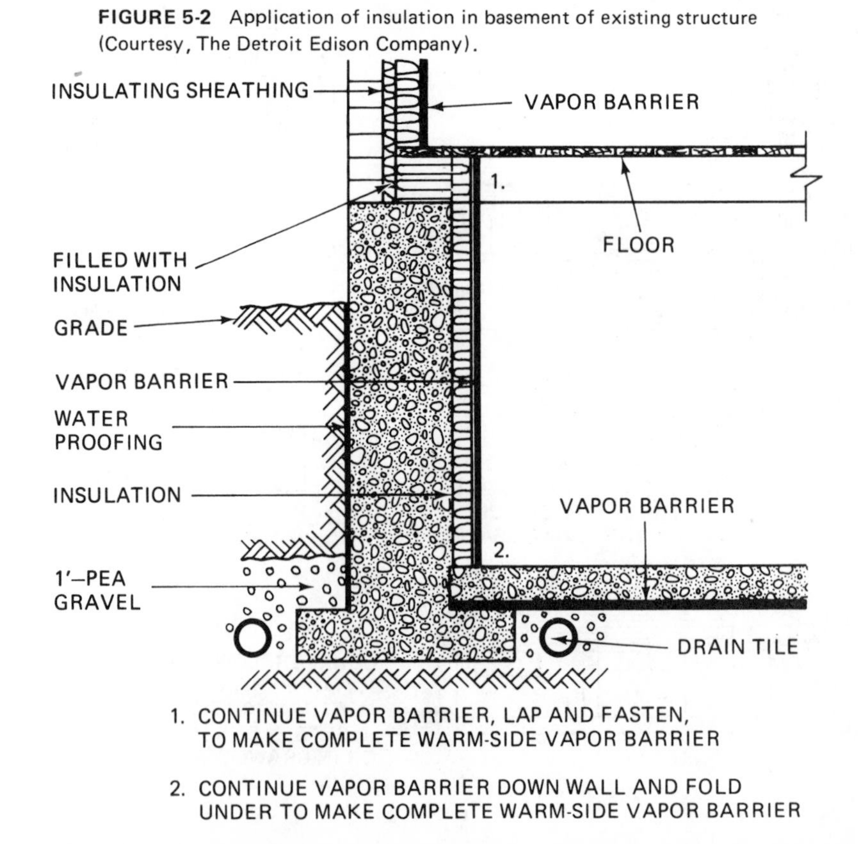

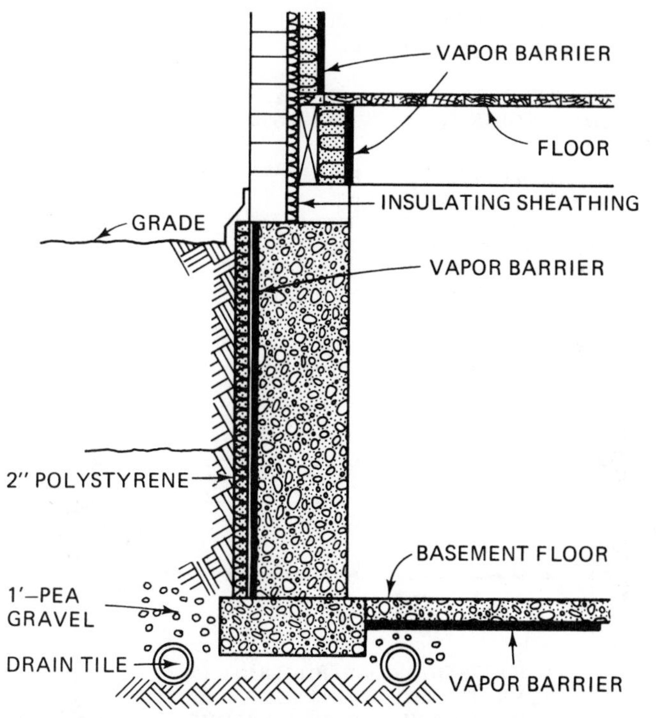

FIGURE 5-3 Application of insulation in basement of new structure (Courtesy, The Detroit Edison Company).

STRUCTURES WITH CRAWL SPACE:

It is important that the crawl space be heated in this type of structure. Heating produces warm floors on the level above. The best construction is similar in many ways to basement construction. Where crawl spaces are built above the bare ground a moisture barrier should be used, see Figure 5-4.

The following are features of crawl space construction:

- A moisture barrier is placed over the ground and extended upward a minimum of 6 in. on the side walls
- The walls are waterproofed on the outside below grade
- Perimeter joist spaces should be insulated
- On existing structures insulation should be applied to the inside of the crawl space walls. While on new structures it should be applied to the outside of the walls
- Ventilation is provided in summer only
- Dampers permit closing the ducts feeding the heat to the crawl space area

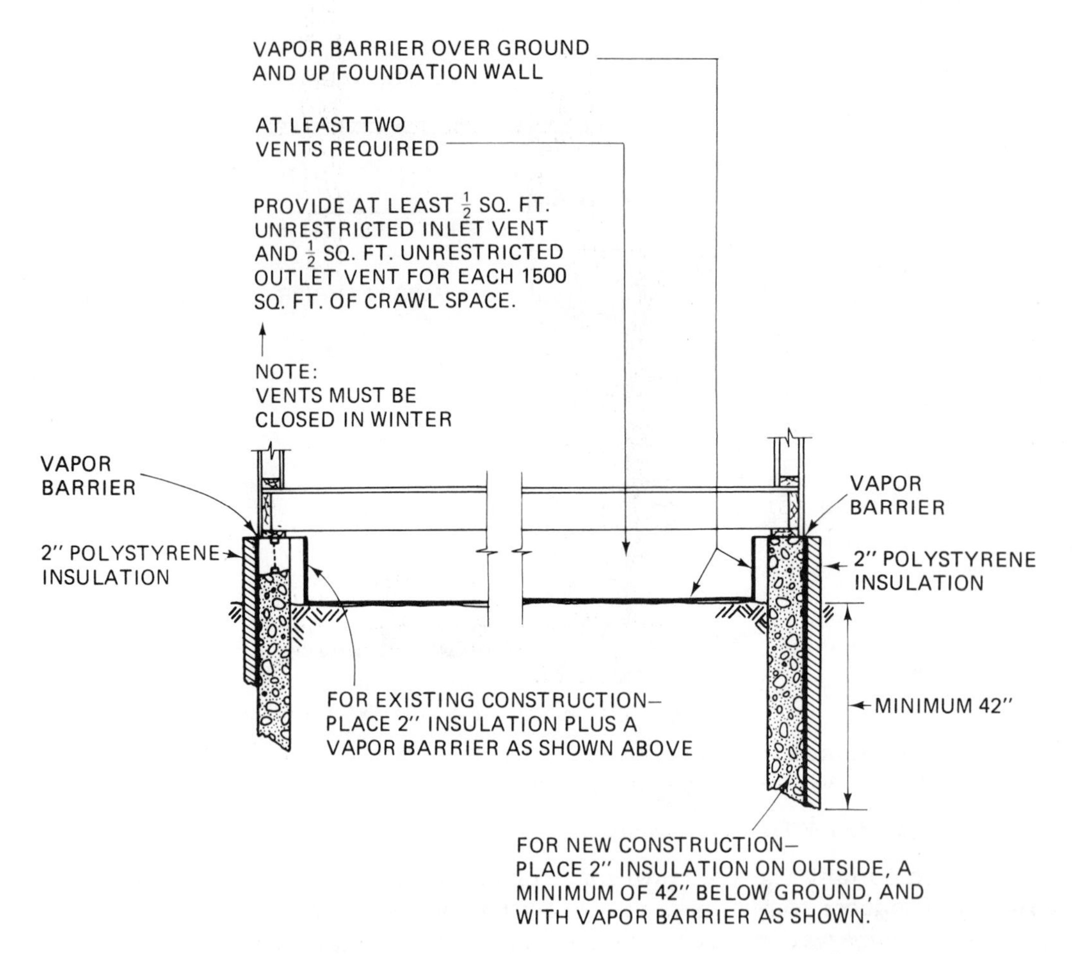

FIGURE 5-4 Recommended crawl space construction and ventilation.

CONCRETE SLAB CONSTRUCTION:

It is important to warm the floor in concrete slab construction, particularly around the perimeter where the greatest heat loss occurs. This can be done by placing the heat-distributing ducts in the concrete floor, as shown in Figures 5-5 through 5-9.

Some of the items included in this construction and shown in the illustrations are:

- Edge insulation is installed around the perimeter of the slab, as shown in Figure 5-5
- Ducts embedded in the floor supply perimeter heating (Figures 5-6 through 5-8)

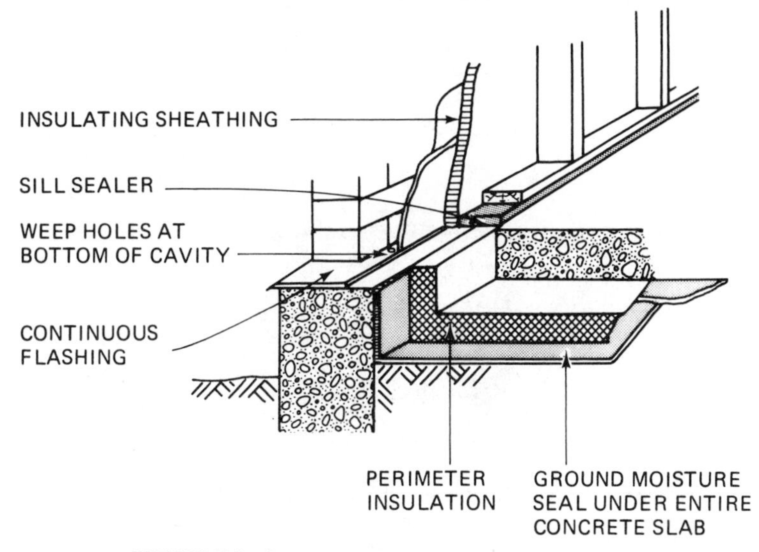

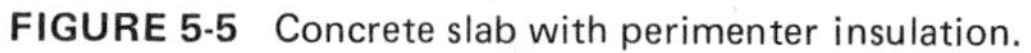

FIGURE 5-5 Concrete slab with perimenter insulation.

FIGURE 5-6 Perimeter-loop system with feeder and loop ducts in concrete slab (Reprinted with permission from the 1976 Systems Volume, ASHRAE Handbook & Product Directory).

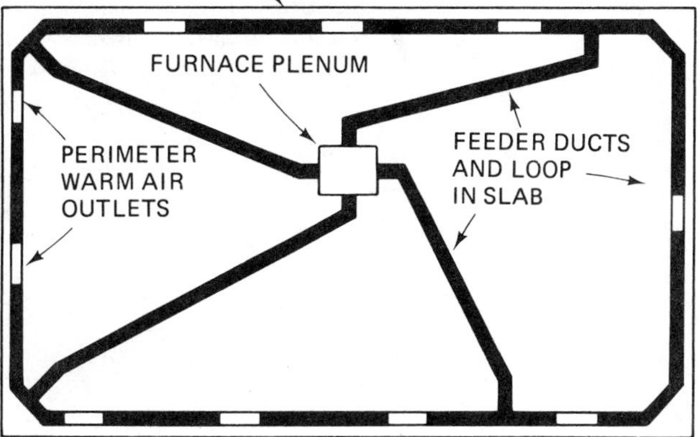

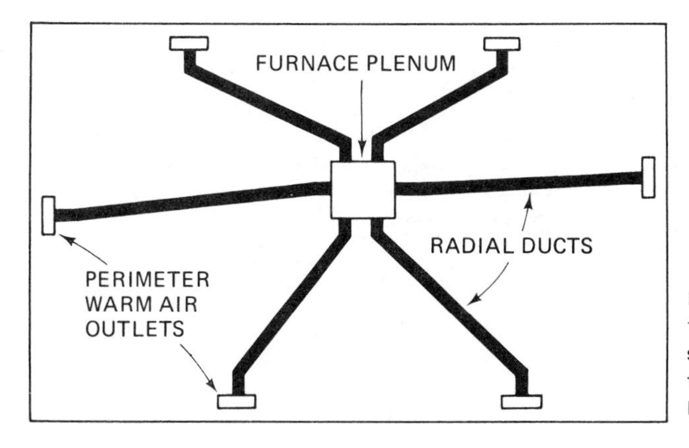

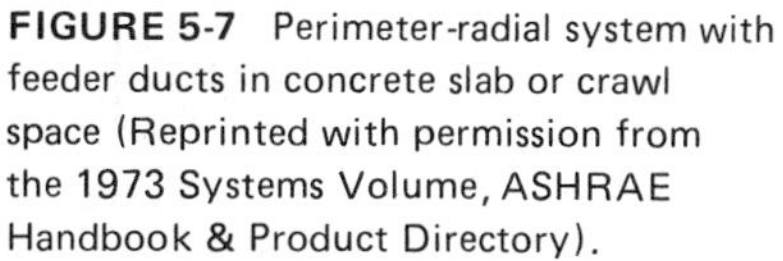

FIGURE 5-7 Perimeter-radial system with feeder ducts in concrete slab or crawl space (Reprinted with permission from the 1973 Systems Volume, ASHRAE Handbook & Product Directory).

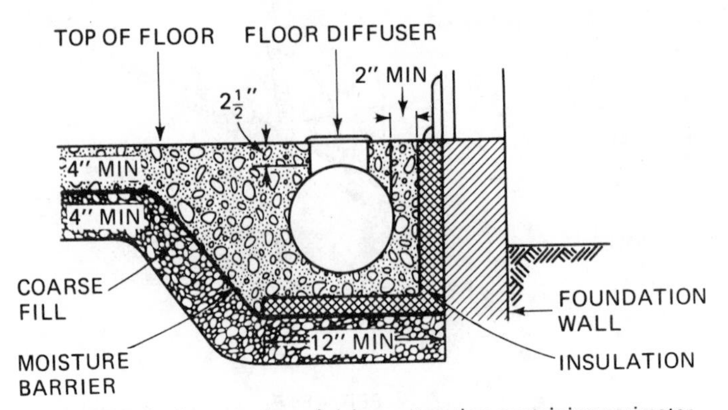

FIGURE 5-8 Cross-section of slab construction containing perimeter duct made of one type of material (Reprinted with permission from the 1976 Systems Volume, ASHRAE Handbook & Product Directory.

FIGURE 5-9 Construction of feeder ducts to plenum pit (Reproduced with permission of Carrier Corporation, ©Copyright 1979, Carrier Corporation).

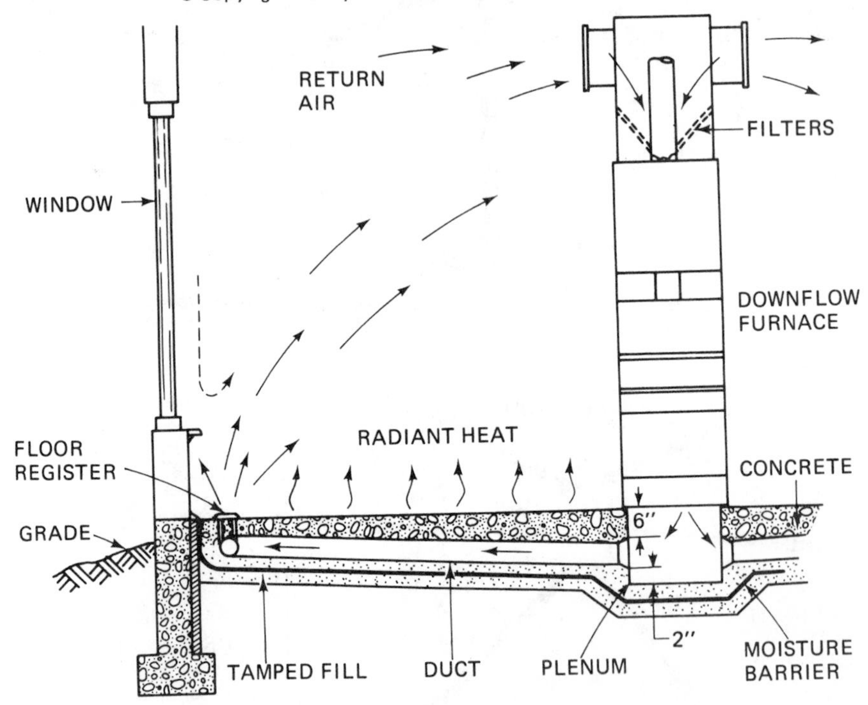

- A concrete pit of proper size is poured below the furnace for the supply air plenum chamber (Figure 5-9)
- A moisture barrier is placed over the ground before the concrete slab is poured
- Feeder ducts slope downward from the perimeter to the plenum pit
- Ducts are constructed of waterproof materials with waterproof joints
- Dampering of individual outlets is provided at the register location

SPLIT-LEVEL STRUCTURES:

This type of construction presents a problem in balancing the heat distribution to provide even heating because each level has its own heat-loss characteristics. Each level must be treated separately from an air distribution and balancing standpoint. Continuous fan operation is strongly recommended. In large structures of this type, a separate unit can be installed for each section.

Furnace Installation

When a service technician inspects a furnace installation for the first time, it should be determined if minimum standards have been met in the original installation. The owner should be advised of any serious faults that may affect safety and performance. Some installation conditions of importance are:

- Clearances from combustible material
- Circulating air supply
- Air for combustion, draft hood dilution, and ventilation
- Vent connections
- Electrical connections
- Gas piping

A service technician should know the local codes and regulations that govern installations. If there are no local codes, the equipment should be installed in accordance with National Electric Code AN 51.C1–1968, and in accordance with the recommendations made by the National Board of Fire Underwriters and the American National Standards Institute ASA 221.30–1964.

CLEARANCES FROM COMBUSTIBLE MATERIAL:

Clearances between the furnace and combustible construction should not be reduced to less than standard unless permissible clearances are indicated on the attached furnace nameplate.

Standard clearances are as follows:

1. Keep 1 in. between combustible material and the top of plenum chamber, 6 in. between sides and rear of unit
2. Keep 9 in. between combustible material and the draft hood and vent pipe in any direction
3. Keep 18 in. between combustible material and the front of the unit

Accessibility clearances take precedence over fire protection clearances (minimum clearances). Figure 5-10 shows recommended minimum clearances in a confined space. Allow at least 24 in. at the front of the furnace if all parts are accessible from the front. Otherwise, allow 24 in. on three sides of the furnace if back must be reached for servicing. When installation is made in a utility room, the door must be of sufficient size to allow replacement of the unit.

FIGURE 5-10 Recommended minimum clearances in a confined space.

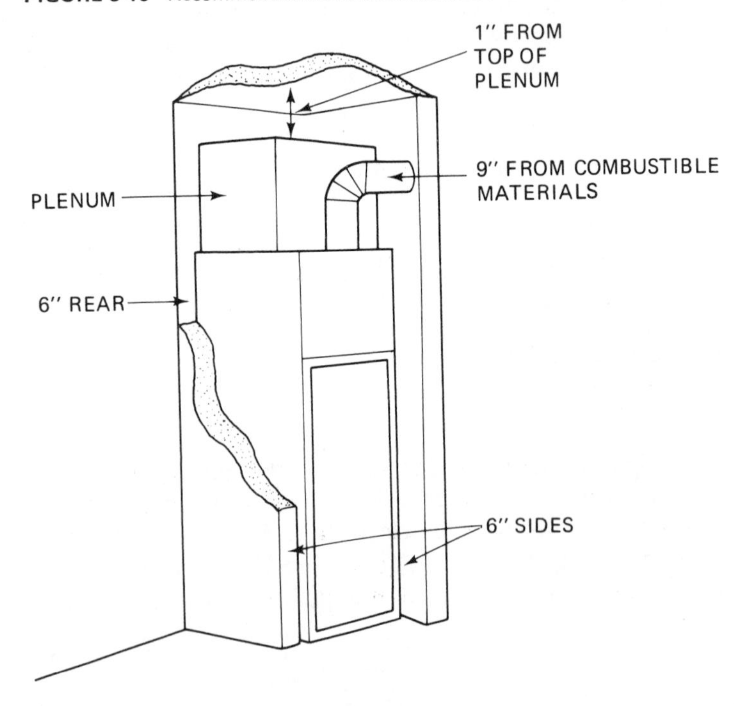

CIRCULATING AIR SUPPLY:

Circulating air supply may be 100% return air or any combination of fresh outside air and return air. It is recommended that return air plenums be lined with an acoustical duct liner to reduce any possible fan noise. This is particularly important when the distance from the return air grille to the furnace is close.

All duct connections to the furnace must extend outside the furnace closet. Return air must not be taken from the furnace room or closet. Adequate return air duct height must be provided to allow filters to be removed and replaced. All return air must pass through the filter after it enters the return air plenum.

AIR FOR COMBUSTION, DRAFT HOOD DILUTION, AND VENTILATION:

Air for combustion, draft hood dilution, and ventilation differs somewhat for two types of conditions:

1. Furnace in confined space
2. Furnace in unconfined space

Confined Space: If furnace is located in a confined space, such as a closet or small room, provisions must be made for supplying combustion and ventilation air (Figures 5-11 and 5-12). Two properly located openings of equal area are required. One opening should be located in the wall, door, or ceiling above the relief opening of the draft diverter. The other opening should be installed in the wall, door, or floor below the combustion air inlet of the furnace. The total free area* of each opening must be at least 1 in.^{2} for each 1000 Btuh input. It is recommended that air to this confined space come from the outside, not from the air in the house.

In closet installations where space is restricted, it is important to properly separate the incoming air to the furnace used for heating from that which is used for combustion. A schematic diagram of the air separation is shown in Figure 5-13.

Unconfined Space: Air for combustion, draft hood dilution, and ventilation must be obtained from outside or from spaces connected with the outside. If the unconfined space is within a building of unusually tight construction, then a permanent opening or openings, having a total free area of not less than 1 in.^{2} per 5000 Btuh of total input rating of all appliances must be provided as specified in AAS1-221.30-1964 and NFDA No. 54-1964. These standards are adopted and approved by both the National Fire Protection Association and the National Board of Fire Underwriters. See appendix for additional information.

* The free area of a grille is the total area of the opening through which the air passes.

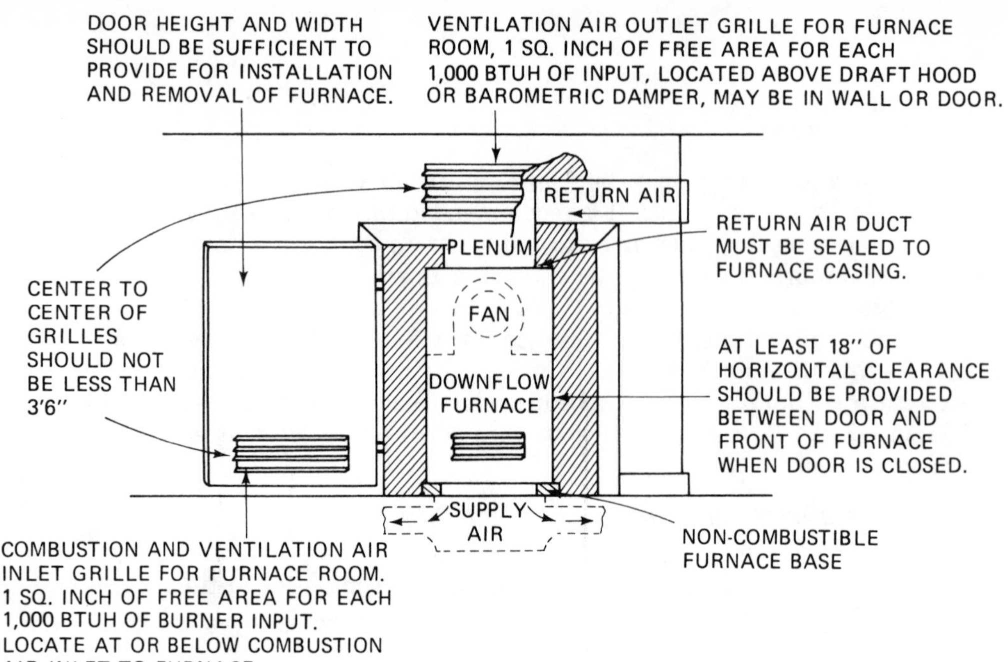

FIGURE 5-11 Provisions for combustion and ventilation air for furnace closet.

FIGURE 5-12 Provisions for combustion air for furnace, from vented attic space.

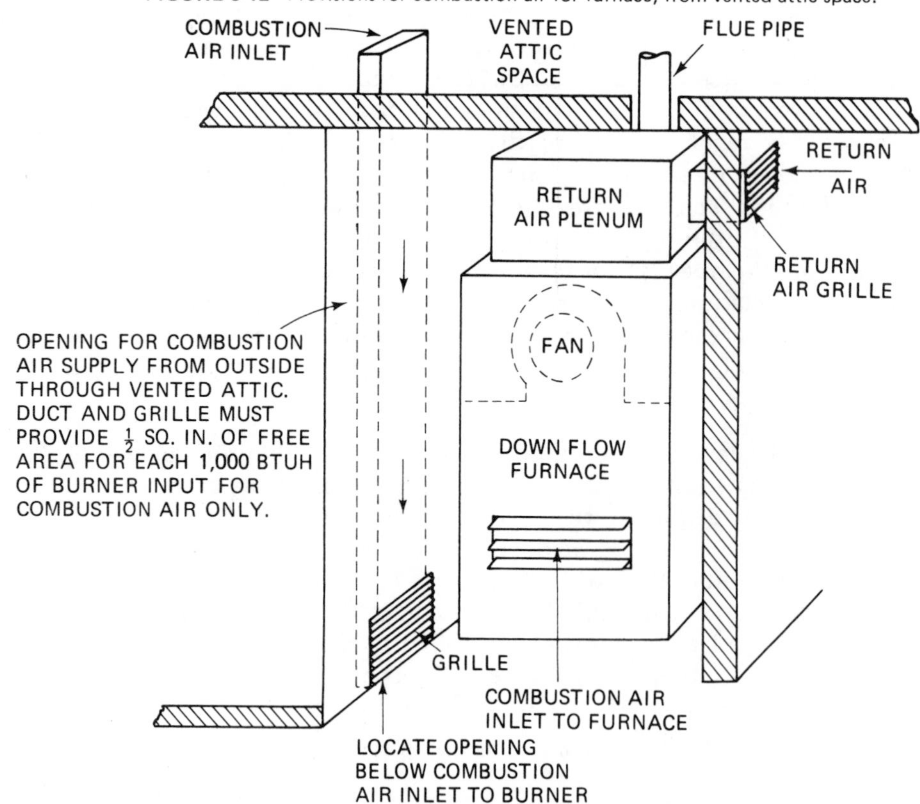

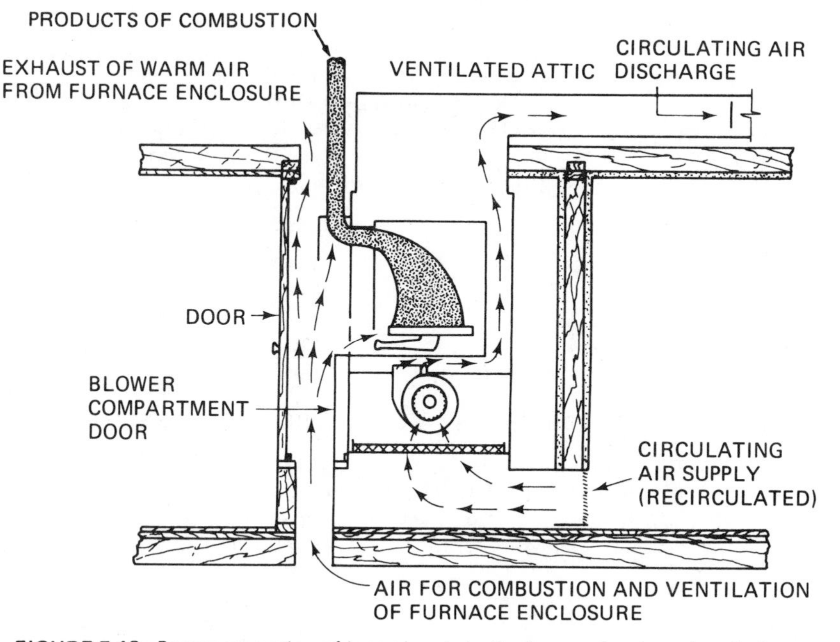

FIGURE 5-13 Proper separation of incoming air to the furnace for closet installations (Courtesy, Robertshaw Controls Company).

VENT CONNECTIONS:

It is important to provide proper venting of flue gases from the standpoint of fire protection as well as for the safety of people in a building. Good practices in venting include the following:

1. A chimney or flue vent outlet must extend above the roof surface and terminate no less than 2 ft above any object within a 10-ft radius or, if an anti-downdraft flue cap is used it is permissible to terminate 2 ft above the roof line, anyplace on the roof. The vent termination should be 1 ft above, or 4 ft away from, any opening or air inlet to the building. Recommended chimney and flue heights are shown in Figure 5-14
2. Horizontal runs of flue should maintain a minimum pitch of $^1/_4$ in./linear foot, and should not exceed 75% of the vertical vent length (Figure 5-15)
3. Support pipes rigidly with hangers or straps
4. Pipe must be the same size as the flue collar on the unit (Figure 5-16)
5. Run pipe as directly as possible with a minimum number of turns
6. Do not connect vent piping to a chimney serving an open fireplace

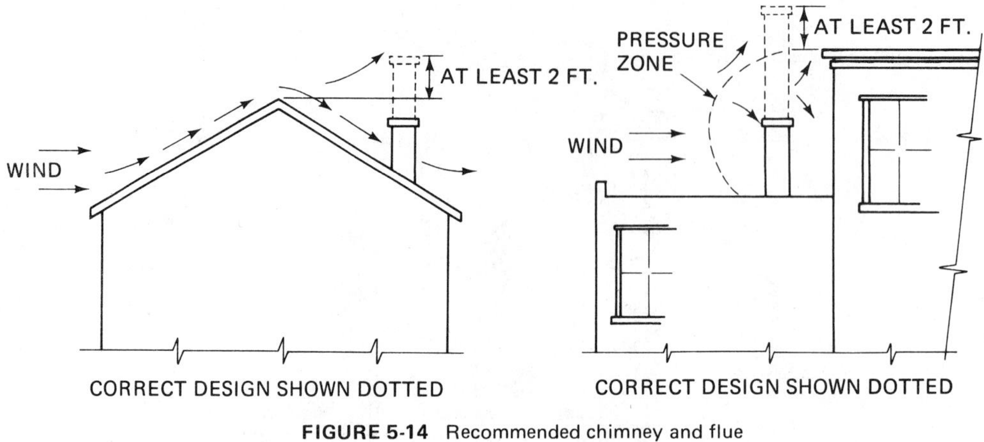

FIGURE 5-14 Recommended chimney and flue heights (Reproduced by permission of Carrier Corporation, ©Copyright 1979, Carrier Corporation).

FIGURE 5-15 Maximum horizontal run of flue.

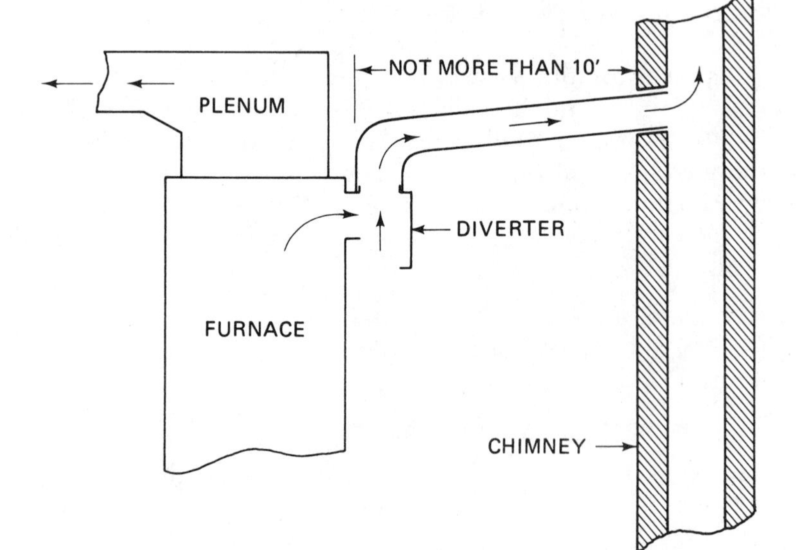

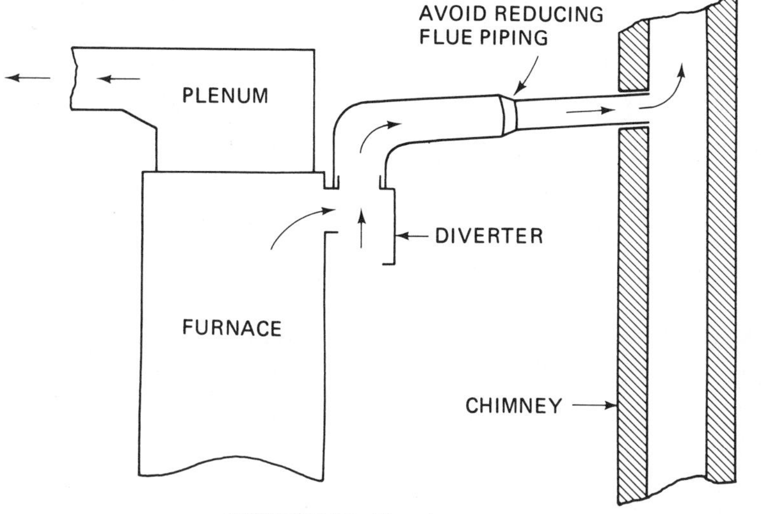

FIGURE 5-16 Flue pipe sizing.

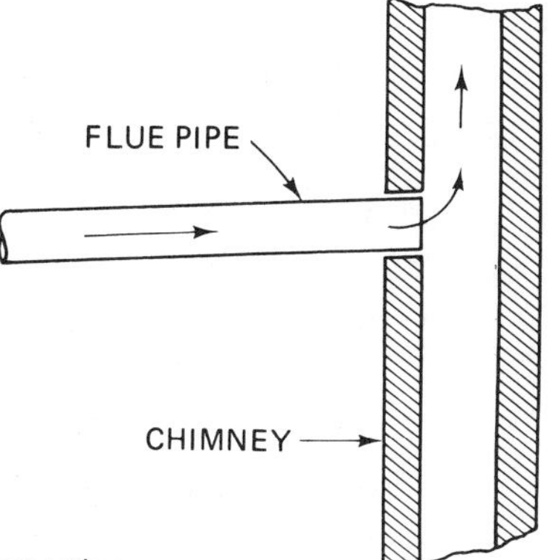

FIGURE 5-17 Chimney flue connection.

7. Extend flue pipe through chimney walls flush with the inner face of the chimney lines (Figure 5-17)

8. When more than one unit is vented into the same flue, the cross-sectional area of the main flue should be equal to the area of the one flue plus one-half the area of the second (Figure 5-18)

9. Observe local ordinances covering vent piping. When local ordinances do not cover the subject, it is recommended that the installer be guided by a book published by the Gas Vent Institute entitled *Gas Vent Capacity Tables* (AIA file 40-D or NFPA No. 54). See appendix for additional information.

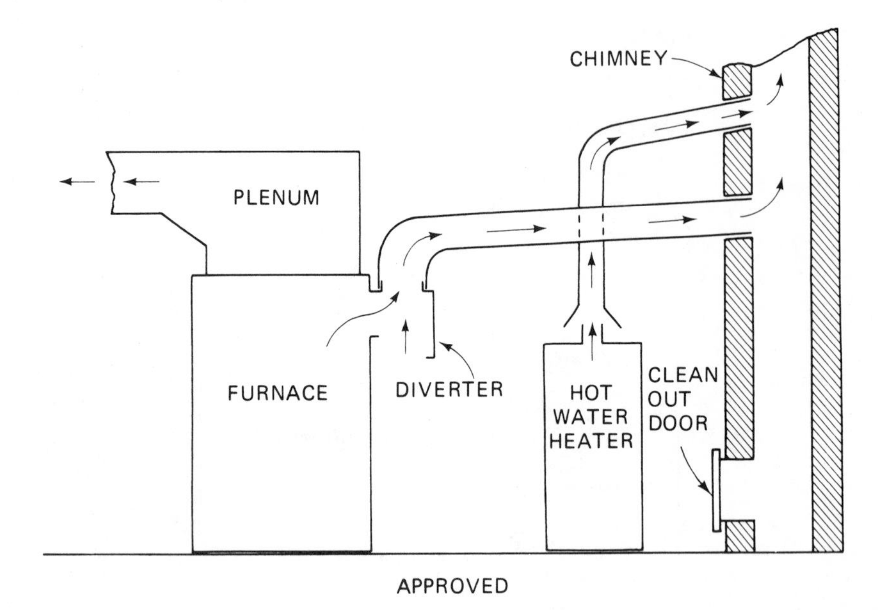

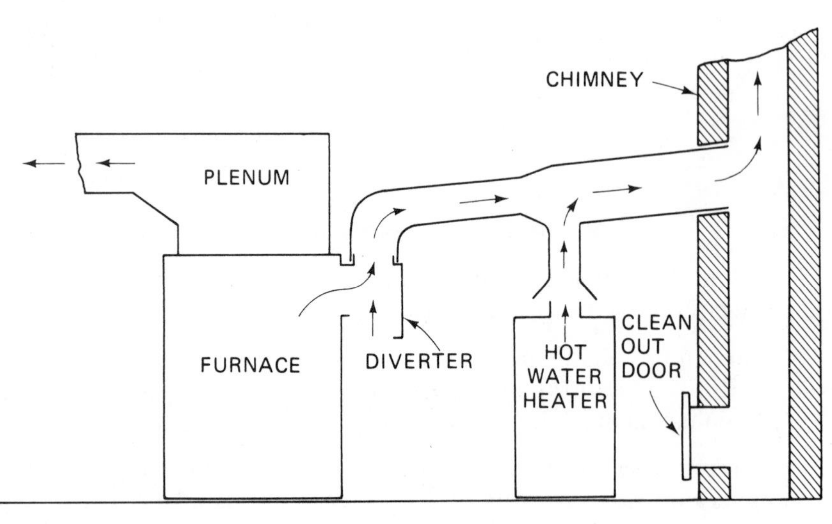

FIGURE 5-18 Arrangement for dual flue connection.

ELECTRICAL CONNECTIONS:

All electrical wiring and connections should be made in accordance with the National Electric Code and with any local ordinances that may apply. Some of the important items to be observed in making the electrical connections are as follows:

1. A separate 120-V power circuit properly fused should be provided, with a manual disconnect at the furnace
2. The fuse size and the wire size is determined by the National Electric Code. The wire size and fuse size are based on 125% of full-load current rating. The length of the wire run will also affect the wire size. The table in Figure 5-19 gives the maximum length of the two-wire runs. The table is based on 240 V and a 3% voltage drop

The following factors apply to other voltages:

Voltage (V)	*Factor*
110	0.458
115	0.479
120	0.500
125	0.521
220	0.917
230	0.966
240	1.000
250	1.042

EXAMPLE: Find the maximum length of run for No. 12 wire carrying a 15-A load at 120 V.

SOLUTION: From the table in Figure 5-19 for No. 12 wire and 15 A, the maximum length of run is 145 ft. Therefore, to find the length for 120 V, the equation is

$$145 \text{ ft} \times 0.50 \text{ factor} = 72.5 \text{ ft}$$

The maximum length of No. 12 wire carrying 15 A at 120 V is 72.5 ft.

3. All replacement wire used on a furnace should be of the same type and size as the original wire and rated from 90 to 105°C. A typical label on furnace wire is shown in Figure 5-20

WIRE SIZE	AMPERES												
	5	10	15	20	25	30	35	40	45	50	55	70	80
14	274	137	91										
12		218	145	109									
10			230	173	138	115							
8					220	182	156	138					
6								219	193	175	159		
4									309	278	253	199	
3										350	319	250	219
2											402	316	276
1												399	349
0												502	439
00													560

FIGURE 5-19 Table indicating maximum length of two-wire run, for voltage drop of three percent, at 240 volts.

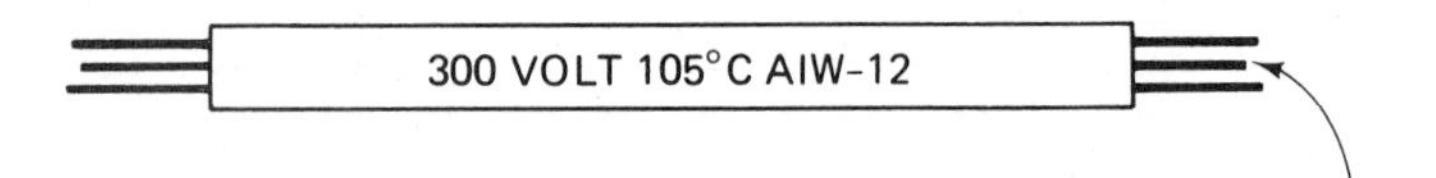

FIGURE 5-20 Typical label on furnace wire (12-2 with ground).

4. Unless a circuit breaker is used, the fuse should be of the time-delay type
5. Waterproof strain relief connectors should be used at the entrance and exits of makeup boxes
6. Control circuit wire (for 24 V) should be No. 18 wire

GAS CONNECTION TO FURNACE

A recommended gas piping arrangement at the furnace is shown in Figure 5-21.

Some of the characteristics of good gas piping are as follows:

1. Piping should include a vertical section to collect scale and dirt
2. A ground joint union should be placed at the connection to the furnace

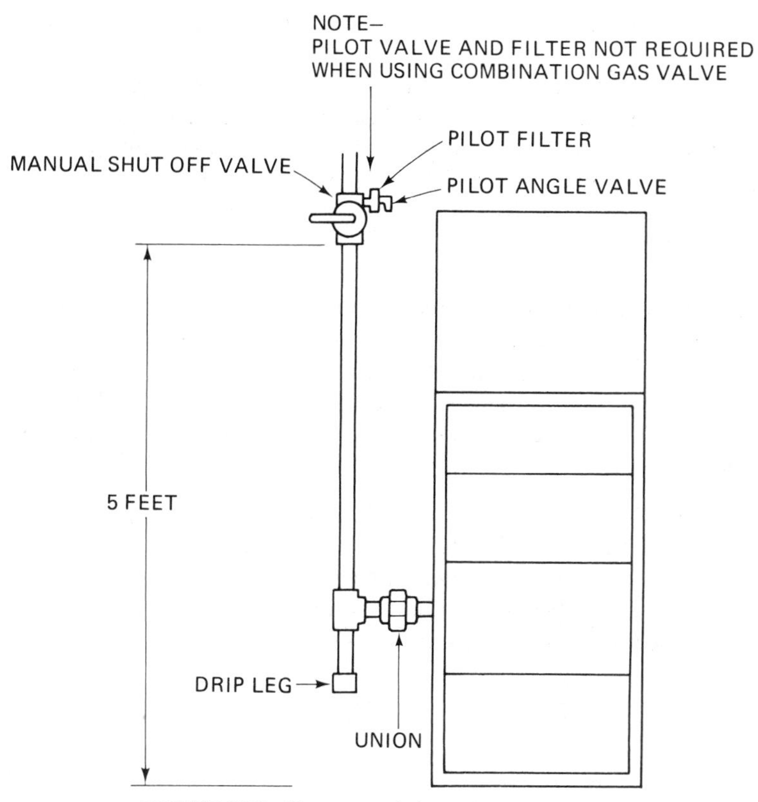

FIGURE 5-21 Recommended gas piping arrangement.

3. A drip leg should be installed at the bottom of the vertical riser
4. The manual shut off valve must be located external to the furnace casing, except where a combination gas valve is used. The manual shut off is not required with a combination gas valve because the manual shut off is part of this valve
5. Natural gas service pressure to the furnace should not exceed 14 in. water column (8.1 ounces). Liquid petroleum (LP) gas is furnished with a tank pressure regulator to provide 11 in. water column (6.3 ounces) to the furnace

GAS PIPE SIZING

Adequate gas supply must be provided to the gas furnace. The installation contractor is usually responsible for providing piping from the gas source to the equipment. Therefore, if other gas appliances are used in the building, the entire gas piping layout must be examined from the gas meter to each piece of equipment to determine the proper gas piping sizes for the systems.

It may also be advisable, if additional gas requirements are being added to an existing building, to be certain that the gas meter supplied is adequate. This can be done by advising the gas company of the total connected load. The utility company will provide the necessary service at the meter.

The procedure for determining the proper gas piping sizes from the meter is as follows:

1. Sketch a layout of the actual location of the gas piping from the meter to each gas appliance. Indicate on the sketch the length of each section of the piping diagram. Indicate on the sketch the Btuh input required for each gas appliance. See Figure 5-22 for an example. Note in this diagram that there are three appliances on the system

2. Convert the Btuh input of each gas appliance to cubic feet per hour (cfh) by dividing the Btuh input by the Btu/ft^3 rating of the gas used. The example shown uses natural gas with a rating of 1000 Btu ft^3. Thus, for the appliances shown in Figure 5-22 the cubic feet of gas used for each appliance is as follows:

Outlet	*Appliance*	*Btuh input*	*Natural gas (cfh)*
A	Water heater	30,000	30
B	Range	75,000	75
C	Furnace	136,000	136

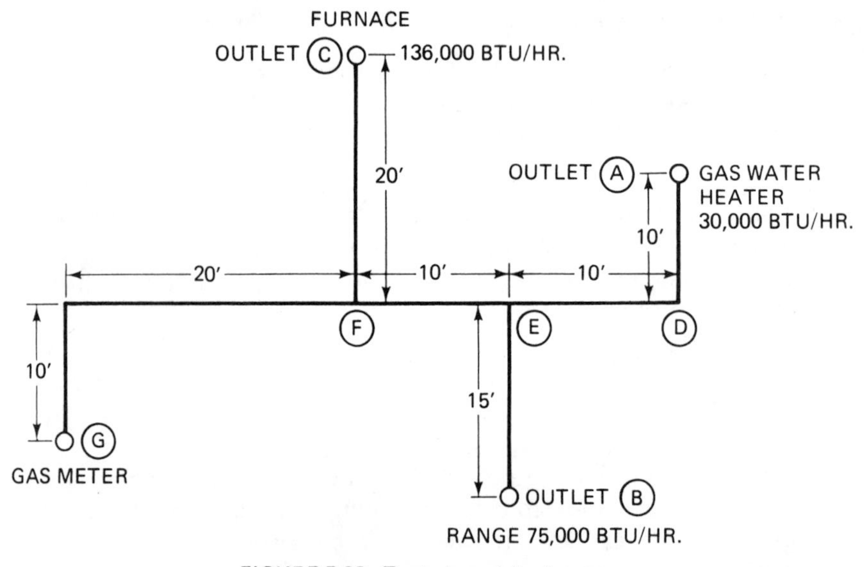

FIGURE 5-22 Typical gas piping layout.

3. Determine the amount of gas required in cfh for each section of the piping system
4. Using the appropriate figure for pipe size (Figures 5-23 through 5-30), select and record the proper pipe sizes

Most residential gas piping systems are sized using Figure 5-23, because most natural gas is delivered from the meter at a pressure of 3/10 (0.3) in. of water column and a specific gravity of 6/10 (0.6). The specific gravity refers to the relative weight of 1 ft^3 of the gas as compared to an equal quantity of air.

In using Figure 5-23, the maximum length of pipe to the appliance the greatest distance from the meter is used for sizing the entire piping system. Thus, in Figure 5-22 the length of pipe to outlet (A) is 60 ft, and this number is used for all pipe sizes. Using the data above with the piping layout in Figure 5-22 and Figure 5-23, the pipe sizes are determined as follows:

1. Since 60 ft is the greatest distance from the meter, point A to point G (10 ft + 10 ft + 10 ft + 20 ft + 10 ft), the 60 "Length of Feet" line is used in Figure 5-23
2. Staying on the 60 "Length of Feet" line, move to the right until the cfh of gas carried in AD is covered. A $\frac{1}{2}$-in. pipe is used because it can carry up to 50 cfh
3. DE is the same, since it is also 30 cfh

FIGURE 5-23 Maximum capacity of pipe in cubic feet of gas per hour for gas pressures of 0.5 psig or less and a pressure drop of 0.5 inch water column (Courtesy, American Gas Association).

Nominal Iron Pipe Size, Inches	Internal Diameter, Inches	Length of Pipe, Feet													
		10	20	30	40	50	60	70	80	90	100	125	150	175	200
¼	.364	32	22	18	15	14	12	11	11	10	9	8	8	7	6
⅜	.493	72	49	40	34	30	27	25	23	22	21	18	17	15	14
½	.622	132	92	73	63	56	50	46	43	40	38	34	31	28	26
¾	.824	278	190	152	130	115	105	96	90	84	79	72	64	59	55
1	1.049	520	350	285	245	215	195	180	170	160	150	130	120	110	100
1¼	1.380	1,050	730	590	500	440	400	370	350	320	305	275	250	225	210
1½	1.610	1,600	1,100	890	760	670	610	560	530	490	460	410	380	350	320
2	2.067	3,050	2,100	1,650	1,450	1,270	1,150	1,050	990	930	870	780	710	650	610
2½	2.469	4,800	3,300	2,700	2,300	2,000	1,850	1,700	1,600	1,500	1,400	1,250	1,130	1,050	980
3	3.068	8,500	5,900	4,700	4,100	3,600	3,250	3,000	2,800	2,600	2,500	2,200	2,000	1,850	1,700
4	4.026	17,500	12,000	9,700	8,300	7,400	6,800	6,200	5,800	5,400	5,100	4,500	4,100	3,800	3,500

Based on a 0.60 specific gravity gas.

FIGURE 5-24 Maximum capacity of pipe in cubic feet of gas per hour for gas pressures of 0.5 psig or less and a pressure drop of 0.5 inch water column (Courtesy, American Gas Association).

Nominal Iron Pipe Size, Inches	Internal Diameter, Inches	Length of Pipe, Feet													
		10	20	30	40	50	60	70	80	90	100	125	150	175	200
1/4	.364	43	29	24	20	18	16	15	14	13	12	11	10	9	8
3/8	.493	95	65	52	45	40	36	33	31	29	27	24	22	20	19
1/2	.622	175	120	97	82	73	66	61	57	53	50	44	40	37	35
3/4	.824	360	250	200	170	151	138	125	118	110	103	93	84	77	72
1	1.049	680	465	375	320	285	260	240	220	205	195	175	160	145	135
1 1/4	1.380	1,400	950	770	660	580	530	490	460	430	400	360	325	300	280
1 1/2	1.610	2,100	1,460	1,180	990	900	810	750	690	650	620	550	500	460	430
2	2.067	3,950	2,750	2,200	1,900	1,680	1,520	1,400	1,300	1,220	1,150	1,020	950	850	800
2 1/2	2.469	6,300	4,350	3,520	3,000	2,650	2,400	2,250	2,050	1,950	1,850	1,650	1,500	1,370	1,280
3	3.068	11,000	7,700	6,250	5,300	4,750	4,300	3,900	3,700	3,450	3,250	2,950	2,650	2,450	2,280
4	4.026	23,000	15,800	12,800	10,900	9,700	8,800	8,100	7,500	7,200	6,700	6,000	5,500	5,000	4,600

Based on a 0.60 specific gravity gas.

FIGURE 5-25 Maximum capacity of semi-rigid tubing in cubic feet of gas per hour for gas pressures of 0.5 psig or less and a pressure drop of 0.3 inch water column (Courtesy, American Gas Association).

Outside Diameter, Inch	Length of Tubing, Feet													
	10	20	30	40	50	60	70	80	90	100	125	150	175	200
3/8	20	14	11	10	9	8	7	7	6	6	5	5	4	4
1/2	42	29	23	20	18	16	15	14	13	12	11	10	9	8
5/8	86	59	47	40	36	33	30	28	26	25	22	20	18	17
3/4	150	103	83	71	63	57	52	49	46	43	38	35	32	30
7/8	212	146	117	100	89	81	74	69	65	61	54	49	45	42

Based on a 0.60 specific gravity gas.

FIGURE 5-26 Maximum capacity of semirigid tubing in cubic feet of gas per hour for gas pressures of 0.5 psig or less and a pressure drop of 0.5 inch water column (Courtesy, American Gas Association).

Outside Diameter, Inch	Length of Tubing, Feet													
	10	20	30	40	50	60	70	80	90	100	125	150	175	200
3/8	27	18	15	13	11	10	9	9	8	8	7	6	6	5
1/2	56	38	31	26	23	21	19	18	17	16	14	13	12	11
5/8	113	78	62	53	47	43	39	37	34	33	29	26	24	22
3/4	197	136	109	93	83	75	69	64	60	57	50	46	42	39
7/8	280	193	155	132	117	106	98	91	85	81	71	65	60	55

Based on a 0.60 specific gravity gas.

FIGURE 5-27 Multipliers to be used only with Figures 5-23, 5-24, 5-25 and 5-26 when applying different specific gravity factors (Courtesy, American Gas Association).

Specific Gravity	Multiplier	Specific Gravity	Multiplier
.35	1.31	1.00	.78
.40	1.23	1.10	.74
.45	1.16	1.20	.71
.50	1.10	1.30	.68
.55	1.04	1.40	.66
.60	1.00	1.50	.63
.65	.96	1.60	.61
.70	.93	1.70	.59
.75	.90	1.80	.58
.80	.87	1.90	.56
.85	.84	2.00	.55
.90	.82	2.10	.54

FIGURE 5-28 Maximum capacity of pipe in thousands of Btu per hour of undiluted liquefied petroleum gases (Courtesy, American Gas Association).

Nominal Iron Pipe Size, Inches	Length of Pipe (Feet)											
	10	20	30	40	50	60	70	80	90	100	125	150
1/2	275	189	152	129	114	103	96	89	83	78	69	63
3/4	567	393	315	267	237	217	196	185	173	162	146	132
1	1071	732	590	504	448	409	378	346	322	307	275	252
1 1/4	2205	1496	1212	1039	913	834	771	724	677	630	567	511
1 1/2	3307	2299	1858	1559	1417	1275	1181	1086	1023	976	866	787
2	6221	4331	3465	2992	2646	2394	2205	2047	1921	1811	1606	1496

Based on a pressure drop of 0.5 inch water column.

FIGURE 5-29 Maximum capacity of semirigid tubing in thousands of Btu per hour of undiluted liquefied petroleum gases (Courtesy, American Gas Association).

Outside Diameter (Inches)	Length of Tubing (Feet)									
	10	20	30	40	50	60	70	80	90	100
3/8	39	26	21	19	—	—	—	—	—	—
1/2	92	62	50	41	37	35	31	29	27	26
5/8	199	131	107	90	79	72	67	62	59	55
3/4	329	216	181	145	131	121	112	104	95	90
7/8	501	346	277	233	198	187	164	155	146	138

Based on a pressure drop of 0.5 inch water column.

FIGURE 5-30 Flow capacity of copper tubing in cubic feet of gas per hour.

LENGTH (FEET)	3/8" O.D. (1/4" NOM)	1/2" O.D. (3/8" NOM)	5/8" O.D. (1/2" NOM)	3/4" O.D. (5/8" NOM)	7/8" O.D. (3/4" NOM)	1 1/8" O.D. (1" NOM)
10	21.1	49.3	94.0	162.4	254.1	525.7
20	14.7	33.1	63.1	109.0	170.6	352.9
30	11.2	26.2	50.0	86.3	135.1	279.5
40	9.5	22.2	42.4	73.2	114.5	236.9
50	8.4	19.5	37.3	64.4	100.7	208.4
60	7.5	17.6	33.6	58.0	90.7	187.6
70	6.9	16.1	30.7	53.0	83.0	171.7
80	6.4	14.9	28.4	49.1	76.9	159.0
90	6.0	13.9	26.6	45.9	71.8	148.6
100	5.6	13.1	25.0	53.2	67.6	139.9

COPPER TUBE SIZE – TYPE L

SPECIFIC GRAVITY 0.6

PRESSURE DROP 0.3" W.C.

4. BE requires a $^3/_4$-in. pipe, which can carry up to 105 cfh
5. EF is also $^3/_4$ in., since 105 cfh is needed
6. CF requires a 1-in. pipe, which can carry up to 195 cfh
7. FG requires a $1^1/_4$-in. pipe, which can carry up to 400 cfh

Pipe section	*Total cfh natural gas carried in section*	*Iron pipe size (in.)*
AD	30	$\frac{1}{2}$
DE	30	$\frac{1}{2}$
BE	75	$\frac{3}{4}$
EF	105 (30 + 75)	$\frac{3}{4}$
CF	136	1
FG	241 (136 + 105)	$1\frac{1}{4}$

Figures 5-23, 5-24, 5-25, 5-26, 5-30 and 5-31 are for natural gas. Figures 5-28 and 5-29 are for liquid petroleum gas. Figure 5-27 provides multipliers to use when converting the natural gas tables to specific gravities other than 6/10 (0.6). Factors relating to gas piping systems in trailer parks are shown in Figure 5-32.

For example, using Figure 5-27, assume that the natural gas has a specific gravity of 0.75. From Figure 5-27, the multiplier would be 0.90. This multiplier would be used to redetermine the cfh of gas handled by each part of the piping system.

When the 0.90 multiplier is applied to the above example, the results are as follows:

Pipe section	*Cfh determined in instruction 2*	*Cfh using multiplier of 0.90*	*Iron pipe size (in.)*
AD	30	27 (0.90 × 30)	$\frac{1}{2}$
DE	30	27 (0.90 × 30)	$\frac{1}{2}$
BE	75	68 (0.90 × 75)	$\frac{3}{4}$
EF	105	94 (0.90 × 105)	$\frac{3}{4}$
CF	136	122 (0.90 × 136)	1
FG	241	217 (0.90 × 241)	$1\frac{1}{4}$

FIGURE 5-31 Flow capacity of various sizes and lengths of standard straight steel pipe and fittings, in cubic feet of gas per hour.

LENGTH IN FEET	CUBIC FEET PER HOUR 3/8"	1/2"	3/4"	1"	1 1/4"	1 1/2"	2"	3"
10	62	109	222	420	890	1300	2515	7680
20	43	78	157	295	625	940	1775	5430
30	36	63	128	235	510	775	1450	4430
40	31	55	111	205	445	660	1260	3840
50	27	48	100	183	400	590	1120	3430
60	24	45	91	168	365	538	1030	3140
70	22	42	85	154	340	495	950	2900
80	21	39	79	145	315	470	940	2720
90	20	37	74	135	295	440	890	2570
100	19	35	70	129	282	420	800	2440
150	15	26	57	105	230	340	650	1990
200	13	24	49	92	200	293	565	1710
300	11	20	40	75	165	240	460	1410

FITTING	EQUIVALENT FEET OF STRAIGHT PIPE 3/8"	1/2"	3/4"	1"	1 1/4"	1 1/2"	2"	3"
ELL	0.6	0.9	1.2	1.6	2.2	2.6	3.6	5.7
GATE VALVE	0.2	0.3	0.5	0.6	0.8	1.0	1.3	2.1
GLOBE VALVE	1.7	2.5	3.5	4.7	6.5	7.8	10.6	17.1
TEE (SIDE)	1.1	1.7	2.3	3.1	4.4	5.2	7.1	11.4

Based on pressure drop 0.3" W.C.

FIGURE 5-32 Demand factors for use in calculating gas piping systems in trailer parks (Courtesy, American Gas Association).

Demand Factors for use in Calculating Gas Piping Systems in Trailer Parks

No. of Trailer Sites	Btu Per Hour Per Trailer Site
1	125,000
2	117,000
3	104,000
4	96,000
5	92,000
6	87,000
7	83,000
8	81,000
9	79,000
10	77,000
11 - 20	66,000
21 - 30	62,000
31 - 40	58,000
41 - 60	55,000
Over 60	50,000

REVIEW PROBLEM

Using the sketch in Figure 5-22, assume the following loads:

Outlet A	50,000
Outlet B	60,000
Outlet C	150,000

Determine the iron pipe sizes for natural gas using a pressure drop of 0.3 in. of water column and 0.60 specific gravity gas.

REVIEW QUESTIONS

Select the letter representing your choice of the correct answer.

5-1. LP gas pressure at the furnace should be:

(a) 3.5 in.

(b) 7.0 in.

(c) 11.0 in.

(d) 14.0 in.

5-2. In basement construction a moisture barrier should be placed:

(a) Between the ceiling and first floor

(b) On the inside of the walls

(c) Below the floor

(d) Around all plumbing

5-3. In a crawl space ventilation is provided:

(a) All year round

(b) In summer only

(c) In winter only

(d) Only if there is a furnace in the space

5-4. A split-level house presents a problem in heating because:

(a) Each level must be treated separately

(b) It is difficult to find locations for outlets

(c) The partition walls do not line up with each other

(d) All of the return air goes to the basement

5-5. What is the minimum distance from combustible material that a furnace (sides and rear) should be placed?

(a) 24 in.

(b) 18 in.

(c) 12 in.

(d) 1 in.

5-6. Access to the front of the furnace should be not less than:

(a) 12 in.

(b) 24 in.

(c) 36 in.

(d) 48 in.

5-7. If the furnace is placed in a confined space, provision must be made for:

(a) Insulating all walls of the room

(b) Using a fireproof floor

(c) Humidity control

(d) Combustion and ventilation air

5-8. The opening area to provide air from the outside to the furnace, where the unit is placed in a house of tight construction, should be not less than:

(a) 1 $in.^2$ per 5000 Btuh input

(b) 2 $in.^2$ per 5000 Btuh input

(c) 2 $in.^2$ per 10,000 Btuh input

(d) 1 $in.^2$ per 15,000 Btuh input

5-9. A chimney or flue vent outlet must extend above the highest point on the roof at least:

(a) 1 ft

(b) 2 ft

(c) 3 ft

(d) 4 ft

5-10. The electric power supply for a furnace:

(a) Must be placed on a separate circuit

(b) Can plug into any 120-V outlet

(c) Should be combined with other appliances

(d) Should always be on a 240-V circuit

6

COMBUSTION AND FUELS

OBJECTIVES

After studying this chapter, the student will be able to:

- Compare the heating qualities of various fuels
- Determine the conditions necessary for efficient utilization of fuels

COMBUSTION

Combustion is the chemical process in which oxygen is combined rapidly with a fuel to release the stored energy in the form of heat. There are three conditions necessary for combustion to take place:

1. **Fuel:** Consisting of a combination of carbon and hydrogen
2. **Heat:** Sufficient to raise the temperature of the fuel to ignition (burning) point
3. **Oxygen:** From the air, combined with the elements in the fuel

Figure 6-1 illustrates the conditions necessary for combustion.

The fuel can be gas (such as natural gas), liquid (such as fuel oil), or solid (such as coal). Two elements all fuels have in common are hydrogen and carbon.

Fuel must be heated to burn. For example, a pilot burner (small flame) is used to ignite gas burners; electric ignition (an electric spark) is used to ignite oil; and, usually, a wood-burning fire is used to ignite coal. An example of a pilot burner and spark igniter is shown in Figure 6-2.

Air containing oxygen must be present for burning of fuel to take place. As an example, a burning candle can be extinguished by placing a glass jar around it to enclose it (Figure 6-3). The candle goes out when it no longer has oxygen to burn.

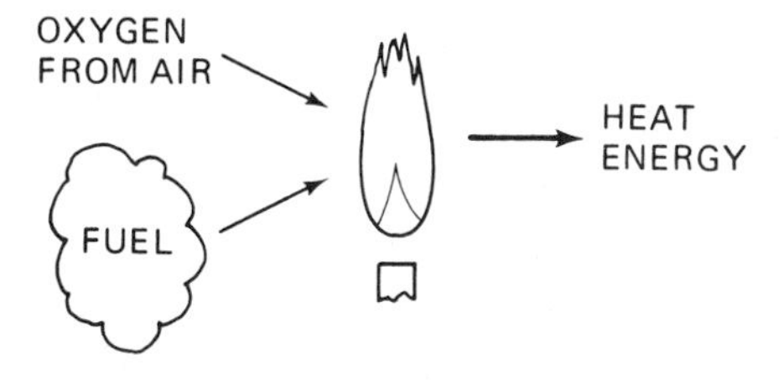

FIGURE 6-1 Conditions necessary for combustion.

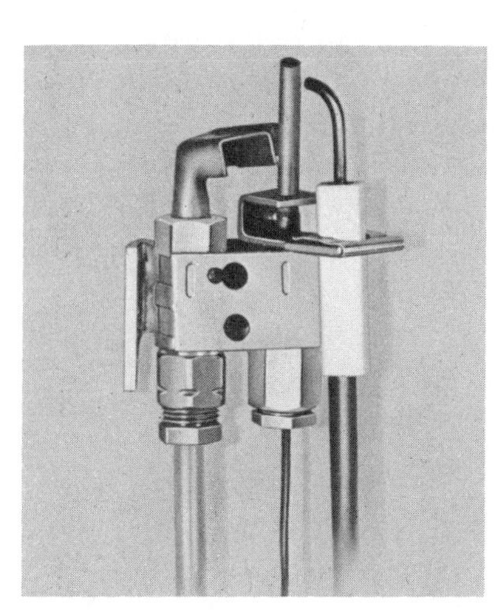

FIGURE 6-2 Example of pilot burner and spark igniter (Courtesy, White-Rodgers, Division of Emerson Electric Company).

FIGURE 6-3 Candle flame extinguished by lack of oxygen.

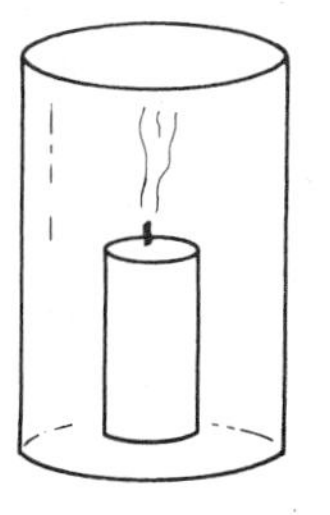

TYPES OF COMBUSTION

There are two types of combustion: complete combustion and incomplete combustion. Complete combustion must be obtained in all fuel-burning devices because incomplete combustion is dangerous.

Complete combustion results when carbon combines with oxygen to form carbon dioxide (CO_2), which is nontoxic and can be readily exhausted to the atmosphere. The hydrogen combines with oxygen to form water vapor (H_2O), which also can be harmlessly exhausted to the atmosphere.

Incomplete combustion results when a lack of sufficient oxygen causes the formation of undesirable products, including:

- Carbon monoxide (CO)
- Pure carbon or soot (C)
- Aldehyde, a colorless volatile liquid with a strong unpleasant odor (CH_3CHO)

Both carbon monoxide and aldehyde are toxic and poisonous. Soot causes coating of the heating surface of the furnace and reduces heat transfer (useful heat). Thus, the heating service technician must so adjust the fuel-burning device to produce complete combustion of the fuel.

During complete combustion the fuel combines with oxygen in the air to produce carbon dioxide and water vapor:

$$CH_4 + O_2 \longrightarrow CO_2 + 2H_2O$$

CAUTION: Sufficient air must be provided for proper combustion to take place to prevent the dangers of incomplete combustion.

Air Requirements

Air consists of about 21% oxygen and 79% nitrogen by volume (Figure 6-4). The nitrogen in the air dilutes the oxygen, which otherwise would be too concentrated to breathe in its pure form. Nitrogen is an *inert* chemical. The inert quality in a chemical means that it remains in a pure state without combining with other chemicals under ordinary conditions. For example, in a furnace when nitrogen is heated to temperatures higher than 2000° F, it does not react with the elements of the fuel. It enters the furnace with the combustion air and leaves through the chimney as pure nitrogen.

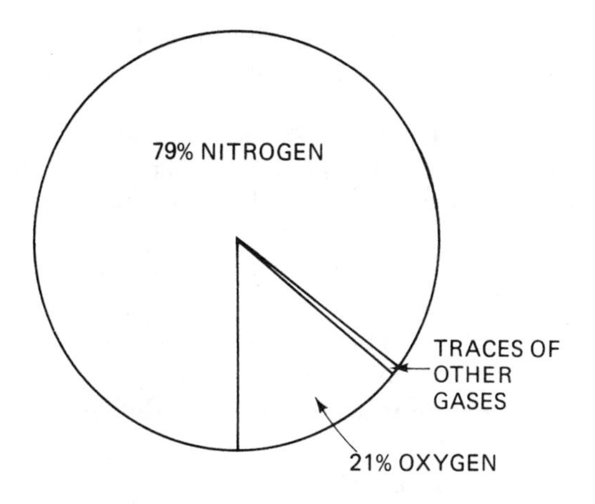

FIGURE 6-4 Chemical composition of air.

THE FLUE GASES:

The flue gases of a furnace operating to produce complete combustion contain:

- Carbon dioxide
- Water vapor
- Nitrogen
- Excess air

Carbon dioxide and water vapor are the products of complete combustion. Nitrogen remains after oxygen in the combustion air is consumed by the fuel. Excess air is supplied to the fuel-burning device to guard against the possibility of producing incomplete combustion. Normally, furnaces are adjusted to use 5 to 50% excess air. The effect of the amount of excess air on the CO_2 in the flue gases is shown in Figure 6-5.

Heating Values

Each fuel when burned is capable of producing a given amount of heat, depending on the constituents of the fuel. This information is useful in determining the heating capacity of a furnace. If the heating value of a unit of fuel produces a given amount of Btu, and the number of units of fuel burned per hour is known, the input rating of the furnace can be calculated. The Btu ratings for units of natural gas are shown in Figure 6-6. The typical gravity and heating value of fuel oil is shown in Figure 6-7, and the approximate Btu ratings for coal are shown in Figure 6-8.

EXAMPLE: An oil furnace burning No. 2 fuel oil uses 0.75 gal/h. What is the input rating of the furnace?

SOLUTION: The heating value of No. 2 oil from Figure 6-7 is between 141,800 and 137,000 Btu/gal. To simplify calculations, rating is rounded off to 140,000 Btu/gal. Therefore, the input rating of the furnace would be

$$140{,}000 \text{ Btu/gal} \times 0.75 \text{ usage rate} = 105{,}000 \text{ Btuh input}$$

NOTE: The Btu ratings for gas are given in Btu/ft^3, for oil in Btu/gal, and for coal in Btu/lb. The units used differ for each fuel because of the form in which they are delivered, (Figure 6-9).

Losses and Efficiency

When fuel is burned in a furnace, a certain amount of heat is lost in the hot gases that rise through the chimney. Although this function is necessary for disposal of the products of combustion, the loss should be minimized to allow the furnace to operate at its highest efficiency. Air entering the furnace at room temperature, or lower, is heated to fuel gas temperatures. These temperatures range from 350° F to 600° F, depending on the design of the furnace and its adjustment by a service technician.

FIGURE 6-5 The effect of excess air on the CO_2 in the flue gasses (Reprinted with permission from the 1977 Fundamentals Volume, ASHRAE Handbook & Product Directory).

APPROXIMATE MAXIMUM CO_2 VALUES FOR VARIOUS FUELS WITH DIFFERENT PERCENTAGES OF EXCESS AIR

TYPE OF FUEL	MAXIMUM THEORETICAL CO_2 PERCENT	PERCENT CO_2 AT GIVEN EXCESS AIR VALUES		
		20%	40%	60%
GASEOUS FUELS				
NATURAL GAS	12.1	9.9	8.4	7.3
PROPANE GAS (COMMERCIAL)	13.9	11.4	9.6	8.4
BUTANE GAS (COMMERCIAL)	14.1	11.6	9.8	8.5
MIXED GAS (NATURAL AND CARBURETED WATER GAS)	11.2	12.5	10.5	9.1
CARBURETED WATER GAS	17.2	14.2	12.1	10.6
COKE OVEN GAS	11.2	9.2	7.8	6.8
LIQUID FUELS				
NO. 1 AND 2 FUEL OIL	15.0	12.3	10.5	9.1
NO. 6 FUEL OIL	16.5	13.6	11.6	10.1
SOLID FUELS				
BITUMINOUS COAL	18.2	15.1	12.9	11.3
ANTHRACITE	20.2	16.8	14.4	12.6
COKE	21.0	17.5	15.0	13.0

NATURAL GAS DISTRIBUTED IN VARIOUS CITIES IN THE UNITED STATES*

NO.	CITY	HEAT VALUE, BTU/CU FT	SPECIFIC GRAVITY
1	ABILENE, TEX.	1121	0.710
2	AKRON, OHIO	1037	0.600
3	ALBUQUERQUE, N.M.	1120	0.646
4	ATLANTA, GA.	1031	0.604
5	BALTIMORE, MD.	1051	0.590
6	BIRMINGHAM, ALA.	1024	0.599
7	BOSTON, MASS.	1057	0.604
8	BROOKLYN, N.Y.	1049	0.595
9	BUTTE, MONT.	1000	0.610
10	CANTON, OHIO	1037	0.600
11	CHEYENNE, WYO.	1060	0.610
12	CINCINNATI, OHIO	1031	0.591
13	CLEVELAND, OHIO	1037	0.600
14	COLUMBUS, OHIO	1028	0.597
15	DALLAS, TEX.	1093	0.641
16	DENVER, COLO.	1011	0.659
17	DES MOINES, IOWA	1012	0.669
18	DETROIT, MICH.	1016	0.616
19	EL PASO, TEX.	1082	0.630
20	FT. WORTH, TEX.	1115	0.649
21	HOUSTON, TEX.‡	1031	0.623
22	KANSAS CITY, MO.	945	0.695
23	LITTLE ROCK, ARK.	1035	0.590
24	LOS ANGELES, CALIF.	1084	0.638

NO.	CITY	HEAT VALUE, BTU/CU FT	SPECIFIC GRAVITY
25	LOUISVILLE, KY.	1034	0.506
26	MEMPHIS, TENN.	1044	0.608
27	MILWAUKEE, WIS.	1051	0.627
28	NEW ORLEANS, LA.	1072	0.612
29	NEW YORK CITY	1049	0.595
30	OKLAHOMA CITY, OKLA.	1080	0.615
31	OMAHA, NEB.	1020	0.669
32	PARKERSBURG, W. VA.	1049	0.592
33	PHOENIX, ARIZ.	1071	0.633
34	PITTSBURGH, PA.	1051	0.595
35	PROVIDENCE, R.I.	1057	0.601
36	PROVO, UTAH	1032	0.605
37	PUEBLO, COLO.	980	0.706
38	RAPID CITY, S.D.	1077	0.607
39	ST. LOUIS, MO.	–	–
40	SALT LAKE CITY, UTAH	1082	0.614
41	SAN DIEGO, CALIF.	1079	0.643
42	SAN FRANCISCO, CALIF.	1086	0.624
43	TOLEDO, OHIO	1028	0.597
44	TULSA, OKLA.	1086	0.630
45	WACO, TEX.	1042	0.607
46	WASHINGTON, D.C.	1042	0.586
47	WICHITA, KAN.	1051	0.690
48	YOUNGSTOWN, OHIO	1037	0.600

*Average analyses obtained from the operating utility company (a) supplying the city; the supply may vary considerably from these data—especially where more than one pipeline supplies the city. Also, as new supplies may be received from other sources, the analyses may change. Peak shaving (if used) is act accounted for in these data.

FIGURE 6-6 The Btu ratings for units of natural gas (Reprinted with permission from the 1972 Fundamentals Volume, ASHRAE Handbook & Product Directory).

GRADE NO.	GRAVITY, API	WEIGHT, LB PER GALLON	HEATING VALUE, BTU PER GALLON
1	38-45	6.95 -6.675	137,000-132,900
2	30-38	7.296-6.960	141,800-137,000
4	20-28	7.787-7.396	148,100-143,100
5L	17-22	7.94 -7.686	150,000-146,800
5H	14-18	8.08 -7.89	152,000-149,400
6	8-15	8.448-8.053	155,900-151,300

(API AMERICAN PETROLEUM INSTITUTE RATING)

FIGURE 6-7 Typical gravity and heating values for standard grades of fuel oil (Reprinted with permission from the 1977 Fundamentals Volume, ASHRAE Handbook & Product Directory).

RANK	BTU PER LB AS RECEIVED
ANTHRACITE	12,700
SEMIANTHRACITE	13,600
LOW-VOLATILE BITUMINOUS	14,350
MEDIUM-VOLATILE BITUMINOUS	14,000
HIGH-VOLATILE BITUMINOUS A	13,800
HIGH-VOLATILE BITUMINOUS B	12,500
HIGH-VOLATILE BITUMINOUS C	11,000
SUBBITUMINOUS B	9,000
SUBBITUMINOUS C	8,500
LIGNITE	6,900

FIGURE 6-8 Approximate Btu ratings for various types of coal (Reprinted with permission from the 1977 Fundamentals Volume, ASHRAE Handbook & Product Directory).

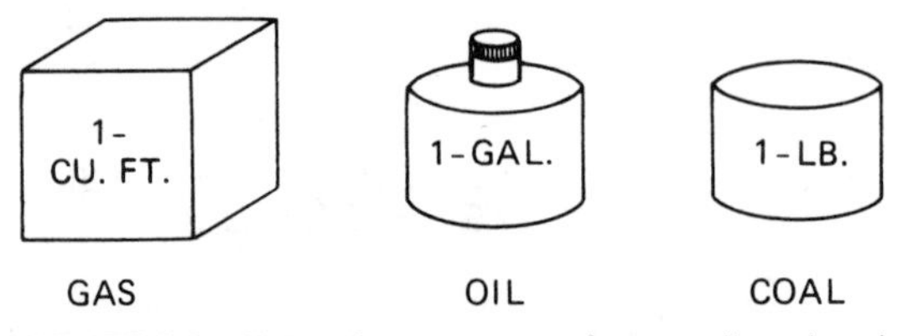

FIGURE 6-9 Units of measurement for gas, oil, and coal.

If the amount of heat lost is 20%, the efficiency of the furnace would be 80%. Charts are available (see Chapter 10) for determining the efficiency of the furnace based on knowing the temperature and the carbon dioxide content of the flue gases.

Knowing the efficiency of the furnace makes it possible for a heating service technician to calculate the output of the furnace. The following formula is used:

$$\text{Btuh input} \times \%\ \text{efficiency} = \text{Btu output}$$

EXAMPLE: The input of an oil furnace is 105,000 Btuh. Its efficiency is 80%. What is its output?

SOLUTION: 105,000 Btu × 0.80 efficiency = 84,000 Btuh output.

Types of Flames

Basically, there are two types of flames: yellow and blue (Figure 6-10). Pressure-type oil burners burn with a yellow flame. Modern Bunsen-type gas burners burn with a blue flame. The difference is due mainly to the manner in which air is mixed with the fuel.

A yellow flame is produced when gas is burned by igniting fuel gushing from an open end of a gas pipe, such as may be seen in fixtures used for ornamental decoration.

A blue flame is produced when approximately 50% of the air requirement is mixed with the gas prior to ignition. This is called primary air. A Bunsen-type burner uses this arrangement. The balance of air, called *secondary air*, is supplied during combustion to the exterior of the flame. Air adjustments are discussed in Chapter 8.

Improper gas flames are the result of inefficient or incomplete combustion and can be caused by:

- An excess supply of primary air
- A lack of secondary air
- The impingement of the flame on a cool surface (Figure 6-11)

FIGURE 6-10 Types of flames produced by pressure type oil burner and Bunsen burner (Courtesy, Robertshaw Control Company [Bunsen Burner]).

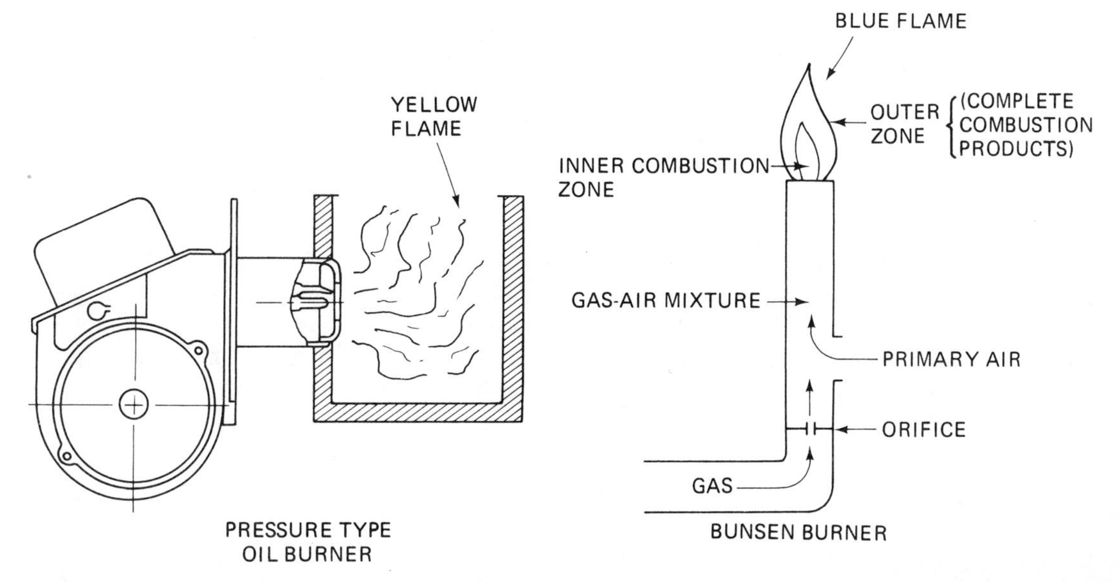

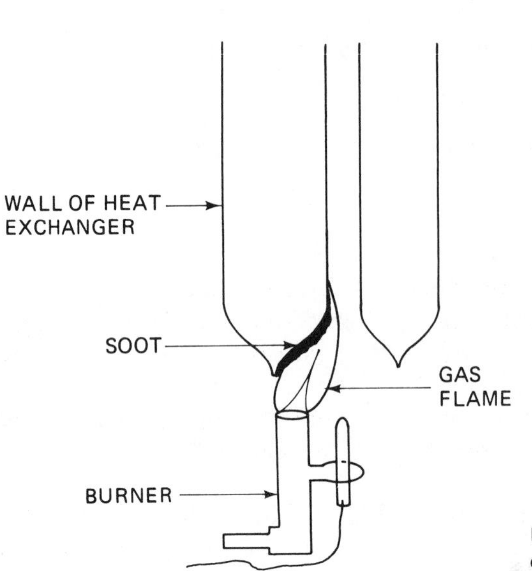

FIGURE 6-11 Soot caused by impingement of flame on cool wall of heat exchanger.

FUELS

Fuels are available in three forms: gases, liquids, and solids. Gases include natural gas and liquid petroleum; fuel oils are rated by grades 1, 2, 4, 5, and 6; and coals are of various types, mainly anthracite and bituminous. Each fuel has its own individual Btu heat content per unit and its own desirable or undesirable characteristics. The fuel selection is usually based on availability, price, and type of application.

Types and Properties of Gaseous Fuels

There are three types of gaseous fuels. They are:

- Natural gas
- Manufactured gas
- Liquid petroleum

Natural gas comes from the earth in the form of gas and often accumulates in the upper part of oil wells. Manufactured gases are combustible gases which are usually produced from solid or liquid fuel and used mainly for industrial processes. Liquid petroleum (LP) is a by-product of the oil refining process. It is so named because it is stored in liquid form. However, LP is vaporized when used.

NATURAL GAS:

Natural gas is nearly odorless and colorless. Therefore, an odorant, such as a mercaptan (any of various compounds containing sulfur and having a disagreeable odor) is added so that a leak can be sensed. The content of gases differs somewhat according to locality. It is recommended that information be obtained from the local gas company relative to the specific gravity and the Btu/ft^3 content of the gas available.

The specific gravity affects piping sizes. *Specific gravity* is the ratio of the weight of a given volume of substance to an equal volume of air or water at a given temperature and pressure. Thus, gas with a specific gravity of 0.60 weighs $^6/_{10}$ or $^3/_5$ as much as air for an equal volume.

The Btu/ft^3 content of gas varies from 900 to 1200 depending on the locality, but it is usually in the range of 1000 to 1050 Btu/ft^3.

The chief constituent of natural gas is methane. Commonly called marsh gas, methane is a gaseous hydrocarbon that is a product of decomposition of organic matter in marshes or mines or of the carbonization of coal. Natural gas is comprised of from 55 to 95% methane combined with other hydrocarbon gas.

MANUFACTURED GAS:

Manufactured gas is produced from coal, oil, and other hydrocarbons. It is comparatively low in Btu/ft^3, usually in the range of 500 to 600. It is not considered an economical space-heating fuel.

LIQUID PETROLEUM:

There are two types of liquid petroleum: propane and butane. Propane is the more useful of the two as a space heating fuel since it boils at -40° F and therefore can be readily vaporized for heating in a northern climate. Butane boils at about 32° F.

Propane has a heating value of 21,560 Btu/lb or about 2500 Btu/ft^3. Butane has a heating value of 21,180 Btu/lb or about 3200 Btu/ft^3. When LP gas is used as a heating fuel, the equipment must be designed to use this type of gas. When ordering equipment, the purchaser must indicate which type of gaseous fuel is being used. Conversion kits are available to convert natural gas furnaces to LP gas when necessary. It is important to follow the manufacturer's instructions with great care.

NOTE: Propane and butane vapors are generally considered more dangerous than those of natural gas, since they have higher specific gravities (propane 1.52, butane 2.01). Because the vapor is heavier it tends to accumulate near the floor, thereby increasing the danger of an explosion upon ignition.

Types and Properties of Fuel Oils

Fuel oils are rated according to their Btu/lb content and API gravity (shown in Figure 6-7). The API gravity is an index selected by the American Petroleum Institute. There are six grades of oil: Nos. 1, 2, 4, 5 (light), 5 (heavy), and 6. Note that the lighter-weight oils have a higher API gravity.

GRADE NO. 1:

A light-grade distillate prepared for vaporizing-type oil burners.

GRADE NO. 2:

A heavier distillate than Grade No. 1 and is manufactured for domestic pressure-type oil burners.

GRADE NO. 4:

Light residue or heavy distillate. It is produced for pressure-type commercial oil burners using a higher pressure than domestic burners.

GRADE NO. 5 (LIGHT):

A residual-type fuel of medium weight. It is used for commercial-type burners that are specially designed for its use.

GRADE NO. 5 (HEAVY):

A residual-type fuel for commercial oil burners. It usually requires preheating.

GRADE NO. 6:

Also called Bunker C; a heavy residue used for commercial burners. It requires preheating in the tank to permit pumping, and additional preheating at the burner to permit atomization (breaking up into fine particles).

Types and Properties of Coals

There are four different types of coal: anthracite, bituminous, subbituminous, and lignite. Coal is constituted principally of carbon with the better grades having as much as 80% carbon. The types vary not only in Btu/lb but also in their burning and handling qualities.

ANTHRACITE:

A clean, hard coal. It burns with an almost smokeless short flame. It is difficult to ignite but burns freely when started. It is noncaking, leaving a fine ash that does not clog grates and ash removal equipment.

BITUMINOUS:

A wide range of coals varying from high grade in the east to low grade in the west. It is more brittle than anthracite coal and readily breaks up into small pieces for grading and screening. The length of the flame is long, but varies for different grades. Unless burning is carefully controlled, much smoke and soot can result.

SUBBITUMINOUS:

Has a high moisture content and tends to break up when dry. It can ignite spontaneously when stored. It ignites easily and burns with a medium flame. It is desirable for its noncaking characteristic and because it forms little soot and smoke.

LIGNITE:

Has a woody consistency, is high in moisture content, low in heating value, and is clean to handle. Lignite has a greater tendency to break up when dry than does subbituminous coal. Because of its high moisture content, it is difficult to ignite. It is noncaking and forms little smoke and soot.

REVIEW QUESTIONS

Select the letter representing your choice of the correct answer.

6-1. Combustion is what type of process?

(a) Mechanical

(b) Chemical

(c) Electrical

(d) Pneumatic

6-2. What are the three conditions necessary for combustion?

(a) Fuel, air, and moisture

(b) Carbon, air, and pressure

(c) Fuel, heat, and oxygen

(d) Carbon, hydrogen, and oxygen

6-3. Complete combustion is obtained when all the carbon and hydrogen in the fuel are:

(a) Oxidized

(b) Heated

(c) Combined

(d) Dissipated

6-4. Incomplete combustion may be caused by:

(a) Too hot a fire

(b) Too much draft

(c) Insufficient supply of fuel

(d) Insufficient supply of air

6-5. What is the percent of oxygen in air?

(a) 79%

(b) 21%

(c) 69%

(d) 31%

6-6. Expressed in percentage, efficiency is the useful heat divided by:

(a) The heat loss of the building

(b) The capacity of the furnace

(c) Heating value of the fuel

(d) Temperature of the flue gases

6-7. The amount of excess air required for complete combustion is within which percentage range?

(a) 5 to 30%

(b) 25 to 60%

(c) 40 to 75%

(d) 5 to 50%

6-8. The percent of CO_2 in the flue gas is not affected by the amount of excess air used by the combustion equipment. True or false?

(a) True

(b) False

6-9. Natural gas is comprised chiefly of:

(a) Methane

(b) Propane

(c) Butane

(d) Ethylene

6-10. Preheating of No. 6 fuel oil is usually necessary. True or false?

(a) True

(b) False

7

PARTS COMMON TO ALL FURNACES

OBJECTIVES

After studying this chapter, the student will be able to:

- Identify the various components of forced warm air heating furnaces
- Adjust the speed of a furnace fan.
- Provide proper service and maintenance for common parts of a heating system

PARTS COMMON TO ALL FURNACES

A warm air furnace is a device for providing space heating. Fuel, a form of energy, is converted into heat and distributed to various parts of a structure.

Fuel-burning furnaces provide heat by combustion of the fuel within a heat exchanger. Air is passed over the outside surface of the heat exchanger, transferring the heat from the fuel to the air. In a fuel-burning furnace, the products of combustion are exhausted to the atmosphere through the flue passages connected to the heat exchanger.

In an electric furnace, air passes directly over the electrically heated elements without the use of a heat exchanger, since no products of combustion are formed.

A forced air furnace uses a fan to propel the air over the heat exchanger and to circulate the air through the distribution system. A furnace is considered to be a residential type when its input is less than 250,000 Btuh.

BASIC COMPONENTS OF A WARM AIR FURNACE

There are seven basic components of a forced warm air furnace:

1. Heat exchanger
2. Fuel-burning device
3. Cabinet or enclosure
4. Fan and motor
5. Air filters
6. Humidifier
7. Controls

The fuel-burning device, item 2, will be discussed in Chapter 8. The controls, item 7, will be discussed in Chapters 16 through 19.

Gravity-type warm air furnaces include all the components listed above with the exception of the fan and motor, and air filters. The basic components of a gravity warm air heating system are shown in Figure 7-1. The gravity furnace is so named because air is circulated over the heat exchanger and through the distribution system by the force of gravity. Air heated by the furnace becomes lighter and rises. The movement of air is caused by the colder (heavier) air replacing the heated air, creating continuous movement as long as heat is applied. Gravity furnaces for residential heating have been replaced to a great extent by the forced air design.

ARRANGEMENT OF COMPONENTS

Four different designs of forced air furnaces are in common use, each requiring a different arrangement of the basic components. The four designs are:

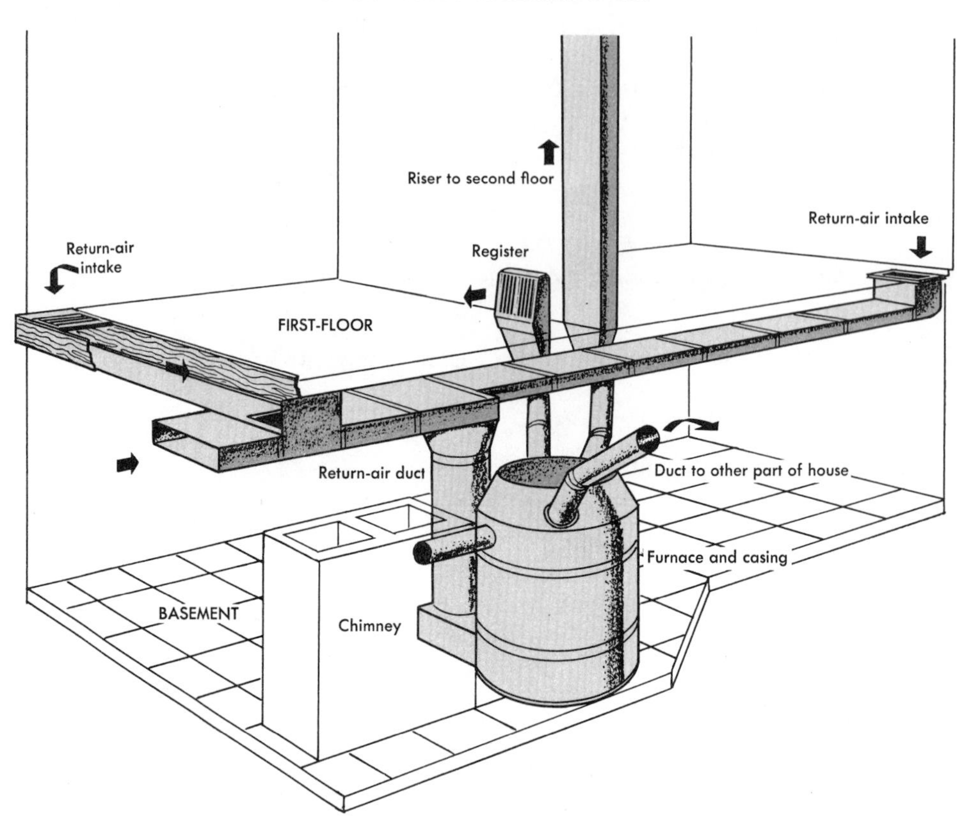

FIGURE 7-1 Basic components of a gravity warm air heating system (Courtesy, Small Homes Council – Building Research Council, University of Illinois).

1. Horizontal
2. Upflow or high-boy
3. Low-boy
4. Downflow or counterflow

The *horizontal* furnace is used in attic spaces or crawl spaces where the height of the furnace must be kept as low as possible (Figures 7-2 and 7-3). Air enters at one end of the unit through the fan compartment and is forced horizontally over the heat exchanger, exiting at the opposite end.

The *upflow* or *high-boy* design is used in the basement or a first-floor equipment room where floor space is at a premium (Figures 7-4 and 7-5). The fan is located below the heat exchanger. Air enters at the bottom or lower sides of the unit and leaves at the top through a warm air plenum.

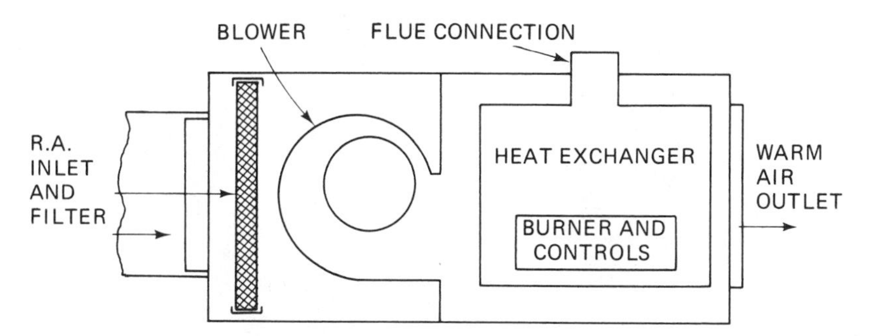

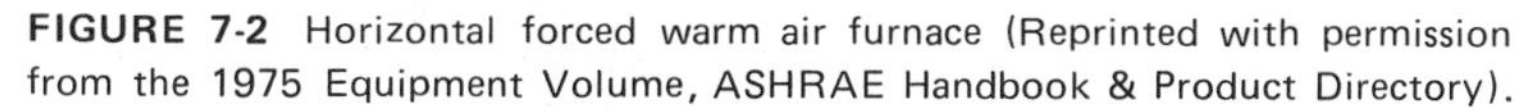

FIGURE 7-2 Horizontal forced warm air furnace (Reprinted with permission from the 1975 Equipment Volume, ASHRAE Handbook & Product Directory).

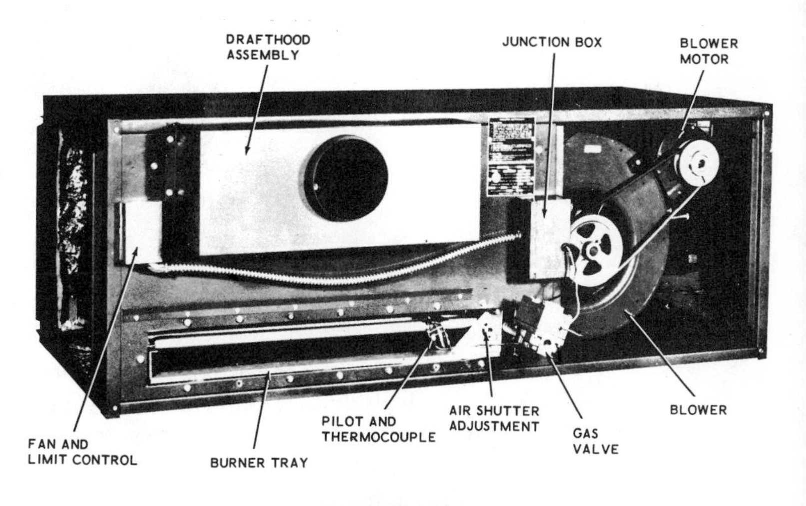

FIGURE 7-3 Horizontal forced warm air furnace, gas fired (Courtesy, Westinghouse Electric Corporation, Heating and Cooling Division).

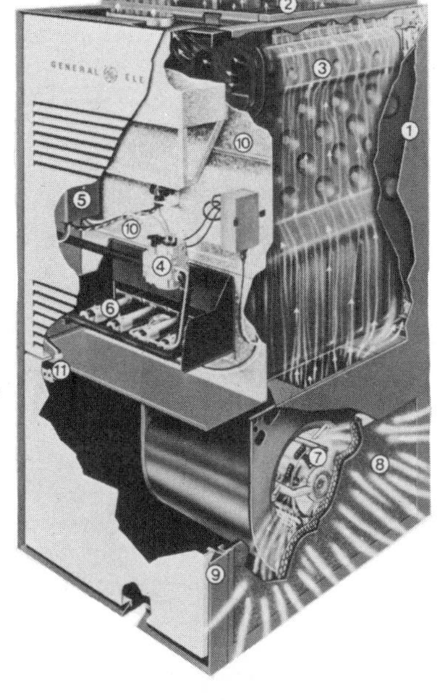

FIGURE 7-4 Upflow (high-boy) forced warm air furnace, gas fired (Courtesy, General Electric Company).

1. Welded double steel sidewall for rigid construction.
2. Shield keeps heat in, jacket cool.
3. Rugged "thermal-trap" heat exchanger.
4. Gas Valve.
5. Temperature sensing fan and limit control protects against circulation of cold air or overheating of the furnace.
6. Linear Type Burner.
7. Direct drive blowers require no noisy belts or pulleys. Automatic speed change.
8. Semi-permanent filter furnished with furnace. When filter is dirty, just wash and replace.
9. All welded steel framework and cabinet
10. Two fusible link protective devices
11. Positive blower door latches to filters.

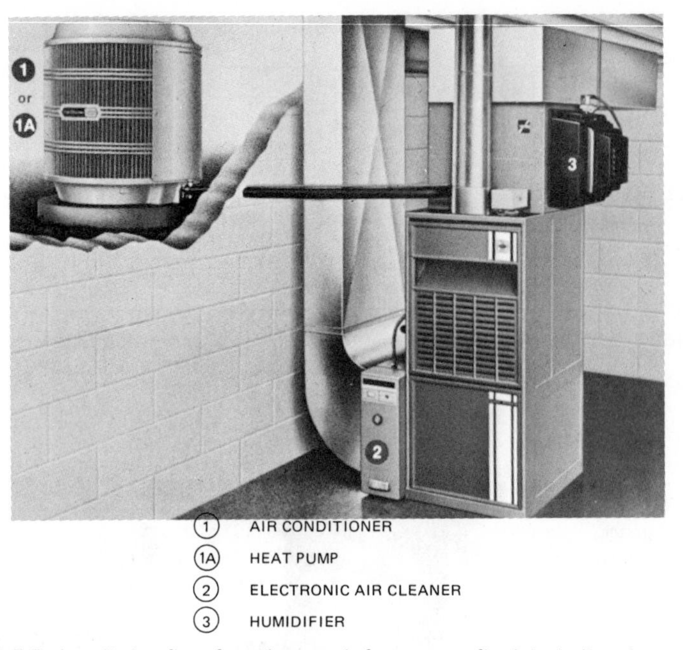

FIGURE 7-5 Installed upflow forced warm air furnace, gas fired; including air conditioner, air cleaner, flue vent damper and humidifier (Reproduced with permission of Carrier Corporation, © Copyright 1979, Carrier Corporation).

The *low-boy* furnace occupies more floor space and is lower in height than the upflow design, making it ideally suited for basement installation (Figure 7-6). The fan is placed alongside the heat exchanger. A return air plenum is built above the fan compartment. A supply air plenum is built above the heat-exchanger compartment.

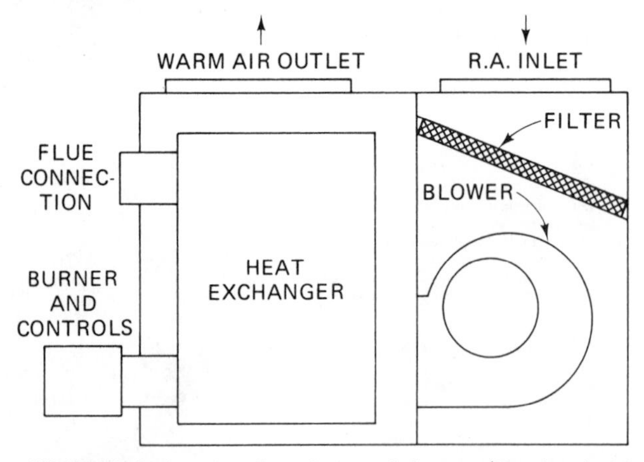

FIGURE 7-6 Low-boy forced warm air furnace (Reprinted with permission from the 1975 Equipment Volume ASHRAE Handbook & Product Directory).

The *downflow* or *counterflow* design is used in houses having an under-the-floor type of distribution system (Figure 7-7). The fan is located above the heat exchanger. The return air plenum is connected to the top of the unit. The supply air plenum is connected to the bottom of the unit.

DESCRIPTION OF COMPONENTS

Although forced warm air furnaces differ in their fuel-burning equipment and many of the controls required for operation, many other components are the same or similar. These components vary somewhat, depending upon the size of the furnace and the arrangement of parts, but basically they have characteristics and functions in common. Components of a typical gas-fired forced warm air furnace are shown in Figure 7-8.

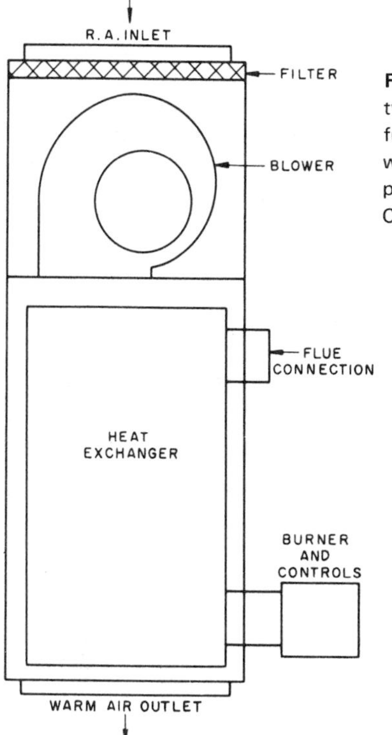

FIGURE 7-7 Downflow (counterflow) forced warm air furnace (Reprinted with permission from the 1975 Equipment Volume, ASHRAE Handbook & Product Directory).

FIGURE 7-8 Components of a typical upflow forced warm air furnace, gas fired (Reproduced with permission of Carrier Corporation, © Copyright 1979, Carrier Corporation).

Heat Exchanger

The heat exchanger is the part of the furnace where combustion takes place using gas, oil, or coal as fuel. The heat exchanger is usually made of cold-rolled, low-carbon steel with welded seams. There are two general types of heat exchangers:

1. Individual section
2. Cylindrical

THE INDIVIDUAL SECTION:

This type of heat exchanger has a number of separate heat exchangers (Figure 7-9). Each section has individual burners (fuel-burning devices). These sections are joined together at the bottom so that a common pilot can light all burners. The sections are joined together at the top so that flue gases are directed to a common flue. The individual section type of heat exchanger is used only on gas-burning equipment.

THE CYLINDRICAL:

This type of heat exchanger has a single combustion chamber and uses a single-port fuel-burning device (Figure 7-10). The cylindrical type of heat exchanger is used on gas, coal, and oil units.

Many heat exchangers for coal, gas, and oil have two types of surface:

1. Primary
2. Secondary

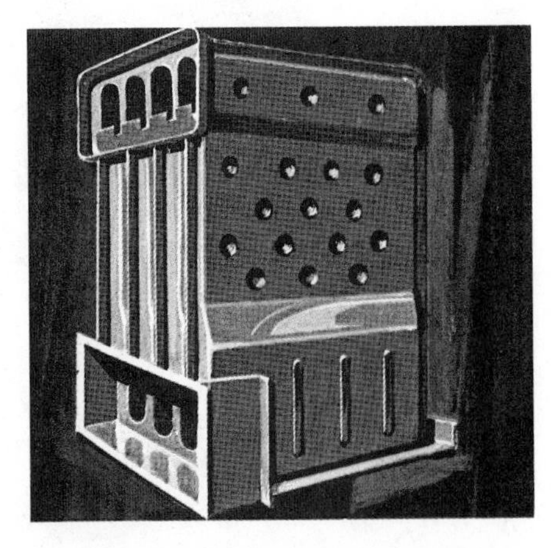

FIGURE 7-9 Individual section heat exchanger (Courtesy, General Electric Company).

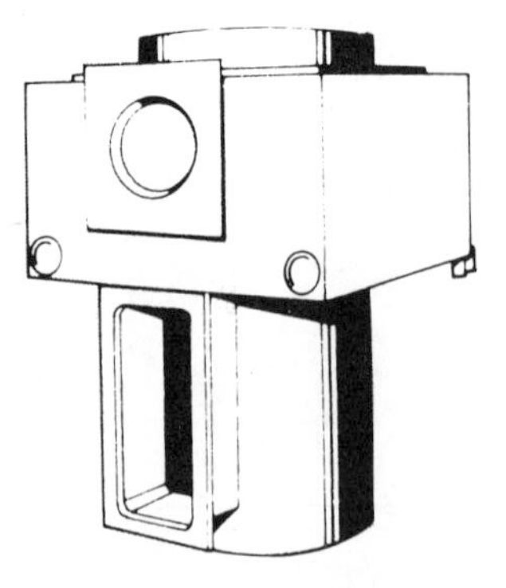

FIGURE 7-10 Cylindrical heat exchanger (Courtesy, Westinghouse Electric Corporation, Heating and Cooling Division).

The primary surface is in contact with or in direct sight of the flame and is located where the greatest heat occurs. The secondary surface follows the primary surface in the path between the burner and the flue, and is used to extract as much heat as possible from the flue gases before the products of combustion are taken out through the flue to the chimney.

The fuel used by the furnace has a definite relation to the amount of chimney draft required. Coal, because of the high resistance of its fuel bed, requires a relatively high draft. Oil requires a relatively low draft. Gas requires a balanced draft which is effective only when the furnace is providing heat.

Cabinet or Enclosure

Cabinets for forced warm air furnaces serve many functions. They provide:

- An attractive exterior
- Insulation in the area around the heat exchanger
- Mounting facilities for controls, fan and motor, filters, and other items which they enclose
- Access panels for parts requiring service
- Connections for the supply and return air duct
- An airtight enclosure for air to travel over the heat exchanger

Cabinets for gas and oil furnaces are usually rectangular in shape and have a baked-enamel finish. Cabinets are insulated on the interior with aluminum-backed mineral wool, fiberglas, or a metal liner so that the surface will not be too hot to the touch. Insulation also makes it possible to place flammable material in contact with the cabinet without danger of fire.

The trend now is to furnish complete units so that a minimum amount of assembly time is required for installation. Usually, only duct and service connections need be made by the installer. Knockouts (readily accessible openings) are provided for gas and electric connections. The bolts used for holding the unit in place during shipment are also used for leveling the unit during installation. On highboy units, return air openings in the fan compartment can be cut in the cabinet sides or bottom to fit installation requirements.

Fan and Motor

Much of the successful performance of a heating system depends upon the proper operation of the fan and fan motors. The fan moves air through the system, obtaining it from return air and outside air, and forces it over the heat exchanger and through the supply distribution system to the space to be heated.

Fans used are the centrifugal type with forward curved blades (Figure 7-11). Air enters through both ends of the wheel (double inlet) and is pumped or compressed (the buildup of air pressure) at the outlet. Fans and motors must be the proper size to deliver the required amount of air against the total resistance produced by the system. Depending upon the quantity of air moved and the resistance of the system, a motor of the proper size is selected.

It is usually necessary to make some adjustment of the air quantities at the time of installation, because the resistance of the system cannot be fully determined beforehand. Generally, the higher the speed of the fan the greater the cfm output, thereby requiring a higher-horsepower motor.

TYPES OF FANS:

Two types of fan motor arrangements are used:

1. Belt-drive
2. Direct-drive

The *belt-drive* arrangement uses a fixed pulley on the fan shaft and a variable-pitch pulley on the motor shaft, Figure 7-12. The speed of the fan is in direct proportion to the ratio of the pulley diameters. The following equation is used:

$$\begin{matrix}\text{speed of fan,}\\ \text{revolutions per}\\ \text{minute (rpm)}\end{matrix} = \frac{\begin{matrix}\text{diameter of the motor}\\ \text{pulley, inches}\end{matrix}}{\begin{matrix}\text{diameter of the fan}\\ \text{pulley, inches}\end{matrix}} \times \begin{matrix}\text{revolutions}\\ \text{per minute}\\ \text{(rpm) of}\\ \text{the motor}\end{matrix}$$

EXAMPLE: What is the fan speed using a 3-in. motor pulley, a 9-in. fan pulley, and an 1800 rpm motor?

SOLUTION:

$$\text{rpm of fan} = \frac{3\text{ in.}}{9\text{ in.}} \times 1800 = 600\text{ rpm}$$

The diameter of a variable-pitch pulley can be changed by adjusting the position of the outer flange of the pulley. This arrangement permits adjustment of the speed up to 30%. For changes greater than this, a different pulley must be used.

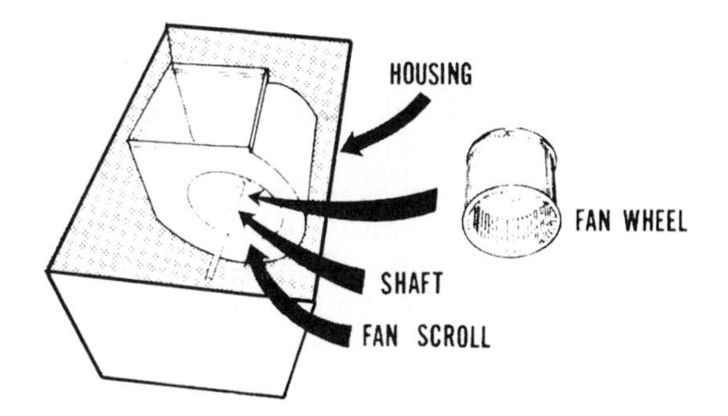

FIGURE 7-11 Centrifugal type fan (Reproduced with permission of Carrier Corporation, © Copyright 1979, Carrier Corporation).

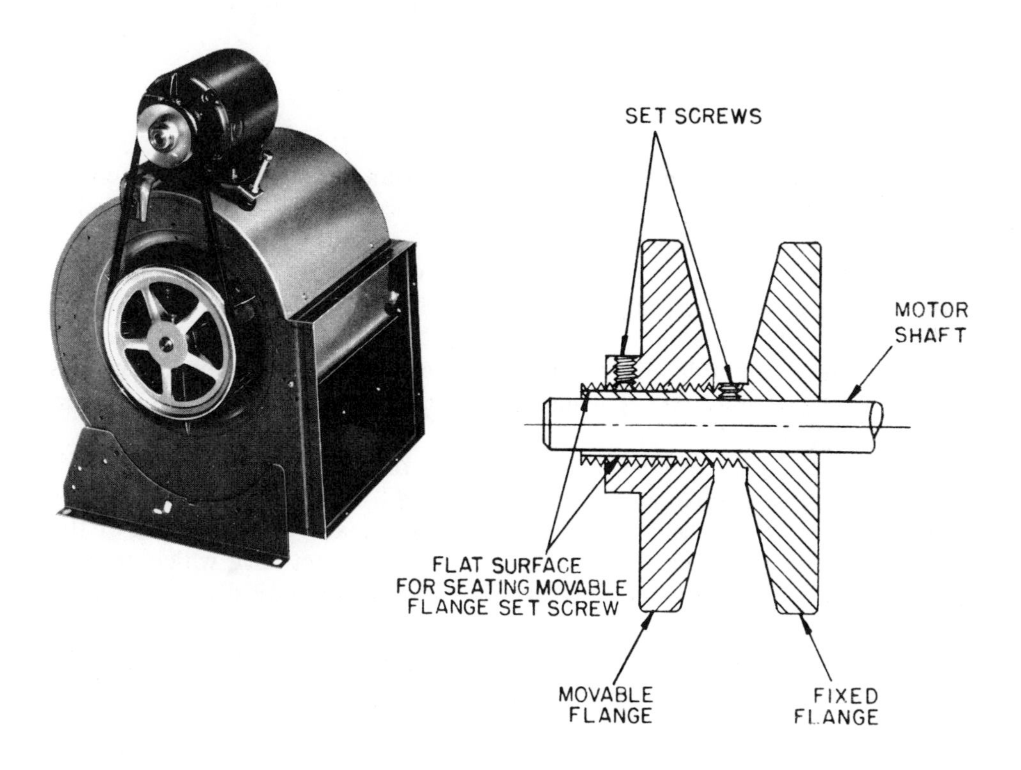

FIGURE 7-12 Belt-drive centrifugal fan with adjustable pitch motor pulley (Courtesy, Lau Division, Phillips Industries, Inc. [Fan and Motor]).

On the belt-drive arrangement, provision is made for regulating the belt tension and isolating the motor from the metal housing. The belt tension should be sufficient to prevent slippage. The motor mounts (usually rubber) provide isolation to prevent sound transmission.

The amperage draw of the motor is an indication of the amount of loading of the motor. The maximum loading is indicated on the motor nameplate as full load amperes (FLA). If the full load amperes are exceeded, a larger motor must be used. Additional information on measuring amperage draw is given in Chapter 14. If the motor is not drawing its full capacity in FLA it is an indication that the motor is underloaded. This may be caused by the additional resistance of dirty filters.

The *direct-drive* arrangement has the fan mounted on an extension of the motor shaft. Fan speeds can be changed only by altering the motor speed, which is accomplished by the use of extra windings on the motor, sometimes called a *tap-wound* or *multispeed* motor (Figure 7-13).

A tap-wound motor has a series of connections, each providing a different speed. Often one speed is selected for heating and another speed is selected for cooling.

Air Filters

Air filters are located on the suction side of a fan. They serve to remove fumes and smoke, as well as undesirable airborne particles such as lint, fly ash, dust, pollen, fungus, spores, and bacteria.

The desirability of air filters in a forced warm air heating system in relation to human comfort, and the three types of filters used, is discussed in Chapter 2. Air filters are also beneficial to health and cleanliness.

Both the disposable and permanent types of air filters remove particles down to 10 μm (microns). The electronic type removes particles down from 10 μm to 0.1 μm.

Disposable types of air filters are usually made of fibrous material and, in the average residence, are normally effective for 3 to 4 months. Some disposable types of air filters have an electrostatic quality. The friction of the air stream through them creates a small charge of electricity which attracts particles. Air filters must be cleaned or replaced periodically to maintain unit efficiency.

Permanent types of air filters are usually made of polyurethane or metal mesh. These filters can be washed and reused. Some require a special adhesive coating to maintain their dust holding capacity.

Electronic air filters create a strong, direct current, electrical field. A charge is applied to the airborne particles and they are collected on a plate with the opposite charge. The collected particles are removed by washing the filter.

FIGURE 7-13 Tap-wound motor used for altering speed of direct-drive centrifugal fan (Courtesy, Lau Division, Phillips Industries, Inc.).

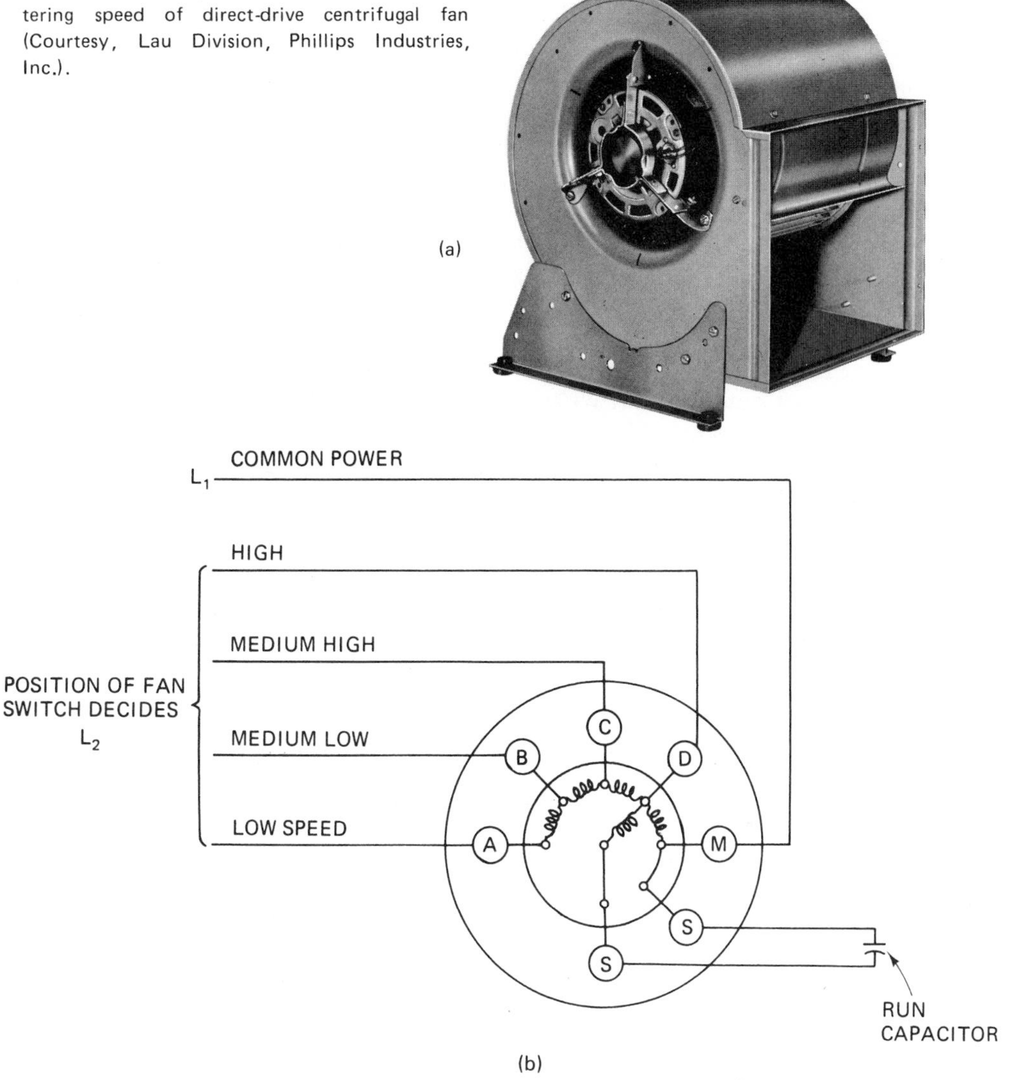

Humidifiers

Humidifiers are used to add moisture to indoor air. The amount of moisture required depends upon:

- The outside temperatures
- The house construction
- The amount of relative humidity that the interior of the house will withstand without condensation problems

It is desirable to maintain 30 to 50% relative humidity. Too little humidity may cause furniture to crack. Too much humidity can cause condensation problems. From a comfort and health standpoint, a range as wide as 20 to 60% relative humidity is acceptable.

The colder the outside temperature, the greater the need for humidification. The amount of moisture the air will hold depends upon its temperature. The colder the outside air, the less moisture it contains. Air from the outside enters the house through infiltration, as discussed in Chapter 3. When outside air is warmed, its relative humidity is lowered unless moisture is added by humidification, Figure 7-14.

For example, air at 20° F and 60% relative humidity contains 8 grains of moisture per pound of air. When air is heated to 72° F, it can hold 118 grains of moisture per pound. Thus, to maintain 50% relative humidity in a 72° F house, the grains of moisture per pound must be increased to 59 (118 ÷ 2 = 59). One pound of air entering a house from the outside will require

FIGURE 7-14 The effect of temperature on relative humidity (Courtesy, Humid-Aire Corporation).

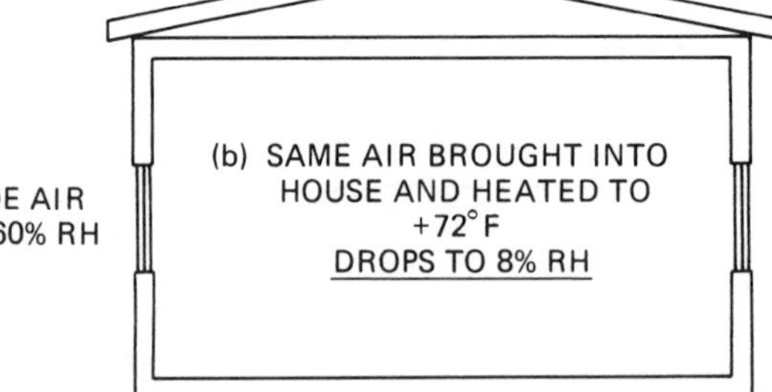

INDOOR RELATIVE HUMIDITY TABLE

OUTDOOR RELATIVE HUMIDITY	0°	10°	20°	30°	40°	50°
100%	6	9	14	21	31	46
80%	5	7	11	17	25	37
60%	3	5	(8%)	13	19	28
40%	2	4	6	8	12	18
20%	1	2	3	4	6	10

OUTDOOR TEMPERATURE

THE ABOVE CHART SHOWS WHAT HAPPENS TO THE HUMIDITY IN YOUR HOME WHEN YOU HEAT COLD WINTER AIR TO 72° ROOM TEMPERATURE. EXAMPLE: WHEN OUTDOOR TEMPERATURE IS 20° AND OUTDOOR RELATIVE HUMIDITY IS 60%, INDOOR RELATIVE HUMIDITY DROPS TO ONLY 8%. YOU NEED 30% TO 45% FOR COMFORT.

the addition of 51 grains of moisture (59 - 8 = 51). The amount of infiltration depends upon the tightness of the windows and doors and other parts of the construction.

It is impractical in most buildings to maintain high relative humidity when the outside temperature is low. Condensation forms on the inside of the window when the surface temperature reaches the dew point temperature of the air. The *dew point temperature* is the temperature at which moisture begins to condense when the air temperature is lowered.

Therefore, it is desirable to control the amount of humidity in the house by the use of a humidistat. A *humidistat* is a device that regulates the "on" and "off" periods of humidification (Figure 7-15). The setting of a humidistat can be changed to comply with changing outside temperatures.

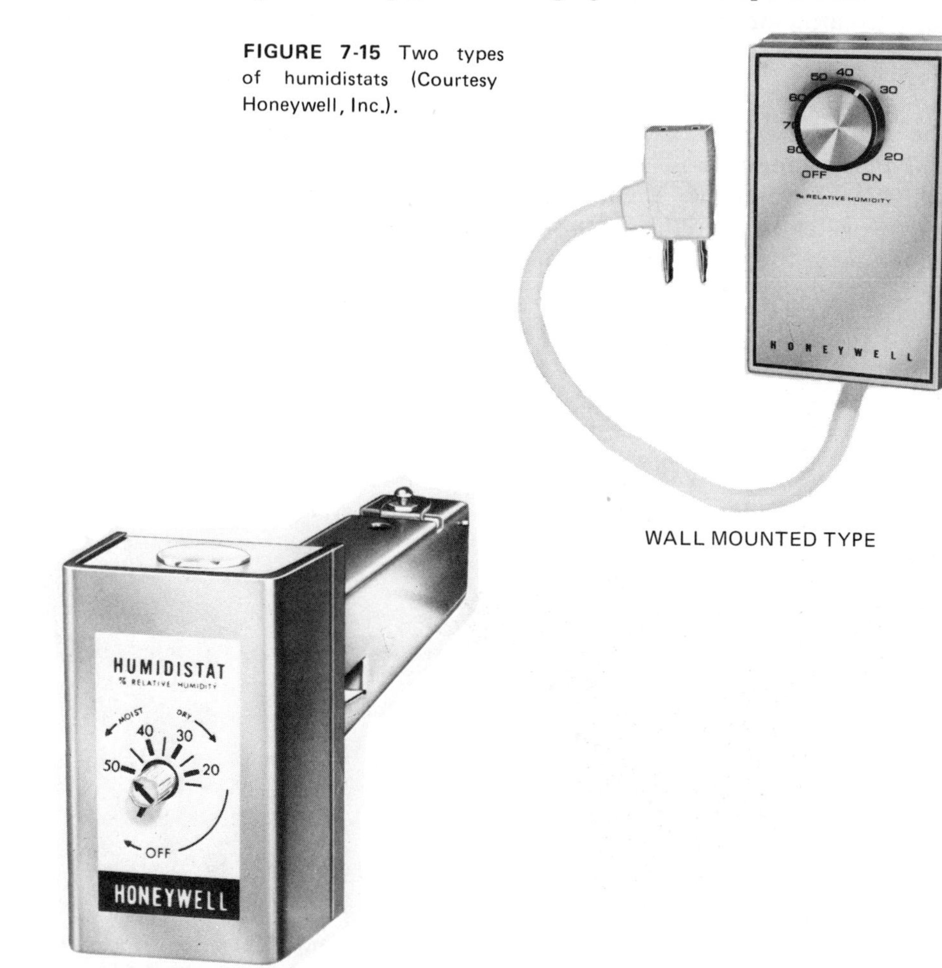

FIGURE 7-15 Two types of humidistats (Courtesy Honeywell, Inc.).

TYPES OF HUMIDIFIERS:

There are three general types of humidifiers:

1. Evaporative
2. Atomizing
3. Vaporizing

There are also three types of evaporative humidifiers:

1. Plate
2. Rotating drum or plate
3. Fan-powered

The *plate* type of evaporative humidifier (Figure 7-16) has a series of porous plates mounted in a rack. The lower section of the plates extends down into the water that is contained in the pan. A float valve regulates the supply of water to maintain a constant level in the pan. The pan and plates are mounted in the warm air plenum.

FIGURE 7-16 Plate type evaporative humidifier (Courtesy, Autoflo, a Division of Masco Corporation).

The *rotating drum* type of evaporative humidifier (Figure 7-17) has a slowly revolving drum covered with a polyurethane pad partially submerged in water. As the drum rotates, it absorbs water. The water level in the pan is maintained by a float valve. The humidifier is mounted on the side of the return air plenum. Air from the supply plenum is ducted into the side of the humidifier. The air passes over the wetted surface, absorbs moisture, then goes into the return air plenum.

The *rotating plate* type of evaporative humidifier (Figure 7-18) is similar to the drum type in that the water-absorbing material revolves. However, this type is normally mounted on the underside of the main warm air supply duct.

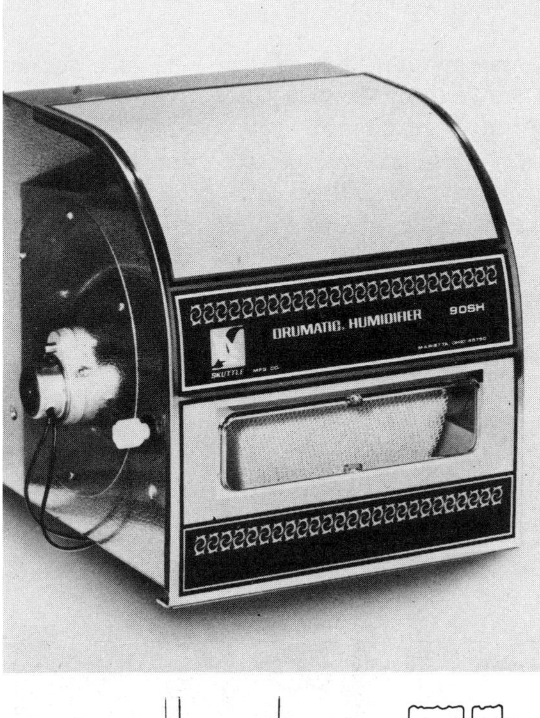

FIGURE 7-17 Rotating drum type evaporative humidifier (Courtesy, Skuttle Manufacturing Company).

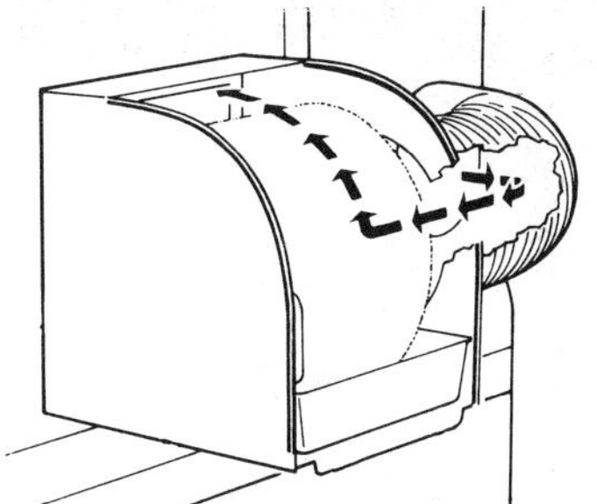

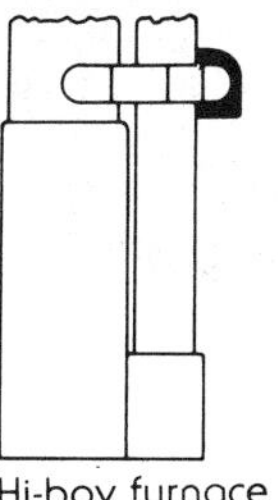

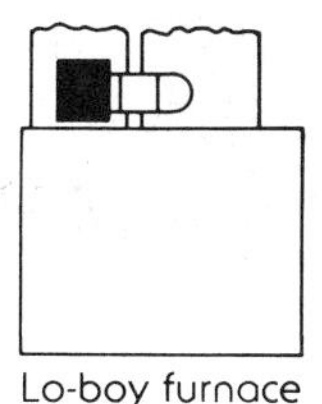

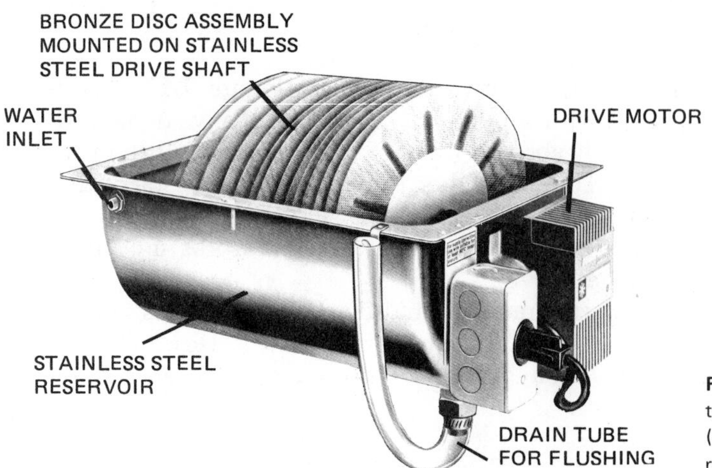

FIGURE 7-18 Rotating plate type evaporative humidifier (Courtesy, Humid-Aire Corporation).

The *fan-powered* type of evaporative humidifier (Figure 7-19) is mounted on the supply air plenum. Air is drawn in by the fan, forced over the wetted core, and delivered back into the supply air plenum. The water flow over the core is controlled by a water valve. A humidistat is used to turn the humidifier on and off, controlling both the fan and the water valve. The control system is set up so that the humidifier can operate only when the furnace fan is running.

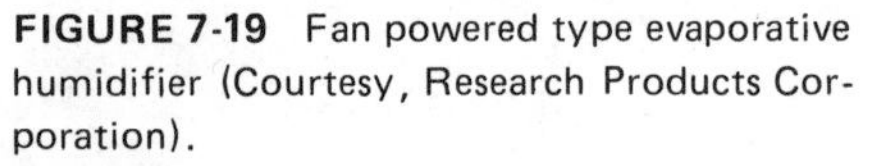

FIGURE 7-19 Fan powered type evaporative humidifier (Courtesy, Research Products Corporation).

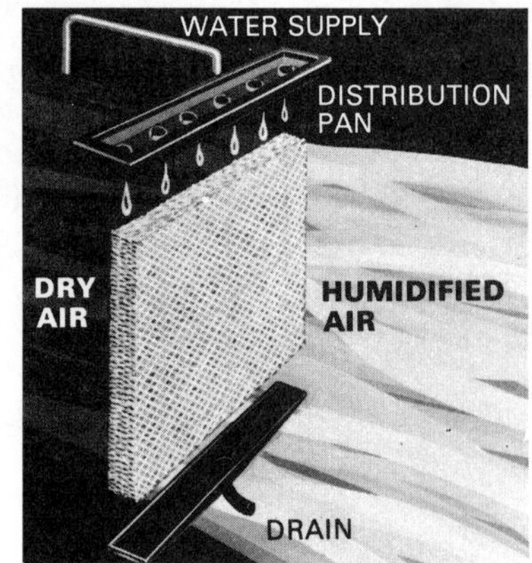

FIGURE 7-20 Atomizing type of humidifier (Courtesy, Humid-Aire Corporation).

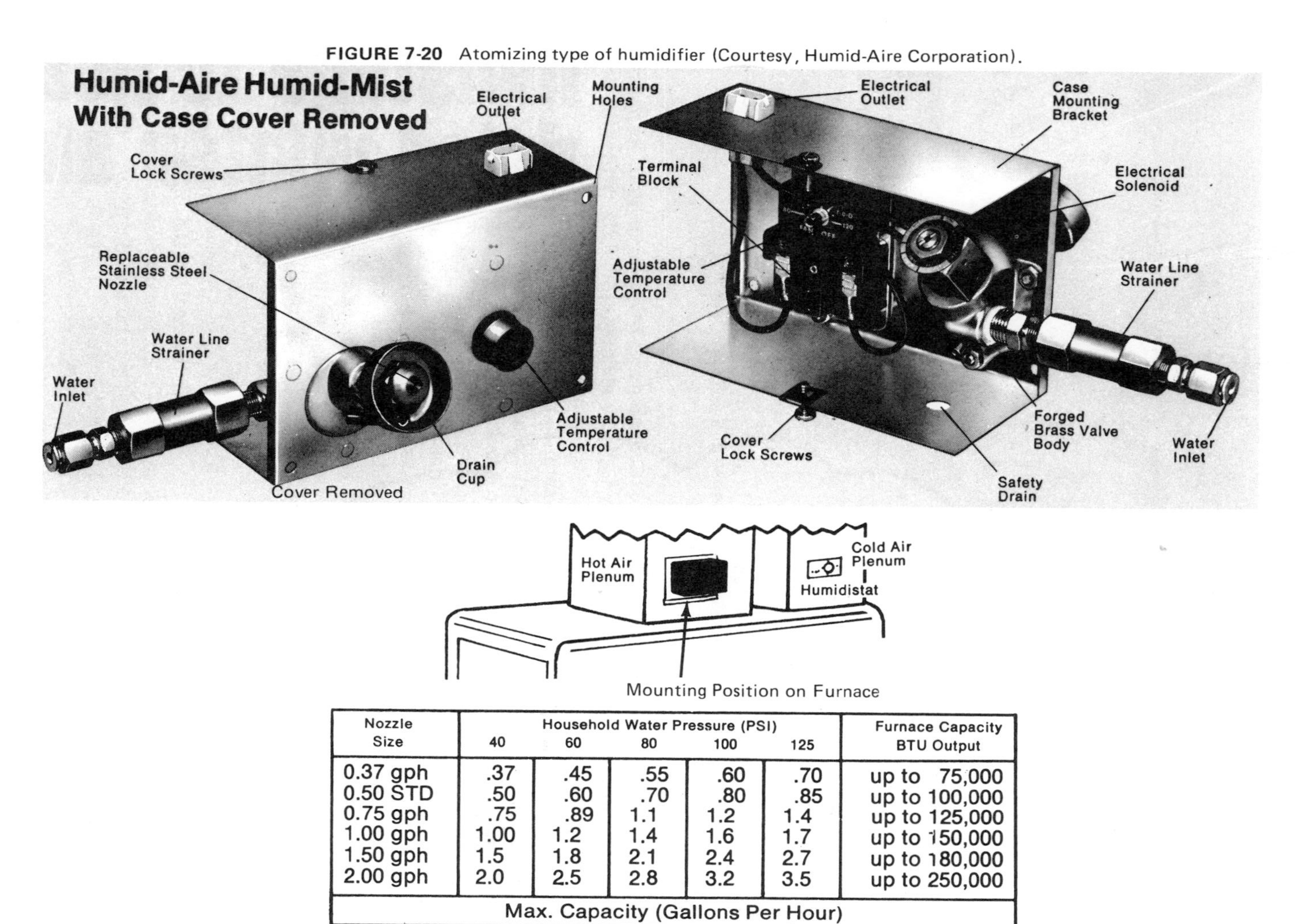

Nozzle Size	Household Water Pressure (PSI) 40	60	80	100	125	Furnace Capacity BTU Output
0.37 gph	.37	.45	.55	.60	.70	up to 75,000
0.50 STD	.50	.60	.70	.80	.85	up to 100,000
0.75 gph	.75	.89	1.1	1.2	1.4	up to 125,000
1.00 gph	1.00	1.2	1.4	1.6	1.7	up to 150,000
1.50 gph	1.5	1.8	2.1	2.4	2.7	up to 180,000
2.00 gph	2.0	2.5	2.8	3.2	3.5	up to 250,000
	Max. Capacity (Gallons Per Hour)					

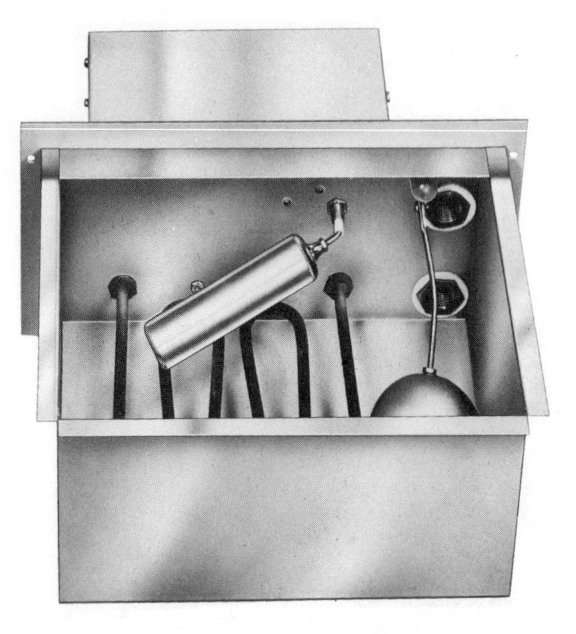

FIGURE 7-21 Vaporizing types of humidifier (Courtesy, Autoflo, a Division of Masco Corporation).

The *atomizing* type of humidifier (Figure 7-20) consists of a metal enclosure containing a stainless steel water atomizing nozzle. With the use of normal household water pressure this precision nozzle produces a finely dispersed water mist capable of instant evaporation in the furnace ductwork. A minimum of 40 psi is required for efficient atomization; therefore, it is not recommended for well water systems. The humidifier operates only when the furnace fan is running.

The *vaporizing* type of humidifier (Figure 7-21) uses an electrical heating element immersed in a water reservoir to evaporate moisture into the furnace supply air plenum. A constant level of water is maintained in the reservoir. A timed flush cycle is included so that the accumulated solids from the water will not remain in the reservoir. These humidifiers can operate even though the furnace is not supplying heat. The humidistat not only starts the water heater but also turns on the furnace fan if it is not running.

NOTE: Problems can arise in the use of humidifiers when the water hardness (mineral content) is too great. Typical water hardness from major U.S. cities is shown in Figure 7-22. An accessory-type water treatment unit can be added to both spray-type and drum-type humidifiers when the water hardness is more than 10 grains of dissolved mineral particles per gallon. A kit is available for testing the hardness of water.

City Water Data

STATE AND CITY	SOURCE OF SUPPLY	MAXIMUM WATER TEMP. F	HARDNESS PPM
ALABAMA			
ANNISTON	W	70	104
BIRMINGHAM	S	85	43
ALASKA			
FAIRBANKS	W	46	120
KETCHIKAN	S	44	4
ARIZONA			
PHOENIX	W	81	210
TUCSON	W	80	222
ARKANSAS			
LITTLE ROCK	WS	89	26
CALIFORNIA			
FRESNO	W	72	87
LOS ANGELES	WS	79	195
SACRAMENTO	S	83	76
SAN FRANCISCO	S	66	181
COLORADO			
DENVER	S	74	123
PUEBLO	S	77	279
CONNECTICUT			
HARTFORD	S	73	12
NEW HAVEN	S	76	46
DELAWARE			
WILMINGTON	S	83	48
DISTRICT OF COLUMBIA			
WASHINGTON	S	84	162
FLORIDA			
JACKSONVILLE	WS	90	305
MIAMI	W	82	78
GEORGIA			
ATLANTA	S	87	14
SAVANNAH	W	85	120
HAWAII			
HONOLULU	S	70	57
IDAHO			
BOISE	WS	65	71
ILLINOIS			
CHICAGO	S	73	125
PEORIA	W	67	386
SPRINGFIELD	S	84	164

City Water Data

STATE AND CITY	SOURCE OF SUPPLY	MAXIMUM WATER TEMP. F	HARDNESS PPM
INDIANA			
EVANSVILLE	S	87	140
FORT WAYNE	S	84	95
INDIANAPOLIS	WS	85	279
IOWA			
DES MOINES	S	77	340
DUBUQUE	W	60	324
SIOUX CITY	W	62	548
KANSAS			
KANSAS CITY	S	92	230
KENTUCKY			
ASHLAND	S	85	93
LOUISVILLE	S	85	104
LOUISIANA			
NEW ORLEANS	S	93	150
SHREVEPORT	S	90	36
MAINE			
PORTLAND	S	70	12
MARYLAND			
BALTIMORE	S	75	50
MASSACHUSETTS			
CAMBRIDGE	S	74	46
HOLYOKE	S	77	23
MICHIGAN			
DETROIT	S	78	100
MUSKEGON	S	71	153
MINNESOTA			
DULUTH	S	58	54
MINNEAPOLIS	S	83	172
MISSISSIPPI			
JACKSON	S	85	38
MERIDIAN	WS	89	7
MISSOURI			
SPRINGFIELD	WS	80	187
ST. LOUIS	S	88	83
MONTANA			
BUTTE	WS	54	63
HELENA	WS	57	96

FIGURE 7-22 Typical water hardness for major U.S. cities (Courtesy, The Trane Company).

MAINTENANCE AND SERVICE

Air Flow Adjustment

For heating, the volume of air handled by the fan should be adjusted to maintain an air temperature rise through the furnace of 85°F (or the air temperature rise indicated on the nameplate).

Adjustments vary according to the type of fan drive as follows:

BELT-DRIVE:

Changes in fan speed are made by altering the position of the adjustable motor pulley flanges. Opening up the flanges decreases the speed of the fan. Closing the flange opening increases the speed of the fan.

City Water Data

STATE AND CITY	SOURCE OF SUPPLY	MAXIMUM WATER TEMP. F	HARDNESS PPM
NEBRASKA			
LINCOLN	W	63	188
OMAHA	S	85	135
NEW HAMPSHIRE			
BERLIN	S	69	10
NASHUA	W	70	25
NEVADA			
RENO	S	63	114
NEW JERSEY			
ATLANTIC CITY	WS	73	12
NEWARK	S	75	29
NEW MEXICO			
ALBUQUERQUE	W	72	155
NEW YORK			
ALBANY	S	70	42
BUFFALO	S	76	118
NEW YORK	WS	73	30
NORTH CAROLINA			
ASHEVILLE	S	79	4
WILMINGTON	S	89	34
NORTH DAKOTA			
BISMARCK	S	80	172
OHIO			
CINCINNATI	S	85	120
CLEVELAND	S	77	121
OKLAHOMA			
OKLAHOMA CITY	S	83	100
TULSA	S	85	80
OREGON			
PORTLAND	S	65	10
PENNSYLVANIA			
PHILADELPHIA	S	83	98
PITTSBURGH	S	84	95
RHODE ISLAND			
PROVIDENCE	S	71	26

City Water Data

STATE AND CITY	SOURCE OF SUPPLY	MAXIMUM WATER TEMP. F	HARDNESS PPM
SOUTH CAROLINA			
CHARLESTON	S	85	18
GREENVILLE	S	79	4
SOUTH DAKOTA			
SIOUX FALLS	W	60	489
TENNESSEE			
CHATTANOOGA	S	84	87
JACKSON	W	73	49
TEXAS			
DALLAS	WS	87	75
EL PASO	W	88	160
HOUSTON	W	89	120
SAN ANTONIO	W	78	221
UTAH			
SALT LAKE CITY	WS	62	140
VERMONT			
BURLINGTON	S	71	60
VIRGINIA			
LYNCHBURG	S	76	12
NORFOLK	S	83	14
RICHMOND	S	84	57
WASHINGTON			
SPOKANE	W	55	147
WEST VIRGINIA			
CHARLESTON	S	85	34
WHEELING	S	81	125
WISCONSIN			
EAU CLAIRE	W	65	54
LA CROSSE	W	55	278
MILWAUKEE	S	69	126
WYOMING			
CHEYENNE	S	60	91

S = Surface: river, reservoir, lake
W = Well

FIGURE 7-22 Continued

DIRECT-DRIVE (TAP-WOUND MOTOR):

The motor speed is adjusted by changing the motor leads. Different speeds use different motor leads. Consult the wiring diagram for the connections necessary to increase or decrease fan speed.

> **CAUTION:** Increasing the fan speed will increase the power required to drive the fan. Check the amperage draw of the motor to be certain that it is within nameplate amperage. Too high an amperage can cause overheating, overload dropout, or motor damage.

Lubrication

For proper lubrication refer to the manufacturer's instructions, if available. Check first to see if the fan or motor has sealed bearings, in which case no lubrication is required. If a unit does not have sealed bearings, the following information is useful:

FAN:

Some fan bearings are grease-type with screw-on caps. These require No. 2 consistency neutral mineral grease. A full cap should be screwed down one revolution for every 6 months of operation.

Other fan bearings have cups with a spring cap. Lubricate at startup, and every 6 months, with SAE No. 10 nondetergent oil. Add oil slowly to saturate the wick and packing.

MOTORS:

Belt-driven motors generally use SAE No. 10 oil at the beginning of operation and every 6 months thereafter.

Direct-drive motors are usually prelubricated with the lubrication lasting about 3 years. After this period of time, add oil slowly to each bearing using about 3 drops of SAE No. 10 oil.

FAN SHAFT END PLAY

Inspect fan for excessive shaft end play. Normal end play on units with sleeve bearings is $^1/_{32}$ to $^1/_{16}$ in. If adjustment is necessary, reposition thrust collar, which is located on the shaft next to the bearing.

Pulley Alignment (Belt-Drive)

Use a straightedge along both motor and fan pulleys. Move pulleys on shafts or adjust motor mount to attain proper alignment (Figure 7-23).

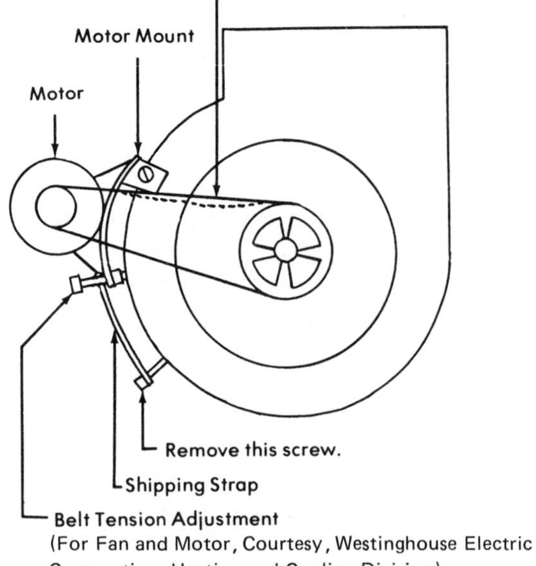

(For Fan and Motor, Courtesy, Westinghouse Electric Corporation, Heating and Cooling Division).

Always re-adjust belt tension after adjusting motor pulley so that it is not loose enough to cause slippage nor tight enough to cause bearing wear. To adjust tension, remove the screw which fastens the shipping strap to the blower housing (the entire strap may be removed if desired.) Adjust tension by means of the bolt in the motor tail piece. A properly adjusted belt can be deflected approximately one inch with moderate pressure of the hand.

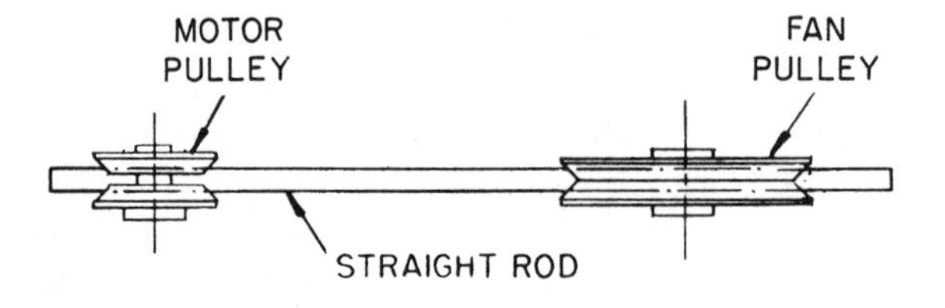

CHECKING PULLEY ALIGNMENT

FIGURE 7-23 Checking pulley alignment and belt tension (Reprinted with permission of Carrier Corporation, © 1979 Copyright, Carrier Corporation [Pulley alignment]).

Adjusting Belt Tension

Adjust belt tension just tight enough to prevent slippage (Figure 7-23). Allow belt to be depressed 1 in. minimum.

Heat Exchanger

Inspect flue passages for soot collection on the heat exchangers. The accumulation of soot can be caused by defective equipment or improper

FIGURE 7-24 Cleaning flue passage in furnace heat exchanger.

burner adjustment. After the cause is determined and corrected, clean the passages with a brush, (Figure 7-24). Remove draft hood and burner assembly for access to flue passages, heat exchangers, and flue baffles.

Humidifier

Refer to manufacturers' instructions for maintaining humidifiers. Each type has specific service requirements. While in use, humidifiers require inspection once a month. Periodic cleaning to remove minerals is important. The frequency of cleaning depends upon the mineral content of the water being supplied to the unit. When a humidifier is not is use, it should be thoroughly cleaned, and the evaporative media pad replaced, the water shut off, and the float assembly cleaned and adjusted. Water is turned on again when the humidifier is put back into service.

REVIEW QUESTIONS

Select the letter representing your choice of the correct answer.

7-1. Which of the following items is not included in a gravity-type furnace?

(a) Heat exchanger

(b) Cabinet

(c) Air filter

(d) Humidifier

7-2. The location of the fan on a counterflow furnace is:

(a) Above the heat exchanger

(b) Below the heat exchanger

(c) At the side of the heat exchanger

(d) In a separate unit

7-3. The material used to construct the heat exchanger is:

(a) Galvanized iron

(b) Copper

(c) Aluminum

(d) Cold-rolled, low-carbon steel

7-4. Individual-section-type heat exchangers are used on what type of furnaces?

(a) Gas

(b) Oil

(c) Coal

(d) Electric

7-5. The type of fuel that requires the greatest chimney draft is:

(a) Gas

(b) Oil

(c) Coal

(d) Electric

7-6. The blades of fans on forced air furnaces are:

(a) Backward-curved

(b) Forward-curved

(c) Flat

(d) Propeller-shaped

7-7. If a motor has a 3-in. pulley, the fan a 6-in. pulley, and the motor runs at 1200 rpm, the speed of the fan is:

(a) 600 rpm

(b) 1200 rpm

(c) 1800 rpm

(d) 2400 rpm

7-8. If the fan is mounted on the shaft of the motor, the type of drive is called:

(a) Unified

(b) V-belt

(c) Indirect

(d) Direct

7-9. A downflow furnace is also called a:

(a) High-boy

(b) Low-boy

(c) Horizontal

(d) Counterflow

7-10. The colder the outside temperature, the greater or lesser the need for humidification?

(a) Greater

(b) Lesser

8

COMPONENTS OF GAS-BURNING FURNACES

OBJECTIVES

After studying this chapter, the student will be able to:

- Identify the components of a gas-burning assembly
- Adjust a gas burner for best performance
- Provide service and maintenance for a gas burner assembly

GAS-BURNING ASSEMBLY

The function of a gas-burning assembly is to produce a proper fire at the base of the heat exchanger. To do this, the assembly must

- Control and regulate the flow of gas
- Assure the proper mixture of gas with air
- Ignite the gas under safe conditions

To accomplish these actions, a gas burner assembly must consist of four major parts or sections. These parts are the

1. Gas valve
2. Safety pilot
3. Manifold and orifice
4. Gas burners and adjustment

An illustration of a complete gas burner manifold assembly is shown in Figure 8-1.

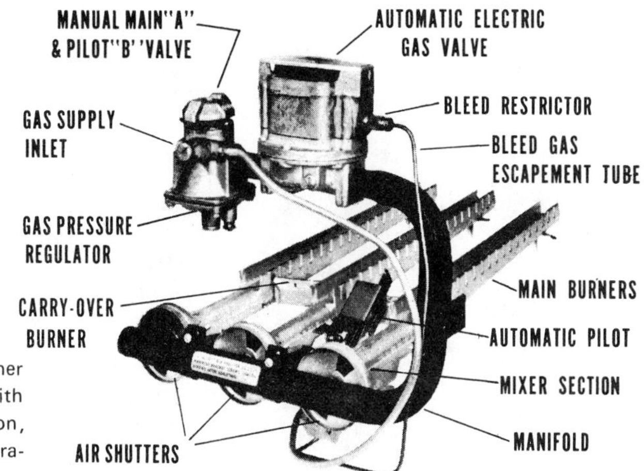

FIGURE 8-1 Complete gas burner manifold assembly (Reprinted with permission of Carrier Corporation, © Copyright 1979 Carrier Corporation).

Gas Valve

The gas valve section consists of a number of parts, with each performing a different function. On older units, these operations were entirely separate. On more modern units, these parts are all contained in a combination gas valve (CGV), and include:

- Hand shutoff valve
- Pressure reducing valve
- Safety shutoff equipment
- Operator or automatic gas valve

Figure 8-2 shows the separate parts of the gas valve section installed along the manifold, sometimes called a *gas regulation train.* Figure 8-3 shows the combination valve (CGV) which includes these separate parts, each of which is replaceable if trouble should develop.

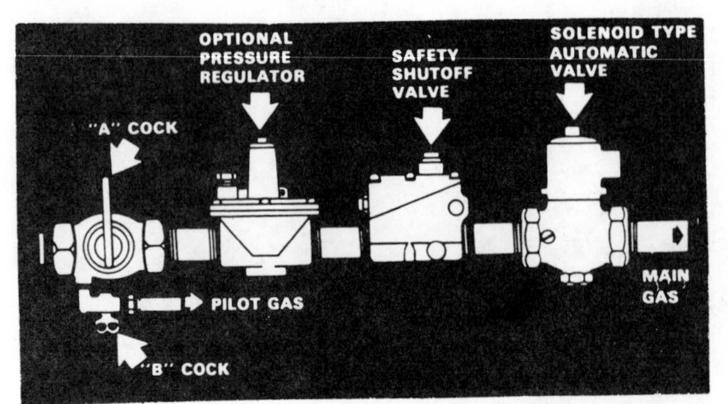

FIGURE 8-2 Older style gas manifold showing separate components (Courtesy, Honeywell, Inc.).

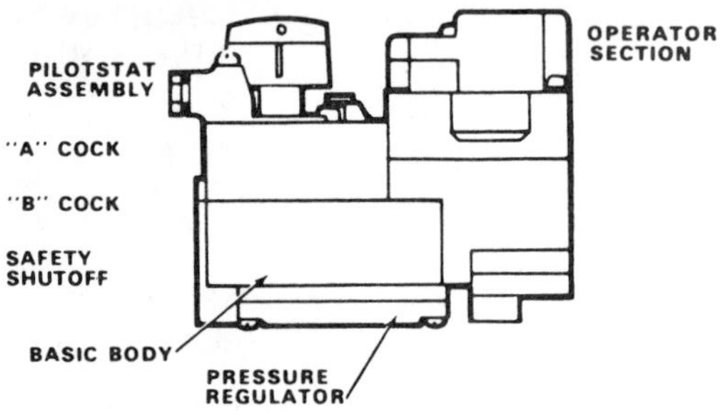

FIGURE 8-3 Combination gas valve (CGV) (Courtesy, Honeywell, Inc.).

HAND SHUTOFF VALVES:

Referring to Figure 8-2, note that the main shutoff valve or gas cock is composed of two parts; an A cock and a B cock. The A cock is used to manually turn the main gas supply on and off. The B cock is used to manually turn the gas supply to the pilot on and off.

These gate or ball-type valves are open when the handle is parallel to the length of the pipe, and closed when the handle is perpendicular to the length of the pipe. They should be either "on" or "off" and not placed in an intermediate position. If it is desirable to regulate the supply of gas, it should be done by adjusting the pressure regulator.

PRESSURE-REDUCING VALVE:

The pressure-reducing valve, or pressure regulator, decreases the gas pressure supplied from the utility meter at approximately 7 in. of water column to 3½ in. of water column. The regulator maintains a constant gas pressure at the furnace to provide a constant input of gas to the furnace. On an LP system the regulator is supplied at the tank and, therefore, an additional regulator is not necessary at the furnace.

As shown in Figure 8-4, the spring exerts a downward pressure on the diaphragm, which is connected to the gas valve. The spring pressure tends to open the valve. When the gas starts to flow through the valve, the down-

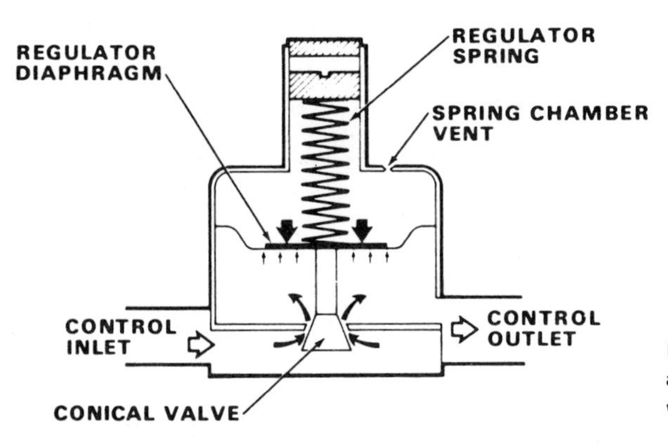

FIGURE 8-4 Pressure reducing valve and its operation (Courtesy, Honeywell, Inc.).

stream pressure of the gas creates an upward pressure on the diaphragm tending to close it. Thus, the spring pressure and the downstream gas pressure oppose each other and must equal each other for the pressure to remain in equilibrium. To adjust the gas pressure, the cap is removed and the spring tension is adjusted. Also note that the vent maintains atmospheric pressure on the top of the diaphragm.

SAFETY SHUTOFF EQUIPMENT:

The safety shutoff equipment consists of a

- Pilot burner
- Thermocouple or thermopile
- Pilotstat power unit

The pilot burner and the thermocouple or thermopile are assembled into one unit as shown in Figures 8-12, 8-13, and 8-14. This unit is placed near the main gas burner. The thermocouple is connected to the pilotstat power unit. The pilotstat power unit may be located in the CGV or installed as a separate unit in the gas regulation train as shown in Figure 8-2.

The pilot burner has two functions. It

1. Directs the pilot flame for proper ignition of the main burner flame
2. Holds the thermocouple or thermopile in correct position with relation to the pilot flame

The thermocouple or thermopile extends into the pilot flame and generates sufficient voltage to hold in the pilotstat. The thermopile generates additional voltage, which also operates the main gas valve.

The pilotstat power unit is held in by the voltage generated by the thermocouple. If the voltage is insufficient, indicating a poor or nonexistent pilot light, the power unit drops out, shutting off the main gas supply. These units are of two types: one that shuts off the gas supply to the main burner only, and the other that shuts off the supply of gas to both the pilot and the main burner. Units burning LP gas always require 100% shutoff.

The Principle of the Thermocouple: Two dissimilar metal wires are welded together at the ends to form a thermocouple (Figure 8-5). When one junction is heated and the other remains relatively cool, an electrical current is generated which flows through the wires. The voltage generated is used to operate the pilotstat power unit.

The Thermopile: The thermopile is a number of thermocouples connected in series (Figure 8-6). One thermocouple may generate 25 to 30 millivolts (mV). A thermopile may generate as high as 750 or 800 mV (1000 mV equals 1 V).

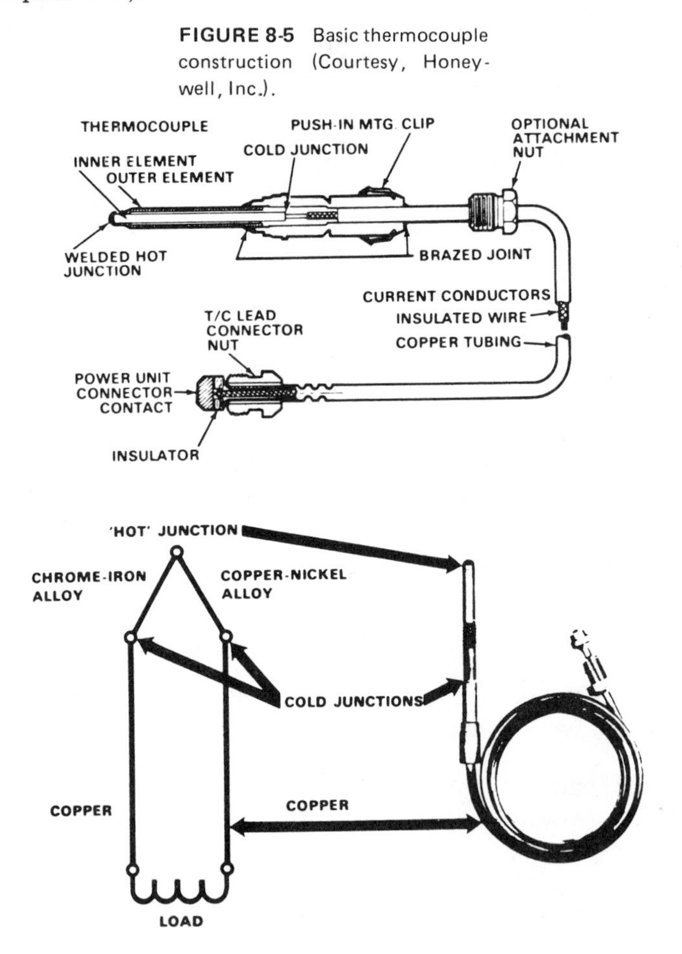

FIGURE 8-5 Basic thermocouple construction (Courtesy, Honeywell, Inc.).

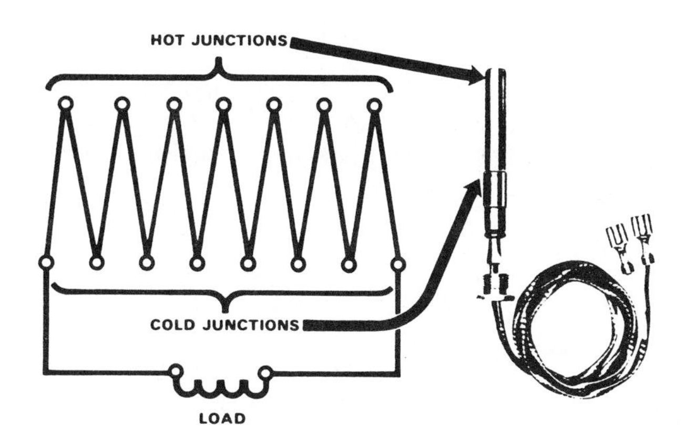

FIGURE 8-6 Basic thermopile construction (Courtesy, Honeywell, Inc.).

The Power Unit: The power unit is energized by the voltage generated by the thermocouple or thermopile. This voltage operates through an electromagnet. The voltage generated is sufficient to hold in the plunger against the pressure of the spring. The position of the plunger against the electromagnet must be manually set. If the thermocouple voltage drops due to a poor or nonexistent pilot, the plunger is released by the spring. This action either causes the plunger itself to block the flow of gas or operates an electrical switch that shuts off the supply of gas. The plunger must be manually reset when the proper pilot is restored. Figure 8-7 shows the pilotstat mechanism installed in a CGV.

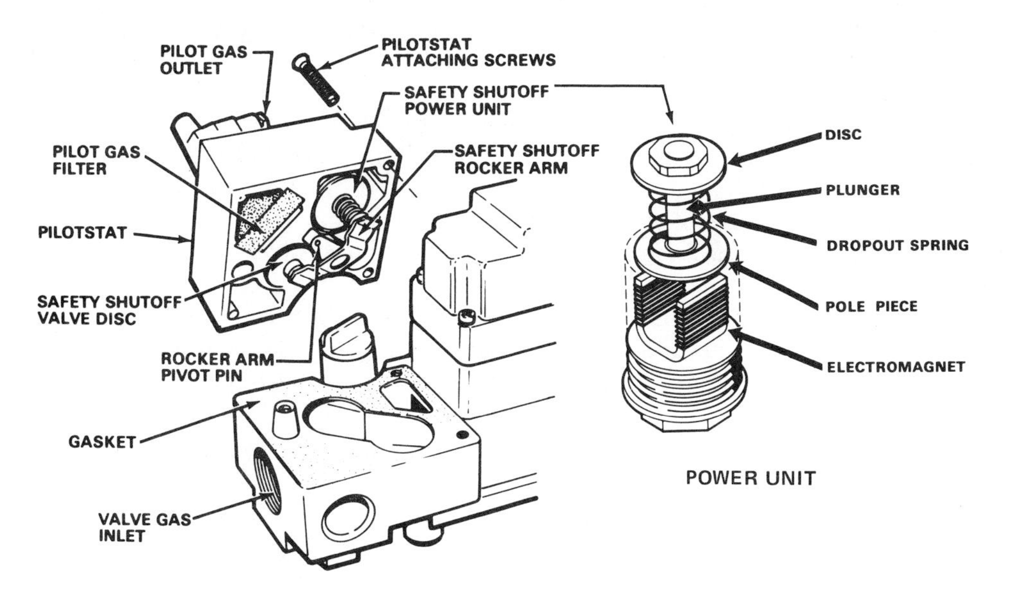

FIGURE 8-7 Pilotstat mechanism in a Combination gas valve (CGV) [Courtesy, Honeywell, Inc.].

OPERATOR OR AUTOMATIC GAS VALVE:

The main function of the operator is to control the gas flow. Some operators control the gas flow directly; some regulate the pressure on a diaphragm, which, in turn, regulates the gas flow. The various types of operators are

- Solenoid
- Diaphragm
- Bimetal
- Bulb

Solenoid Valve Operators: These units (Figure 8-8) employ electromagnetic force to operate the valve plunger. When the thermostat calls for heat, the plunger is raised, opening the gas valve. These valves are often filled with oil to eliminate noise and to serve as lubrication.

Diaphragm Valve Operators: On these operators (Figure 8-9) gas is used to control the pressure above and below the diaphragm. The diaphragm is attached to the valve. To close the valve, gas pressure is released above the diaphragm. With equal pressure above and below the diaphragm, the weight of the valve closes it. To open the valve, the supply of gas is cut off above the diaphragm and gas pressure is vented to atmosphere. The pressure of the gas below the diaphragm opens the valve. These valves also can be filled with oil.

Bimetal Operators: Bimetal operators (Figure 8-10) have a high-resistance wire wrapped around the blade. When the thermostat calls for heat, current is supplied to the wire, causing it to heat and warp the blade. This warping action opens the valve. The valve action is slow and therefore this valve is sometimes referred to as a *delayed-action valve.*

Bulb-Type Operators: Bulb-type operators (Figure 8-11) use the expansion of a liquid-filled bulb to provide the operating force. A snap-action disk is also incorporated to speed up the opening and closing action.

Safety Pilot

The function of the safety pilot is to provide an ignition flame for the main burner and to heat the thermocouple to provide safe operation.

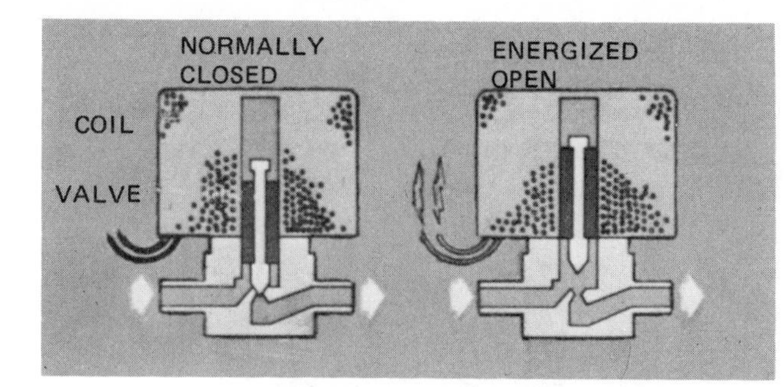

FIGURE 8-8 Solenoid valve operator (Courtesy, Honeywell, Inc.).

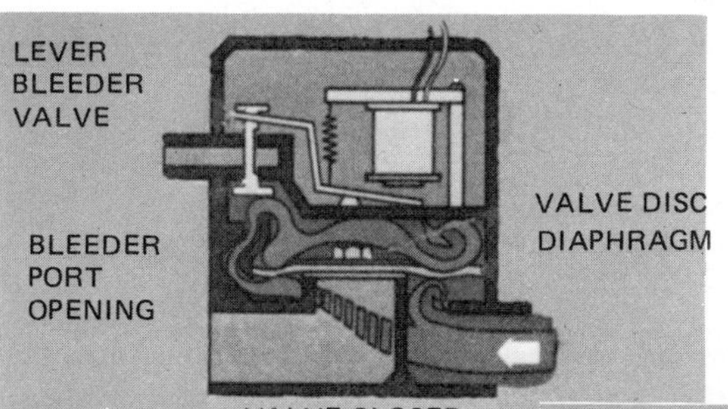

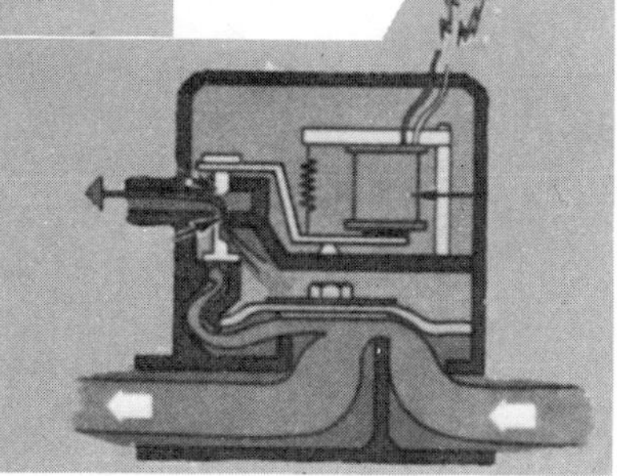

FIGURE 8-9 Magnetic diaphragm valve operator (Reprinted with permission of Carrier Corporation, © Copyright 1979, Carrier Corporation).

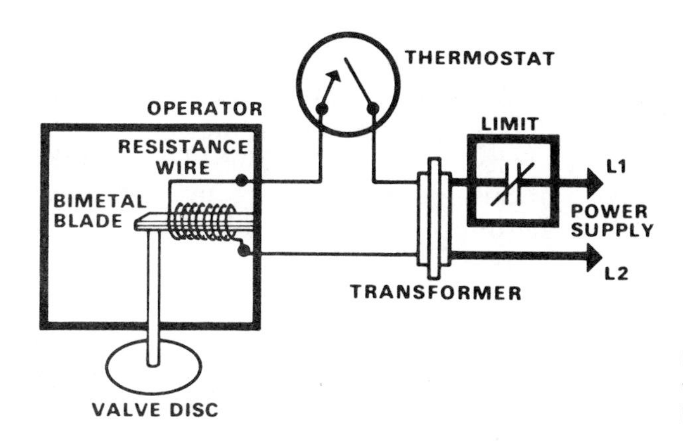

FIGURE 8-10 Bimetal type valve operator (Courtesy, Honeywell, Inc.).

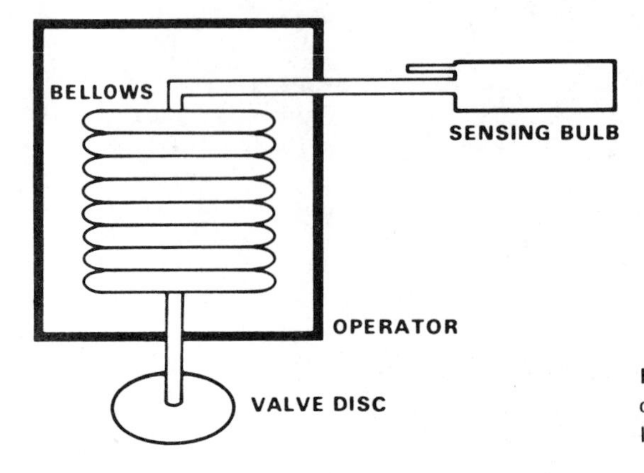

FIGURE 8-11 Bulb type valve operator (Courtesy, Honeywell, Inc.).

PILOT BURNERS:

There are two types of pilot burners. They are

1. Primary-aerated
2. Non-primary-aerated

On the *primary-aerated pilot* (Figure 8-12), the air is mixed with the gas before it enters the pilot burner. The disadvantage of this is that dirt and lint tend to clog the screened air opening. The air opening must be periodically cleaned.

On the *non-primary-aerated pilot* (Figure 8-13), the gas is supplied directly to the pilot, without the addition of primary air. All the necessary air is supplied as secondary air. This eliminates the need for cleaning the air passages.

PILOT BURNER ORIFICE:

The orifice is the part that controls the supply of gas to the burner. The drilled opening permits a small stream of gas to enter the burner. The amount that enters is dependent upon the size of the drilled hole and the manifold pressure. Some cleaning of this opening may be required. This can be done by blowing air through the orifice or by using a suitable nonoily solvent.

PILOT FLAME:

The pilot flame should envelop the thermocouple $^3/_8$ to $^1/_2$ in. at its top, as shown in Figure 8-14. The gas pressure at the pilot is the same as in the burner manifold, $3^1/_2$ to 4 in. of water column for natural gas; 11 in. for LP gas. Too high a pressure decreases the life of the thermocouple. Too low a pressure provides unsatisfactory heating of the thermocouple tip. Poor flame conditions are shown in Figure 8-15.

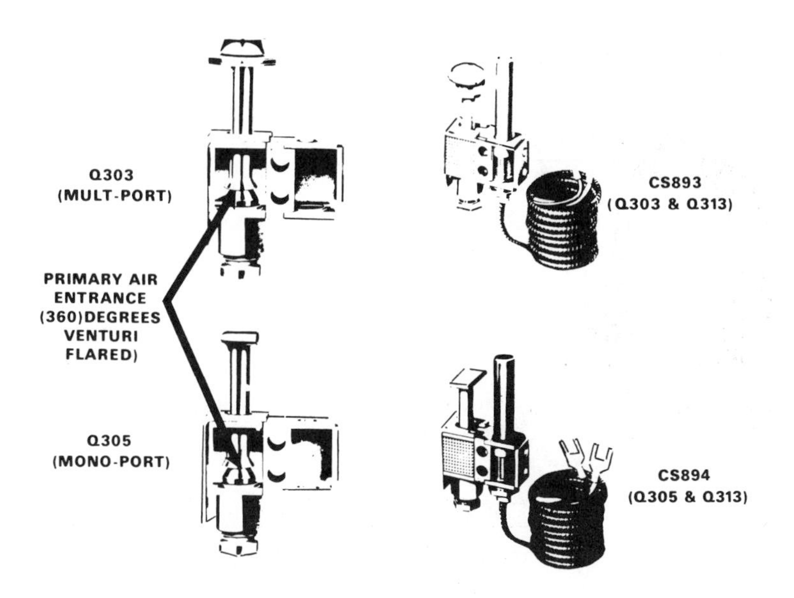

FIGURE 8-12 Primary-aerated pilot designs (Courtesy, Honeywell, Inc.).

FIGURE 8-13 Non-primary-aerated pilot designs (Courtesy, Honeywell, Inc.).

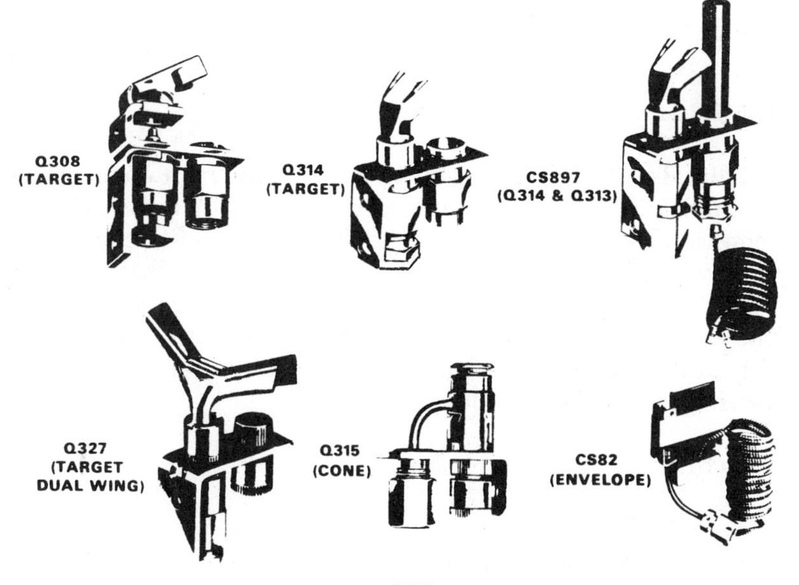

ADJUST PILOT FLOW ADJUSTMENT SCREW TO GIVE A SOFT, STEADY FLAME ENVELOPING 3/8 TO 1/2 INCH OF THE TIP OF THE THERMOCOUPLE OR GENERATOR.

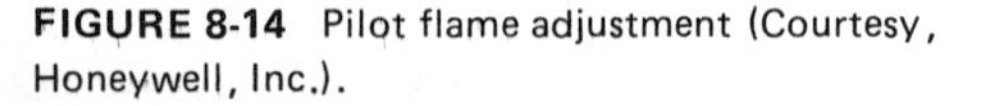
FIGURE 8-14 Pilot flame adjustment (Courtesy, Honeywell, Inc.).

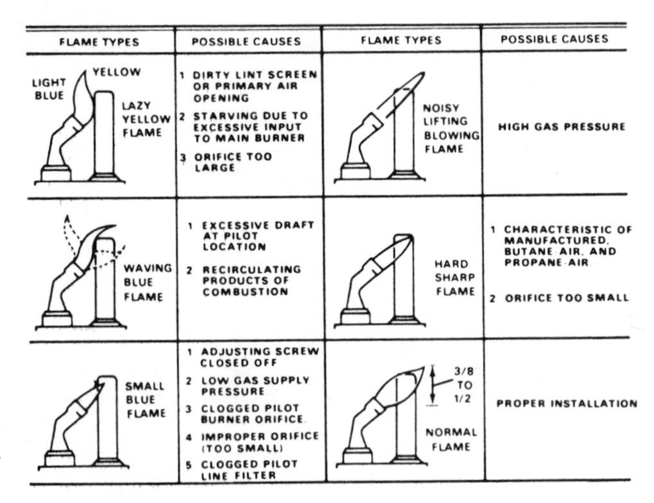

FLAME TYPES	POSSIBLE CAUSES	FLAME TYPES	POSSIBLE CAUSES
LIGHT BLUE, YELLOW, LAZY YELLOW FLAME	1 DIRTY LINT SCREEN OR PRIMARY AIR OPENING 2 STARVING DUE TO EXCESSIVE INPUT TO MAIN BURNER 3 ORIFICE TOO LARGE	NOISY LIFTING BLOWING FLAME	HIGH GAS PRESSURE
WAVING BLUE FLAME	1 EXCESSIVE DRAFT AT PILOT LOCATION 2 RECIRCULATING PRODUCTS OF COMBUSTION	HARD SHARP FLAME	1 CHARACTERISTIC OF MANUFACTURED, BUTANE AIR, AND PROPANE AIR 2 ORIFICE TOO SMALL
SMALL BLUE FLAME	1 ADJUSTING SCREW CLOSED OFF 2 LOW GAS SUPPLY PRESSURE 3 CLOGGED PILOT BURNER ORIFICE 4 IMPROPER ORIFICE (TOO SMALL) 5 CLOGGED PILOT LINE FILTER	3/8 TO 1/2 NORMAL FLAME	PROPER INSTALLATION

FIGURE 8-15 Poor pilot flame conditions (Courtesy, Honeywell, Inc.).

Manifold and Orifice

The manifold delivers gas equally to all the burners. It connects the supply of gas from the gas valve to the burners. It is usually made of $\frac{1}{2}$- to 1-in. pipe. Connections can be made either to the right or left, depending on the design, as shown in Figure 8-16.

The orifice used for the main burners is similar to the pilot orifice only larger. The drilled opening permits a fast stream of raw gas to enter the burner. The orifice is sized to permit the proper flow of gas. Items to be considered in sizing the orifice are:

- Type of gas
- Pressure in manifold
- Input of gas required for each burner

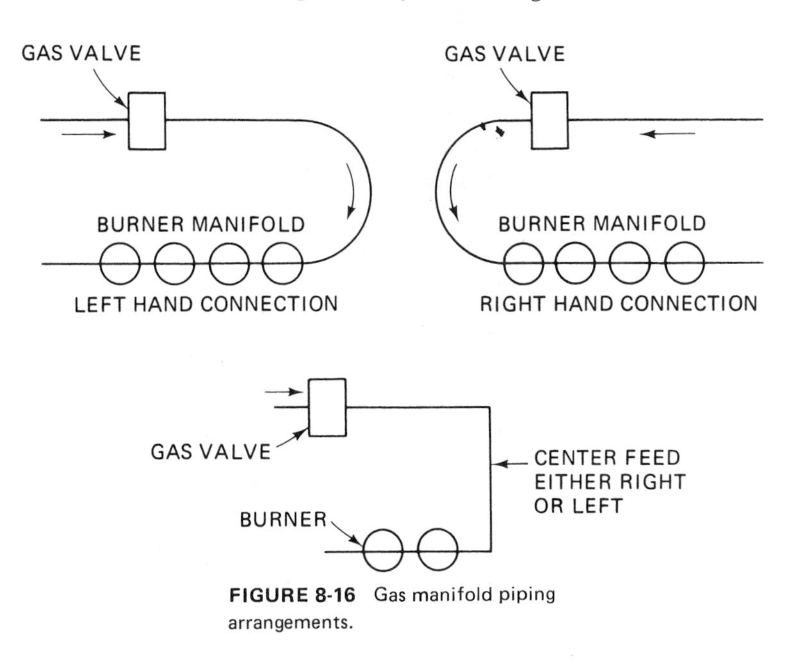

FIGURE 8-16 Gas manifold piping arrangements.

Gas Burners and Adjustment

PRIMARY AND SECONDARY AIR:

Most gas burners require that some air be mixed with the gas before combustion. This air is called primary air (Figure 8-17). Primary air constitutes approximately one-half of the total air required for combustion. Too much primary air causes the flame to lift off the burner surface. Too little primary air causes a yellow flame.

Air that is supplied to the burner at the time of combustion is called secondary air (Figure 8-18). Too little secondary air causes the formation of carbon monoxide. To be certain that enough secondary air is provided, most units operate on an excess of secondary air. An excess of about 50% is con-

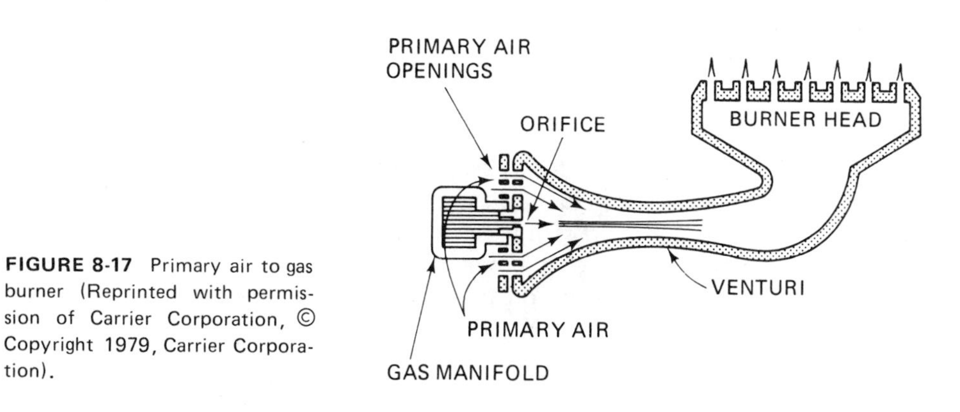

FIGURE 8-17 Primary air to gas burner (Reprinted with permission of Carrier Corporation, © Copyright 1979, Carrier Corporation).

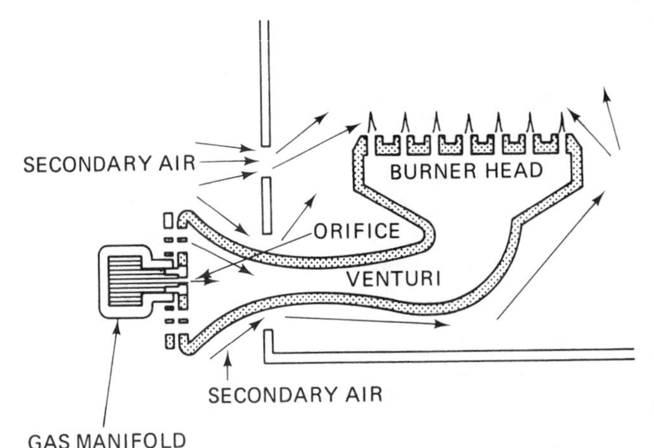

FIGURE 8-18 Secondary air to gas burner (Reprinted with permission of Carrier Corporation, © Copyright 1979, Carrier Corporation).

sidered good practice. To produce a good flame it is essential to maintain the proper ratio between primary and secondary air.

THE VENTURI:

In a multiport burner, gas is delivered into a venturi (mixing tube) creating a high gas velocity. The increase in gas velocity causes a sucking action which causes the primary air to enter the tube and mix with the raw gas.

CROSSOVER IGNITORS:

The crossover or carryover ignitor is a projection on each burner near the first few ports that permits the burner which lights first to pass flame to the other burners. There is a baffle, as shown in Figure 8-21, which directs gas to the first few ports of the burner to assure the proper supply of fuel to the area of the ignitor.

NOTE: Burners that fail to light properly can cause rollout, a dangerous condition where flame emerges into the vestibule.

DRAFT DIVERTER:

The draft diverter (Fig. 8-19) is designed to provide a balanced draft (slightly negative) over the flame in a gas fired furnace. The bottom of the diverter is open to allow air from the furnace room to blend with the products of combustion. In case of blockage or downdraft in the chimney, the flue gases vent to the area around the furnace.

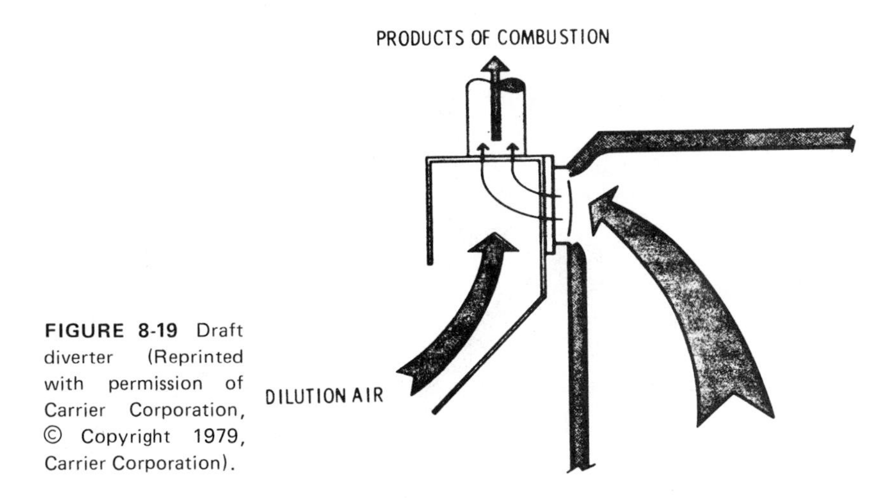

FIGURE 8-19 Draft diverter (Reprinted with permission of Carrier Corporation, © Copyright 1979, Carrier Corporation).

NOTE: The American Gas Association (AGA) requires a rigid test to be certain that under these conditions complete combustion occurs and there is no danger of suffocation of the flame.

TYPES OF BURNERS:

There are four types of burners, classified according to the type of burner head:

1. Single-port or inshot
2. Drilled port
3. Slotted port
4. Ribbon

Single-Port or Inshot Burner: The single port burner (Figure 8-20) is the simplest type. The outlet is an extension of the burner tube. The flame is directed toward a metal plate that spreads the flame. Although the burner is relatively troublefree, it is noisy and less efficient than the other types.

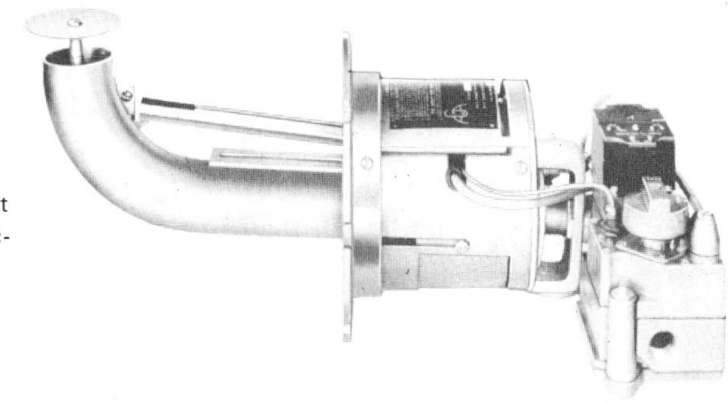

FIGURE 8-20 Single-port or upshot burner (Courtesy, Adams Manufacturing Company).

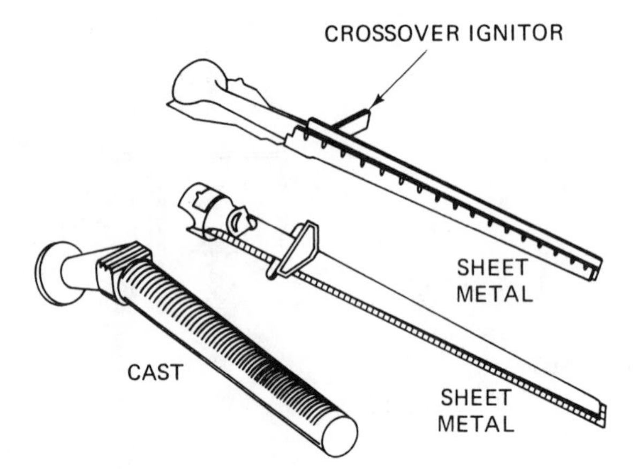

FIGURE 8-21 Slotted-port burners (Reprinted with permission of Carrier Corporation, © Copyright 1979, Carrier Corporation).

Drilled Port Burner: The drilled port burner is usually made of cast iron with a series of small drilled holes. The size of the hole varies for different types of gas.

Slotted Port Burners: The slotted port burner (Figure 8-21) is similar to the drilled port burner, but in place of holes, it uses elongated slots. It has the advantage of being able to burn different types of gas without change of slot size. It is also less susceptible to clogging with lint.

Ribbon Burner: The ribbon burner has a continuous opening down each side and, when it is lit, the flame has the appearance of a "ribbon." The burner itself is made either of cast iron or fabricated metal. The ribbon insert is usually corrugated stainless steel.

MAINTENANCE AND SERVICE*

The main items requiring service on gas-burning equipment are the:

- Gas burner assembly
- Pilot assembly
- Automatic gas valve

After the equipment is in use for a period of time, each of these items may require service, depending on the conditions of the installation.

* For service on electric ignition, see Chapter 21.

Problems that can arise include:

- Flashback
- Carbon on the burners
- Dirty air mixer
- Noise of ignition
- Gas flames lifting from ports
- Appearance of yellow tips in flames

When the velocity of the gas/air mixture is reduced below a certain speed, the flame will flash back through the ports. This is undesirable and must be corrected. The solution is to close the primary air shutter as much as possible and still maintain a clear blue flame, (Figure 8-22).

FIGURE 8-22 Primary air and its adjustment (Courtesy, Westinghouse Electric Corporation, Heating and Cooling Division).

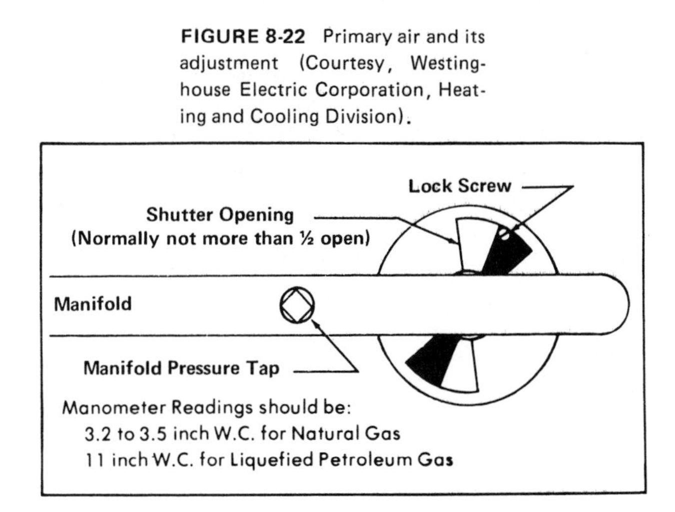

PRIMARY AIR ADJUSTMENT
To adjust primary air supply, turn main gas supply **ON** and operate unit for 15 to 20 minutes. Then loosen set screw and rotate shutter on burner bell. Open shutter until the yellow tips just disappear. Lock shutter in position by tightening set screw. Turn main gas supply **OFF**, let cool completely, and re-check operation and flame characteristics from a cold start.

Burners that become clogged with carbon due to flashback conditions must be cleared so as not to interfere with the proper flow of gas.

If the burner flame becomes soft and yellow-tipped, it is an indication that cleaning is required. Items requiring cleaning are the mixer face, venturi, and the burner head itself. In some cases these can be cleaned with the suction of a vacuum cleaner. In other cases, a thin, long-handled brush may be used to thoroughly reach the inside of the mixing tube.

Noise of ignition is usually caused by delayed or faulty ignition. Conditions that must be checked and corrected, if necessary, are: location of the pilot, poor flame travel, or poor distribution of the flame on the burner itself.

Lifting flames are usually caused by too high a gas pressure or improper primary air adjustment. The gas pressure for natural gas at the manifold should be 3.5 in. of water column. Primary air should be adjusted to provide a blue flame.

The appearance of yellow tips on the flame indicate incomplete combustion, which can cause sooting of the heat exchanger. The following items should be checked and corrected, if necessary:

(a) **Gas Pressure:** Should be 3.5 in. of water column for natural gas

(b) **Clogged Orifices:** Inspect and clean if required

(c) **Air Adjustment:** Use minimum primary air required for blue flame

(d) **Alignment:** Gas stream should move down center of venturi tube

(e) **Flues:** Air passages must not be clogged

(f) **Air Leaks:** Air from fan or strong outside air current can cause yellow flame

Pilot Assembly

Pilot problems can be caused by:

- Clogged air openings
- Dirty pilot filters
- Clogged orifice
- Defective thermocouples

Dirt and lint in the air opening is the most common problem. Cleaning solves this problem.

Pilots for manufactured gases use pilot filters. The cartridge must be replaced periodically.

Occasionally, it is necessary to clean the pilot orifice, which may become clogged from an accumulation of dirt, lint, carbon, or condensation in the lines.

If the pilot flame does not adequately heat the thermocouple, not enough dc voltage will be generated and the electromagnet will not energize sufficiently to permit the opening of the automatic gas valve. The thermocouple can be checked, as shown in Figure 14-10, using a millivoltmeter.

Automatic Gas Valve

Most gas valves are of the diaphragm type with an electromagnetic or heat motor-operated controller. The two principal problems that can occur with gas valves are:

1. The valve will not open
2. Gas leakage through the valve

First, the pilot must be checked to be certain that the proper power is being produced by the thermocouple. If the gas valve will not open after following the lighting procedure shown on the nameplate, proceed as follows:

Check to determine if power is available at the valve. If power is available, connect and disconnect the power and listen for a muffled click which indicates that the lever arm is being actuated. Check to be sure that the vent above the diaphragm is open. If the valve still does not operate, replace the entire operator or valve top assembly.

If gas continues to leak through the valve outlet or the vent opening after the valve is deenergized, the entire valve head assembly should be replaced.

REVIEW QUESTIONS

Select the letter representing your choice of the correct answer.

8-1. The function of the gas burner assembly is to:

(a) Produce proper fire at base of heat exchanger

(b) Heat the air that goes to the rooms

(c) Mix gas and air

(d) Provide safe lighting of the burner

8-2. A hand shutoff valve is used to turn gas on and off to the:

(a) Burner

(b) Pilot

(c) Burner and pilot

8-3. The proper gas manifold pressure for a residential gas unit is:

(a) $2\frac{1}{2}$ in. of water column

(b) $3\frac{1}{2}$ in. of water column

(c) 7 in. of water column

(d) 11 in. of water column

8-4. A thermocouple produces how many millivolts of electric power?

(a) 25 to 30

(b) 35 to 50

(c) 500 to 600

(d) 750 to 800

8-5. CGV stands for:

(a) Continuous gas valve

(b) Correct gas volume

(c) Cast gas vent

(d) Combination gas valve

8-6. The type of gas valve operator with a delayed action feature is a:

(a) Solenoid

(b) Diaphragm

(c) Bimetal

(d) Bulb

8-7. The type of pilot where gas is mixed with air before it enters the burner is:

(a) Primary-aerated

(b) Non-primary-aerated

8-8. What type of fuel requires 100% shutoff at the gas valve?

(a) Natural gas

(b) LP gas

8-9. Air that is mixed with gas before burning is called:

(a) Outside air

(b) Total combustion air

(c) Secondary air

(d) Primary air

8-10. A draft diverter:

(a) Creates a draft of 0.02 in. of water column over fire

(b) Connects directly to the gas burner

(c) Is closed at the bottom

(d) Is really not necessary on a modern furnace

9

COMPONENTS OF OIL-BURNING FURNACES

OBJECTIVES

After studying this chapter, the student will be able to:

- Identify the components of an oil burner
- Adjust an oil burner for best performance
- Provide service and maintenance for an oil burner

OIL-BURNING UNITS

Oil must be vaporized to burn. There are two ways to vaporize oil:

1. **By Heat Alone:** For example, heating an iron pot containing oil and igniting it with a flame or spark
2. **By Atomization and Heating:** Oil is vaporized by forcing it under pressure through an orifice, causing it to break up into small droplets. The atomized oil is then ignited by an electric spark

TYPES OF BURNERS

There are four types of oil burners:

Type	*Method of vaporizing oil*
1. Pot	Heating
2. Rotary	Atomizing
3. Low pressure	Atomizing
4. High pressure	Atomizing

Pot-Type Oil Burners

In a pot-type oil burner (Figure 9-1), oil enters near the bottom of an open pot. The oil is ignited by a pilot flame or by electric ignition. The starting flame heats a plate in the lower portion of the pot which then serves to vaporize oil and mix it with combustion air. The flame increases in size, heating not only the plate but the sides of the chamber. The high fire consumes enough vaporized oil for the full capacity of the burner.

FIGURE 9-1 Pot-type oil burner (Courtesy, Shell Oil Company).

Under low-fire conditions, such as on startup, the pot-type burner forms carbon in the pot, which tends to clog the air passages. The pot burner also is sensitive to changes in draft conditions, affecting combustion. These problems in operation have restricted its use.

Horizontal-Rotary-Type Oil Burner

A horizontal-rotary-type oil burner (Figure 9-2) uses centrifugal force to atomize the oil. Oil is fed into a rapidly rotating cup located in the center of the burner. The oil is thereby forced through an orifice, throwing the atomized oil toward the furnace walls. Forced air for combustion is mixed with the atomized oil.

The mixture is usually ignited at the base of the heat exchanger by an electric spark. The flame rises along the outside surface of the heat exchanger, evenly heating its surface.

This type of burner has excellent combustion efficiency. However, it requires considerable maintenance. The amount of service required by the horizontal type of burner has limited its residential use. A vertical model is in common use for commercial- and industrial-size burners.

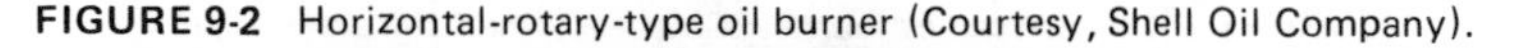

FIGURE 9-2 Horizontal-rotary-type oil burner (Courtesy, Shell Oil Company).

Low-Pressure Gun-Type Oil Burner

In a low-pressure gun-type burner (Figure 9-3), oil and primary air are mixed prior to being forced through an orifice or nozzle. A pressure of 1 to 15 psig on the mixture, plus the action of the orifice, causes the atomization of the oil. Secondary air is drawn into the spray mixture after it is released from the nozzle. Electric spark ignition is used to light the combustible mixture.

Although the low-pressure-type burner has achieved some success, it has been replaced to a great extent by the high-pressure burner. Higher pressure produces better atomization and high efficiencies.

High-Pressure Gun-Type Oil Burner

A high-pressure gun-type burner (Figure 9-4) forces oil at 100 psig pressure through the nozzle, breaking the oil into fine, mistlike droplets.

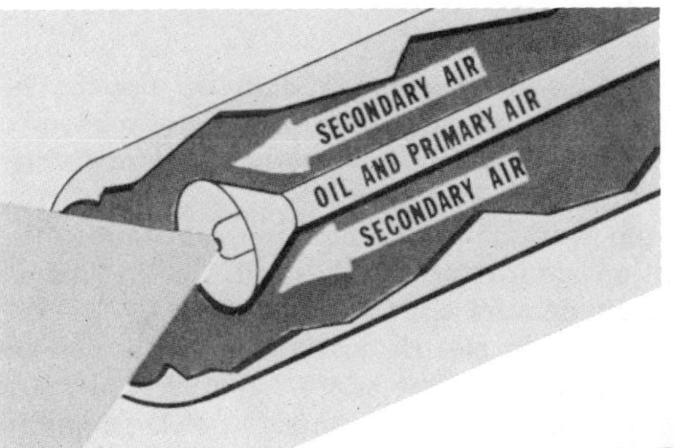

FIGURE 9-3 Low-pressure gun-type oil burner (Reproduced by permission of Carrier Corporation; Copyright 1979, Carrier Corporation).

FIGURE 9-4 High-pressure gun-type oil burner (Courtesy, R. W. Beckett Corporation).

The atomized oil spray creates a low-pressure area into which the combustion air flows. Combustion air is supplied by a fan through vanes, creating turbulence and complete mixing action.

The high-pressure gun-type burner is the most popular domestic burner. It is simple in construction, relatively easy to maintain, and efficient in operation. Therefore, the balance of this unit will concentrate exclusively on providing information relative to the high-pressure burner.

HIGH-PRESSURE BURNER COMPONENTS

The high-pressure oil burner is actually an assembled unit. The component parts are made by a few well-known manufacturers. The parts are mass-produced, low in cost, and readily available for servicing requirements. The component parts of the high-pressure burner are shown in Figure 9-5.

Power Assembly

The power assembly consists of the motor, fan, and fuel pump. The nozzle assembly consists of the nozzle, the electrodes, and parts related to the oil/air mixing action. The ignition system consists of the transformer and electrical parts.

The motor drives the fan and fuel pump. The fan forces air through the blast tube to provide combustion air for the atomized oil. The fuel pump draws oil from the storage tank and delivers it to the nozzle.

The oil/air mixture is ignited by an electric spark which is formed between two properly positioned electrodes in the nozzle assembly. The spark is created by a transformer that increases the 120-V primary power supply to 10,000 V secondary to form a low-current spark arc.

FUEL PUMPS:

All fuel pumps are of the rotary type using cams or gears or a combination of both. The principal parts of the fuel pump, in addition to the gears are: the shaft seal, the pressure-regulating valve, and the automatic cutoff valve.

The shaft seal is necessary since the pump is driven by an external source of power.

The pressure-regulating valve has an adjustable spring that permits regulation of the oil pressure. The pump actually delivers more oil than the burner can use. The excess oil is dumped back into the supply line or returns to the tank. All pumps also have an adjustment screw for regulating the pressure of the oil delivered to the nozzle.

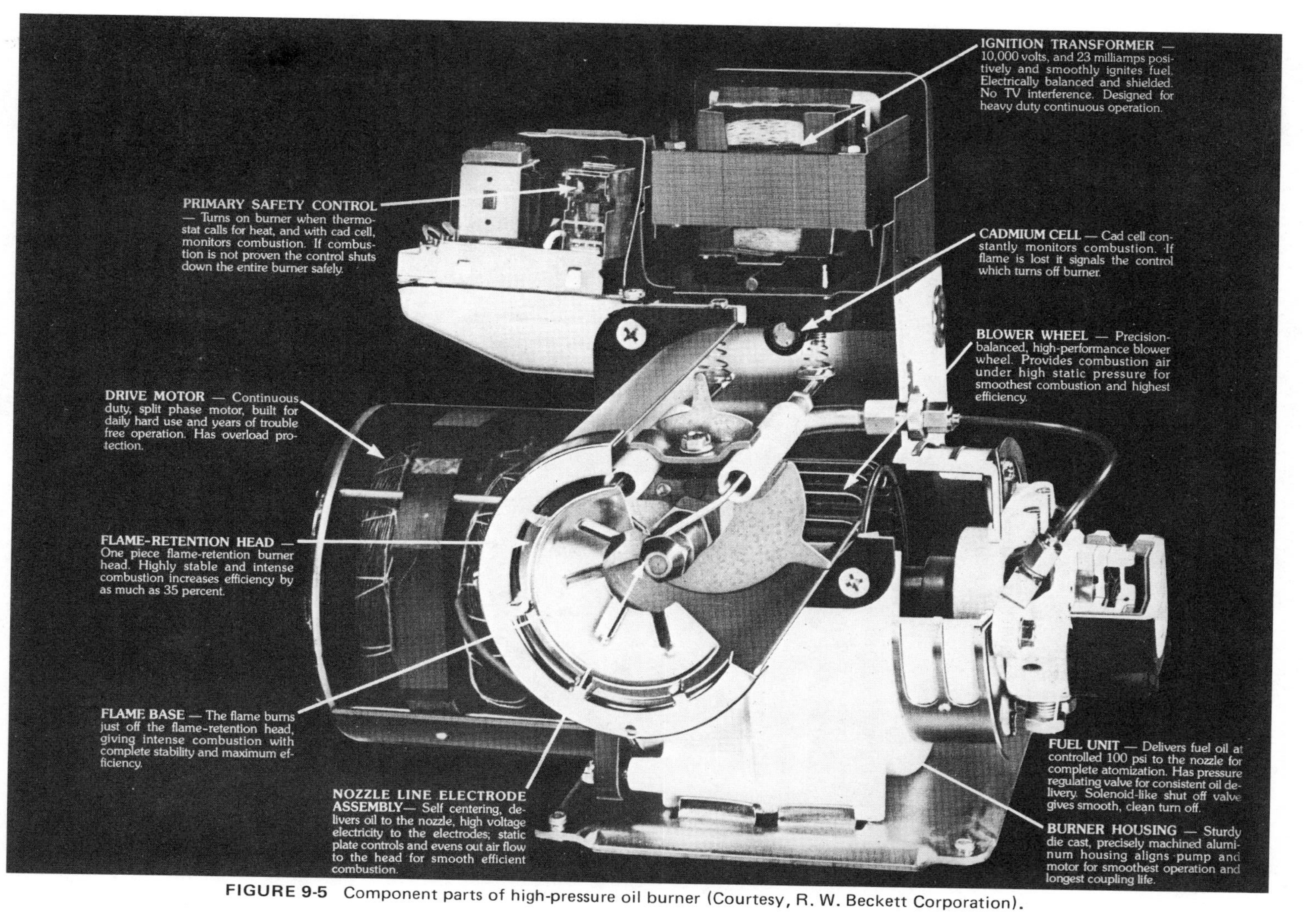

FIGURE 9-5 Component parts of high-pressure oil burner (Courtesy, R. W. Beckett Corporation).

The automatic cutoff valve stops the flow of oil as soon as the pressure drops. Thus, when the burner is stopped, the oil is quickly cut off to the nozzle to prevent oil dripping into the combustion chamber.

Fuel oil pumps are designed for single-stage (Figures 9-6 and 9-7) or two-stage operation (Figures 9-8 and 9-9). The single-stage unit is used where the supply of oil is above the burner and the oil flows to the pump by gravity. The two-stage unit is used where the storage tank is below the burner. The first stage is used to draw the oil to the pump. The second stage is used to provide the pressure required by the nozzle. The suction on the pump should not exceed a 15-in. vacuum.

Piping connections to the fuel pump are of two types: single-pipe and two-pipe. In the single-pipe system there is only one pipe from the storage tank to the burner. Plug B must be removed from the pump so that the unused oil can return to the low-pressure side of the pump. On the two-pipe system, two pipes are run from the storage tank to the burner. One carries supply oil and the other returns oil. Plug B must be left in place and plug A removed so that unused oil can return to the storage tank. The bleed plug is used on a single-pipe system to remove the air from the line (Figure 9-10).

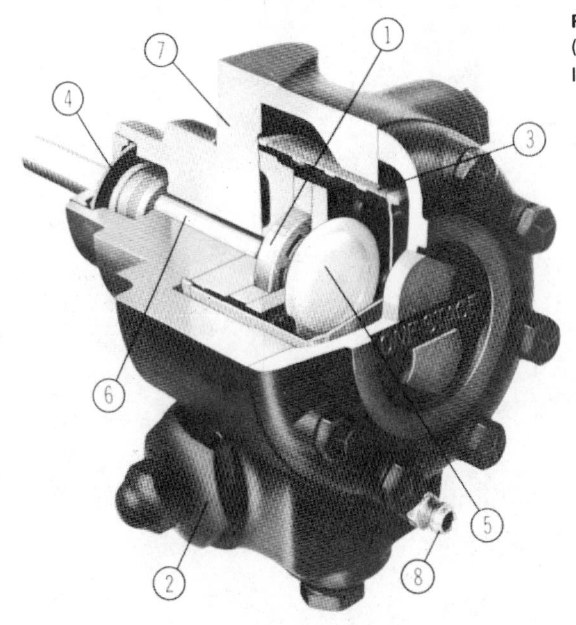

FIGURE 9-6 Single-stage oil pump (Courtesy, Sundstrand Hydraulics, Inc.).

1. GEARS
2. CUTOFF VALVE
3. STRAINER
4. SHAFT SEAL
5. ANTI-HUM DEVICE
6. SHAFT BEARING
7. BODY
8. BLEED VALVE

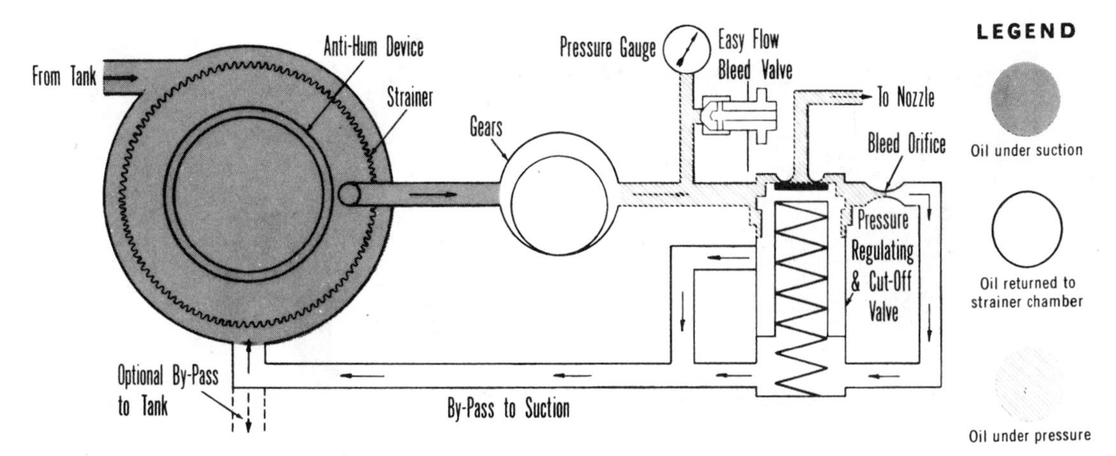

FIGURE 9-7 Circuit diagram for single-stage oil pump (Courtesy, Sundstrand Hydraulics, Inc.).

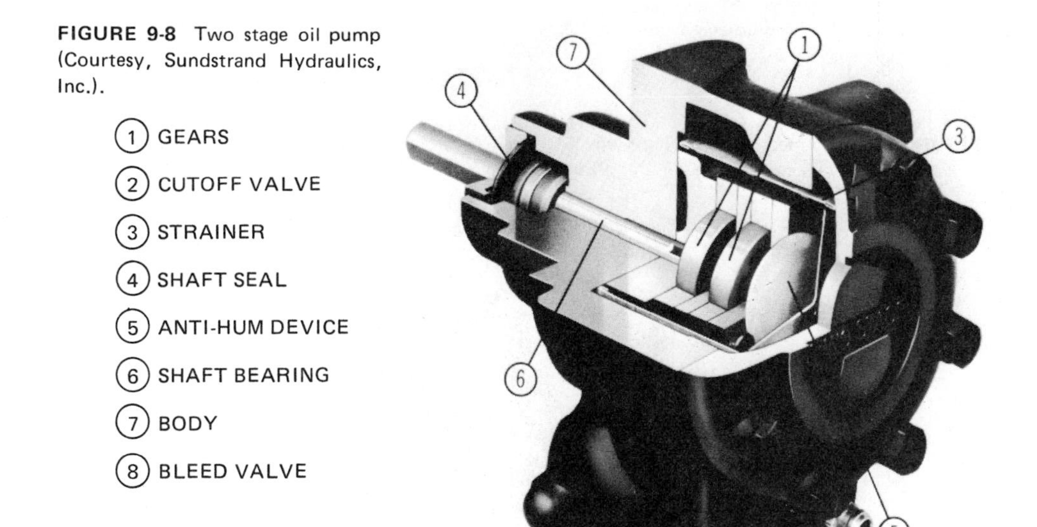

FIGURE 9-8 Two stage oil pump (Courtesy, Sundstrand Hydraulics, Inc.).

1. GEARS
2. CUTOFF VALVE
3. STRAINER
4. SHAFT SEAL
5. ANTI-HUM DEVICE
6. SHAFT BEARING
7. BODY
8. BLEED VALVE

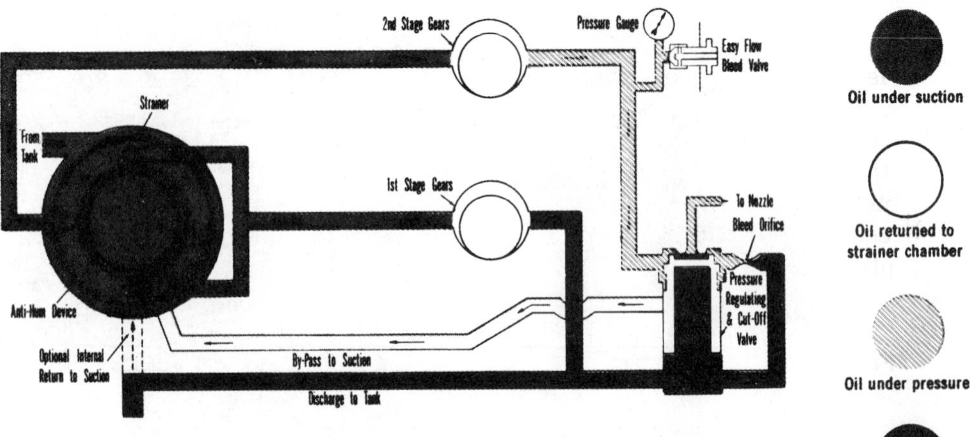

Note that the first stage has two inlets. When Model H is mounted in any of the three recommended positions, one inlet is always above the inlet to the second stage. For this reason, air in the system is drawn off by the first stage, and the second stage pumps only solid, air-free oil to the nozzle.

FIGURE 9-9 Circuit diagram for two-stage oil pump (Courtesy, Sundstrand Hydraulics, Inc.).

FIGURE 9-10 Connections showing plugs "A" and "B" used to convert a single-pipe to a two-pipe system (Courtesy, Sundstrand Hydraulics, Inc.).

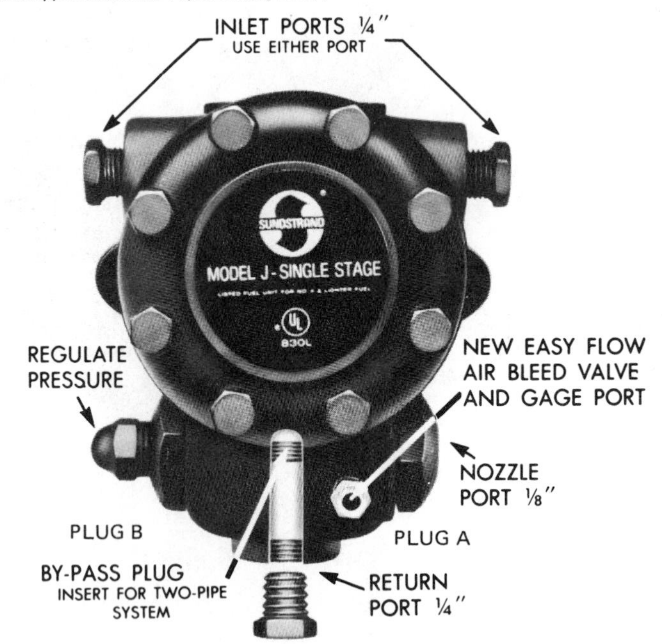

Various pump models are made by a number of manufacturers. Some pumps have clockwise rotation, some counterclockwise rotation. It is important when changing a pump to use an identical replacement, since oil pumps vary in the location of connections, as shown in Figures 9-11 and 9-12. Always refer to the manufacturer's information and specification sheets for connection details.

Some other variations that are available from pump manufacturers are high-speed pumps (3450 rpm) and pumps equipped with electrical solenoid oil supply cutoff valves (Figure 9-13).

SIMPLIFIED TEST PROCEDURE FOR OIL BURNER PUMPS

A simplified test procedure for oil burner pumps is shown in Figure 9-14. Note that the connections shown are for a specific model pump. However, similar connections can be made to other models by referring to the manufacturer's connection details on the specification sheets.

MOTOR:

The motor supplies power to rotate the pump and fan. Usually, a split-phase motor (with a start and run winding) is used (Figure 9-15). This type of motor provides enough torque (starting power) to move the connected components. In the event a motor fails, it must be replaced by substituting one with the same horsepower, direction of rotation, mounting dimension, revolutions per minute, shaft length, and shaft diameter.

MULTIBLADE FAN AND AIR SHUTTERS:

The multiblade (squirrel cage) centrifugal fan delivers air for combustion. It is enclosed within the fan housing. The inlet to the fan has an adjustable opening so that the amount of air volume handled by the fan can be manually controlled. The outlet of the fan delivers the combustion air through the blast tube of the burner (see Figure 9-5).

SHAFT COUPLING:

The shaft coupling connects the motor to the fuel pump (Figure 9-16). This coupling:

- Provides alignment between the pump and motor shafts
- Absorbs noise that may be created by the rotating parts
- Is strongly constructed to endure the starting and stopping action of the motor

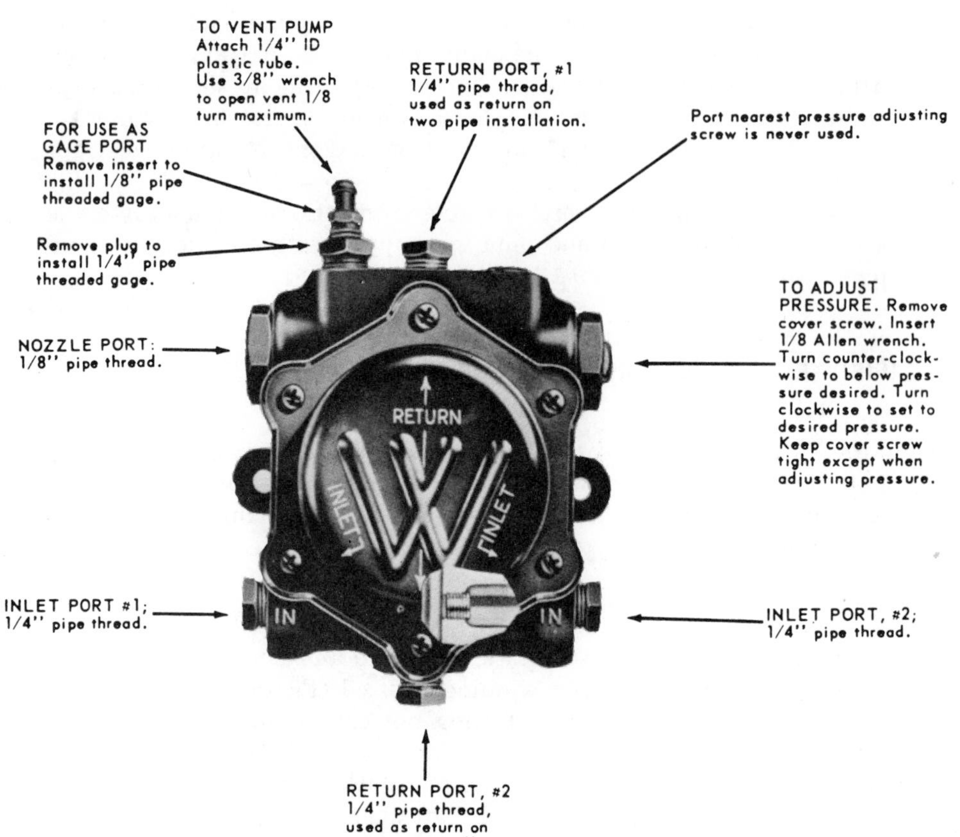

NOTE: For maximum performance INLET VACUUM, when measured at unused INLET PORT, should not exceed 10'' Hg on single stage pumps, and 15'' Hg on two stage pumps.

SINGLE PIPE INSTALLATION

Recommended only when bottom of tank is above fuel unit, unless pump code ends in 15.

1. Remove BYPASS PLUG if installed, through INLET PORT #2.
2. Connect inlet line to preferred INLET PORT.
3. Plug all unused ports securely.
4. Start burner and bleed all air from the system by opening VENT PLUG. Close VENT securely when oil flow in-tube is clear.

TWO-PIPE INSTALLATION

1. Insert BYPASS PLUG if not installed, through INLET PORT #2.
2. Connect inlet line to preferred INLET PORT.
3. Connect return line to preferred RETURN PORT.
4. Plug all unused ports securely.
5. Start burner. Two stage pumps will self-vent. If single stage and code ends in 3 or 4, bleed all air from system by opening VENT PLUG. Close VENT securely when oil flow in tube is clear.

FIGURE 9-11 Oil pump showing connection locations (Courtesy, Webster Electric Company, Inc.).

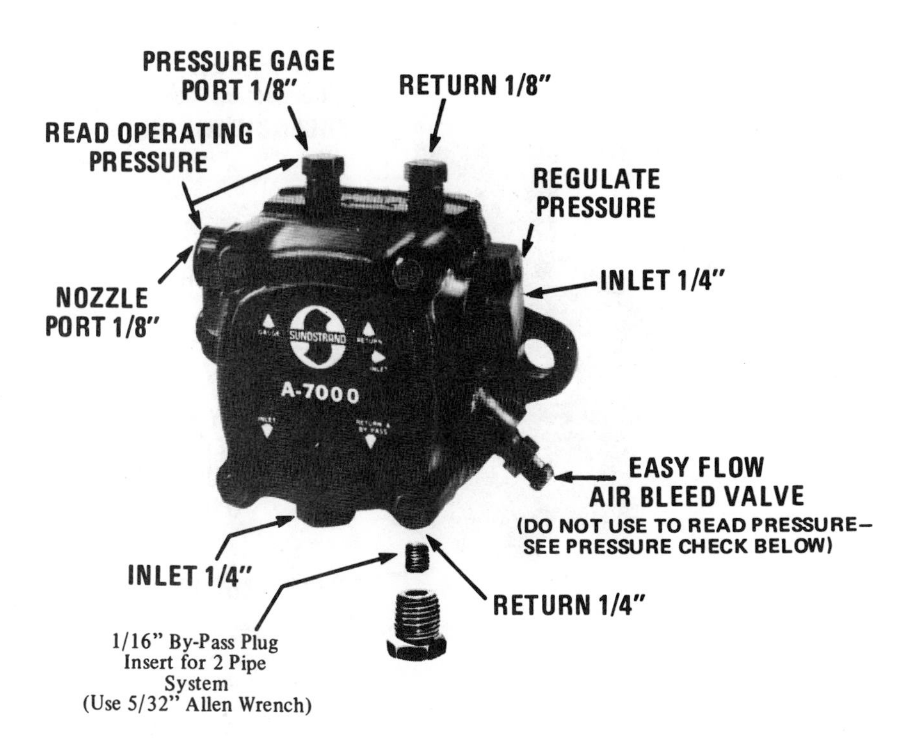

ONE-PIPE SYSTEM

Connect inlet line to pump inlet. Start burner. Arrange primary burner control for continuous operation during purging. Open easy flow bleed valve 1 turn CCW. Bleed unit until all air bubbles disappear – HURRIED BLEEDING WILL IMPAIR EFFICIENT OPERATION OF UNIT. Tighten easy flow bleed valve securely. (Fig. 4)

TWO-PIPE SYSTEM

Remove 1/16" by pass plug from plastic bag attached to unit. Remove 1/4" plug from return port. Insert by pass plug (See Figure 1 or 2). Attach return and inlet lines.
Start burner—Air bleeding is automatic. Opening Easy Flow Air Bleed Valve will allow a faster bleed if desired.
Return line must run to within 3" of the bottom of the tank (See Figure 5). Failure to do this may introduce air into the system and could result in loss of prime.

FIGURE 9-12 Oil pump showing connection locations (Courtesy, Sundstrand Hydraulics, Inc.).

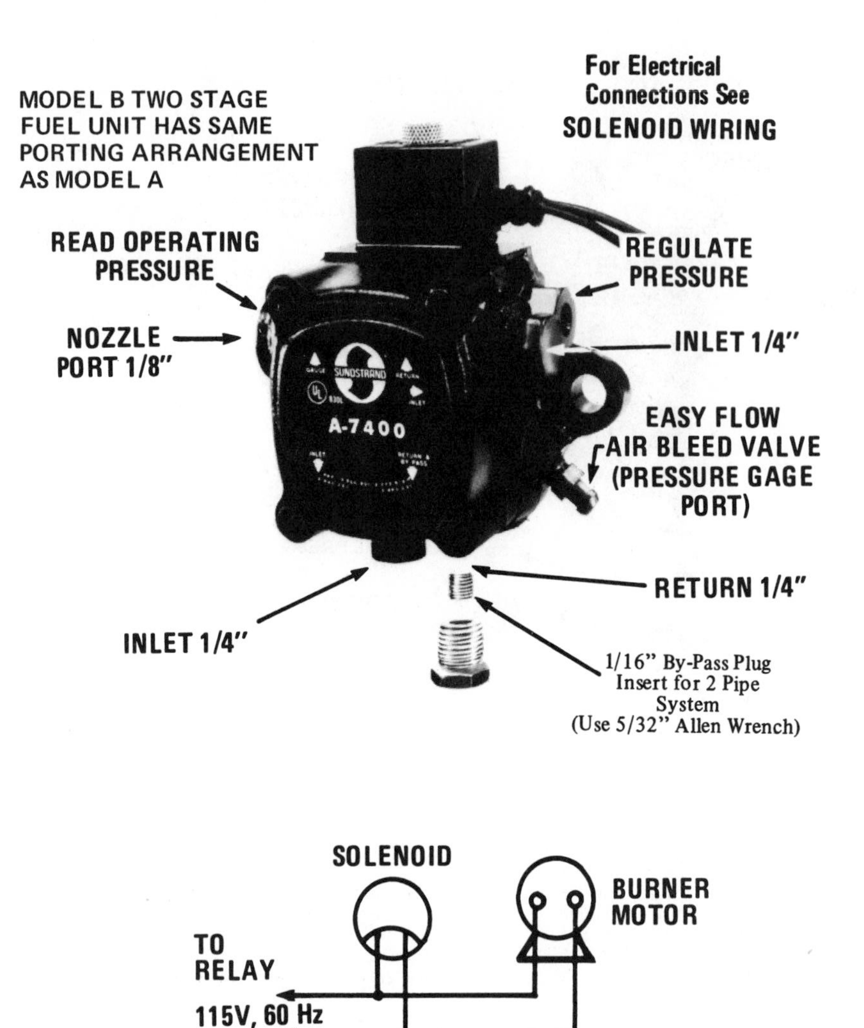

SOLENOID WIRING

DISCONNECT POWER SUPPLY BEFORE WIRING TO PREVENT ELECTRICAL SHOCK OR EQUIPMENT DAMAGE. Lead wires on these devices are long enough to reach the junction box on most burner installations. Wire solenoid in parallel with burner motor (See Fig. 3). All electrical work should be done according to local and national codes. (Solenoid 115V, 0.1A, 60 Hz)

FIGURE 9-13 Oil pump with mounted solenoid locations (Courtesy, Sundstrand Hydraulics, Inc.).

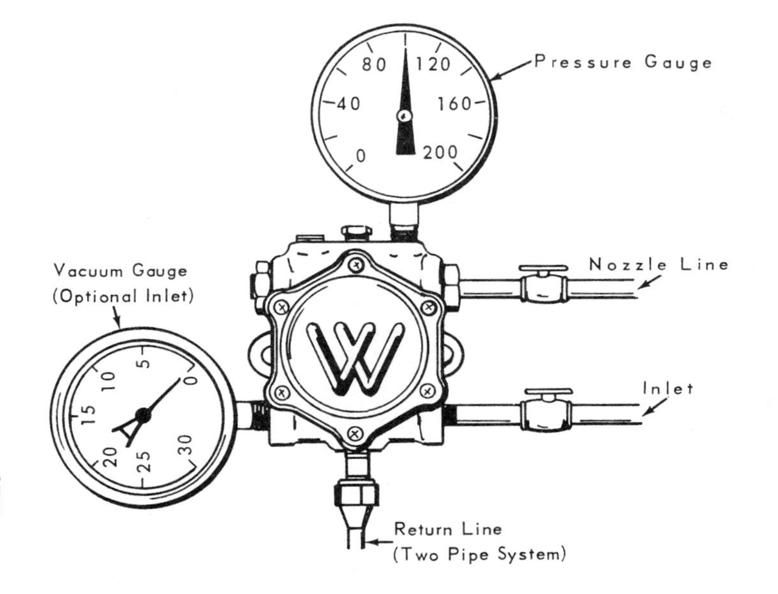

SET–UP–A

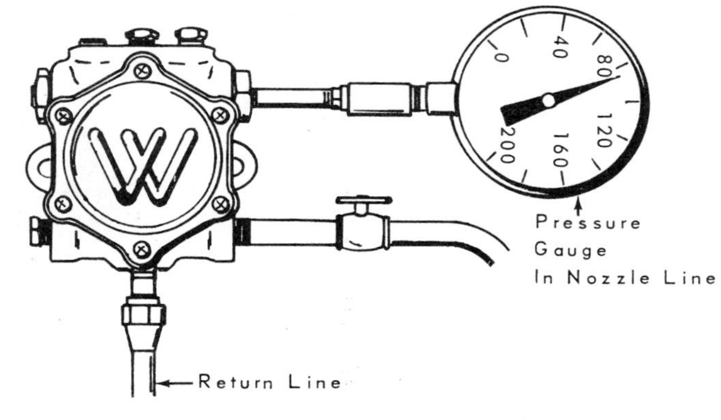

SET–UP–B

Test All Units With By-Pass Installed For Two Pipe System

Delivery and Vacuum Test - - Set-Up A Above

1. Set pressure adjusting screw to required pressure.
2. Measure nozzle delivery at set pressure.
3. Close inlet valve to check vacuum.

Cut Off Test - - Set-Up B Above

1. Set pressure at 100 PSI.
2. Shut off motor. Pressure should hold at between 75 and 90 PSI. If pressure drops to 0 PSI cut off leaks.

FIGURE 9-14 Test procedure for oil burner pump (Courtesy, Webster Electric Company, Inc.).

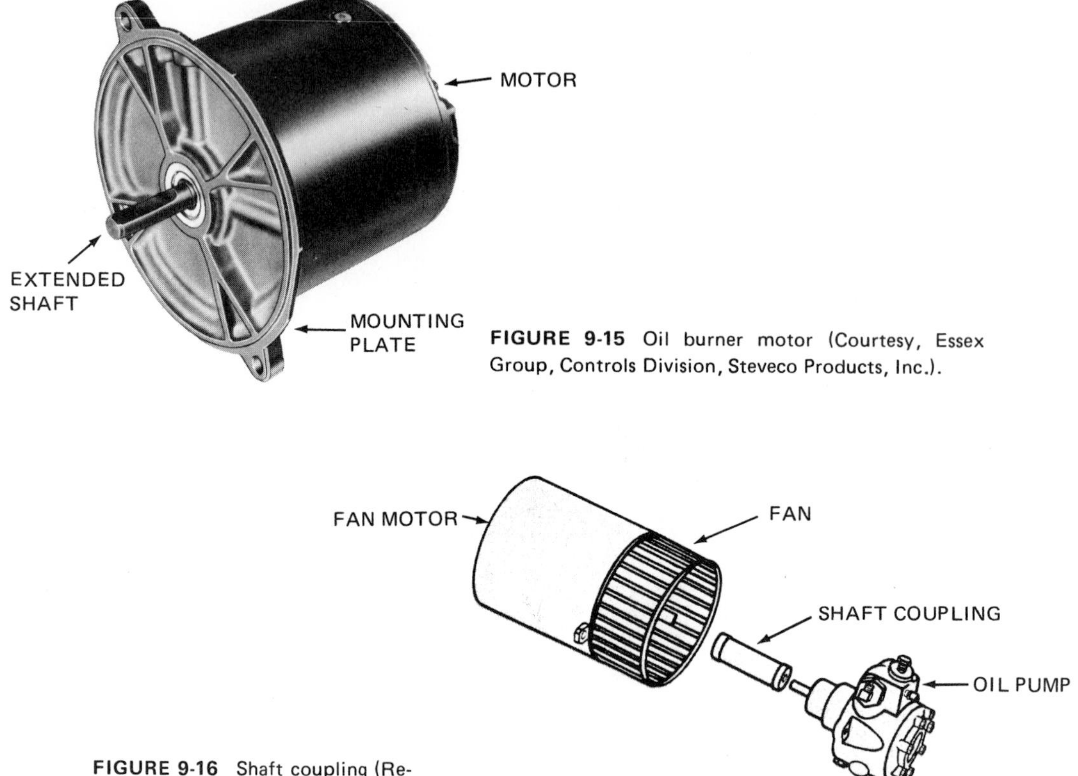

FIGURE 9-15 Oil burner motor (Courtesy, Essex Group, Controls Division, Steveco Products, Inc.).

FIGURE 9-16 Shaft coupling (Reproduced by permission of Carrier Corporation; Copyright 1979, Carrier Corporation).

STATIC DISC CHOKE, AND SWIRL VANES:

As shown in Figure 9-17, the static disc which is located in the center of the draft tube, causes the air from the fan to build up velocity at the inside surface of the tube. The choke is located at the end of the tube and restricts the area, thus further increasing the air velocity. The swirl vanes located near the choke give turbulence to the leaving air, which assists the mixing action.

Nozzle Assembly

The nozzle assembly consists of the oil feed line, the nozzle, electrodes, and transformer connections (Figure 9-18). This assembly serves to position the electrodes in respect to the nozzle opening, and provides a mounting for the high potential electrical leads from the transformer. The assembly is located near the end of the blast tube.

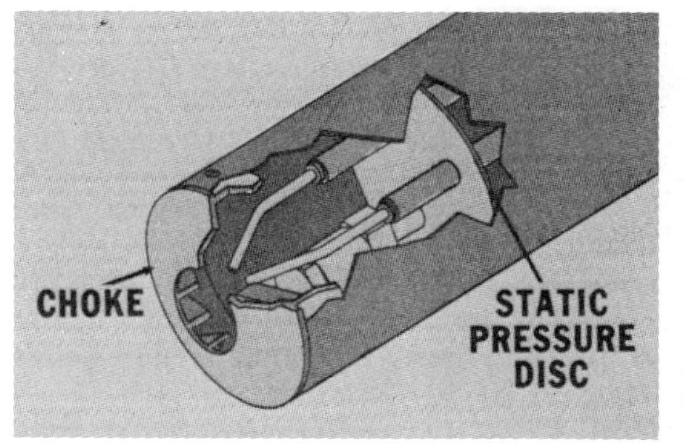

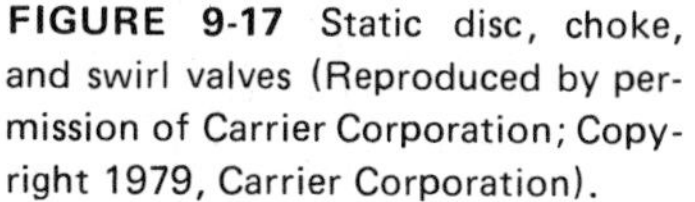

FIGURE 9-17 Static disc, choke, and swirl valves (Reproduced by permission of Carrier Corporation; Copyright 1979, Carrier Corporation).

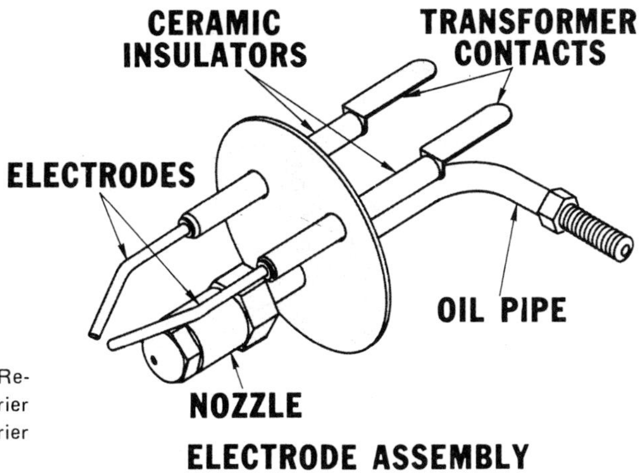

FIGURE 9-18 Nozzle assembly (Reproduced by permission of Carrier Corporation; Copyright 1979, Carrier Corporation).

THE NOZZLE:

A nozzle construction is shown in Figure 9-19. The purpose of the nozzle is to prepare the oil for mixing with the air. The process is called *atomization.*

Oil first enters the strainer. The strainer has a mesh which is finer in size than the nozzle orifice so as to catch solid particles that could clog the nozzle.

From the strainer, the oil enters slots that direct the oil to the swirl chamber. The swirl chamber gives the oil a rotary motion when it enters the nozzle orifice, thus shaping the spray pattern. The nozzle orifice increases the velocity of the oil. The oil leaves in the form of a mist or spray and mixes with the air from the blast tube. Because of the fine tolerances of the nozzle construction, servicing is usually impractical. A defective or dirty nozzle is normally replaced.

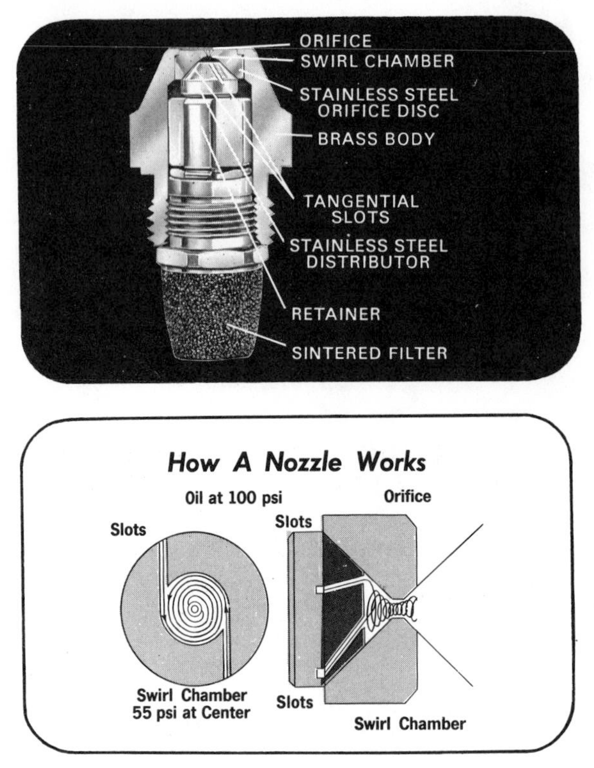

FIGURE 9-19 Nozzle construction (Courtesy, Delavan Corporation).

SPRAY PATTERNS:

The requirements for nozzles vary with the type of application. Nozzles are supplied with two variations: the shape of the spray and the angle between the sides of the spray.

There are three spray shapes, as shown in Figure 9-20: hollow (H), semihollow (SH), and solid (S). The hollow and the semihollow are most popular on domestic burners because they provide better efficiencies when used with modern combustion chambers.

The angle of the spray must correspond to the type of combustion chamber (Figure 9-21). An angle of 70 to 90 degrees is usually best for square or round chambers. An angle of 30 to 60 degrees is best for long, narrow chambers.

THE IGNITION SYSTEM:

All high-pressure burners have electric ignition. The power is supplied by a step-up transformer connected to two electrodes. Types of transformers are shown in Figure 9-22. The transformer supplies high voltage which causes a spark to jump between the two electrodes. The force of the air in the blast tube causes the spark to arc (or bend) into the oil/air mixture, igniting it.

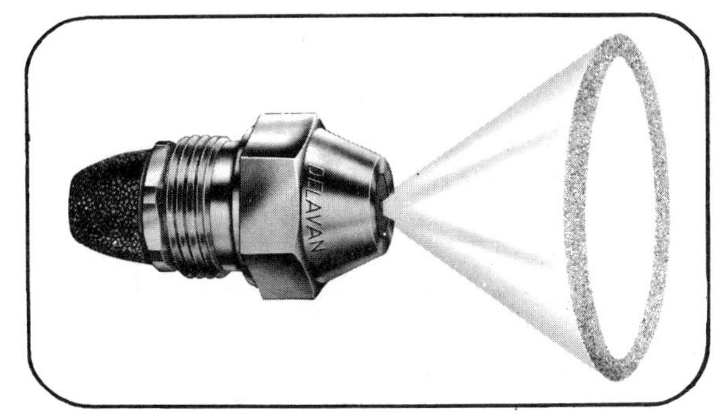

HOLLOW CONE (H)

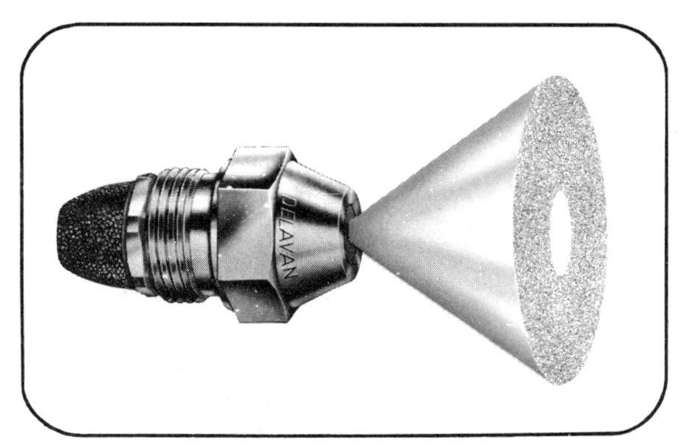

SEMI–HOLLOW CONE (SH)

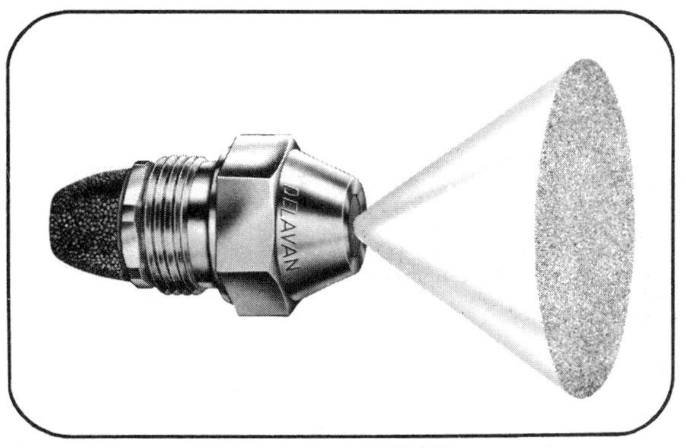

SOLID CONE (S)

FIGURE 9-20 Spray patterns (Courtesy Delavan Corporation).

Ceramic insulators surround the electrodes where they are close to metal parts. Insulators also serve to position the electrodes.

The transformer* increases the voltage from 120 V to 10,000 V, but reduces the amperage (current flow) to about 20 mA. This low amperage is relatively harmless and reduces wear on the electrode tips.

* Because of its high voltage, caution should be used in checking the transformer.

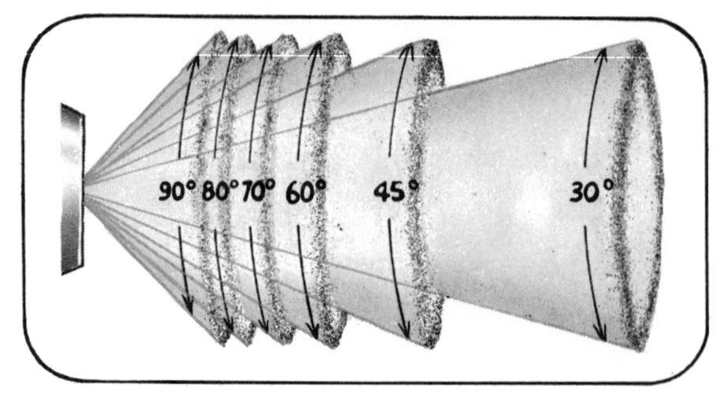

VARIETY OF SPRAY ANGLES

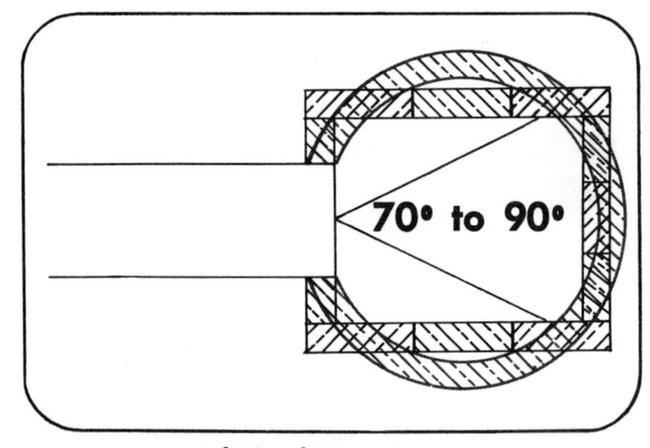

70° TO 90° FOR ROUND
OR SQUARE CHAMBERS

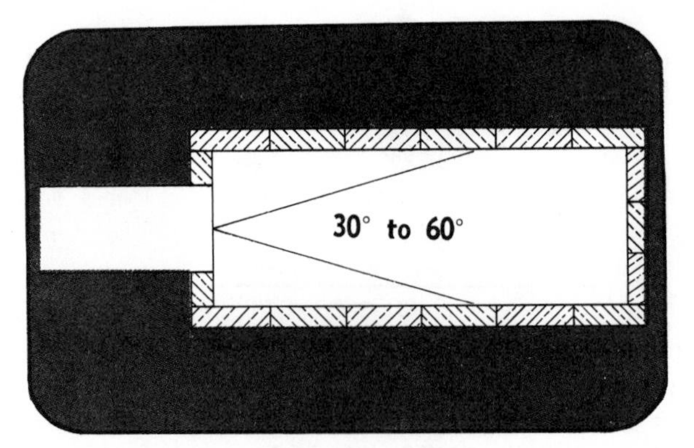

30° TO 60° FOR LONG
NARROW CHAMBERS

FIGURE 9-21 Relationship of spray angle to combustion chamber shape (Courtesy, Delavan Corporation).

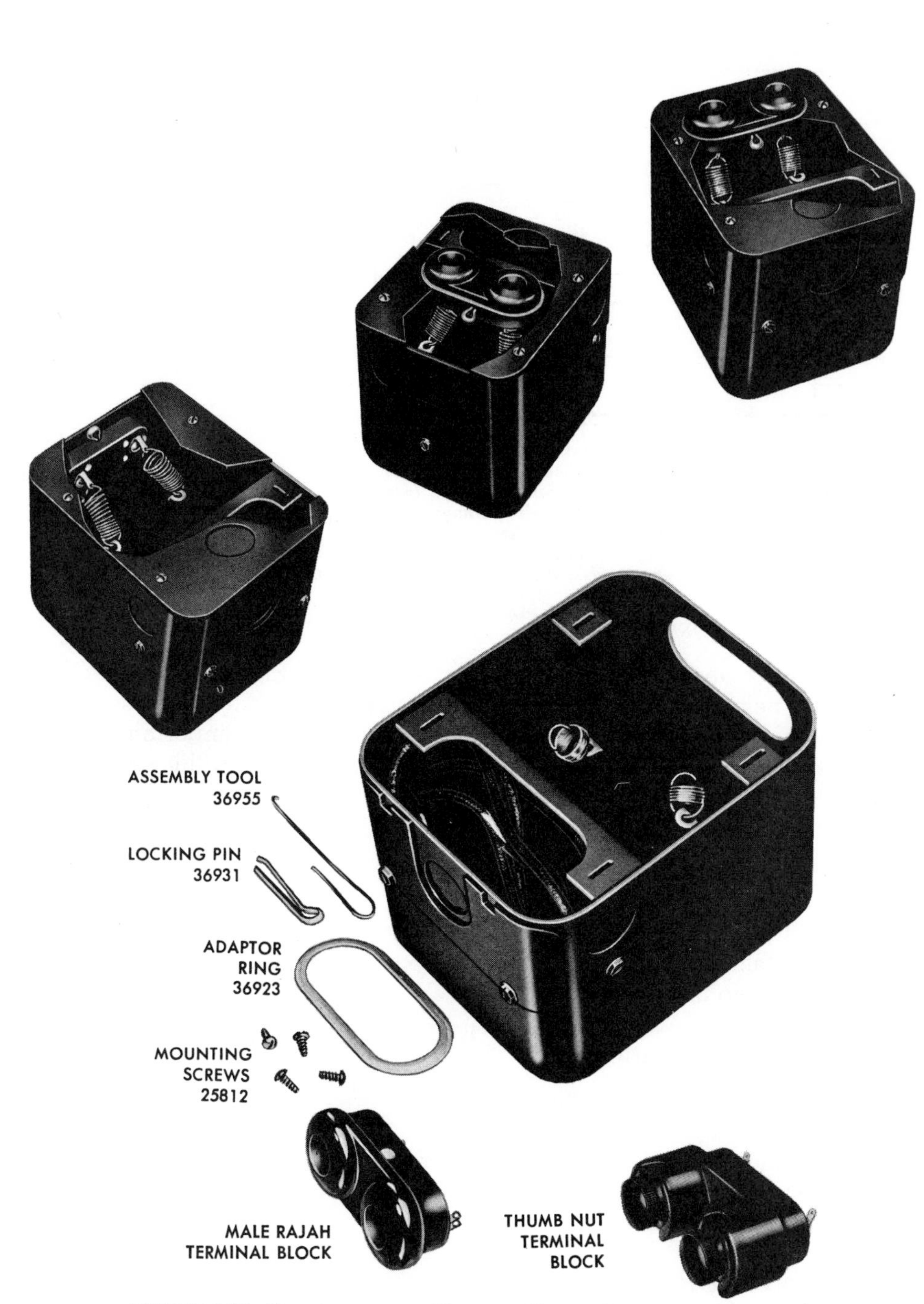

FIGURE 9-22 Transformer types (Courtesy, Webster Electric Company, Inc.).

Nozzle	GPH	A	B	C
45°	(.75 to 4.00)	1/8" to 3/16"	1/2" to 9/16"	1/4"
60°	(.75 to 4.00)	1/8" to 3/16"	9/16" to 5/8"	1/4"
70°	(.75 to 4.00)	1/8" to 3/16"	9/16" to 5/8"	1/8"
80°	(.75 to 4.00)	1/8" to 3/16"	9/16" to 5/8"	1/8"
90°	(.75 to 4.00)	1/8" to 3/16"	9/16" to 5/8"	0

Recommended Electrode Settings. NOTE: Above 4.00 GPH, it may be advisable to increase dimension C by 1/8" to insure smooth starting. When using double adapters: (1) Twin ignition is the safest and is recommended, with settings same as above. (2) With single ignition, use the same A and B dimensions as above, but add 1/4" to dimension C. Locate the electrode gap on a line midway between the two nozzles.

FIGURE 9-23 Electrode positioning (Courtesy, Delavan Corporation).

ELECTRODES POSITION:

The correct position of the electrodes is shown in Figure 9-23. The electrodes must be located out of the oil spray, but close enough for the spark to arc into the spray. The spark gap between the electrodes is important. If the electrodes are not centered on the nozzle orifice, the flame will be one-sided and cause carbon to form on the nozzle.

Oil Storage and Piping

Most cities have rules and regulations governing the installation and piping of oil tanks which must be adhered to by the installation mechanic.

There are two tank locations:

1. Outside (underground)
2. Inside (usually in the basement)

An outside tank can be 550, 1000, or 1500 gal. in capacity. An inside tank usually has a capacity of 275 gal. Tanks must be approved by Underwriters Laboratories (UL).

OUTSIDE UNDERGROUND TANK:

As shown in Figure 9-24, an outside underground tank should be installed at least 2 ft below the surface. The fill pipe should be 2 in. international pipe size (IPS) and the vent pipe $1\frac{1}{4}$ in. IPS. The vent pipe must be of

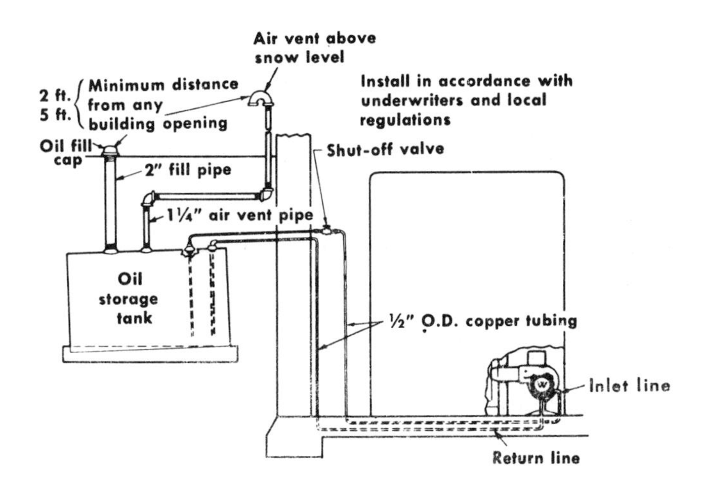

Suction lines should not be less than 3/8'' O.D. Copper Tubing when tank is *above* burner, or 1/2'' O.D. Copper Tubing when tank is *below* burner.

FUEL-UNIT BELOW TANK, TWO-PIPE SYSTEM
(No. 2 oil 50 SSU at 68°F.)

Head in Feet	Total length of tubing in ft.	
	3/8Tube	1/2Tube
10	80	292
9	77	279
8	73	266
7	70	253
6	66	240
5	63	227
4	59	214
3	55	201
2	52	188
1	48	175
0	45	162

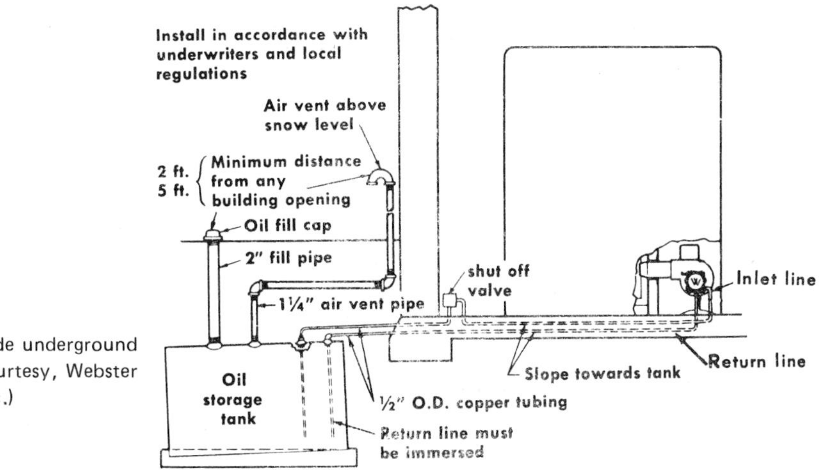

FIGURE 9-24 Outside underground tank installation (Courtesy, Webster Electric Company, Inc.)

a greater height than the fill pipe. The bottom of the tank should slope away from the end having the suction connection. A slope of 3 in. in its length is adequate. Piping connections to the tank should use swing connections to prevent breaking the pipe when movement occurs. Oil suction and return lines are usually made of $^1/_2$-in.-outside diameter (OD) copper tubing, positioned within 3 in. of the bottom of the tank.

A UL-approved suction line filter should be used. A globe valve should be installed between the tank and the filter. A check valve should be installed in the suction line at the pump to keep the line filled at all times. A suitable oil level gauge should be installed. The tank should be painted with at least two coats of tar or asphaltum paint.

INSIDE TANK:

An inside tank installation is shown in Figure 9-25. Many of the regulations and installation conditions for inside tanks are the same as those for outside tanks. The fill pipe should be 2 in. IPS, and the vent pipe $1^1/_4$ in. IPS. An oil level gauge should be installed. The oil piping is usually $^3/_8$-in.-OD copper tubing. A filter should be installed in the oil suction lines.

Inside tank installation varies in some ways from outside tank installation. The slope of the tank bottom should be about 1 in. and directed toward the oil suction connection. Two globe type shutoff valves should be used, one at the tank and one at the burner. The valve at the tank should be a fuse type. The oil line between the tank and the burner should be buried in concrete. The tank should be at least 7 ft from the burner.

ACCESSORIES

Two important accessories are the combustion chamber and the draft regulator. The combustion chamber is placed in the lower portion of the heat exchanger, surrounding the flame on all sides with the exception of the top. A barometric damper is used as a draft regulator. The draft regulator is placed in the flue pipe between the furnace and the chimney.

Combustion Chamber

The purpose of a combustion chamber is to protect the heat exchanger and to provide reflected heat to the burning oil. The reflected heat warms the tips of the flame, assuring complete combustion.

No part of the flame should touch the surface of the combustion chamber. If the combustible mixture does touch, the surface will be cooled and incomplete combustion will result. It is essential that the chamber fit the flame. The nozzle must be located at the proper height above the floor. The

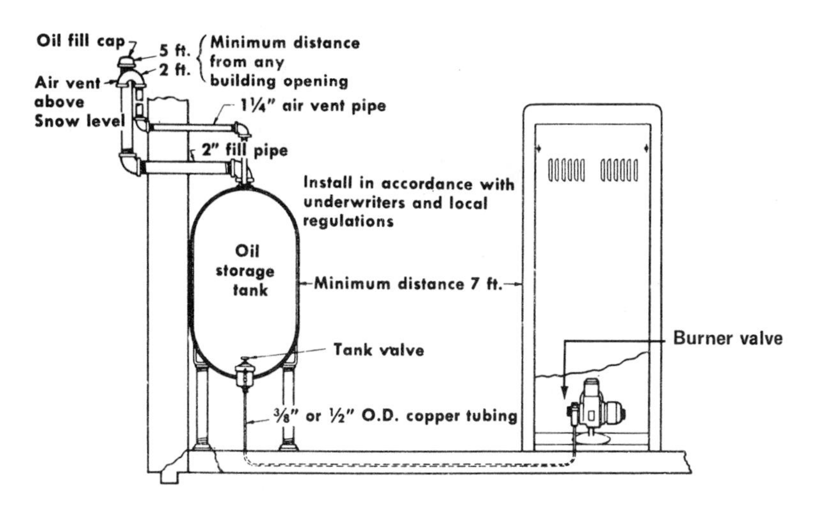

FUEL-UNIT BELOW TANK, ONE-PIPE SYSTEM

Firing rates for this type installation do not normally exceed 1-1/2 G.P.H. A head of oil up to 20 feet may be maintained, if necessary. Under these circumstances maximum length of suction line should not exceed 140 feet.

Any increase in firing rates or reduction in head pressures would necessitate a corresponding reduction in maximum length of the supply line.

NOTE: In pressurized suction lines the pressure should never exceed 10 psi – preferably less.

FIGURE 9-25 Inside tank installation (Courtesy, Webster Electric Company, Inc.)

bottom area of the combustion chamber is usually 80 in.^{2}/gal for nozzles between 0.75 and 3.00 gal/min, 90 in.^{2}/gal for nozzles between 3.50 and 6.00 gal/min, and 100 in.^{2}/gal for nozzles between 6.50 gal/min and higher.

Three types of material are generally used for combustion chambers:

1. Metal (usually stainless steel)
2. Insulating firebrick
3. Molded ceramic

Metal is used in factory-built, self-contained furnaces and is used without backfill (material in space between chamber and heat exchanger). Insulated firebrick is used for conversion burners to fit almost any type of heat-

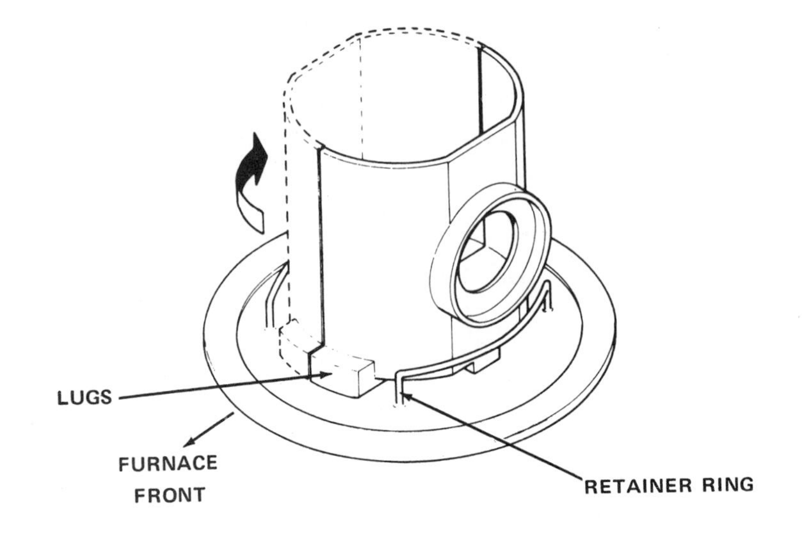

FIGURE 9-26 Molded ceramic combustion chamber (Courtesy, Luxaire, Inc.).

exchanger shape, and uses fiberglas-type backfill. The molded ceramic chamber is prefabricated to the proper shape and size, and uses fiberglas-type backfill (Figure 9-26).

DRAFT REGULATOR:

A draft regulator maintains a constant draft over the fire, usually 0.01 to 0.03 in. water column. Too high a draft causes undue loss of heat through the chimney. Too little draft causes incomplete combustion.

(a)

FIGURE 9-27 Draft regulator (Courtesy, Field Control Division, Heico, Inc.).

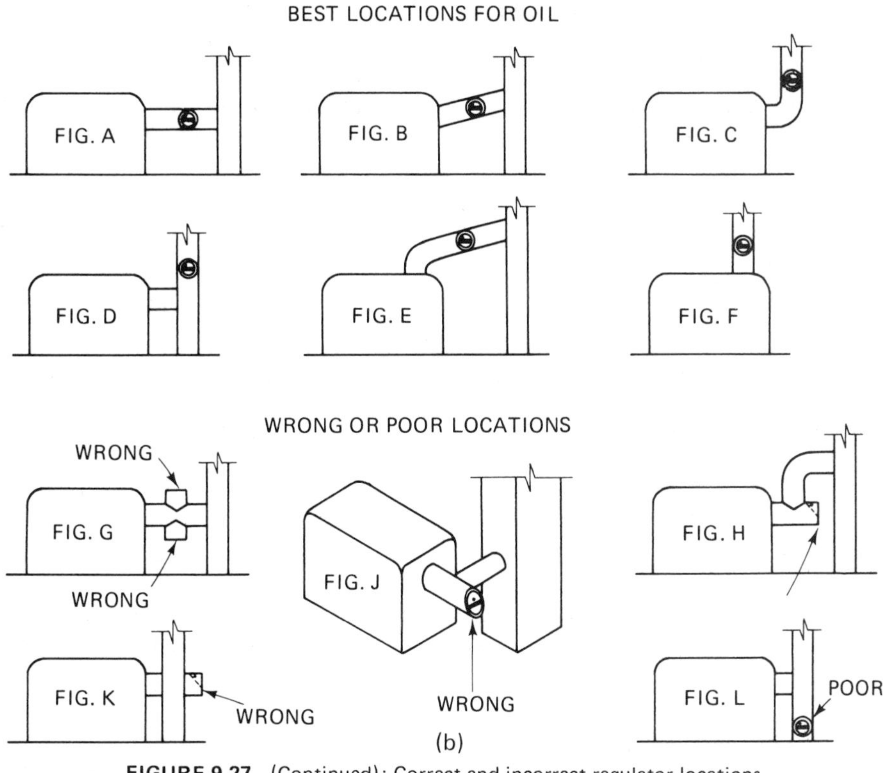

FIGURE 9-27 (Continued): Correct and incorrect regulator locations.

A draft regulator (Figure 9-27) consists of a small door in the side of the flue pipe. The door is hinged near the center and controlled by adjustable weights. Basement air is admitted to the flue pipe as required to maintain a proper draft over the fire.

MAINTENANCE AND SERVICE

Annual maintenance of oil burner equipment is essential to good operation. Service procedures are outlined below.

Power Assembly

1. Clean fan blades, fan housing, and screen
2. Oil motor with a few drops of SAE No. 10 oil
3. Clean pump strainer

4. Adjust oil pressure to 100 psig
5. Check oil pressure cutoff
6. Conduct combustion test and adjust air to burner for best efficiency (see Chapter 10)

Nozzle Assembly

1. Replace nozzle
2. Clean nozzle assembly
3. Check ceramic insulators for hairline cracks and replace, if necessary
4. Check location of electrodes and adjust, if necessary
5. Replace cartridges in oil line strainers

Ignition System and Controls

1. Test transformer spark. The spark should jump $\frac{1}{2}$ in. or better and can be checked with a screw driver with an insulated handle as shown in Figure 9-28
2. Clean thermostat contacts
3. Clean control elements that may become contaminated with soot, especially those that protrude into furnace or flue pipe
4. Check system electrically (see Chapter 18)

Furnace (Figure 9-29)

1. Clean combustion chamber and flue passages
2. Clean furnace fan blades
3. Oil fan motor
4. Replace air filter

After this work is completed, run the furnace through a complete cycle and check all safety controls (Chapter 18). Clean up exterior of furnace and area around furnace.

CAUTION: If the unit runs out of oil or if there is an air leak in the lines, the fuel pump can become air-bound and not pump oil. To cor-

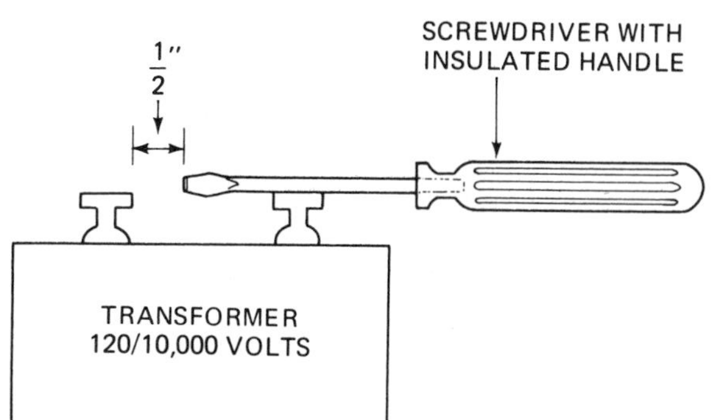

FIGURE 9-28 Checking the ignition transformer (Reproduced by permission of Carrier Corporation; Copyright 1979, Carrier Corporation).

FIGURE 9-29 Components of a typical forced warm air furnace—oil fired.

rect this, the air must be bled from the pump and replaced with oil. On a one-pipe system this is done by loosening or removing the plug on the port opposite the intake. Start the furnace and run until oil flows out of the opening; then turn it off and replace the plug. The system can then be put back into operation. A two-pipe pump is considered to be self priming. However, if it does fail to prime follow the procedure described above.

REVIEW QUESTIONS

Select the letter representing your choice of the correct answer.

9-1. Oil is vaporized by:

(a) Pressurization

(b) Heat

(c) Centrifugal force

(d) Atomization

9-2. The most popular oil burner is:

(a) Pot type

(b) Rotary type

(c) Low-pressure type

(d) High-pressure type

9-3. A high-pressure burner mixes oil and air:

(a) Before entering the nozzle

(b) After oil passes through the nozzle

9-4. The location of the electrodes on a gun-type burner is:

(a) Within the air/oil spray

(b) Outside the air/oil spray

9-5. The type of pump used with an outside tank located below the oil burner is called:

(a) Two-stage

(b) Single-stage

9-6. The type of motor used on an oil burner is a:

(a) Capacitor start

(b) Split-phase

(c) Capacitor run

(d) Shaded pole

9-7. The purpose of the nozzle on a high-pressure burner is:

(a) To blow air

(b) To mix the oil and the air

(c) To atomize the oil

(d) To burn the oil

9-8. The shape of the oil/air spray must:

(a) Correspond to the type of combustion chamber

(b) Fit the burner supplied

(c) Make a good-appearing fire

(d) Fit the heating load

9-9. The oil capacity of a single inside tank in gallons is usually:

(a) 275

(b) 550

(c) 1000

(d) 1500

9-10. What size of copper tubing is usually used for inside tank oil lines?

(a) $\frac{1}{4}$ in. OD

(b) $\frac{3}{8}$ in. OD

(c) $\frac{1}{2}$ in. OD

(d) $\frac{5}{8}$ in. OD

10

COMBUSTION TESTING

OBJECTIVES

After studying this chapter, the student will be able to:

- Use combustion test instruments
- Evaluate the results of combustion testing
- Determine the changes that must be made to reach maximum combustion efficiency

EFFICIENCY OF OPERATION

Combustion is a chemical reaction resulting in the production of a flame. The fuels used for heating consist chiefly of carbon. In combustion, carbon and oxygen are combined to produce heat. Instruments are used to measure how well the equipment performs the process of combustion. The heating service technician makes whatever adjustments are necessary to produce the most efficient operation.

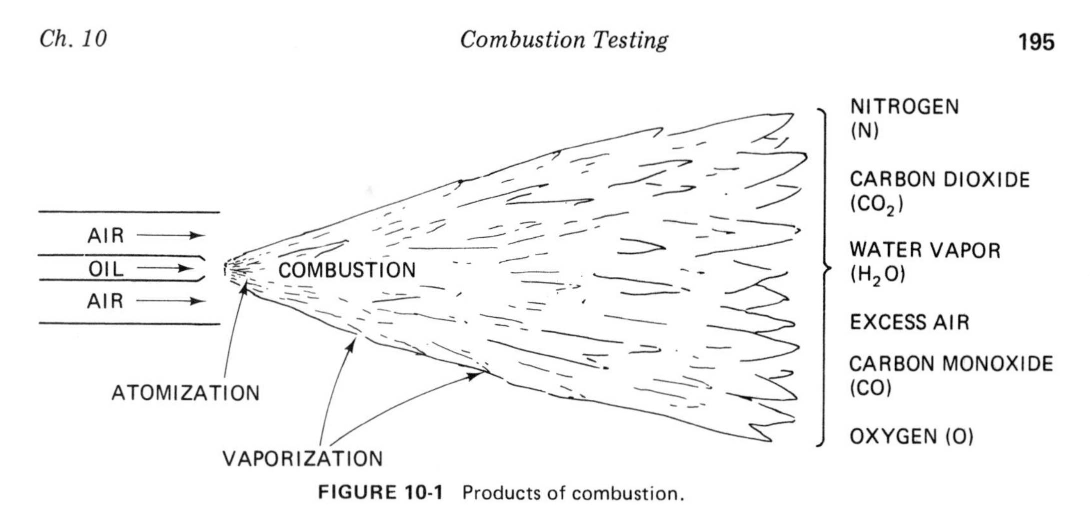

FIGURE 10-1 Products of combustion.

The products of combustion that leave the oil-heating unit through the flue are:

- Carbon dioxide (CO_2)
- Carbon monoxide (CO)
- Oxygen (O)
- Nitrogen (N)
- Water vapor (H_2O)

A description of the products of combustion is shown in Figure 10-1.

The measurement of the CO_2 in the flue gases is an indication of the amount of air used by the fuel-burning equipment. It is desirable to have a high CO_2 measurement, as this indicates a hot fire. The maximum CO_2 measurement for oil with no excess air is 15.6%; for natural gas, 11.8%. However, it is not practical with field-installed equipment to reach these high CO_2 readings. In practice, oil furnaces should have a CO_2 reading of between 10 and 12%; natural gas furnaces should read between $8\frac{1}{4}$ and $9\frac{1}{2}$%.

The CO_2 in the flue gases is the indicator used (along with stack temperature) to measure the efficiency of combustion.

INSTRUMENTS USED IN TESTING

Instruments used for combustion testing are available from a number of manufacturers. While the illustrations for the instruments described below have been provided by specific companies, instruments that perform similar functions can be obtained from various other manufacturers.

The instruments and their functions are:

- **Draft Gauge:** used for measuring draft (Figure 10-2)
- **True Spot Smoke Tester:** used for determining smoke scale (Figure 10-3)
- **Flue Gas Analyzer:** used for testing CO_2 content (Figure 10-4)
- **Stack Thermometer:** used for determining stack temperature (Figure 10-5)

Complete combustion kits (Figure 10-6) are available with many of the items listed above. The location of the various parts of the heating plant equipment, and sample holes for testing, are shown in Figure 10-7.

TESTING PROCEDURE

Tests should be made in the following order:

1. Draft measurement
2. CO_2 content
3. Stack temperature
4. Smoke sample

The equipment should be run for a minimum of 5 to 10 min before testing to stabilize operating conditions. A record should be kept both before adjustments are made and after test results.

FIGURE 10-2 Draft gauge (Courtesy, Bacharach Instrument Company).

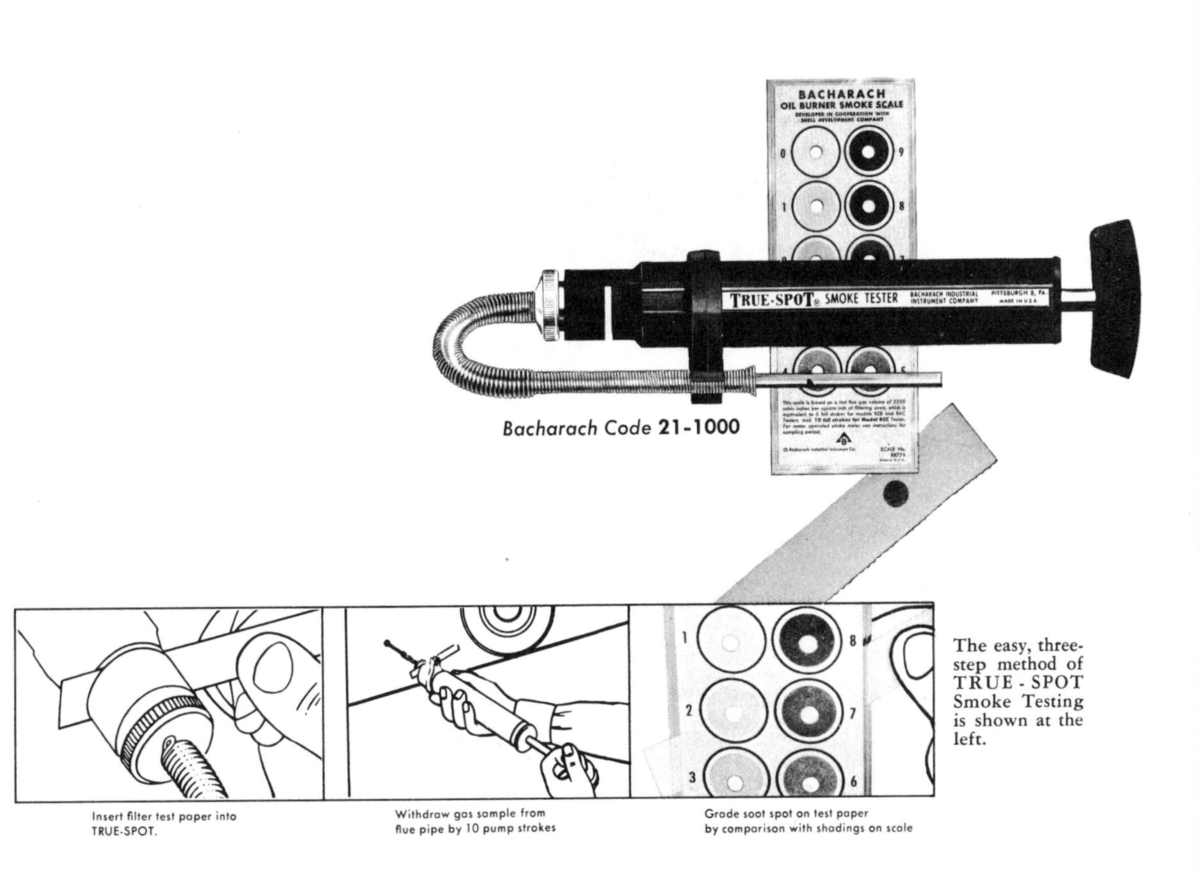

FIGURE 10-3 Smoke tester (Courtesy, Bacharach Instrument Company).

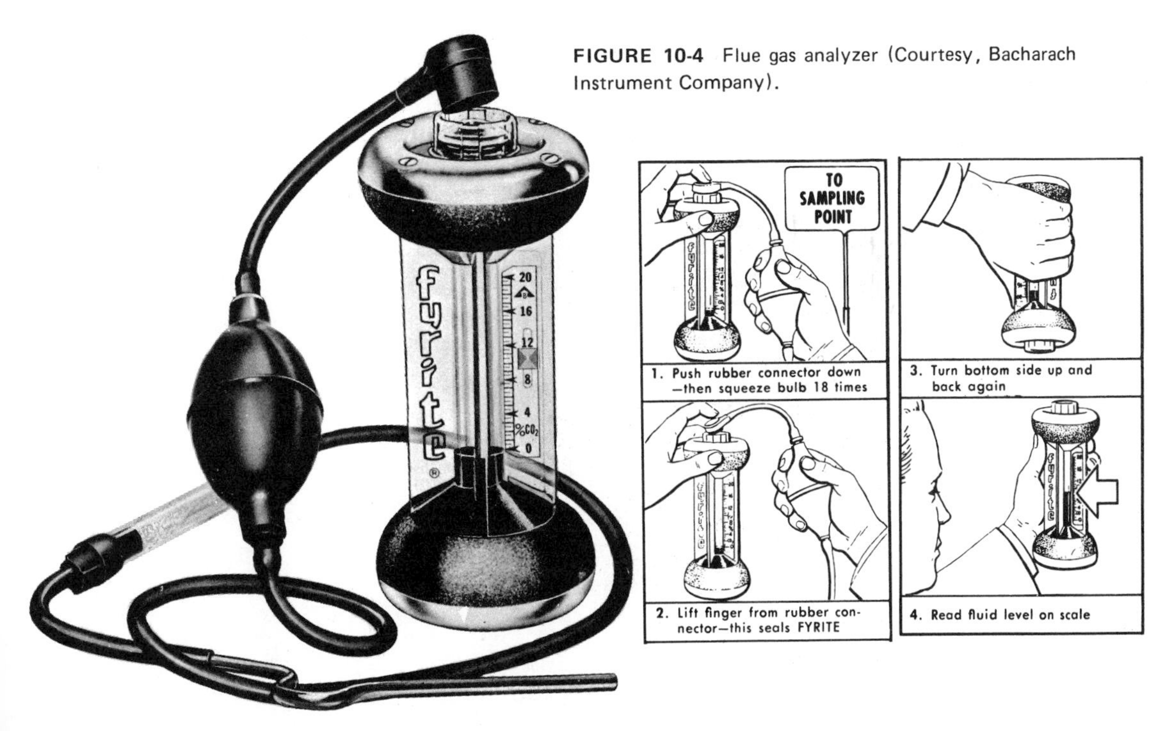

FIGURE 10-4 Flue gas analyzer (Courtesy, Bacharach Instrument Company).

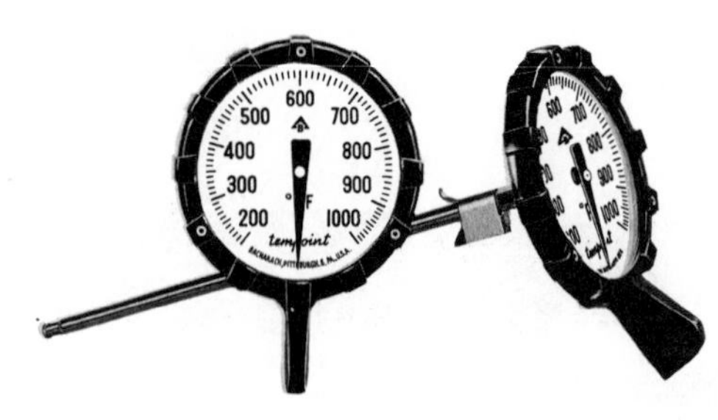

FIGURE 10-5 Stack thermometer (Courtesy, Bacharach Instrument Company).

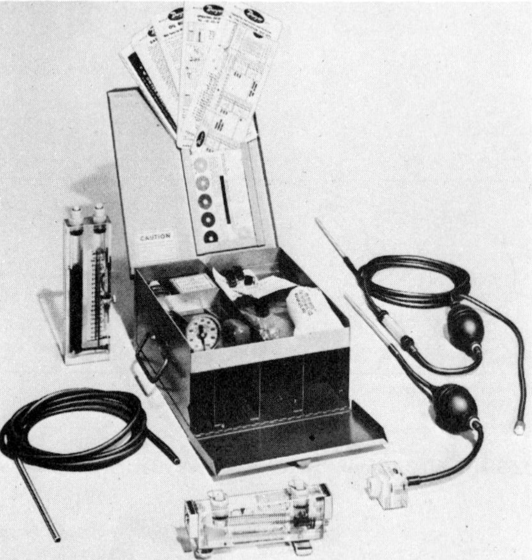

FIGURE 10-6 Complete combustion kits (Courtesy, Dwyer Instruments, Inc.).

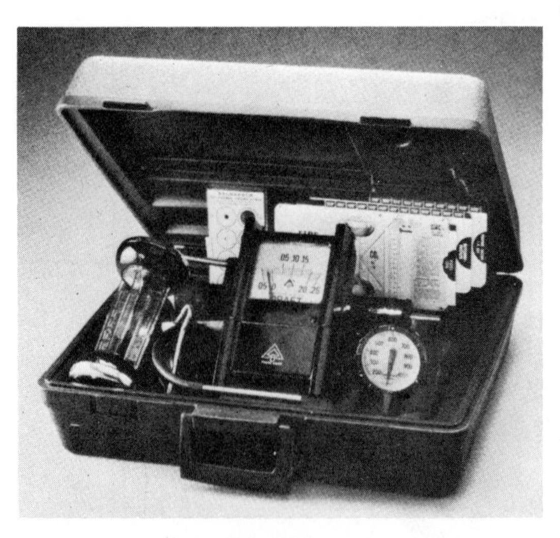

Courtesy, Bacharach

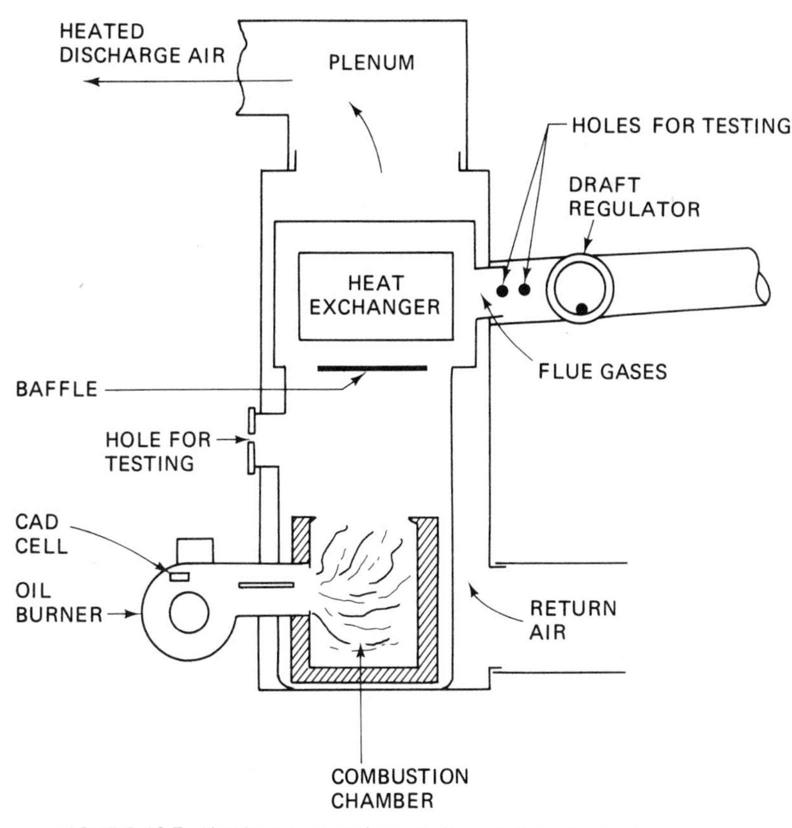

FIGURE 10-7 Heating plant equipment showing holes for testing

Draft Test

The draft gauge reads in inches of water column. For example, a reading of 0.02 in. water column means 2 one-hundredths of an inch of water column.

Always calibrate the instrument before using it. Place the instrument on a level surface and adjust the indicator needle to read zero.

Drill a $\frac{1}{4}$-in. hole in the fire box door (Figure 10-8) and two $\frac{1}{4}$-in. holes in the flue at the point just after it leaves the heat exchanger. The holes in the flue must be placed between the heat exchanger and the draft regulator on an oil furnace, and between the heat exchanger and the draft diverter on a gas furnace. One of the two holes in the flue is used for the stack thermometer; the other is used for the CO_2 and smoke tests.

Occasionally, an additional $\frac{1}{4}$-in. hole is drilled in the flue pipe near the chimney to measure chimney draft. This is normally used only when it is necessary to troubleshoot draft problems.

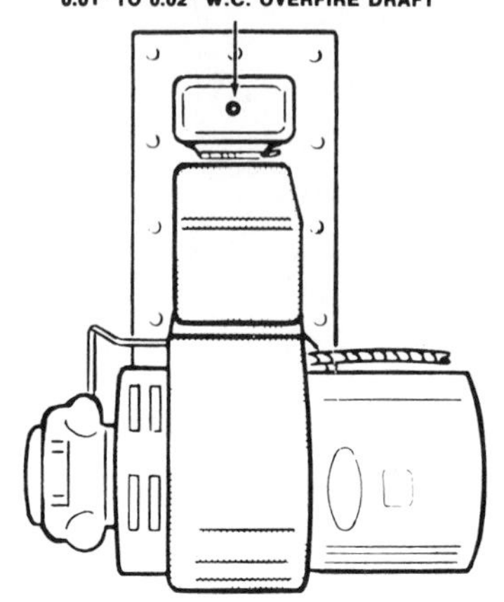

FIGURE 10-8 Location of sample hole for testing (Courtesy, Westinghouse Electric Corporation, Heating and Cooling Division).

The draft should always be negative when measured with the draft gauge, since the flue gases are moving away from the furnace. The draft-over-the-fire, measured at the firebox door, should be -0.01 to -0.02 in. of water column. The difference between the draft reading in the flue at the furnace outlet, and the draft-over-the-fire reading indicates heat-exchanger leakage.

Smoke Test

The burner flame should not produce excessive smoke. Smoke causes soot (carbon) to collect on the surfaces of the heat exchanger, reducing its heat-transfer rate. A soot deposit of $^1/_8$ in. can cause a reduction of 10% in the rate of heat absorption.

Measurement of smoke is performed by using a smoke tester. A sample is collected by pumping 2200 cm^3 (10 full strokes with the sampling pump) through a 0.38-cm^2 area filter paper. The color of the sample filter paper spot is compared to a standard graduated smoke scale (Figure 10-3). Spot zero (0) is white and indicates no smoke. Spot number 9 is darkest and represents the most extreme smoke condition. Between zero and 9 the scale has 10 different grades. It is most desirable to produce smoke at a scale of number 1 or number 2, but not zero. Older style or conversion units could read as high as number 4. Zero indicates excess air supply to the burner and low efficiency. Readings higher than number 2 indicate the production of excessive carbon and soot. Normally, this test is performed only on oil-burning equipment.

Carbon Dioxide Test

An analyzer for CO_2 uses potassium hydroxide (KOH), a chemical that has the property of being able to absorb large quantities of CO_2. A known volume of flue gas is run into the tester. Since KOH will absorb only CO_2, the reduction in volume of the flue gas is an indication of the amount of CO_2 absorbed by the KOH solution.

Instructions for using the tester shown in Figure 10-4 are as follows:

1. Set the instrument to zero by adjusting the sliding scale to the level of the fluid in the tube
2. Insert sample tube in flue opening. Place rubber connector on top of instrument and depress to open valve. Collapse bulb 18 times to fill instrument with flue gas and to release valve
3. Tip instrument over and back twice and hold at 45-degree angle for 5 s
4. Hold instrument upright and read CO_2 content on scale. Release pressure by opening valve when test is complete

Stack Temperature

A stack thermometer is used to determine stack temperature. Insert thermometer in flue hole. Operate burner unit until the temperature rise, as read on the thermometer, is no more than 3°F/min; then read the stack temperature.

COMBUSTION EFFICIENCY

Combustion efficiency, expressed in percentage, is a measure of the useful heat produced in the fuel compared to the amount of fuel available. Thus, a furnace that operates at 80% efficiency is one that loses 20% of the fuel value in the heat that goes out through the chimney.

A combustion efficiency slide rule is a useful tool (Figure 10-9). After determining the stack (flue) temperature and CO_2 content, and by using the appropriate slide on the rule, the combustion efficiency can be determined.

Note that combustion efficiency is based on net stack temperature. Net stack temperature is the reading taken on the stack thermometer minus the room temperature (the temperature of the air entering the furnace).

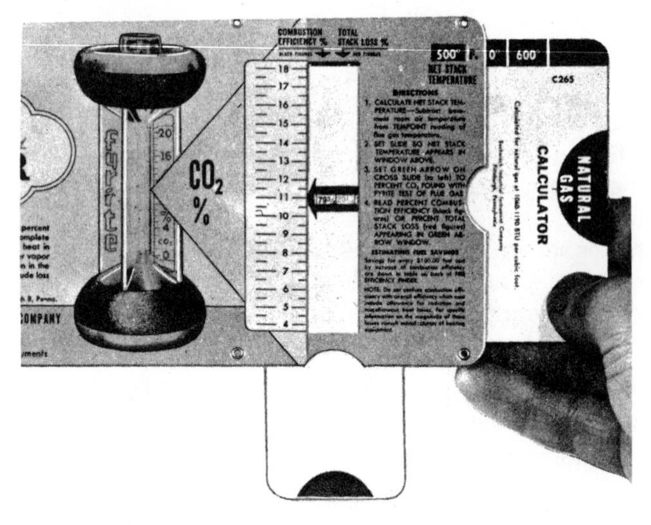

FIGURE 10-9 Combustion efficiency slide rule (Courtesy, Bacharach Instrument Company).

For example, assume that the net stack temperature for an oil burner is found to be 500° F and the CO_2 content determined as 9%. Using the slide rule, move the large slide to the right so that 500 appears in the small window at the upper right marked "net stack temperature." Then move the small vertical slide until the arrow points to reading 9 on the CO_2 scale. Through the window in the arrow, read the figures 80 (black) and 20 (red). This means that the combustion efficiency is 80% and the stack loss is 20%. The table in Figure 10-10 shows, for example, how increasing the combustion efficiency from 80% to 85% would save $5.90 of every $100 of fuel cost.

FIGURE 10-10 Fuel saved by increasing efficiency (Courtesy, R. W. Beckett Corporation).

FROM AN ORIGINAL EFFICIENCY OF	TO AN INCREASED COMBUSTION EFFICIENCY OF:							
	55%	60%	65%	70%	75%	80%	85%	90%
50%	9.10	16.70	23.10	28.60	33.30	37.50	41.20	44.40
55%	–	8.30	15.40	21.50	26.70	31.20	35.30	38.90
60%	–	–	7.70	14.30	20.00	25.00	29.40	33.30
65%	–	–	–	7.10	13.30	18.80	23.50	27.80
70%	–	–	–	–	6.70	12.50	17.60	22.20
75%	–	–	–	–	–	6.30	11.80	16.70
80%	–	–	–	–	–	–	5.90	11.10
85%	–	–	–	–	–	–	–	5.60

ANALYSIS AND ADJUSTMENTS

Usually, the original tests are made after routine maintenance has been performed. The combustion tests indicate the condition of the equipment and point to further service or adjustments that may be necessary.

Oil-Burning Equipment

On oil-burning equipment, adjustments are made in the air supply to the burner. Air is reduced to maintain a smoke test within the number 1 to number 2 limits while providing the highest possible CO_2 content in the flue gases.

If an efficiency rating of 75% or better cannot be obtained as a result of adjustments in the air supply, further service on the burner is required.

Listed below are some common causes of low CO_2 content and smoky fires (Figure 10-11).

1. Incorrect air supply
2. Combustion chamber air leaks
3. Improper operation of draft regulator
4. Insufficient draft
5. Burner "on" periods too short
6. Oil does not conform to burner requirements
7. Air-handling parts defective or incorrectly adjusted
8. Firebox cracked
9. Spray angle of nozzle unsuitable
10. Nozzle worn, clogged, or incorrect type
11. Gallons/hour rate too high for size of combustion chamber
12. Ignition delayed due to defective stack control
13. Nozzle spray or capacity unsuited to type of burner
14. Oil pressure to nozzle improperly adjusted
15. Nozzle loose or not centered
16. Electrodes dirty, loose, or incorrectly set
17. Cutoff valve leaks
18. Rotary burner motor running under speed

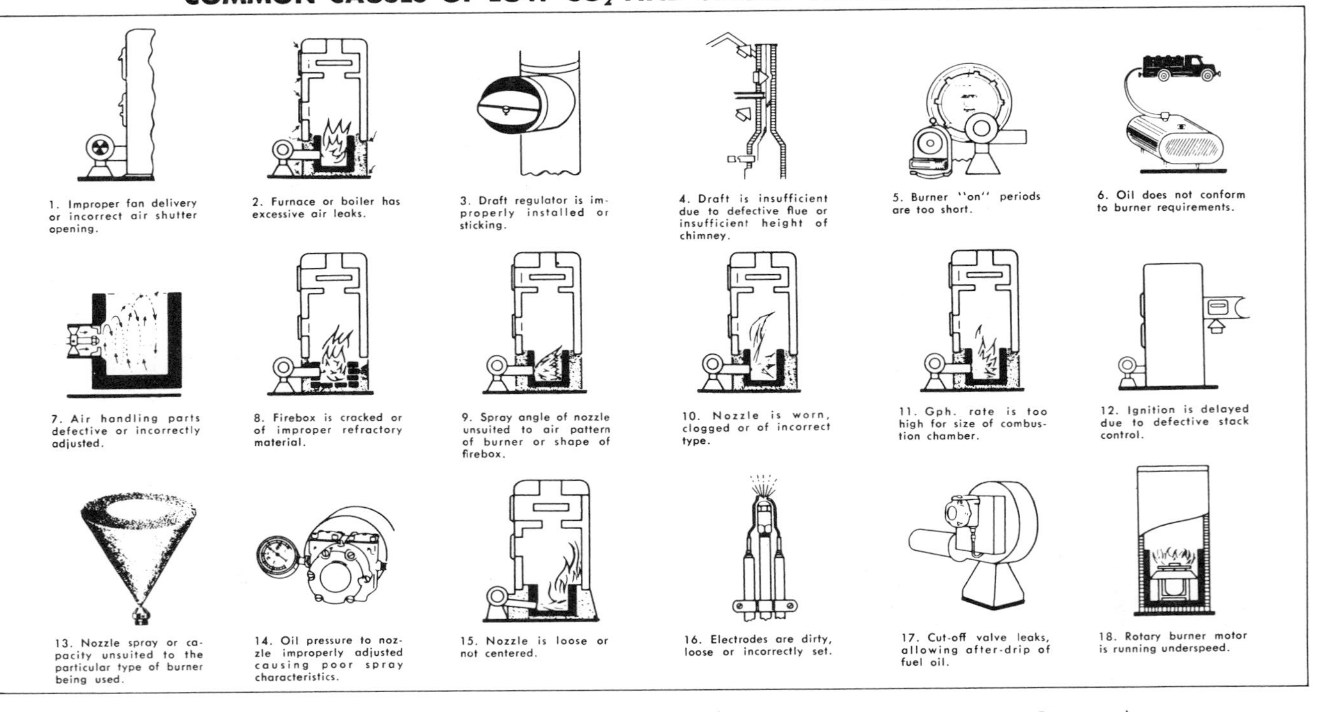

FIGURE 10-11 Causes of low CO_2 content and smoky fires (Courtesy, Bacharach Instrument Company).

Some conditions that produce poor efficiency, such as those indicated below, can be determined by visual inspection.

1. Check burner shutdown. A flame should last no longer than 2s after burner shuts down
2. Check the flame. It should be symetrically shaped and centered in the combustion chamber. It should not strike the walls or floor of the combustion chamber
3. Check for air leaks. There should be no leakage around burner tube where it enters the combustion chamber
4. Check the burner operating period. Burning periods of less than 5 min duration usually do not produce efficient operation

Gas-Burning Equipment

On gas-burning equipment, the primary air is reduced by adjusting the air shutter to provide a blue flame. The secondary air is a fixed quantity on most furnaces. However, inspection should be made to determine if the secondary air restrictor has become loose or, perhaps, removed. Correction should be made, if required.

If the flame does not adjust properly to produce a blue flame, and an efficiency rating of 75% or better cannot be obtained, further service is required.

Some causes of poor efficiency are:

1. **Too Much Excess Air:** It may be necessary to place a restrictor in the flue pipe to reduce the draft. The method of flue restrictor placement is shown in Figure 10-12

There should be a neutral point in all gas furnaces. Also, there should be a slight positive pressure in the top of the heat exchanger and a slight negative pressure near the bottom. The neutral point should be in a location slightly above the fire (observation) door.

2. **Unclean Burner Ports:** To clean burner ports, remove lint and dirt
3. **Insufficient Gas Pressure:** The draft diverter should be checked with the burner in operation. Air should be moving into the opening, thus indicating a negative pressure at the inlet. This can be verified by holding a lighted match at the opening.
4. **Improper Gas Pressure:** The gas pressure should be 3½ in. w.c. at the manifold for natural gas. Check the gas meter for input into the

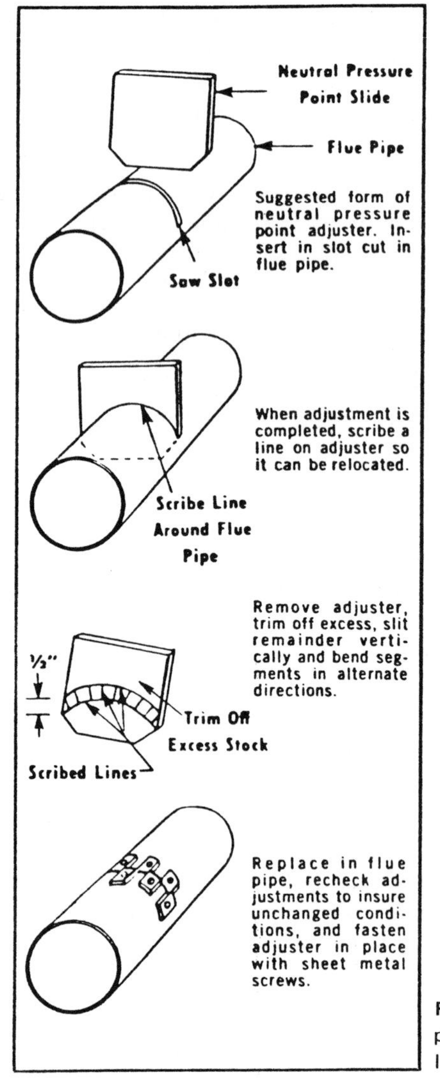

FIGURE 10-12 Flue restrictor placement (Courtesy, Bacharach Instrument Company).

furnace. It should read the same as the input indicated on the furnace nameplate. If the gas pressure reads between 3 and 4 in. w.c. and still does not produce the rated input, the orifice should be cleaned or replaced with one of proper size. For liquid petroleum (LP) the gas pressure should be 11 in. w.c. at the manifold on the furnace.

5. **Check the "On" and "Off" Periods of the Furnace:** "On" periods of less than 5 min are too brief to produce high efficiency

REVIEW QUESTIONS

Select the letter representing your choice of the correct answer.

10-1. The products of combustion are:

(a) C, SO_2, H_2O, N_2, H

(b) CO_2, NaCe, $AgNO_3$

(c) CO_2, CO, O, N, H_2O

(d) AgCl, KCl, C_6, H_5

10-2. In practice, a gas burner should be adjusted to produce what percent of CO_2 in the flue gas?

(a) 10 to 12

(b) $8\frac{1}{4}$ to $9\frac{1}{2}$

(c) 4 to $7\frac{1}{2}$

(d) 20 to $25\frac{1}{2}$

10-3. An instrument to check CO_2 is a(an):

(a) Flue gas analyzer

(b) True spot tester

(c) Micrometer

(d) Psychrometer

10-4. A burner should be run for a minimum of how many minutes before combustion tests are made?

(a) 1 to 2

(b) 5 or 10

(c) 15 to 20

(d) 30 to 45

10-5. What is the maximum number of holes that should be made in the flue for combustion tests?

(a) One

(b) Two

(c) Three

(d) Four

10-6. Draft at the burner (operating) should be:

(a) Negative

(b) Positive

(c) Neutral

10-7. The proper reading for the smoke test for a new oil furnace should be:

(a) Number 1 or number 2

(b) Number 3 or number 4

(c) Number 5 or number 6

(d) Number 7 or number 8

10-8. In making the CO_2 test on an oil furnace, the bulb should be collapsed how many times?

(a) Six

(b) Eight

(c) Twelve

(d) Eighteen

10-9. Combustion efficiency should be:

(a) 50% or better

(b) 60% or better

(c) 75% or better

(d) 85% or better

10-10. If the combustion efficiency is 80%, what should the stack loss be?

(a) 80%

(b) 20%

(c) 70%

(d) 30%

11

BASIC ELECTRICITY DIAGRAMS AND ELECTRICAL LOADS

OBJECTIVES

After studying this chapter, the student will be able to:

- Identify the basic electrical symbols used in wiring diagrams for electrical load devices

ELECTRICAL DIAGRAMS

The greatest advances in the design and use of forced warm air heating equipment came after the invention of automatic control systems. It was then no longer necessary to operate the equipment manually. Automatic controls regulate the furnace to maintain the desired temperature conditions. Perhaps forced air heating owes its very beginning to the use of electricity to drive the fan motor. A knowledge of electricity and its application to domestic heating is essential for both the installer and service mechanic. Many technicians in the field today say that the majority of service problems are electrical in nature.

Along with an understanding of how electricity works, there is a need to understand the special terminology and symbols used in schematic wiring diagrams. A wiring diagram describes an electrical system. There are generally two types of diagrams for warm air heating units:

1. Connection
2. Schematic

A connection diagram shows the electrical devices in much the same way that they are actually positioned on the equipment (Figure 11-1). Straight lines are connected to the electrical terminals to show the paths the electric current will follow.

The schematic diagram (Figure 11-1) uses symbols to represent the electrical devices and the legend describes these symbols (Figure 11-2). The arrangement of the symbols in the diagram, using "ladder"-type connecting lines, indicates how the system works. It shows the sequence of operation in numerical order as it occurs. This diagram is essential to the service technician in troubleshooting problems in heating systems.

This unit prepares students to read a schematic wiring diagram.

SCHEMATIC WIRING DIAGRAM

To read a schematic wiring diagram, it is necessary to have a knowledge of:

- Basic electricity
- Electrical circuits
- Types of electrical devices and the symbols representing them

Basic Electricity

Students must have a knowledge of basic electricity for an understanding of its use in a forced warm air heating system. A service technician must have electrical as well as mechanical training in order to diagnose and correct service problems.

CURRENT:

Current is a term used in electricity to describe the rate of electrical flow. Electricity is believed to be the movement of *electrons*, small electrically charged particles, through a *conductor*. A conductor is a type of metal through which electricity will flow under certain conditions. These conditions will be described later in the unit. Current flow is measured in amperes with an electrical instrument called an ammeter.

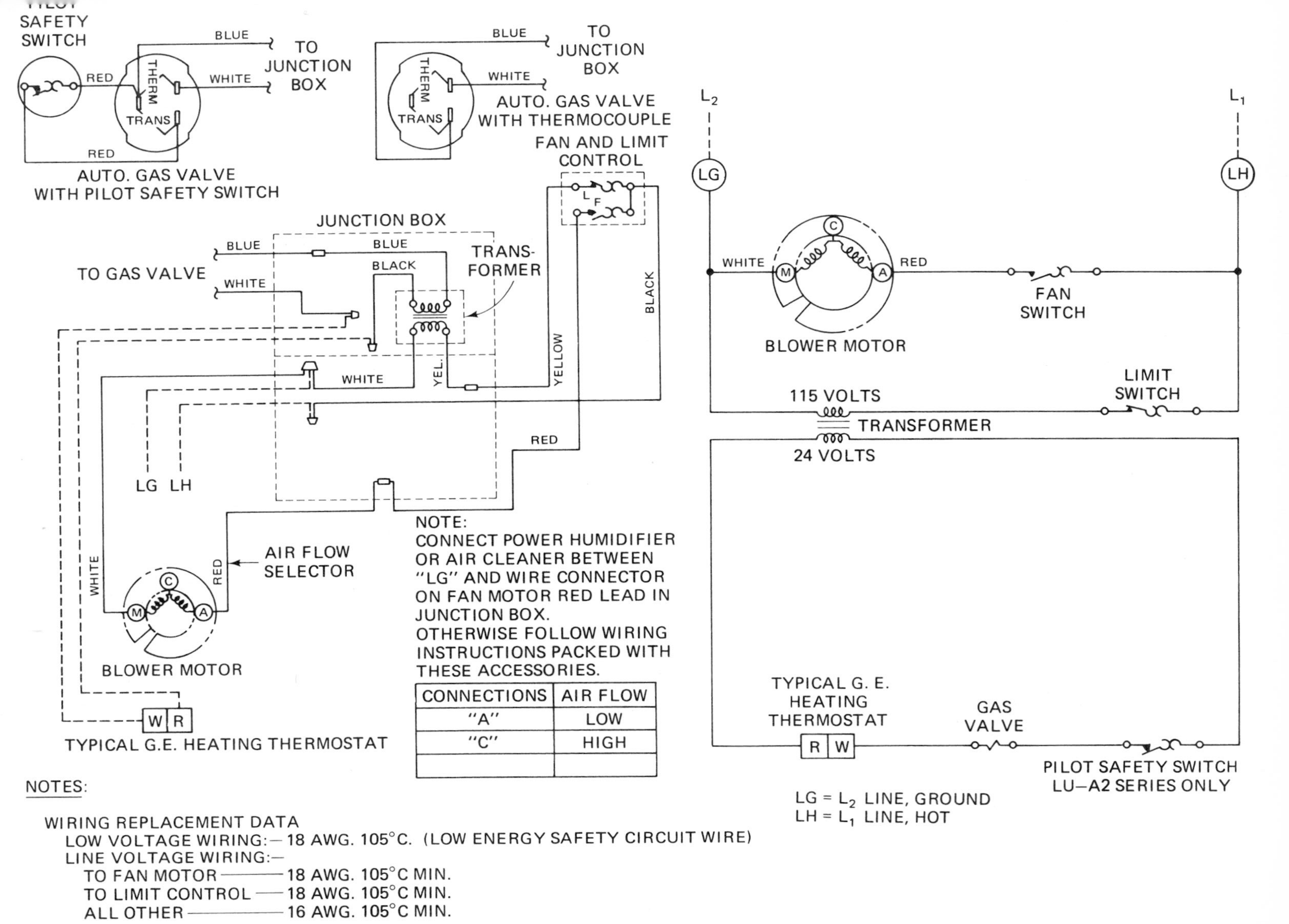

CONNECTIONS	AIR FLOW
"A"	LOW
"C"	HIGH

FIGURE 11-1 Connection and schematic wiring diagrams (Courtesy, General Electric Company).

SYMBOLS

Symbol	Meaning
Millivolt, 24V., Line V.	Factory Wiring
24V., Line V.	Field Wiring
	Magnetic Coil
	Motor Winding
	Transformer
	Capacitor
	Temperature Sensing Switch (Limit)
	Temperature Sensing Switch (Fan)
	Resistor or Heating Element
	Tapped Resistor
	Temperature Sensing Switch
	Relay Contact (S.P.S.T.)
	Relay Contact (S.P.D.T.)
	Wire Connector (Permanent)
	Wire Connector (Pull Apart)
	Wire Nut
	Ground
	Thermocouple
	Fusible Link
	Fuse
	Chassis Connection
	Circuit Breaker

LEGEND

AGV	Automatic Gas Valve
AR	Auxiliary Relay
AS, BS, CS	Safety Relay
BK-1	Wire is coded "Black 1"
C	Common
CB	Circuit Breaker
CF	Fan Capacitor
CN	Wire Connector
CN-2	Other end goes to "CN 2"
F	Indoor Fan Relay
FL	Fusible Link
FR	Relay, Fan Relay
FST	Fan Switch – Thermal
FU	Fuse
FU-2	Other end goes to "FU-2"
GRD	Ground
HTR	Heater
HVTB	High Voltage Terminal Board
ICB-A	Other end goes to no. 1 circuit breaker "A" side
ICB-B	Other end goes to no. 2 circuit breaker "B" side.
LG	Line Ground (Neutral)
LH	Line Hot
LTS	Thermal Sequencer
LVTB	Low Voltage Terminal Board
MCB	Modulating Control Board
MT	Motor
N.C.	Normally Closed
N.O.	Normally Open
PF	Plug Female
PM	Plug Male
RD-2	Wire is coded "Red 2"
RD-3	Wire is coded "Red 3"
RD-7	Wire is coded "Red 7"
SCR	Silicon Controlled Rectifier
SR	Staging Relay
TC	Thermocouple
TCO	Temperature Limit Switch
TCO-1	Other end goes to "TCO 1"
TNS	Transformer
TS	Heating & Cooling Thermostat

COLOR CODES

COLOR OF WIRE
BK/BL Black Wire With Blue Marker
COLOR OF MARKER

BK	Black	GR	Green	RD	Red
BL	Blue	GY	Gray	WH	White
BR	Brown	OR	Orange	YL	Yellow

FIGURE 11-2 Symbols-legend-color codes for Figure 11-1 (Courtesy, General Electric Company).

There are two types of electric current: *direct current* and *alternating current.* In *direct current* (dc), the electrons move through the conductor in only one direction. This is the type of current produced by a battery. In *alternating current* (ac), the electrons move first in one direction and then in the other, alternating their movement usually 60 times per second (called 60 cycles or 60 hertz). Alternating current is the type available for residential use and is supplied by an electric power company.

ELECTRICAL POTENTIAL:

Electrical potential is the force that produces the flow of electricity. It is similar to the action of a pump in a water system which causes the flow of water. The battery in a flashlight is a source of electrical potential. The unit of electrical potential is a *volt* (V). Volts are measured with a voltmeter.

RESISTANCE:

Resistance is the pressure exerted by the conductor in restricting the flow of current. In a water system, the flow would be limited by the size of the pipe. In an electrical system, the flow of current is limited by the size of the conducting wire or by the electrical devices in the circuit. Resistance in an electrical circuit can be a means of converting electrical energy to other forms of energy, such as heat, light, or mechanical work. The unit of resistance is the *ohm* (Ω). Ohms are measured with an ohmmeter.

POWER:

Power is the use of energy to do work. An electrical power company charges its customers for the amount of energy used. The unit of power is the *watt* (W). Watts are measured with a wattmeter.

ELECTROMAGNETIC ACTION:

Magnetic action is the force exerted by a magnet (Figure 11-3). A magnet has two poles: north and south. Like poles repel while unlike poles attract. The force of a magnet is dependent on its strength and the distance between the magnet and the metal it is affecting.

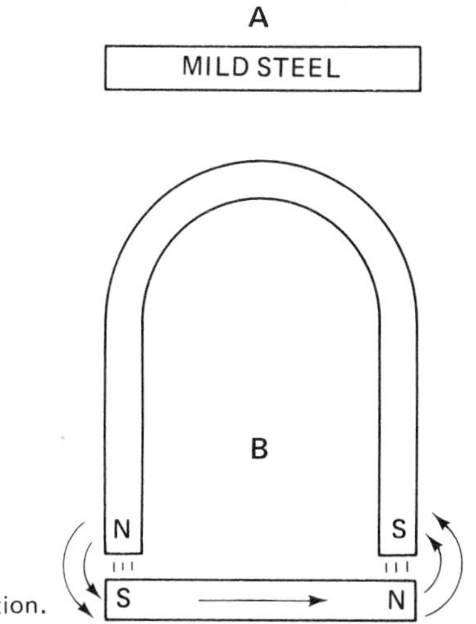

FIGURE 11-3 Magnetic action.

Electromagnetic action is the force exerted by the magnetic field of a magnet (Figure 11-4). Current flowing through a conductor creates a magnetic field around the conductor. This action is extremely useful in the construction of electrical devices.

If an insulated conductor is coiled and alternating current is passed through it, two important effects are produced:

1. A magnetic effect with directional force is produced in the field inside the coil. This force can move a separate metal plunger. The resultant action is called the *solenoid effect.* A solenoid is shown in Figure 11-5
2. A magnetic field effect can transfer current to a nearby circuit, causing current to flow. This is called *electromagnetic induction* (Figure 11-6)

The solenoid effect is used to operate switches and valves. The electromagnetic induction effect is used to operate motors and transformers.

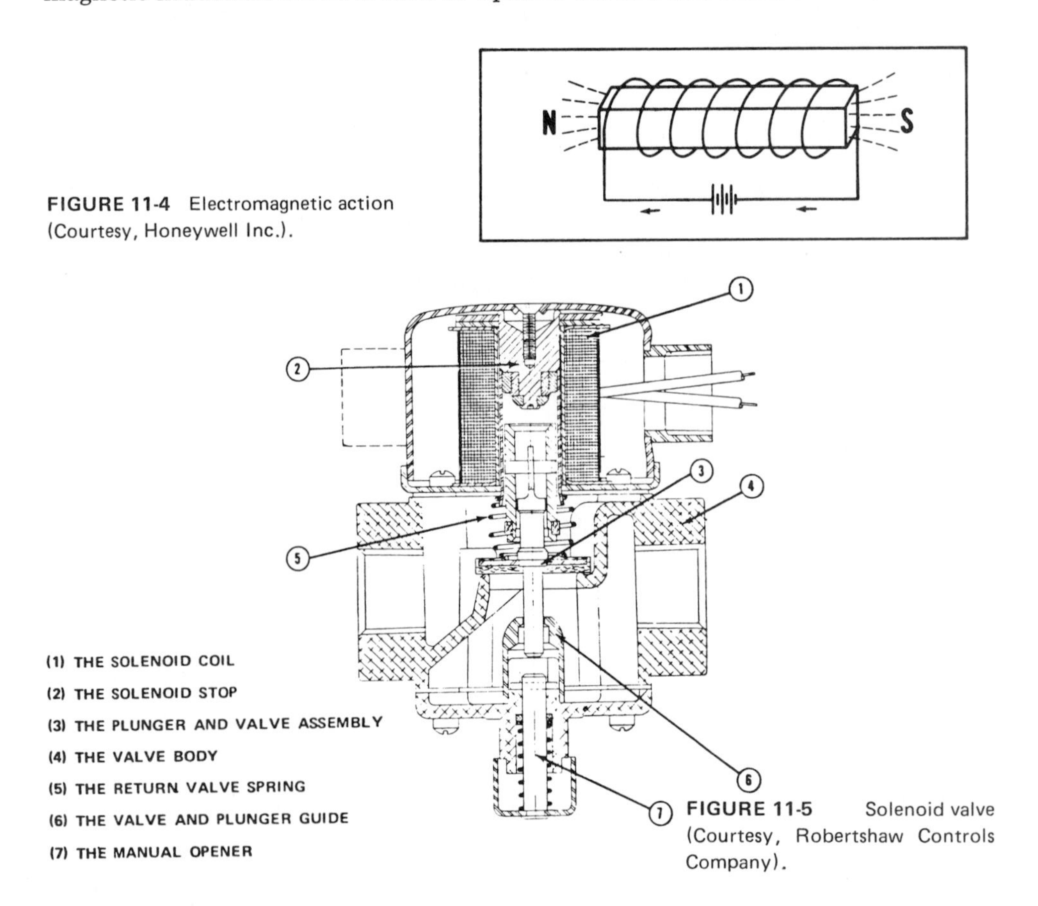

FIGURE 11-4 Electromagnetic action (Courtesy, Honeywell Inc.).

FIGURE 11-5 Solenoid valve (Courtesy, Robertshaw Controls Company).

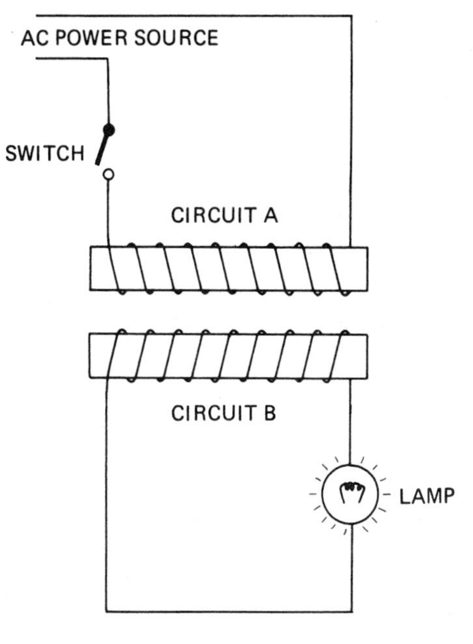

FIGURE 11-6 A flow of current in circuit A induces a flow of current in circuit B by electromagnetic induction.

CAPACITANCE:

Capacitance is the charging and discharging ability of an electrical device, called a *capacitor*, to store electricity. Capacitors are made up of a series of conductor surfaces separated by insulation (Figure 11-7). The unit of capacitance is the *farad* (F). Most capacitors for furnace motors are used to increase the starting power of the motor. These capacitors are rated in microfarads (μF).

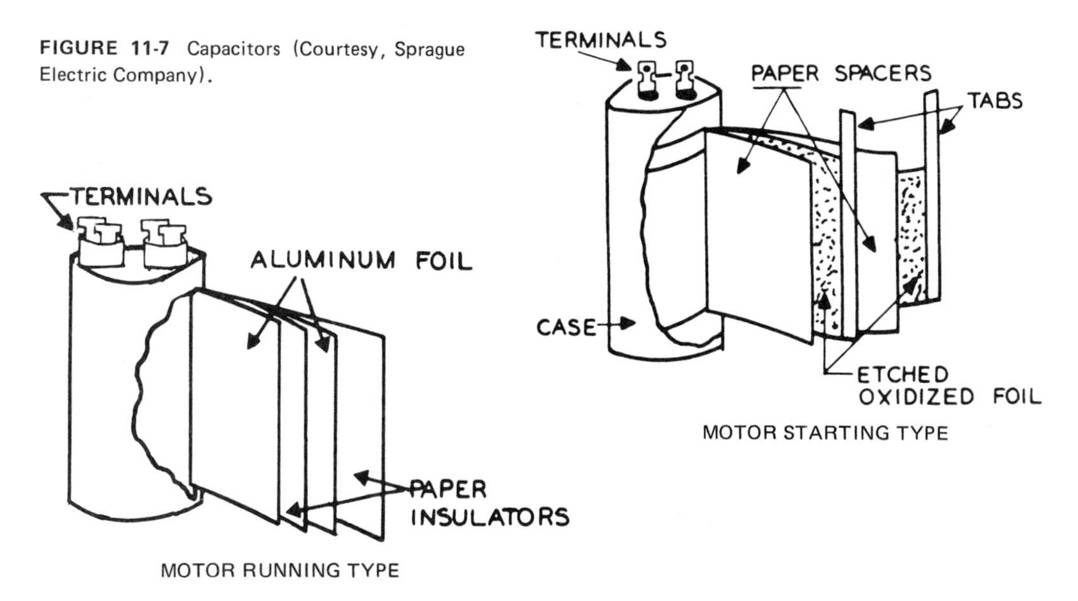

FIGURE 11-7 Capacitors (Courtesy, Sprague Electric Company).

CIRCUITS

Electrical devices are used in *circuit* to perform various functions. A circuit is an electrical system which provides a:

- Source of power
- Path for power to follow
- Place for power to go (load)

An electrical circuit is shown in Figure 11-8.

The source of power can be a battery that stores direct current or a connection to a power company's generator that supplies alternating current. The path is a continuous electrical conductor (such as copper wire) that connects the power supply to the load and the load back to the power supply. The load is a type of electrical resistance that converts power to other forms of energy such as heat, light, or mechanical work.

A switch can be inserted into the circuit to connect (make) or disconnect (break) the flow of current as desired, to control the operation of the load. Thus, a light switch is used to turn a light on and off by making or breaking its connection to the power supply.

A single-phase power supply, such as that used for heating equipment, always consists of two wires. For 120-V ac power, one wire is called *hot* (usually black) and the other wire is called *neutral* (usually white). The hot

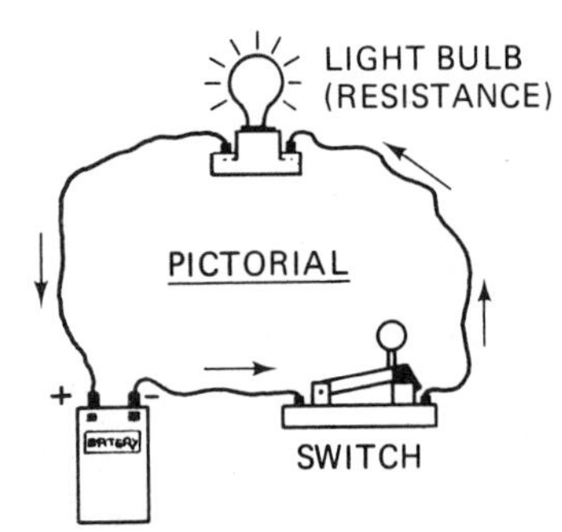

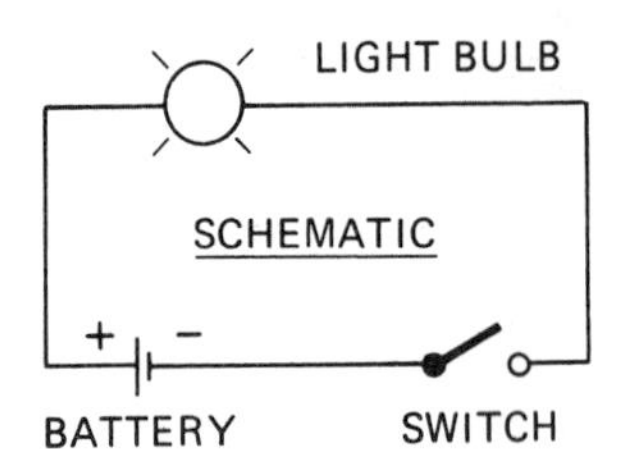

FIGURE 11-8 An electrical circuit.

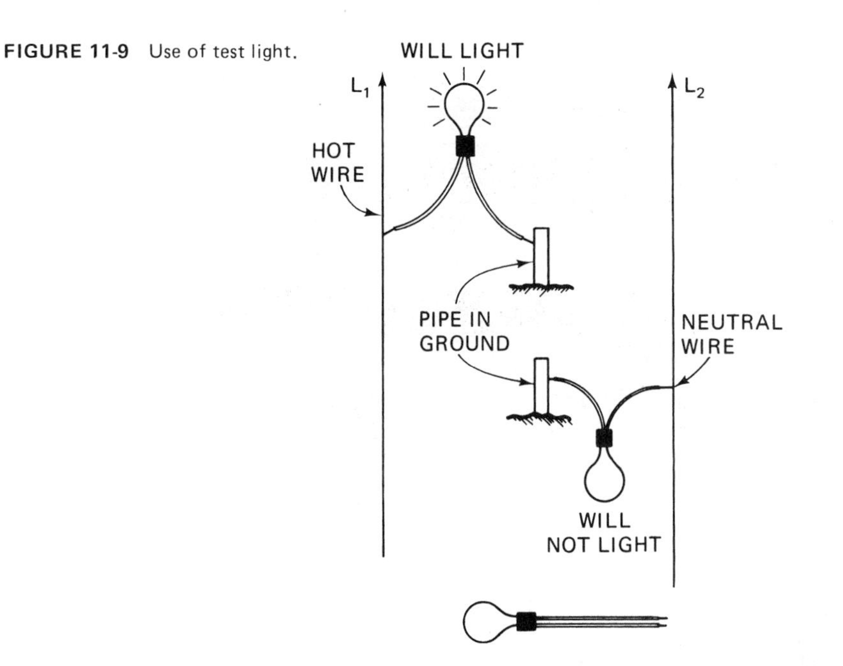

FIGURE 11-9 Use of test light.

wire can be found by using a test light attached to two wire leads (Figure 11-9). Touch one lead to the hot wire and the other to ground (a metal pipe inserted in the ground) and the bulb will light. Touch one lead to the neutral wire and the other to the ground and the bulb will not light.

In a circuit using ac power, the assumption is made that power moves from the hot wire through the load to the neutral wire (although the current does alternate). This assumption is made only to simplify tracing circuits and locating switches.

> **CAUTION:** Switches are always placed between the hot wire and the load; not between the load and the neutral wire. Thus, when the switch is open (disconnecting the circuit) it is safe to work on the load. If the switch were on the neutral side of the load touching the load it could complete the circuit to ground. This dangerous condition could cause current to flow through a person's body with serious, or fatal, results.

ELECTRICAL DEVICES

Electrical devices are chiefly of two types (Figure 11-10):

1. Loads
2. Switches

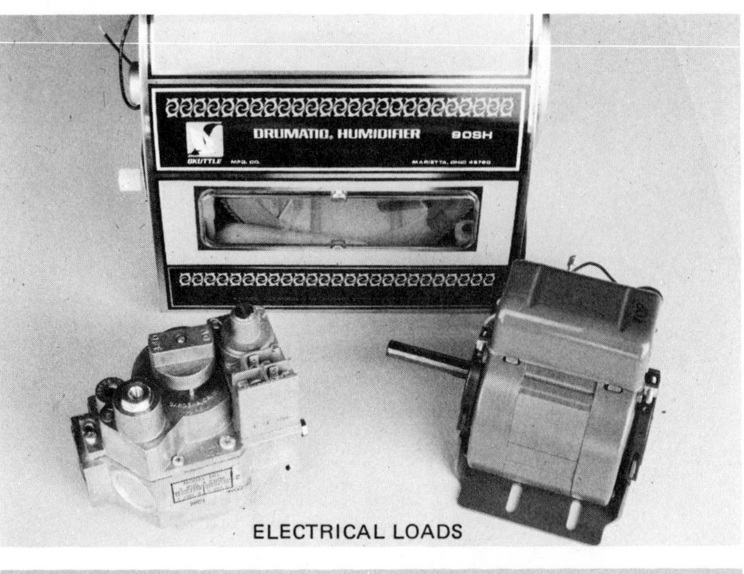

FIGURE 11-10 Loads and switches.

Loads have resistance and consume power. Loads usually transform power into some other form of energy. Examples of loads are motors, resistance heaters, and lights.

Switches are used to connect loads to the power supply or to disconnect them when they are not required.

Loads and Their Symbols

The common loads used on heating equipment are described below, along with the symbol used to represent them in the schematic wiring diagram.

MOTOR:

The electrical motor is usually considered the most important load device in the electrical system. It can be represented in two ways: by a large circle or by the internal wiring (Figure 11-11).

To better understand the electrical motor application, refer to the legend on the wiring diagram. The symbol may represent a fan motor, an oil burner motor, or an automatic humidifier motor. When current flows through the motor, the motor should run.

SOLENOID:

The second most important device is the solenoid (Figure 11-12). When the current flows through the solenoid (coil of wire), magnetism is created. The solenoid is a device designed to harness and use magnetism. It is most frequently used to open and close switches. It is also used to open and close valves.

It is common practice to use letters under the symbol to abbreviate the name of the device and to provide reference to the legend. For example, GV under the symbol would be shown in the legend to mean gas valve.

FIGURE 11-11 Electric motor and symbols (Courtesy, Essex Group, Controls Division, Steveco Products, Inc.).

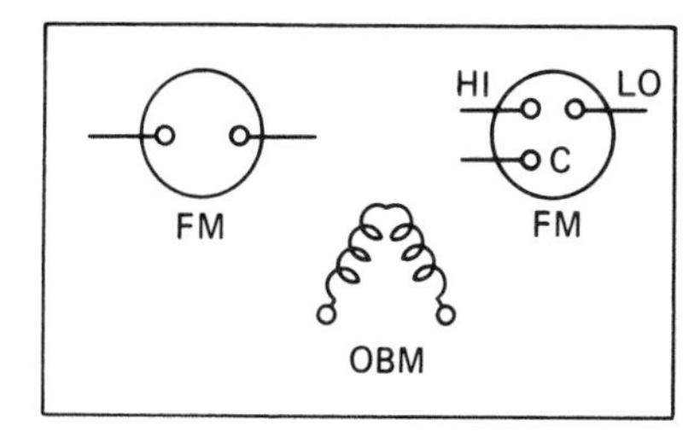

LEGEND

FM = FAN MOTOR

FM = FAN MOTOR, MULTIPLE SPEED

OBM = OIL BURNER MOTOR

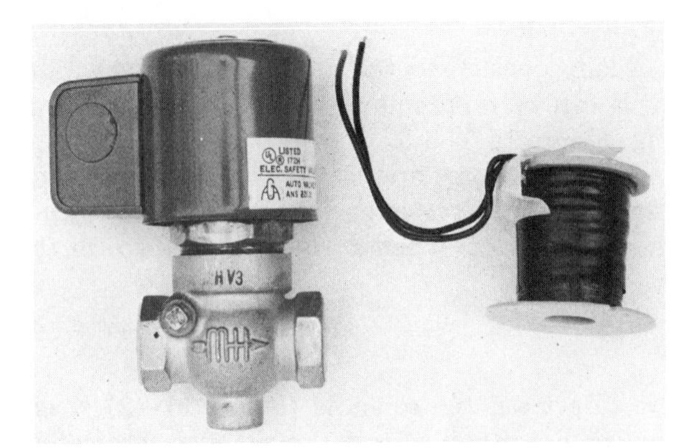

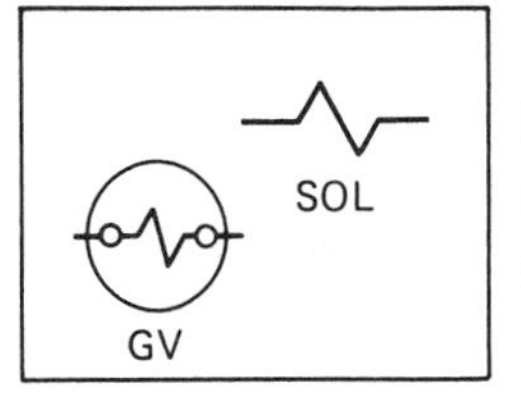

LEGEND
SOL = SOLENOID (RELAY COIL)
GV = SOLENOID (GAS VALVE)

FIGURE 11-12 Solenoid and symbols.

RELAY:

A relay is a useful application of a solenoid. By flowing current through the solenoid coil, one or more mechanically operated switches can be opened or closed. The solenoid coil is located in one circuit, while the switches are usually in separate circuits.

The relay is identified by the symbols shown in Figures 11-13 and 11-14. There at least two parts to the relay: the coil and the switch (or switches). The switch may be in one part of the diagram and the coil in another. The two symbols are identified as belonging to the same electrical device by the letters beneath them.

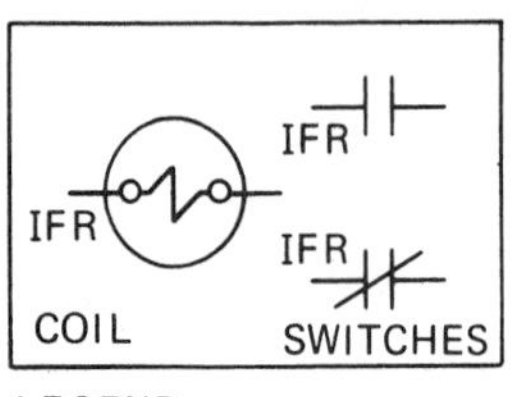

LEGEND
IFR = INDOOR FAN RELAY

FIGURE 11-13 (Courtesy, Essex Group, Controls Division, Steveco Products, Inc.).

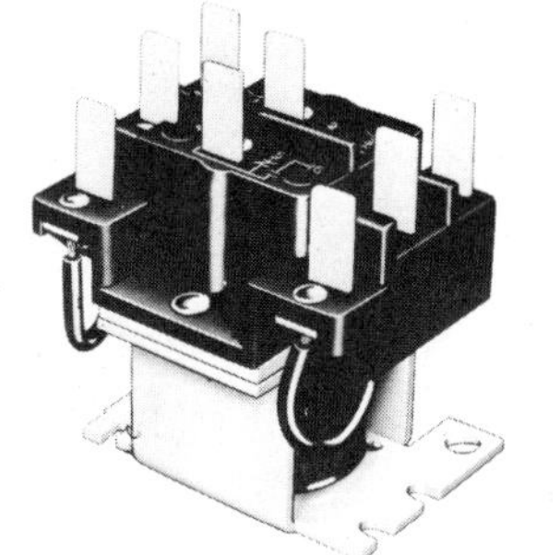

RELAY SWITCH AND SYMBOLS

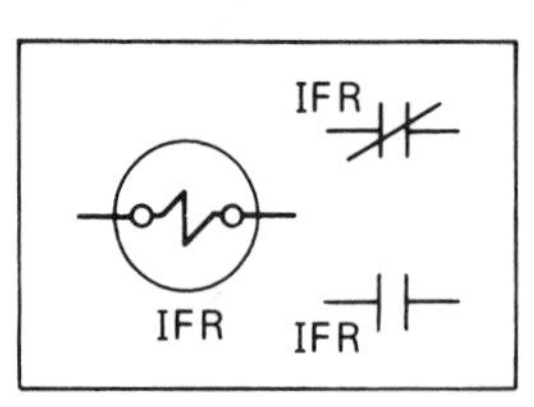

LEGEND
IFR = INDOOR FAN RELAY

FIGURE 11-14 (Courtesy, Essex Group, Controls Division, Steveco Products, Inc.).

When the switch has a diagonal line across it, it is a closed switch; without the diagonal line, the switch is open. All wiring diagrams show relay switches in their normal position, the position when no current is applied to the solenoid coil. Thus, in a diagram, where no current is flowing through the coil, the open switches are called *normally open* (n.o.) and the closed switches are called *normally closed* (n.c.). When current is applied to the coil, all of the related switches change position (Figure 11-14).

There are several types of relays used in heating work. They differ in the number of n.o. and n.c. switches.

RESISTANCE HEATER:

Another form of load device commonly used is the resistance heater. In the resistance heater, electricity is converted to heat. Heat in an electric furnace is produced by electricity. Heat is also used to control switches. The higher the resistance, the greater the amount of heat that is produced. The symbol for a resistance heater is a zigzag line (Figure 11-15). A letter is used under the heater symbol to designate its use as indicated in the legend.

HEAT RELAY:

The symbols and letters for a heat relay are shown in Figure 11-16. In heating circuits, particularly for electric heating, resistance heaters are used to operate switches. The advantage of this type of relay is that it provides a time delay in operating the switch.

FIGURE 11-15 Resistance heater and symbols.

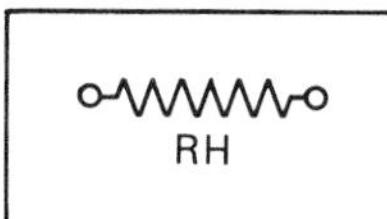

LEGEND

RH = RESISTANCE HEATER

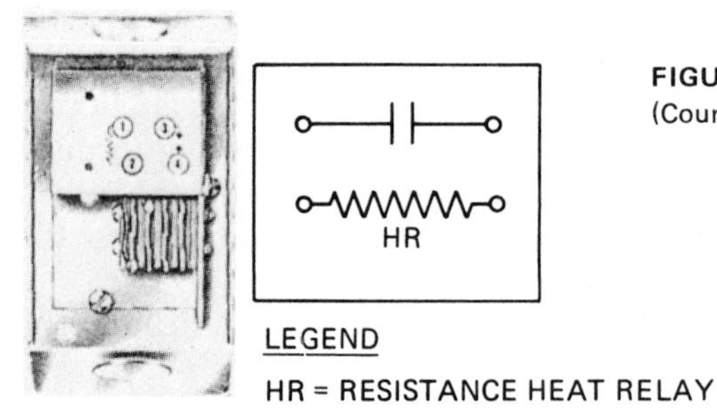

FIGURE 11-16 Heat relay and symbols (Courtesy, Honeywell Inc.).

When current is supplied to the heater, it heats up a bimetal element located in another circuit. Bimetal elements are made by bonding together two pieces of metal that expand at different rates. As the bimetal heats its shape is changed, thereby closing or opening a switch.

LIGHTS:

Lights are a type of load. They have resistance to current flow. A signal light is often used to indicate an electrical condition that cannot otherwise be readily observed. The color of the light is often indicated by a letter on the symbol (Figure 11-17).

TRANSFORMER:

In heating systems, it is often desirable to use two or more different voltages to operate the system. The fan must run on line voltage (usually 120 V), but the thermostat circuit (control circuit) can often best be run on low voltage (24 V). A transformer is used to change from one voltage to another. The legend and symbols for transformers are shown in Figure 11-18.

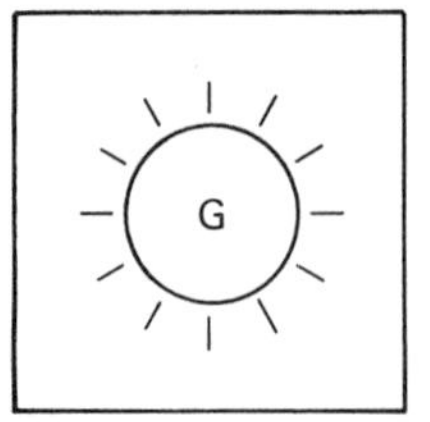

LEGEND

G = GREEN LIGHT

FIGURE 11-17 Light and symbol.

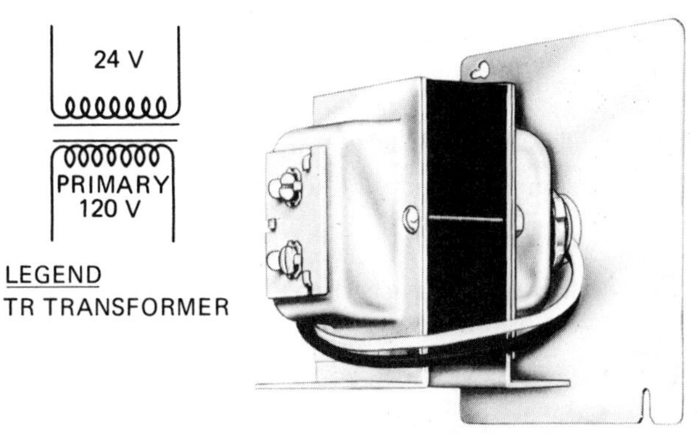

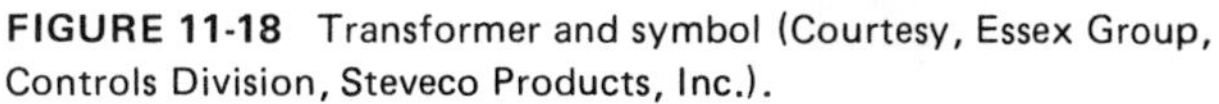

FIGURE 11-18 Transformer and symbol (Courtesy, Essex Group, Controls Division, Steveco Products, Inc.).

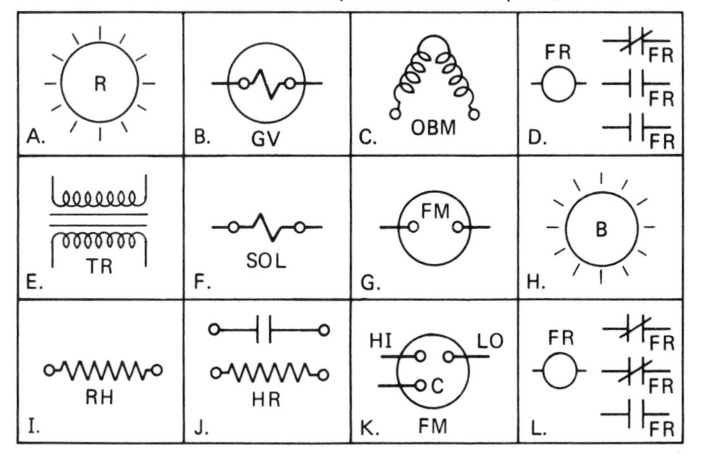

FIGURE 11-19

REVIEW

Match the symbols in Figure 11-19 with the proper descriptions below. (Note that there are more descriptions than there are symbols.)

1. Transformer
2. Heat relay
3. Thermostat
4. Two-speed fan motor
5. Three-speed fan motor
6. Gas valve

7. Red light
8. Blue light
9. Single-speed fan motor
10. Resistance heater
11. Limit control
12. Relay with one n.c. switch and two n.o. switches
13. Relay with two n.c. switches and one n.o. switch
14. Solenoid relay coil
15. Humidistat
16. Oil burner motor

12

BASIC ELECTRICITY AND ELECTRICAL SWITCHES

OBJECTIVES

After studying this chapter, the student will be able to:

- Identify the basic electrical symbols used in wiring diagrams for electrical switch devices

ELECTRICAL SWITCHES

Loads perform many functions. However, switches perform only one: to start and stop the flow of electricity.

Electrical switches are classified according to the force used to operate them: manual, magnetic (solenoid), heat, light, or moisture.

The terms normally open and normally closed refer to the position of the switch with no operating force applied. All wiring diagrams show the position of the switches when the operating solenoid (relay coil) or mechanism is de-energized.

A thermostat is assumed to be a n.o. switch. A cooling thermostat makes (closes) an electrical circuit upon a rise in temperature.

A heating limit switch is considered a n.c. switch since it is normally closed when the system is in operation, and opens only when excessively high temperatures are reached in the furnace.

It is important to identify the type of force that operates a switch. Only then can its normal position be accurately determined.

The simplest type of switch is one that makes (closes) or breaks (opens) a single electrical circuit. Other switches make or break several circuits. The switching action is described by:

- Number of poles (number of electrical circuits through the switch)
- The throw (number of places for the electrical current to go)

The following abbreviations are often used to designate the types of switching action.

SPST : Single-pole single-throw

SPDT : Single-pole double-throw

DPST : Double-pole single-throw

DPDT: Double-pole double-throw

These designations and their symbols are shown in Figure 12-1.

The common types of switches used for heating equipment controls are described, together with the symbols used to represent them.

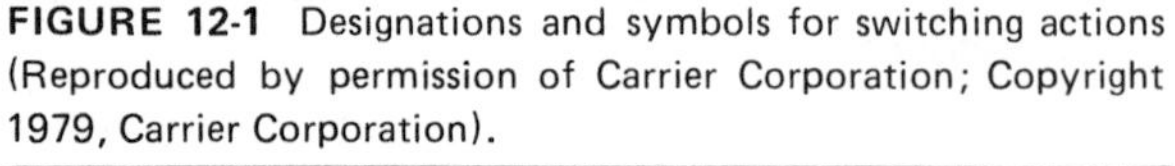

FIGURE 12-1 Designations and symbols for switching actions (Reproduced by permission of Carrier Corporation; Copyright 1979, Carrier Corporation).

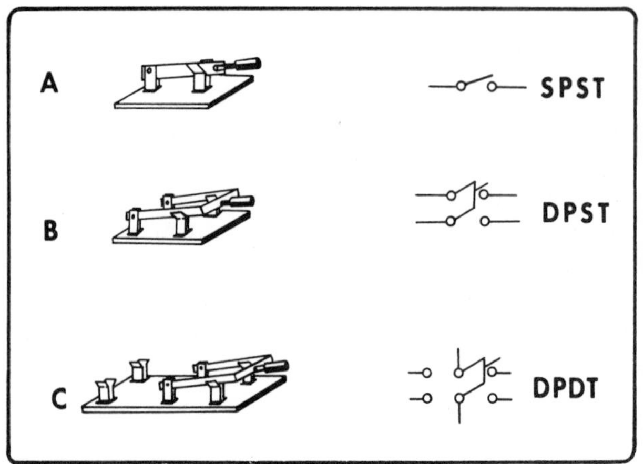

Manual Switches

Manual switches are operated by hand, and are usually used to disconnect the power supply when equipment is shut down for extended periods of time. For example, a room air conditioner is turned off when summer days are cool, or at the end of the cooling season.

Disconnect switches used on heating units are manual switches used to disconnect the power supply to the unit when servicing is required (Figure 12-2). These switches may also have fuses (heat-operated switches), which will be described later.

Manual switches are used when it is undesirable to operate the equipment automatically. For example, a bathroom heater can have a manual switch, thus providing extra bathroom heat only when needed.

Two switches are usually provided on room thermostats so that the occupant has the choice of operating the system on heating or cooling, or turning the unit off. There is a SPDT switch marked "heat–off–cool." A manual fan switch is also provided to permit the choice of fan operation, continuously or automatically. This is a DPST switch usually marked "on–auto." The fan can be run in the summer for ventilation and air movement without operating the unit on cooling. The switches are located on the sub-base of the thermostat.

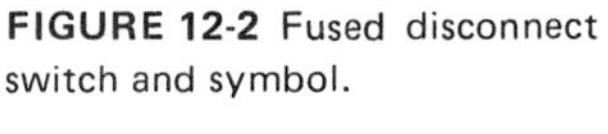

FIGURE 12-2 Fused disconnect switch and symbol.

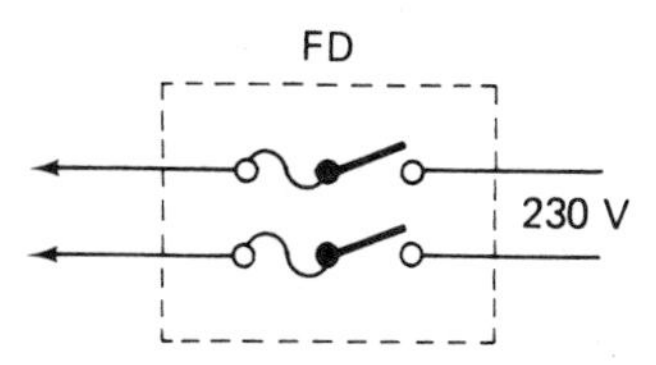

Magnetic (Solenoid) Switches

Magnetic or solenoid switches are electrically operated switches using the force of the magnetic effect to operate the switch. To produce the required amount of power to operate the switch, the wire is coiled, creating a strong magnetic effect on a metal core. This type of electrical mechanism, called a solenoid, is described in Chapter 11.

Magnetic or solenoid switches have various names, depending upon their use. Among these are relays (Figure 12-3), contactors (Figure 12-4), and starters (Figure 12-5).

Relays were described in Chapter 11. Contactors are relatively large electric relays used to start motors. Starters are also relatively large relays that include overload (excess current) protection. The National Electric Code specifies whether a motor requires a manual, contactor, or starter switch.

Solenoid switches may also be used to operate valves which regulate the flow of a fluid (liquid or vapor). An example of this type of valve is the gas valve, described in Chapter 9.

Some types of overloads can also be described as magnetic-type switches (Figure 12-10). An overload device protects a motor against excess current flow. Normal amounts of current will not energize the solenoid, but excess amounts of current will. When the solenoid is energized, it trips a mechanically interlocked switch that cuts off the power to the motor.

FIGURE 12-3 Relay switch, symbols, and switching action (Courtesy, Essex Group, Controls Division, Steveco Products, Inc.).

SINGLE POLE
RELAY (SPST)

DOUBLE POLE
RELAY (DPST)

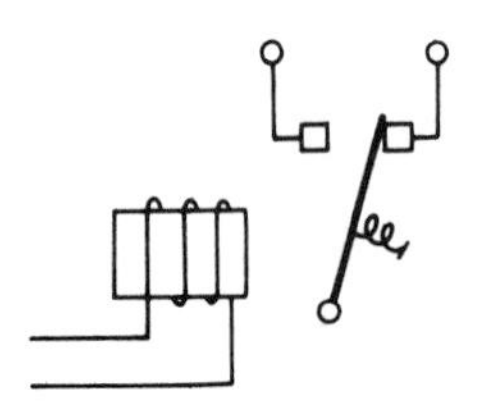

SINGLE POLE DOUBLE
THROW RELAY (SPDT)

DOUBLE POLE DOUBLE
THROW RELAY (DPDT)

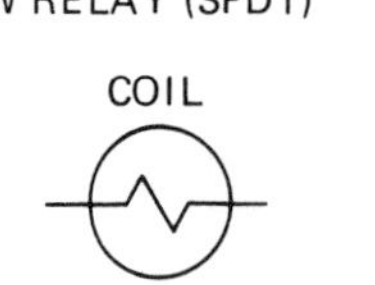

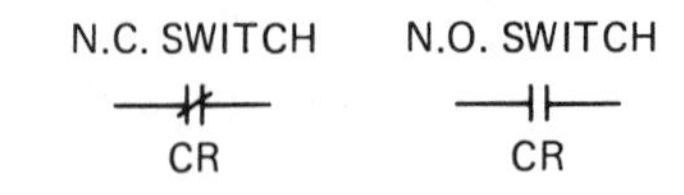

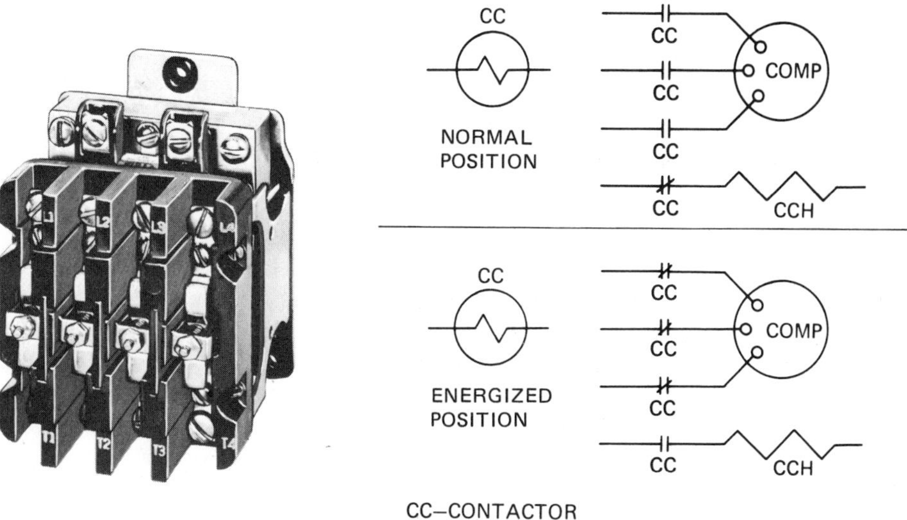

FIGURE 12-4 Contactor switch and symbol (Courtesy, Essex Group, Controls Division, Steveco Products, Inc.).

FIGURE 12-5 Starter switch and symbol.

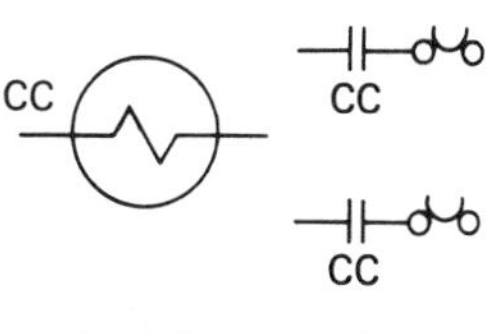

Heat-Operated Switches

There are many types of heat-operated switches. In all cases, heat is the force that operates the switch. These switches include thermostats, fan and limit controls, heat relays, fuses, overloads, and circuit breakers.

THERMOSTATS:

A heating thermostat (Figure 12-6) makes on a drop in room temperature. The intensity of heat reacts on the bimetal changing its shape, which in turn actuates a switch. The thermostat is usually shown as a normally open switch. Thermostats will be discussed in Chapter 16.

FAN AND LIMIT CONTROLS:

Fan and limit controls (Figure 12-7) have a bimetal sensing element that protrudes into the warm air passage of a furnace. The fan control makes on a rise in temperature. It is considered to be a normally open switch because with the furnace shut down, the fan control will have an open switch.

The limit control protects the furnace against excessively high air temperatures, turning off the firing devices when a predetermined temperature is reached. The limit control is diagrammed as a normally closed switch.

FIGURE 12-6 Thermostat switch and symbol (Courtesy, Honeywell Inc.).

CIRCUIT MAKES ON
TEMPERATURE DROP
FOR HEATING

FC L

FIGURE 12-7 Combination fan and limit control switch and symbol.

HEAT RELAYS:

Heat relays (Figure 12-8) are similar to magnetic or solenoid relays in function. However, the mechanical action takes place as a result of heat produced in an electric resistance. When the resistance coil is energized, heat is produced. This heat is applied to one or more bimetal elements. As the bimetal elements expand, they either break an electrical circuit (n.c.) or make an electrical circuit (n.o.).

The heat relay is a delayed-action switch. This means that the switch required some amount of time to change position. Depending on the construction, the time delay may be 15, 30, 45 or more seconds. This feature permits staging or sequencing loads (turning loads on automatically at different times). Since the inrush current to a load when first turned on is usually many times greater than its running current, sequencing is important. Sequencing permits using a smaller power service to the appliance. A sequencer is used on an electric furnace to turn on the electric heating elements in steps.

FUSES:

Fuses (Figure 12-9) are placed in an electrical circuit to cut off the flow of current when there is an overload or a short. An *overload* is current in excess of the circuit design. A *short* is a direct connection between the two wires of a power supply without having the current pass through a load. In either case, the fuse will heat and melt, breaking the circuit continuity and stopping the flow of current. Where the fault is overcurrent, the melting of the fuse takes place slowly. Where the fault is a short, the melting of the fuse takes place quickly.

FIGURE 12-8 Heat relay switch and symbol (Courtesy, Honeywell Inc.).

FIGURE 12-9 Fuse switch and symbol (Courtesy, Bussmann Manufacturing Division, McGraw-Edison Company).

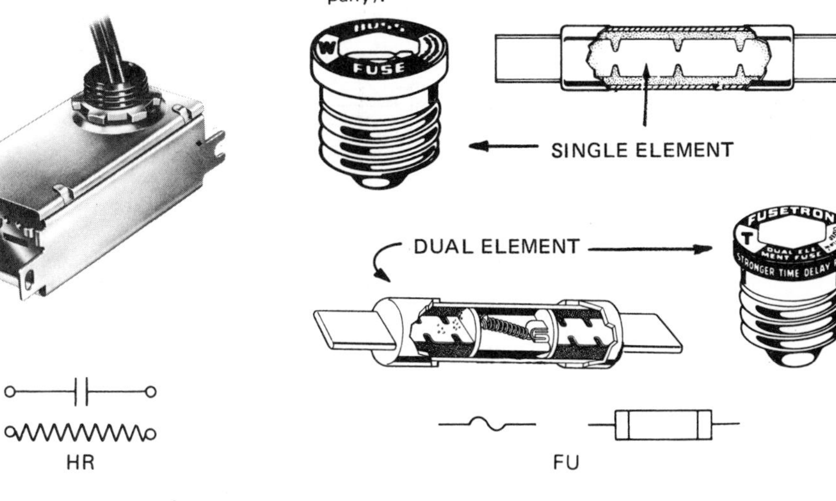

Some circuits require a time-delay fuse. An example is a motor circuit. The in rush of current in starting (a fraction of a second) is so great that an ordinary fuse would "blow" before the motor reached running speed. This initial motor current is called *locked rotor amperes* (LRA). The running amperes or *full-load amperes* (FLA) is much less. Because of the special design of delayed-action fuses, they may be sized on the basis of FLA.

According to the National Electric Code, fuses and wiring can be sized on the basis of 125% of FLA.

OVERLOADS:

Overloads (Figure 12-10) can be constructed in many ways. All overloads are designed to stop the flow of current when safe limits are exceeded. Overloads differ from fuses in that they do not have one-time usage as does a fuse. When a fuse melts, it must be replaced to return the circuit to operation. When an overload senses excess current, a switch is opened, breaking the flow of current. Some of these switches are automatically reset when the current returns to normal, others must be manually reset. An overload is considered a normally closed (n.c.) switch since it is closed when no current is flowing through the circuit and during normal operation of the equipment.

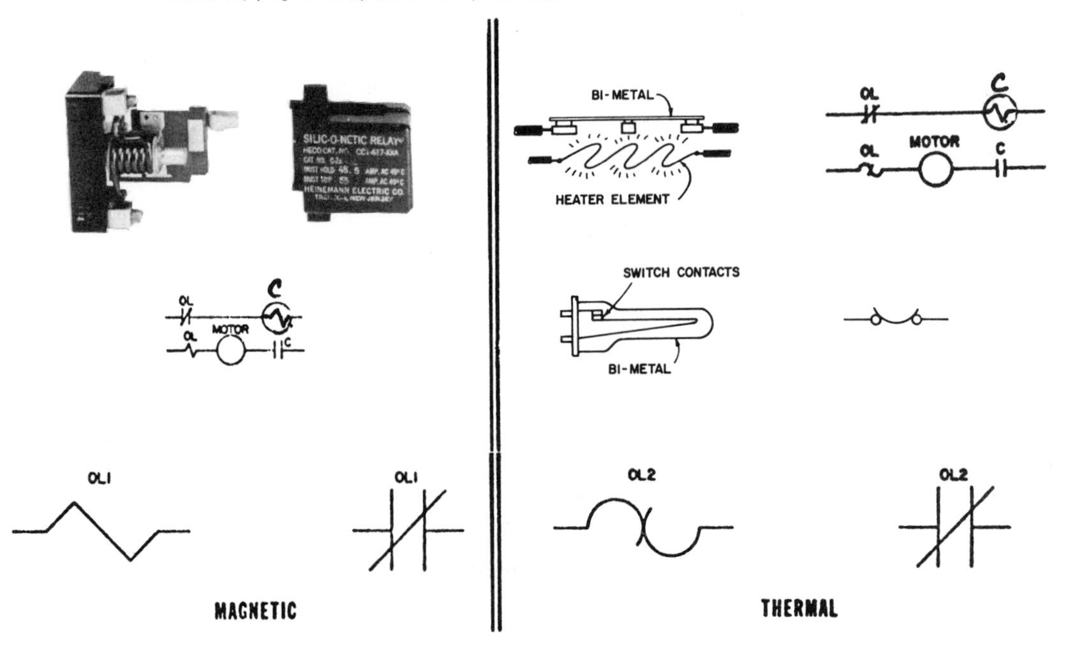

FIGURE 12-10 Overload switches and symbols (Reproduced by permission of Carrier Corporation; Copyright 1979, Carrier Corporation).

CIRCUIT BREAKERS:

The main power circuit to a heating furnace must be protected against excess current flow by a fuse or a circuit breaker (Figure 12-11). Either one of these switches will disconnect the power supply if the equipment draws excess current. The circuit breaker is a type of overload device placed in the power supply which will "trip" (open the circuit) in the event of excessive current flow. When the fault is corrected, the circuit breaker can be manually reset to restore the circuit to its original condition.

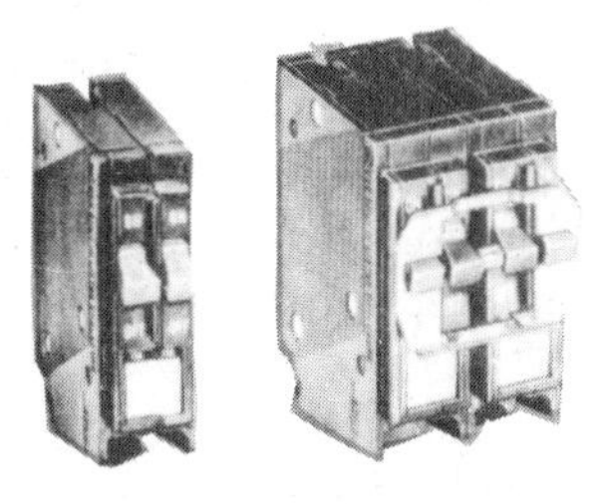

FIGURE 12-11 Circuit breaker switch (Courtesy, Sears, Roebuck and Company).

Light-Operated Switches

Some switches are activated by light. An example is the cad cell used on the primary control of an oil burner. *Cad cell* refers to a cadmium sulfate sensing element. In the presence of light, the electrical resistance of a cad cell is about 1000 Ω; in darkness, its resistance is about 100,000 Ω. The cad cell is located on the draft tube of an oil burner and senses the presence (or proof) of a flame. If the cad cell is in darkness, there is no flame; if the cad cell is lighted, the flame is proof that the burner has been started by the ignition system. It therefore acts as a safety device.

In the electrical circuit the cad cell is placed in series with a relay coil. If the cad cell does not sense adequate light from the flame the relay will not be energized. If the cad cell senses the flame, its resistance is reduced and the relay is energized, which allows the burner to continue running.

Moisture-Operated Switches

The presence of moisture in the air can be used to operate a switch. Certain materials, including human hair and nylon, expand when moist and contract when dry. These two materials are used to sense the relative humidity in the air. The change in the length of a strand of hair or nylon can be used to operate a suitable switch. The switch will turn a humidifier on or off to maintain the desired relative humidity in a space.

ELECTRICAL CIRCUITS

A circuit consists of a source of electrical current, a path for it to follow, and a place for it to go (electrical load). Switches are inserted in the path of the current to control its flow either manually or automatically. Electrical devices are connected in circuits to produce a specific resultant action, such as operating a fan motor or a firing device on a furnace.

There are two basic types of circuits: series and parallel. Electrical devices can be connected in either of these ways, or in a combination of both.

Referring to Figure 12-12, using 120-V alternating current as a power source, the two wires are termed L_1 (hot) and L_2 (neutral). The hot wire is usually black and the neutral wire is usually white. Most 120-V installations include a third wire (color coded green) for the earth ground. These systems require a three-prong plug and outlet.

The fuse inserted in the hot side of the line is usually incorporated in the switching device known as a *fused disconnect*. Where the load is some type of resistance, a thermal fuse is used. In case of a shorted circuit or overload, the fuse melts and automatically disconnects the power. Where the type of load is a motor, a time-delay fuse is used.

FIGURE 12-12 A typical electrical circuit, including a source of power, a path for the current to flow in and a load.

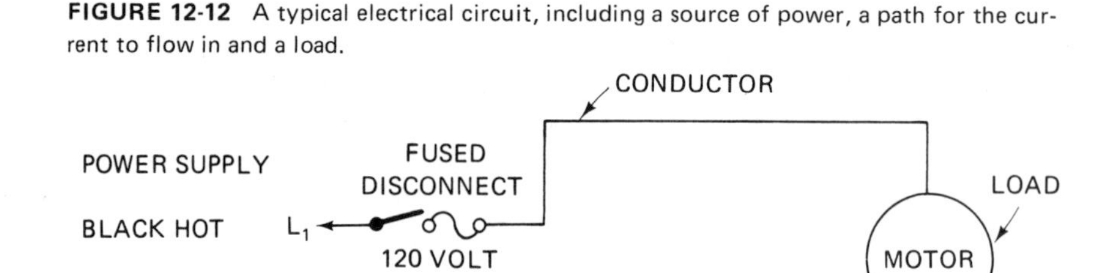

ELECTRICAL TERMS

A knowledge of certain electrical terms is necessary in working with electrical power. Most power used for warm air heating systems is alternating current. Some of the common terms used in describing alternating current are:

Volts: A measure of electromotive force (EMF) or pressure being supplied to cause the electrical current to flow

Amperes: A measure of the flow of current (electrons) through a conductor

Ohms: A measure of the resistance to current flow through a conductor

Watts: A measure of power consumed by an electrical load

Power Factor: The resulting fraction obtained by dividing watts by the product of volts times amperes

OHMS LAW

Ohm's law is used to predict the behavior of electrical current in an electrical circuit. Simply stated,

$$E = IR$$

where: E = electromotive force (EMF), volts

I = intensity of current, amperes

R = resistance, ohms

Ohms law can be stated in three ways (Figure 12-13). An example of the use of Ohm's law is shown in Figure 12-14. Given E = 120-V, R = 10 Ω. Using equation (2), $I = E/R$, the current flow in the circuit, is 12 A:

$$I = \frac{120\text{ V}}{10\ \Omega} = 12\text{ A}$$

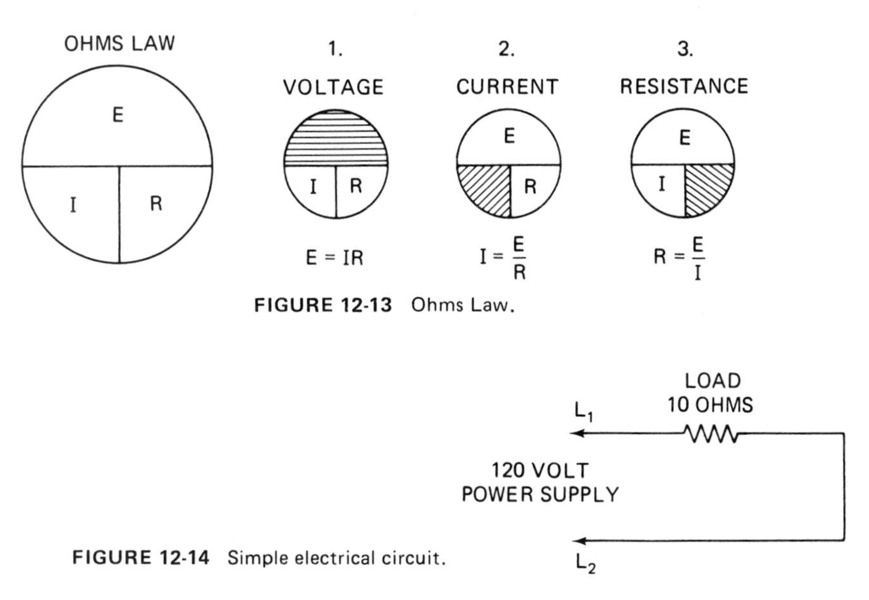

FIGURE 12-13 Ohms Law.

FIGURE 12-14 Simple electrical circuit.

A voltmeter can be used to measure the voltage between L_1 and L_2. An ohmmeter can be used to measure the resistance of the load, but should be used *only when the power is turned off.* An ammeter can be used to measure amperes.

SERIES CIRCUIT

A series circuit is one in which each resistance is wired end-to-end like a string of box cars on a train. In a series circuit the amperage stays the same throughout the circuit. The voltage is divided between the various loads. The total resistance of the circuit is the sum of the various resistances in the circuit. Thus, using Figure 12-15,

$$E_T = E_1 + E_2$$

$$I_T = I_1 = I_2$$

$$R_T = R_1 + R_2$$

Given $E_T = 120, R_1 = 10, R_2 = 10$, then

1. $R_T = R_1 + R_2 = 10 + 10 = 20\ \Omega$
2. $I_T = \dfrac{E_T}{R_T}$ (Ohm's law) $= \dfrac{120}{20} = 6$ A
3. $I_T = I_1 = I_2 = 6$ A
4. $E_1 = I_1 \times R_1 = 10 \times 6 = 60$ V
5. $E_2 = I_2 \times R_2 = 10 \times 6 = 60$ V

FIGURE 12-15 Series circuit.

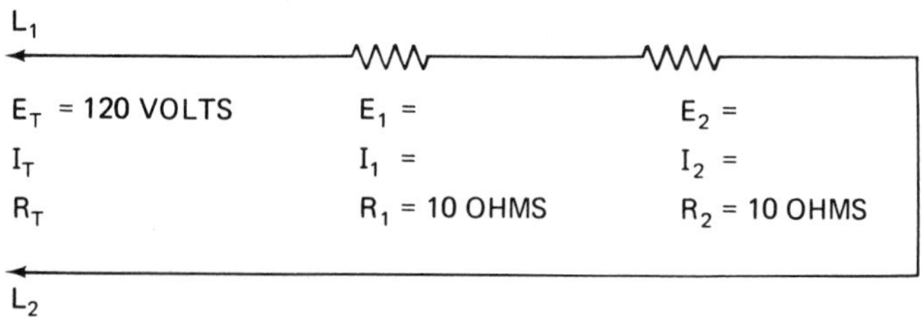

L_1

E_T = 120 VOLTS
I_T
R_T

E_1 =
I_1 =
R_1 = 10 OHMS

E_2 =
I_2 =
R_2 = 10 OHMS

L_2

PARALLEL CIRCUIT

In a parallel circuit each load provides a separate path for electricity and each path may have different current flowing through it. The amount is determined by the resistance of the load. The voltages for each load are the same. The total current is the sum of the individual load currents. The reciprocal of the total resistance is equal to the sum of the reciprocals of the individual resistances. Thus, using Figure 12-16,

$$E_T = E_1 = E_2$$

$$I_T = I_1 + I_2$$

$$\frac{1}{R_T} = \frac{1}{R_1} + \frac{1}{R_2}$$

or for two resistances

$$R_T = \frac{R_1 \times R_2}{R_1 + R_2}$$

Given $E_T = 120$, $R_1 = 10$, $R_2 = 10$, then

1. $E_T = E_1 = E_2 = 120\text{ V}$
2. $I_1 = \frac{E_1}{R_1} \text{(Ohm's law)} = \frac{120}{10} = 12\ \Omega$
3. $I_2 = \frac{E_2}{R_2} = \frac{120}{10} = 12\ \Omega$
4. $I_T = I_1 + I_2 = 12 + 12 = 24\text{ A}$
5. $R_T = \frac{E_T}{I_T} = \frac{120}{24} = 5\ \Omega$

   or

   $$R_T = \frac{R_1 \times R_2}{R_1 + R_2} = \frac{10 \times 10}{10 + 10} = \frac{100}{20} = 5\ \Omega$$

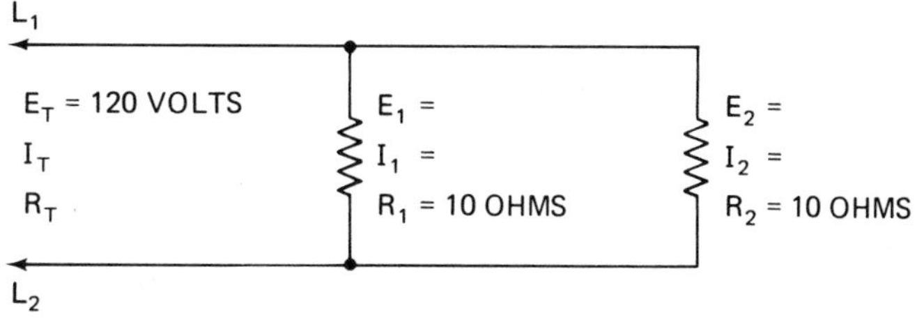

FIGURE 12-16 Parallel circuit.

REVIEW QUESTION

Match the symbols in Figure 12-17 with the proper descriptions below. (Note that there are more descriptions than symbols.)

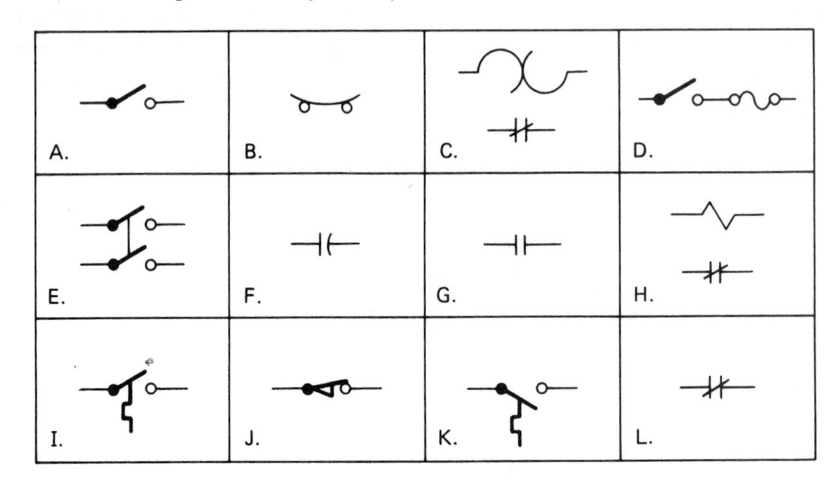

FIGURE 12-17 Symbols for review.

1. DPST switch
2. Magnetic overload (with separate switch)
3. SPST switch
4. Heating thermostat
5. Overload — bimetal type
6. Cooling thermostat
7. Limit switch
8. Thermal overload (with separate switch)
9. Pushbutton
10. Fused disconnect
11. DPDT switch
12. Thermal relay
13. SPDT switch
14. N.O. switch
15. N.C. switch
16. Contactor

13

SCHEMATIC WIRING DIAGRAMS

OBJECTIVES

After studying this chapter, the student will be able to:

- Read and construct a schematic electrical wiring diagram

PURPOSE OF A SCHEMATIC

A schematic wiring diagram (schematic) consists of a group of lines and electrical symbols arranged in "ladder" form to represent the individual circuits controlling or operating a unit. The lines represent the connecting wires. The electrical symbols represent loads or switches. The rungs of the ladder represent individual electrical circuits. The unit can be an electrical-mechanical device such as a furnace.

UNDERSTANDING A SCHEMATIC

To understand a schematic diagram, a student must know:

- The purpose of each electrical component
- Exactly how the unit operates, both mechanically and electrically
- The sequence of operation of each electrical component

Sequence refers to the condition where the operation of one electrical component follows another to produce a final result. It represents the order of events as they occur in a system of electrical controls.

All electrical systems must be wired by making connections to power and to each electrical device. A schematic diagram is important to assist a service technician in wiring the equipment, locating connections for testing circuits, and in analyzing the operation of the control system. Occasionally, however, a schematic diagram may be unavailable, so it is essential that a service technician be able to construct one.

Following is the legend to the schematic diagrams used in this chapter.

CR	Control relay
FC	Fan control
FD	Fused disconnect
FM	Fan motor
FR	Fan relay
FS	Fan Switch
GV	Gas valve
IFR	Indoor fan relay
L	Limit
LA	Limit auxiliary
L_1, L_2	Power supply
N.C.	Normally closed
N.O.	Normally open
T	Thermostat
TR	Transformer
Y, R, G, W	Terminals of thermostat

SCHEMATIC CONSTRUCTION

Connection and schematic wiring diagrams are shown in Figures 13-1 and 13-2.

To draw a schematic and to separate the individual circuits on the unit itself, a service technician must be able to trace (or follow) the wiring of each individual circuit. The method of tracing a circuit is as follows:

1. Start at one side of the power supply (L_1), go through the resistance (load), and return to the other side of the power supply (L_2). This is a *complete circuit*
2. If a technician starts at L_1 and goes through the resistance and cannot reach L_2 or returns back to L_1, this is an *open circuit*
3. If a technician starts at L_1 and reaches L_2 without passing through a resistance (load), this is a *short circuit*

RULES FOR DRAWING A SCHEMATIC (VERTICAL STYLE)

(a) Use letters on each symbol to represent the name of the component

(b) The names of all components represented by letters should be listed in the legend

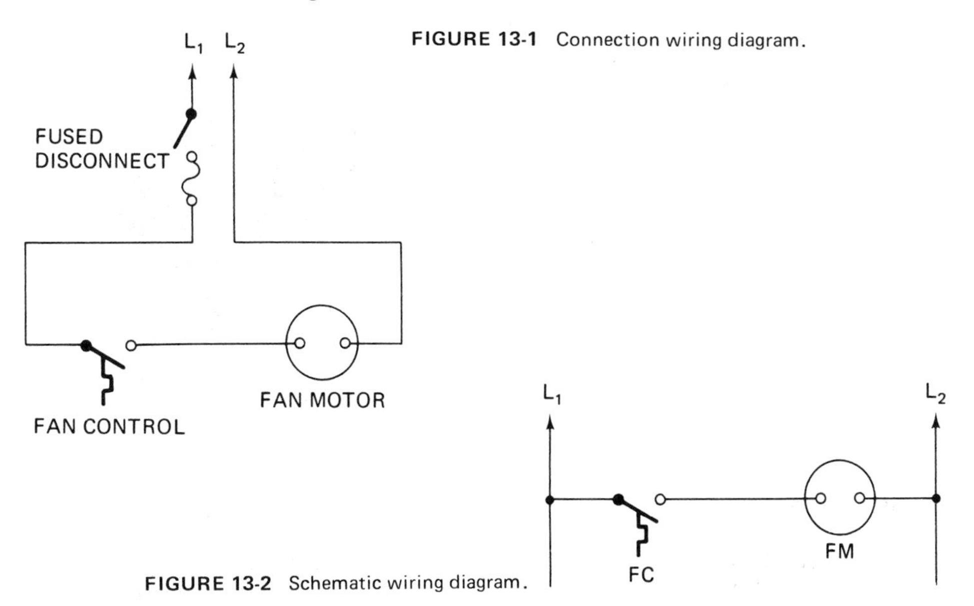

FIGURE 13-1 Connection wiring diagram.

FIGURE 13-2 Schematic wiring diagram.

(c) When using a 120-V-ac power supply,* show the hot line (L_1) on the left side and the neutral line (L_2) on the right side of the diagram

Diagrams can be made with the power supply lines horizontal or vertical.

(d) When using a 120-V power supply, the switches must be placed on the hot side (L_1) of the load

(e) Relay coils and their switches (Figure 13-3) should be marked with the same (matching) symbol letters

(f) Numbers can be used to show wiring connections to controls or terminals

(g) Always show switches in their normal (deenergized) position

(h) Thermostats with switching subbases, primary controls for oil burners, and other more complicated controls can be shown by terminals only in the main diagram. Subdiagrams are used when necessary to show the internal control circuits

(i) It is common practice to start the diagram showing line-voltage circuits first and low voltage (control) circuits second (Figure 13-4)

FIGURE 13-3 Relay coils and switches.

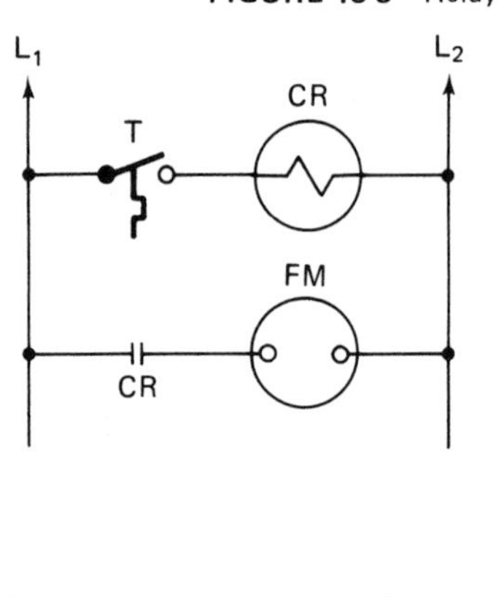

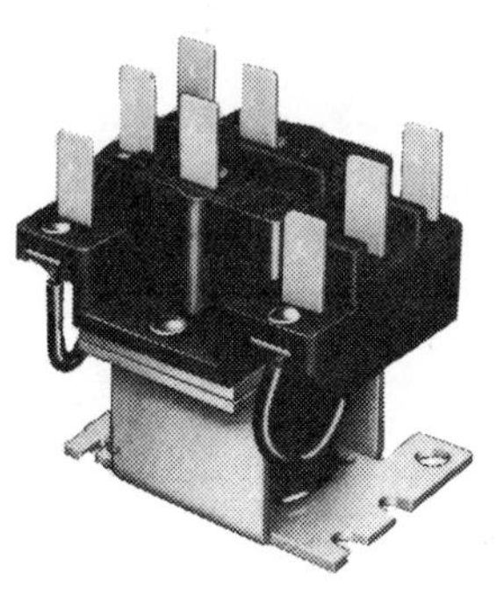

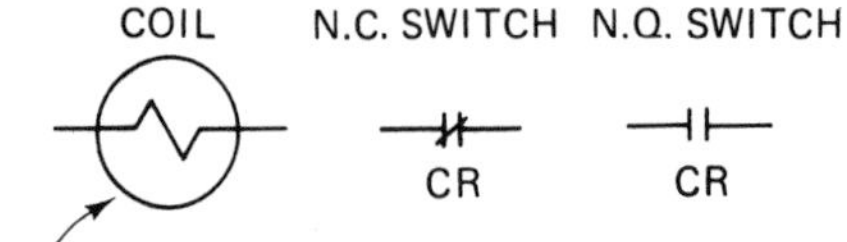

* In 208-, 230-, 240-, or 440-V single-phase circuits, there are two hot wires and no neutral wire. In these circuits switches can be placed on either side of the load.

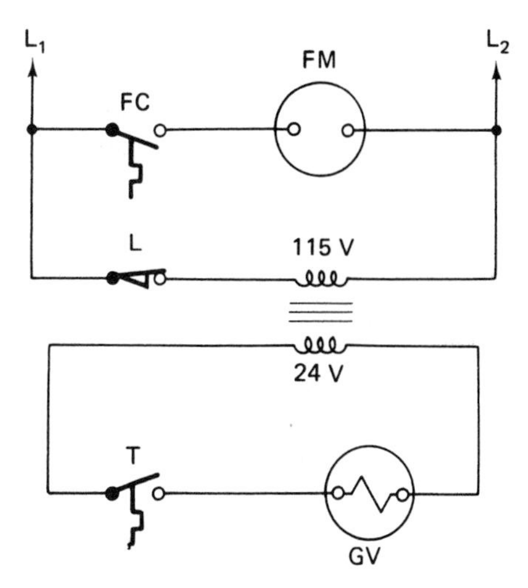

FIGURE 13-4 Line voltage and low voltage circuits schematic diagram.

PROCEDURE FOR DRAWING A SCHEMATIC (HORIZONTAL STYLE)

Assuming that either a completely wired unit or a connection wiring diagram is available, the following is a step-by-step procedure for drawing a schematic:

Referring to the connecting wiring	*Drawing the schematic diagram*
1. Locate the source of power. Trace it back to the disconnect switch.	1. Draw two horizontal lines* representing L_1 and L_2, with sufficient space between them for the diagram. Draw in the disconnect switch.
2. Trace each circuit on the diagram, starting with L_1 and returning to L_2.	2. Draw each circuit on the diagram using a vertical line to connect L_1 through the switches, through the load to L_2.
3. Determine the names of each switch and load.	3. Make a legend by listing the names of the electrical components, together with the letters representing each. The letters are used on the diagram to identify the parts.

* Vertical lines could also be used to represent L_1 and L_2, in which case horizontal lines would then be used to represent the individual circuits.

TYPICAL DIAGRAMS

A connection diagram usually shows the relative position of the various controls on the unit. A schematic diagram makes it possible to indicate the sequence of operation of the electrical devices. A verbal description of the electrical system may also reveal some useful facts about the operation of the unit. Thus, the connection diagram, the schematic diagram, and the verbal description are all useful in understanding how the unit operates electrically. To show how these three are related, a connection diagram and a schematic diagram will be constructed from the description of each circuit.

Power Supply

Description: The power supply is 120-V, 60-H, single-phase (Figure 13-5). A fused disconnect is placed in the hot line to the furnace to disconnect the power when the furnace is not being used. The power supply consists of a hot wire and neutral wire. The circuit has a 20-A fuse.

Circuit 1

Description: A fan motor operating on 120 V is placed in circuit 1 of Figure 13-6. It is controlled by a fan control. This fan control is located in the plenum or in the furnace cabinet by the heat exchanger. It is adjustable

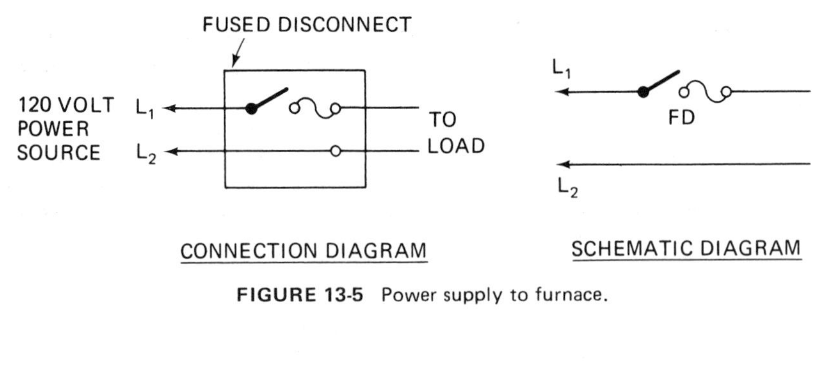

FIGURE 13-5 Power supply to furnace.

FIGURE 13-6 Fan circuit.

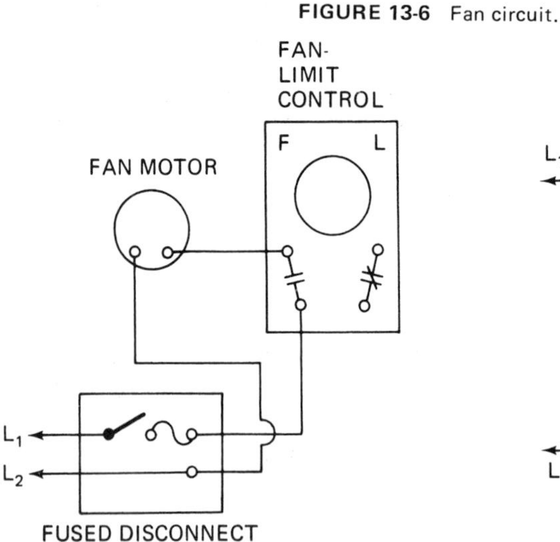

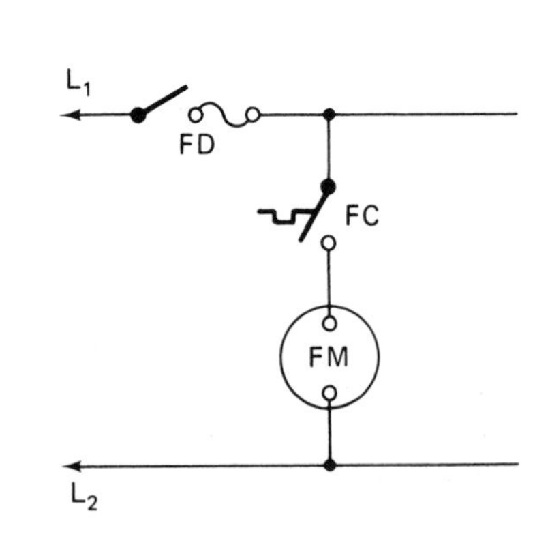

but to conserve energy is usually set to turn the fan on at 110° F (43° C) and off at 90° F (32° C). This permits the furnace to heat up before the fan turns on to deliver heated air to the building.

Note that the fan control is part of a combination fan and limit control. Both use the same bimetal heat element. The limit control settings are higher than the fan control settings.

Circuit 2

Description: The primary of a transformer is placed in a separate 120-V circuit (Figure 13-7). The transformer will be used to supply 24 V (secondary) power to the control circuits.

Circuit 3

Description: This is a 24-V gas valve circuit (Figure 13-8). This circuit includes the thermostat, gas valve, and limit control. When the thermostat calls for heat, the gas valve opens. The limit control will shut off the flow of gas should the air temperature leaving the furnace exceed 200° F (93° C).

FIGURE 13-7 Transformer circuit.

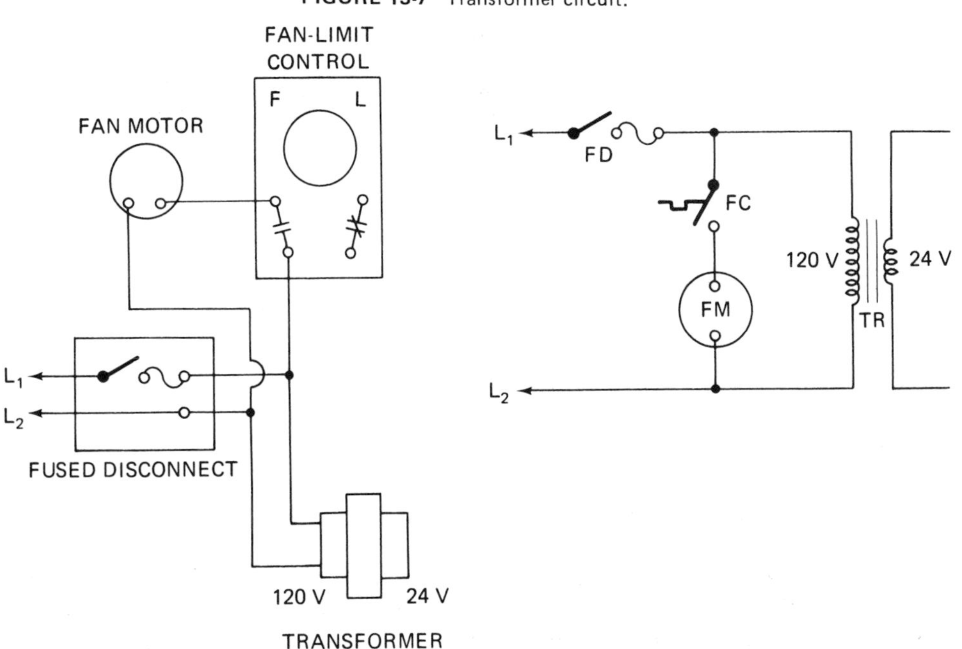

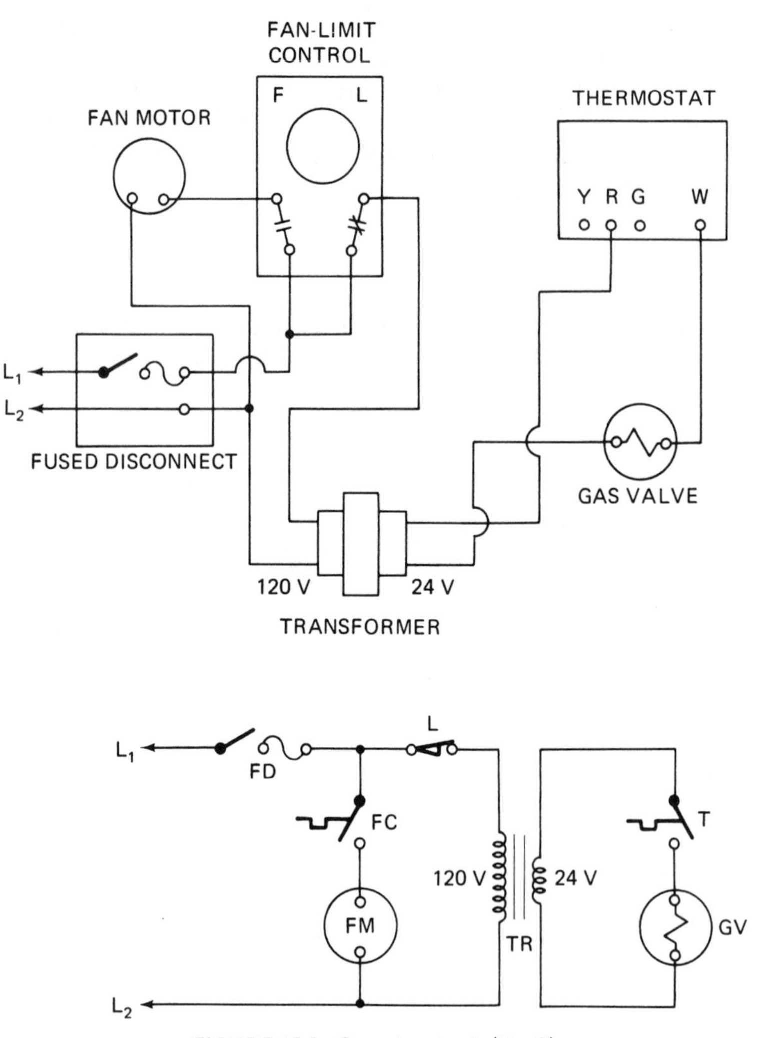

FIGURE 13-8 Gas valve circuit (No. 3).

Circuit 4

Description: A manual switch, located in the subbase of the thermostat, is connected to an indoor fan relay so that the fan can be manually turned on for ventilation even though the gas is off (Figure 13-9). The relay switch is in the 120-V circuit in parallel with the fan control and the solenoid coil of the relay is in the 24-V circuit in series with the fan switch (FS).

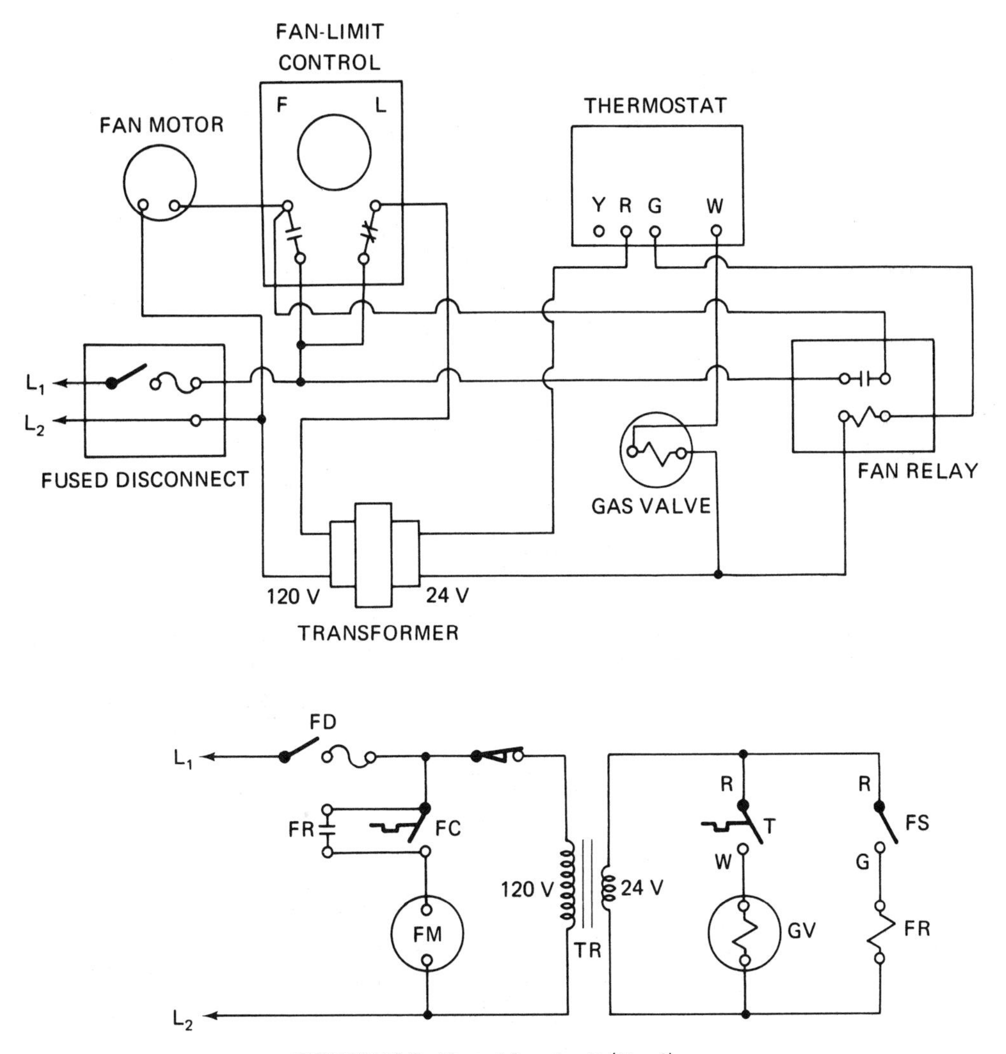

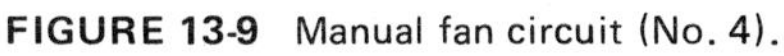

FIGURE 13-9 Manual fan circuit (No. 4).

REVIEW

13-1. Draw a schematic wiring diagram from the connection diagram shown in Figure 13-10.

13-2. Draw a connection diagram from the schematic diagram shown in Figure 13-11.

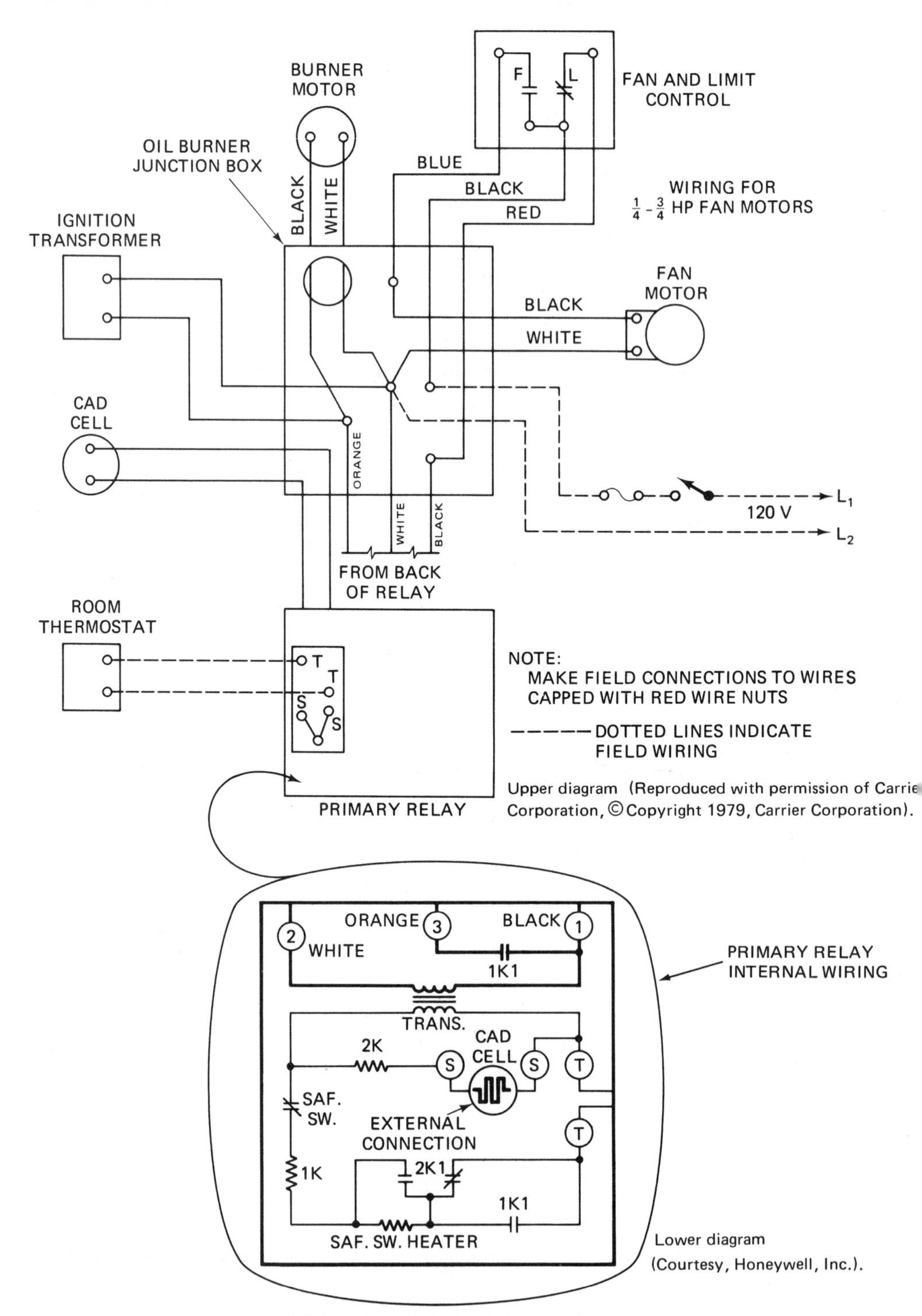

Upper diagram (Reproduced with permission of Carrier Corporation, © Copyright 1979, Carrier Corporation).

Lower diagram (Courtesy, Honeywell, Inc.).

FIGURE 13-10 Oil fired upflow furnace connection diagram.

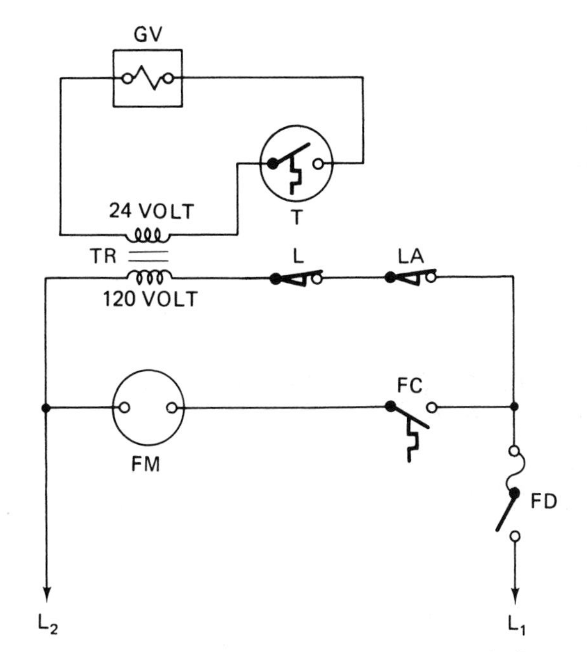

FIGURE 13-11 Gas fired downflow furnace schematic diagram.

14

USING ELECTRICAL TEST INSTRUMENTS

OBJECTIVES

After studying this chapter, the student will be able to:

- Use common electrical test instruments for service or troubleshooting

ELECTRICAL TEST INSTRUMENTS

To test the performance of a heating unit, or to troubleshoot service problems, requires the use of instruments. Since a heating unit has many electrical components a heating technician should:

- Know the unit electrically
- Be able to read and use schematic wiring diagrams
- Be able to use electrical test instruments

The reading and construction of schematic wiring diagrams was discussed in Chapter 13. The use of schematic diagrams is discussed in this unit.

COMPONENT FUNCTIONS

To know the unit electrically means to understand exactly how each component functions. Figure 14-1 shows a simple electrical heating wiring arrangement for a forced air gas system. Two views are shown:

1. The schematic diagram
2. The connection diagram

Three load devices are shown:

1. Fan motor
2. Transformer
3. Gas valve

FIGURE 14-1 Electrical wiring for a 24 volt forced warm air gas-fired furnace.

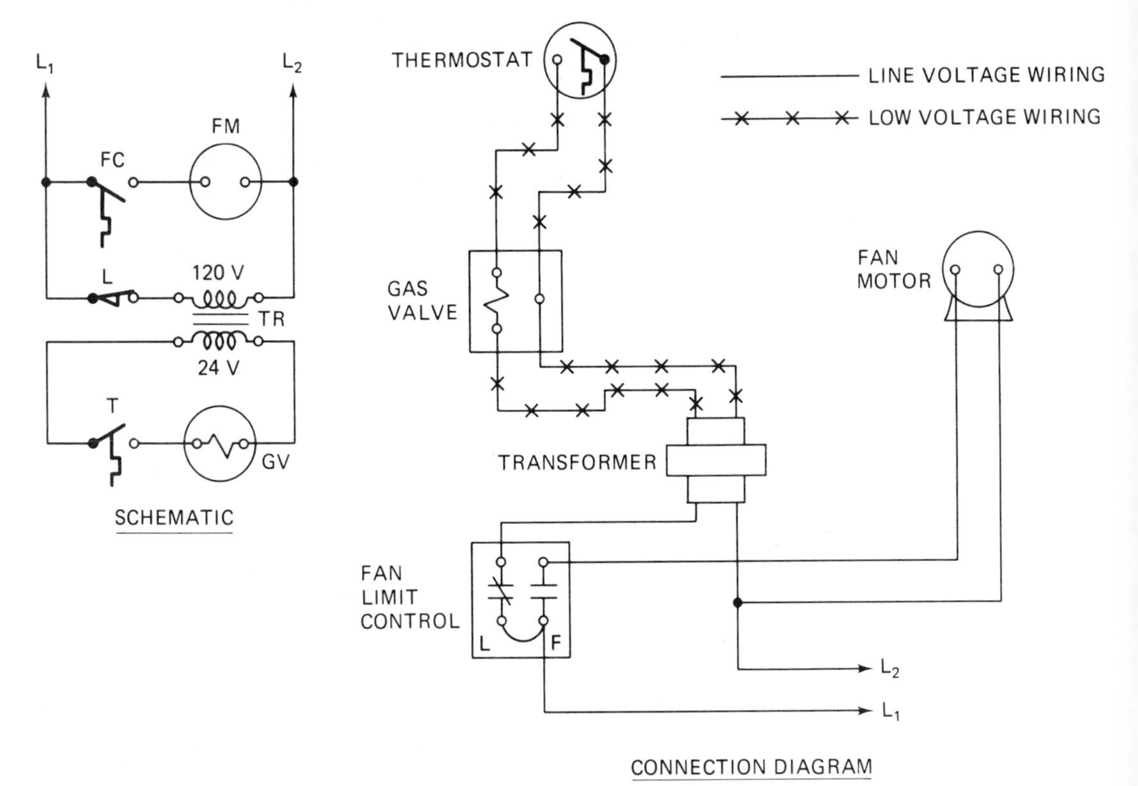

The fan motor turns the fan when it is supplied with power, in this case 120-V, single-phase, 60-Hertz current.

The transformer, which has a primary voltage of 120 V and a secondary voltage of 24 V, is energized whenever power is supplied.

A solenoid type gas valve (a solenoid type) opens when it is energized and operates on 24-V power.

There is at least one automatic switch in each circuit. The fan has one, the transformer has one, and the gas valve has two. The following is a description of the types of switches used:

- The fan control turns on the fan when the bonnet temperature rises to the "cut-in" setting. It turns off the fan when the bonnet temperature drops to the "cut-out" setting.
- The limit control is a safety device that turns off the power to the transformer when the bonnet temperature rises to the cut-out setting. It turns on the power to the transformer automatically when the bonnet temperature drops to the cut-in setting.
- A combination gas valve has a built-in pilotstat. The pilotstat is a safety device operated by a thermocouple (not shown). When the pilot is burning properly, the n.o. contacts are closed, permitting the gas valve to open in response to the thermostat.
- The thermostat is a low-voltage (24-V) switch operated by a bimetal sensing element. When the room temperature drops, and the thermostat reaches its cut-in setting, the contacts close, and the gas valve opens (provided that the pilot flame is proven). When the room temperature rises to the cut-out setting of the thermostat, the contactor opens and the gas valve closes.

INSTRUMENT FUNCTIONS

A number of instruments are required in testing. These include:

- A voltmeter to measure electrical potential.
- An ammeter to measure rate of electrical current flow.
- An ohmmeter to measure electrical resistance.
- A wattmeter to measure electrical power.

Some meters measure a combination of characteristics. For example, a clamp-on ammeter is available which has provisions to measure amperes, volts and ohms. A VOM multimeter is available which can measure volts, ohms, and milliamperes.

Selecting the Proper Instrument

The important thing to consider in selecting a meter is to be certain its scale clearly allows the measured values to be accurately read.

For most heating work, the following scales are adequate:

1. **DC Millivolt Scales:** 0 to 50, 0 to 500, 0 to 1500. These scales are used to read voltages on thermocouples and thermopiles. A millivolt is a thousandth ($1/1000$) of a volt.
2. **AC Voltage Scales:** 0 to 30, 0 to 500. These scales are used to read low-voltage control circuits and line-voltage circuits.
3. **AC Amperage Scales:** 0 to 15, 0 to 75. These scales are used for measuring amperage drawn by various low-voltage and line-voltage loads.
4. **Ohm Scales:** 0 to 400, 0 to 3000. These scales are used for resistance readings on all types of circuits and for continuity checks on all systems.
5. **Wattage Scales:** 0 to 300, 0 to 600, 0 to 1500, 0 to 3000. These scales are used to measure power input to a circuit or system.
6. **Temperature Scale:** 0 to 1200°F. This refers to an electronic temperature instrument. This scale is used to check discharge air temperature, thermocouple hot junction temperature, and surface and ambient temperature of control components (Figure 14-2).

Using Meters

When using meters, the following points should be considered:

1. Always use the highest scale first, then work down until midscale readings are obtained. This prevents damage to the meter.
2. All ohmmeters have batteries. Therefore, when the meter is not being used, it should be turned to the "off" position. This prevents the batteries from running down when not in use.
3. Always check the calibration of a meter before using it. For example, on an ohmmeter the two test leads are shorted together and the zero adjust knob turned until the needle reads zero resistance.
4. When using a clamp-on type ammeter (Figure 14-3), the sensitivity of the instrument can be increased by wrapping the conductor wire around the jaws. The sensitivity will be multiplied by the number of turns taken.

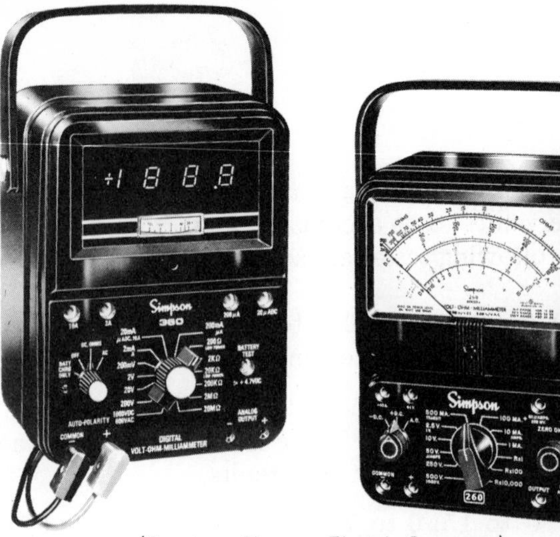

STANDARD AND
DIGITAL VOM MULTI-METERS

(Courtesy, Simpson Electric Company).

(Courtesy, TIF Instruments, Inc.).

FIGURE 14-2 Test instruments.

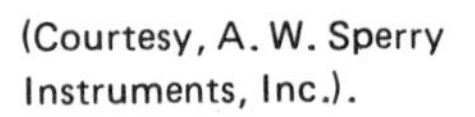

(Courtesy, A. W. Sperry Instruments, Inc.).

COMPACT, PORTABLE MULTI-METERS

COMBINATION VOLT-OHM-AMPERES

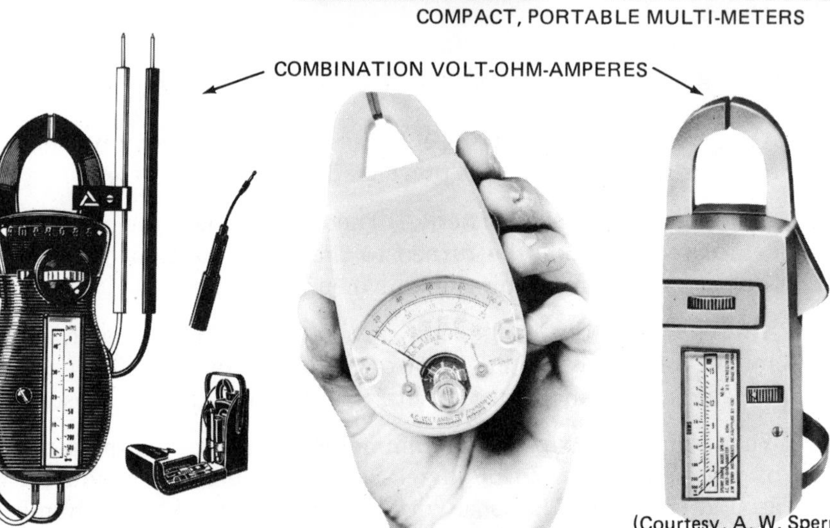

(Courtesy, Amprobe Instruments, Inc.).

(Courtesy, A. W. Sperry Instruments, Inc.).

VOLT-WATTMETER

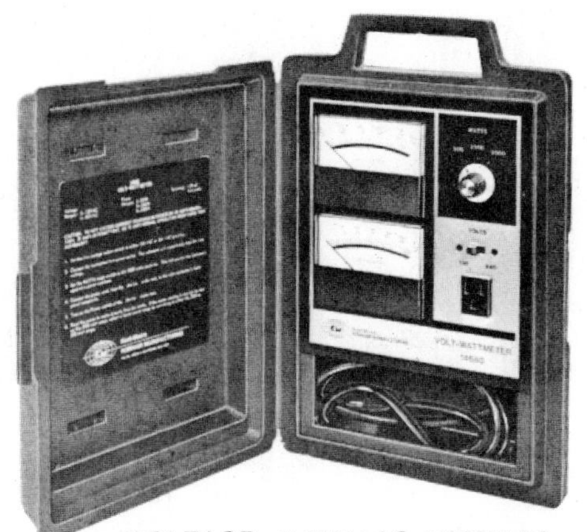

VOLTAGE: 0-130V AC, 0-260V AC
WATTAGE: 0-500, 0-2500, 0-5000
(Courtesy, Robinair Manufacturing Company).

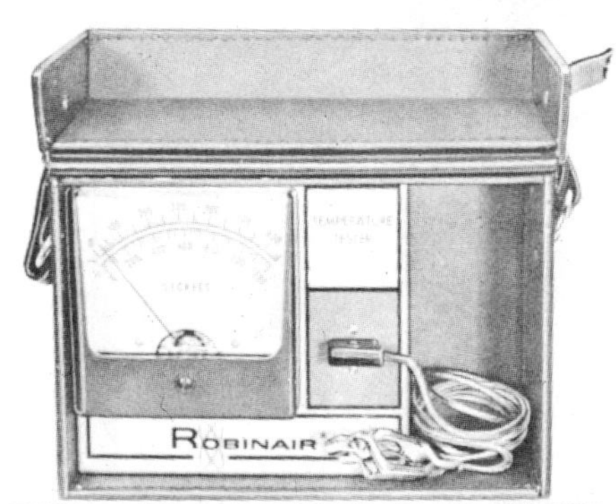

THERMOCOUPLE TEMPERATURE TESTER
RANGE: 0-1200 °F
(Courtesy, Robinair Manufacturing Company).

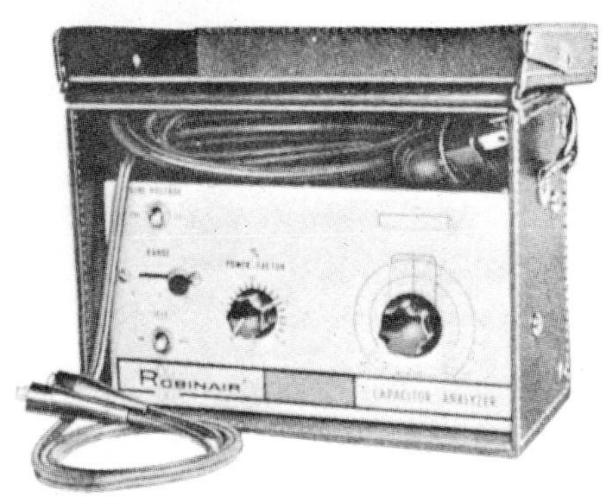

CAPACITOR ANALYSER
SCALES: 1-20, 21-49, 50-200 mfd
(Courtesy, Robinair Manufacturing Company).

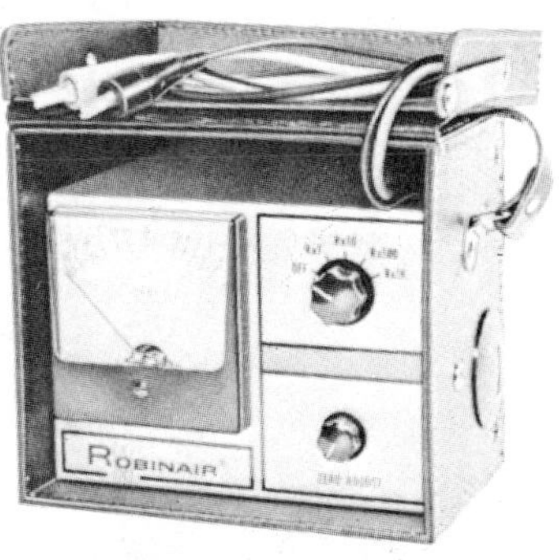

SCALES: 0-500
0-5000
0-50,000
0-500,000

MULTI-RANGE OHMMETER

(Courtesy, Robinair Manufacturing Company).

FIGURE 14-2 Continued

MILLIVOLT METER AND THERMOCOUPLE TESTER SCALES: 0-50, 0-100 μV
(Courtesy, Mechanical Refrigeration Enterprises, Inc.).

PROBE TYPE THERMOMETER RANGE: 0-1200 °F
(Courtesy, Mechanical Refrigeration Enterprises).

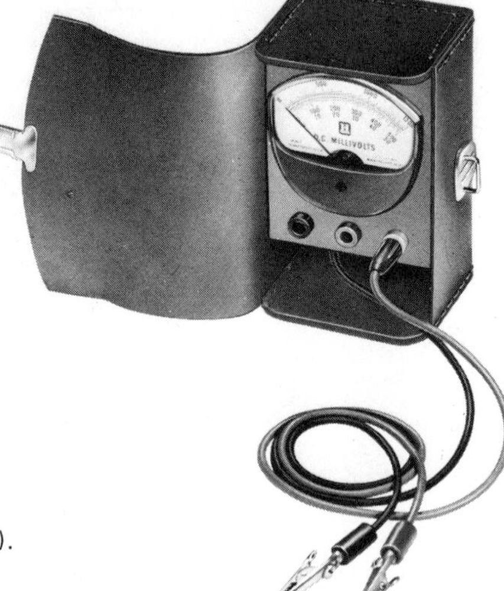

MILLIVOLT METER
SCALES: 0-50, 0-500, 0-1500 μV

(Courtesy, Honeywell, Inc.).

FIGURE 14-2 Continued

FIGURE 14-3 Coiling wire or clamp-on ammeter (Courtesy, Amprobe Instruments, Inc.).

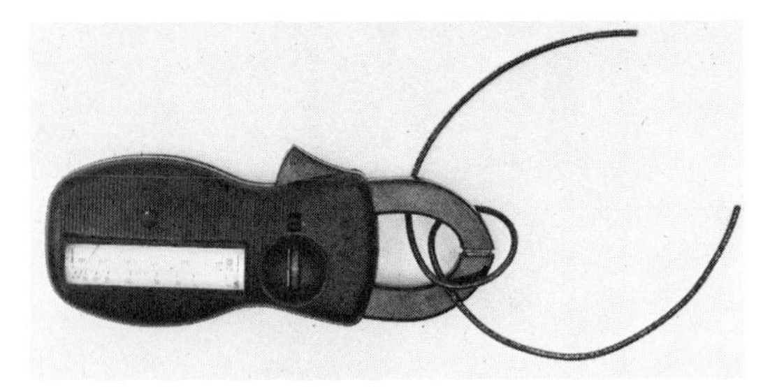

5. When using a clamp-on type of ammeter, be sure the clamp is around only one wire. If it encloses two wires, the current may be flowing in opposite directions, which could cause a zero reading even though current is actually flowing.
6. Always have a supply of the required meter fuses on hand. Certain special fuses may be difficult to obtain on short notice.
7. When measuring dc millivolts being delivered by a thermocouple in a circuit, an adapter is required to provide access to the internal connection (Figure 14-4).

Three types of readings that can be taken with an ohmmeter are shown in Figure 14-5. They are:

1. No resistance ("short"), left illustration. Closed switch is giving full-scale deflection, indicating zero ohms.
2. Measurable resistance, middle illustration. Measurable resistance is providing less deflection or a specific ohm reading on the meter.
3. Infinite resistance ("open"), right illustration. There is no complete path for the ohmmeter current to flow through, so there is no deflection of the meter needle. The resistance is so high that it cannot be measured. The meter shows infinity (∞) representing an open circuit.

FIGURE 14-4 Adapter used for thermocouple measurements.

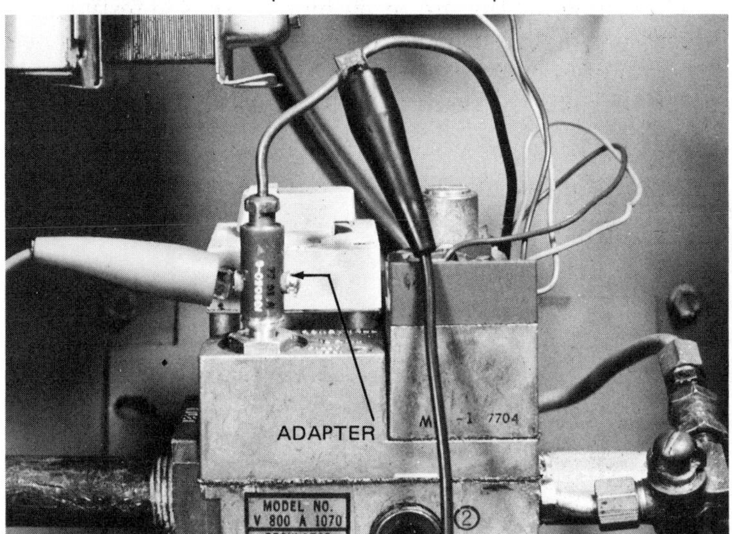

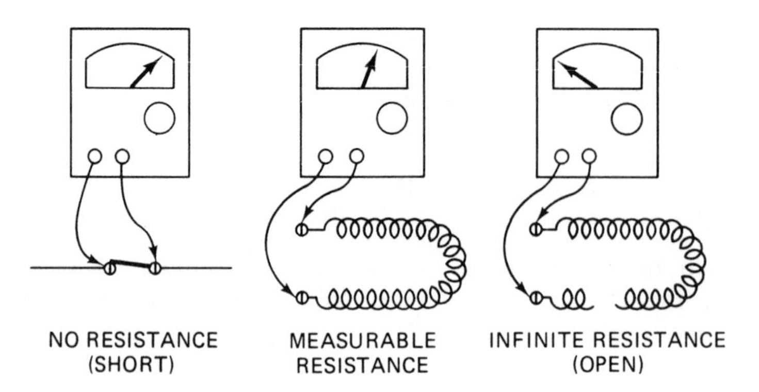

FIGURE 14-5 Three types of readings taken with an ohmmeter.

SELECTING AN INSTRUMENT

It is necessary to decide which meter to use for electrical troubleshooting. Although there is no hard-and-fast rule, there are certain advantages and disadvantages in using one meter in preference to another.

> *NOTE:* It is important to remember when using an ohmmeter that it has its own source of power and must never be used to test a circuit that is "hot" (connected to power). All power must be off or the ohmmeter can be seriously damaged.

One good example of the use of an ohmmeter is in testing fuses (Figure 14-6). The fuse is removed from the power circuit and the ohmmeter is used to make a continuity check. If continuity (zero ohms) is obtained when the meter leads are touching the two ends of the fuse, the fuse is good.

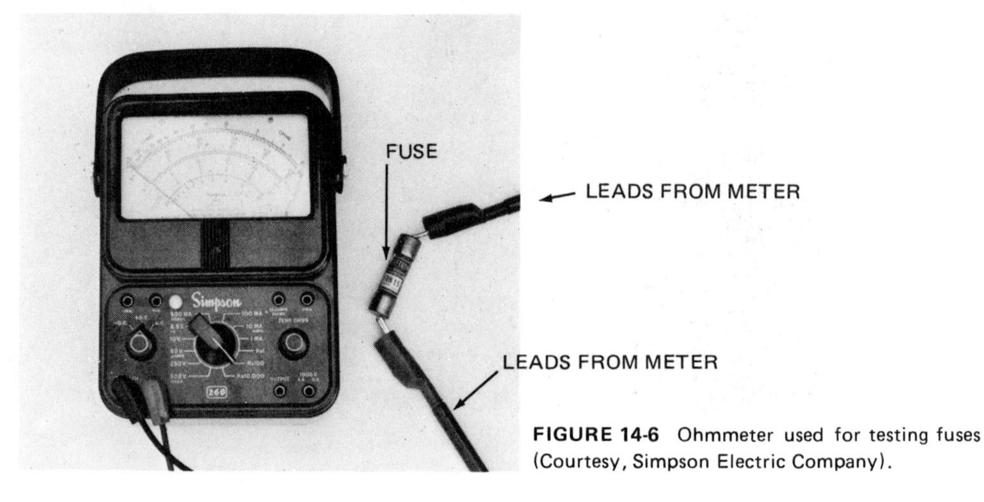

FIGURE 14-6 Ohmmeter used for testing fuses (Courtesy, Simpson Electric Company).

Procedure for Selecting an Instrument

The most helpful procedure for deciding which instrument to use in troubleshooting is to check as follows:

1. If any part of the unit operates, the voltmeter or ammeter are normally used.
2. If no portion of the unit operates, a short circuit could be the problem and the ohmmeter is probably the best instrument to use.

When attempting to measure the resistance of any component, the possibility of obtaining an incorrect reading always exists when a part is wired into the system. In a parallel circuit when testing with an ohmmeter, it is necessary to disconnect one side of the component being tested to avoid the possibility of an incorrect reading of the resistance (Figure 14-7).

A wattmeter measures both amperage and volts simultaneously. This is necessary in an ac circuit because watts take into consideration the power factor. The *power factor* indicates the percent of the volts times the amperes that a consumer actually pays for. Since the peak of the volts and amperes does not usually occur at the same time, wattage is normally less than the product of the volts times the amperes.

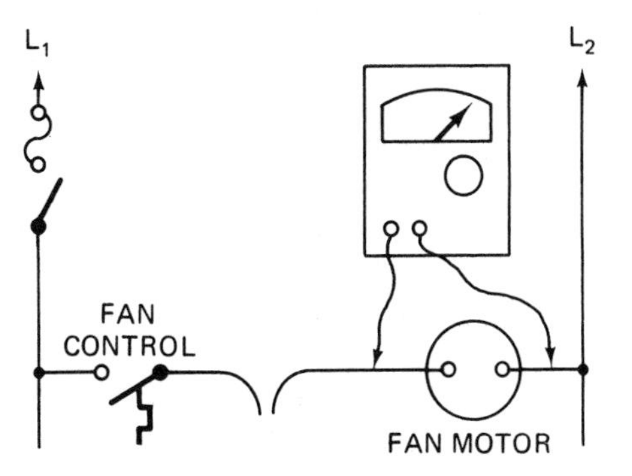

FIGURE 14-7 Disconnecting one side of parallel circuit when testing resistance with Ohmmeter.

POWER CALCULATIONS

Power, measured in watts can be calculated by the equation

$$P = E \times I \times \text{PF}$$

where: P = power, watts

E = EMF, volts

I = current, amperes

PF = power factor

The power factor can be supplied by the electric company or measured by suitable instruments. An example of calculating power, assume that the power factor is 0.90, and using the values for E and I from Figure 12-14,

$$P = 120 \text{ V} \times 12 \text{ A} \times 0.90 \text{ power factor} = 1296 \text{ W}$$

or approximately 1.3 kilowatts (kw), where $K = 1000$ watts.

TROUBLESHOOTING A UNIT ELECTRICALLY

Testing Part of a Unit

To troubleshoot a unit electrically assume, for example, that service is required on a gas-fired forced air unit, with wiring shown in Figure 14-1. The complaint is that the unit does not heat properly. The fan will not run. The gas burner cycles as a result of the limit control cycling on and off.

The first step is to review the schematic wiring diagram and eliminate from consideration any circuits that appear to be operating properly. In this case, the transformer and the gas valve circuits are eliminated. Now the testing can be confined to the fan motor circuit.

Since part of the unit will run, a voltmeter should be used for testing. By the process of elimination the source of the trouble is found.

First, measure the voltage across the ends of the circuit where the connection is made to L_1 and L_2 (Figure 14-8). The voltmeter reads 120 V, which is satisfactory.

Second, with one meter lead on L_2, place the other meter lead between the fan control and the motor (Figure 14-9). The meter reads 120 V. The bonnet temperature is 150° F. The fan control is set to turn the fan on at 120° F. Therefore, the fan control should be made and this checks out as satisfactory.

When the fan motor still does not run after the proper power has been supplied, it is an indication that the fan motor is defective and must be replaced.

The process of elimination can be used on any circuit where troubleshooting is required, regardless of the number of switches. Where power is being supplied to the load and the load does not operate, the load device is defective and needs to be replaced.

Testing a Complete Unit

In troubleshooting a complete unit, there are numerous electrical tests that can be made with the instruments. By the process of elimination, it is important to rule out as many of the circuits as possible that are not causing trouble.

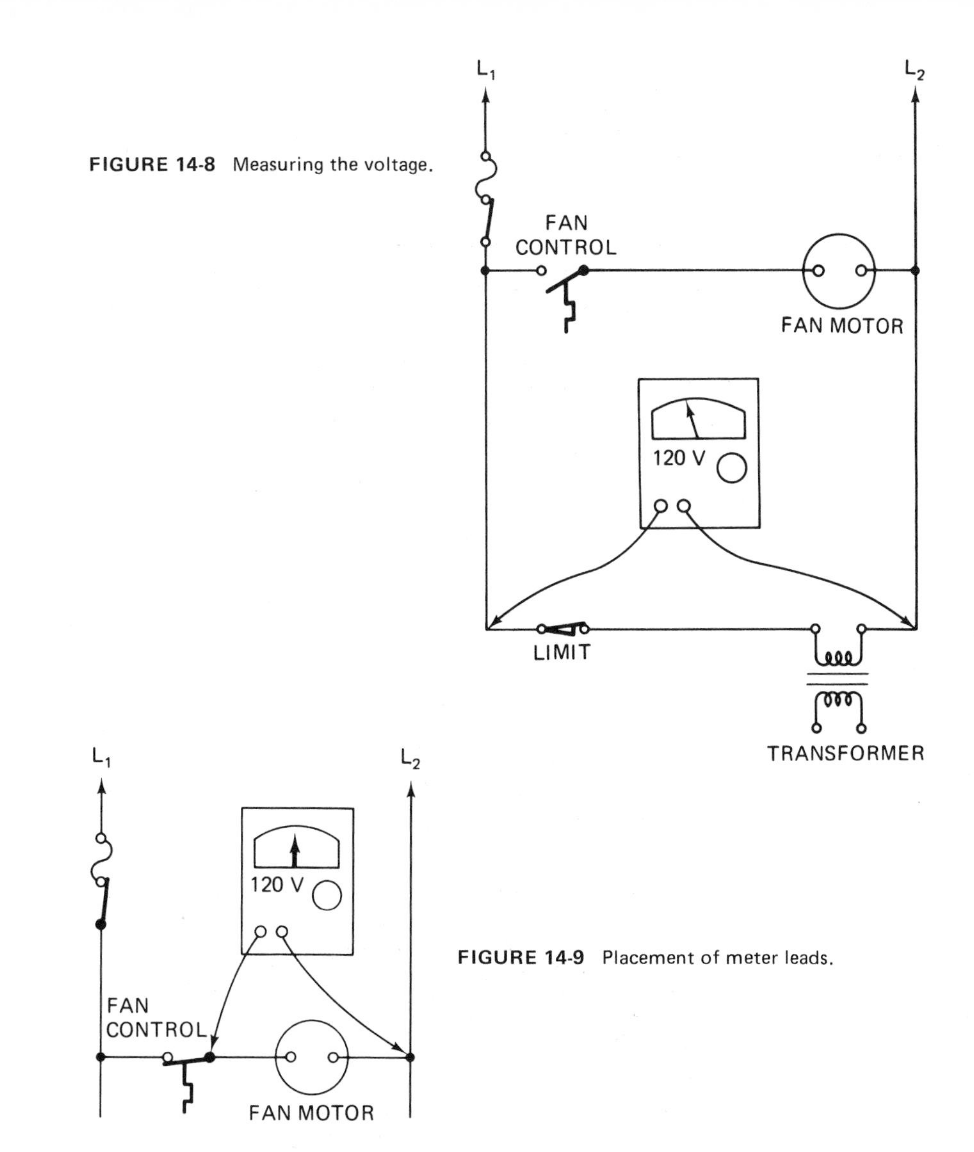

FIGURE 14-8 Measuring the voltage.

FIGURE 14-9 Placement of meter leads.

METER LEADS ON L_2 AND LOAD SIDE OF FAN CONTROL SWITCH TO CHECK POWER AVAILABLE FOR LOAD (MOTOR).

The methods of testing various heating systems are indicated in the following diagrams:

1. Thermopile, self-generating, gas heating control system (Figure 14-10)
2. 24-V gas heating control system (Figure 14-11)
3. Line-voltage gas heating control system (Figure 14-12)
4. Oil burner control system (Figure 14-13)
5. Electric heating system (Figure 14-14)

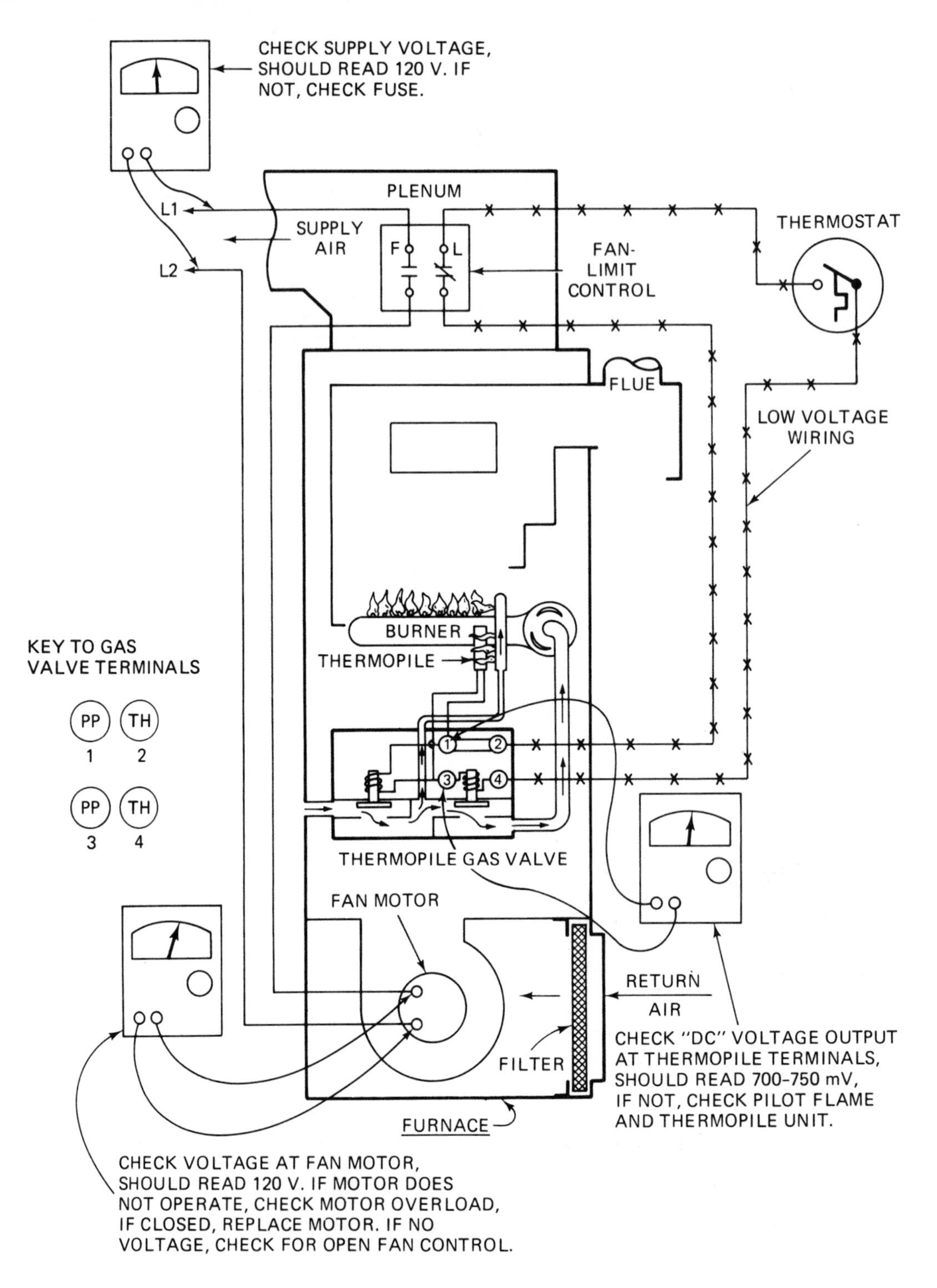

FIGURE 14-10 Thermopile system, gas-fired, troubleshooting.

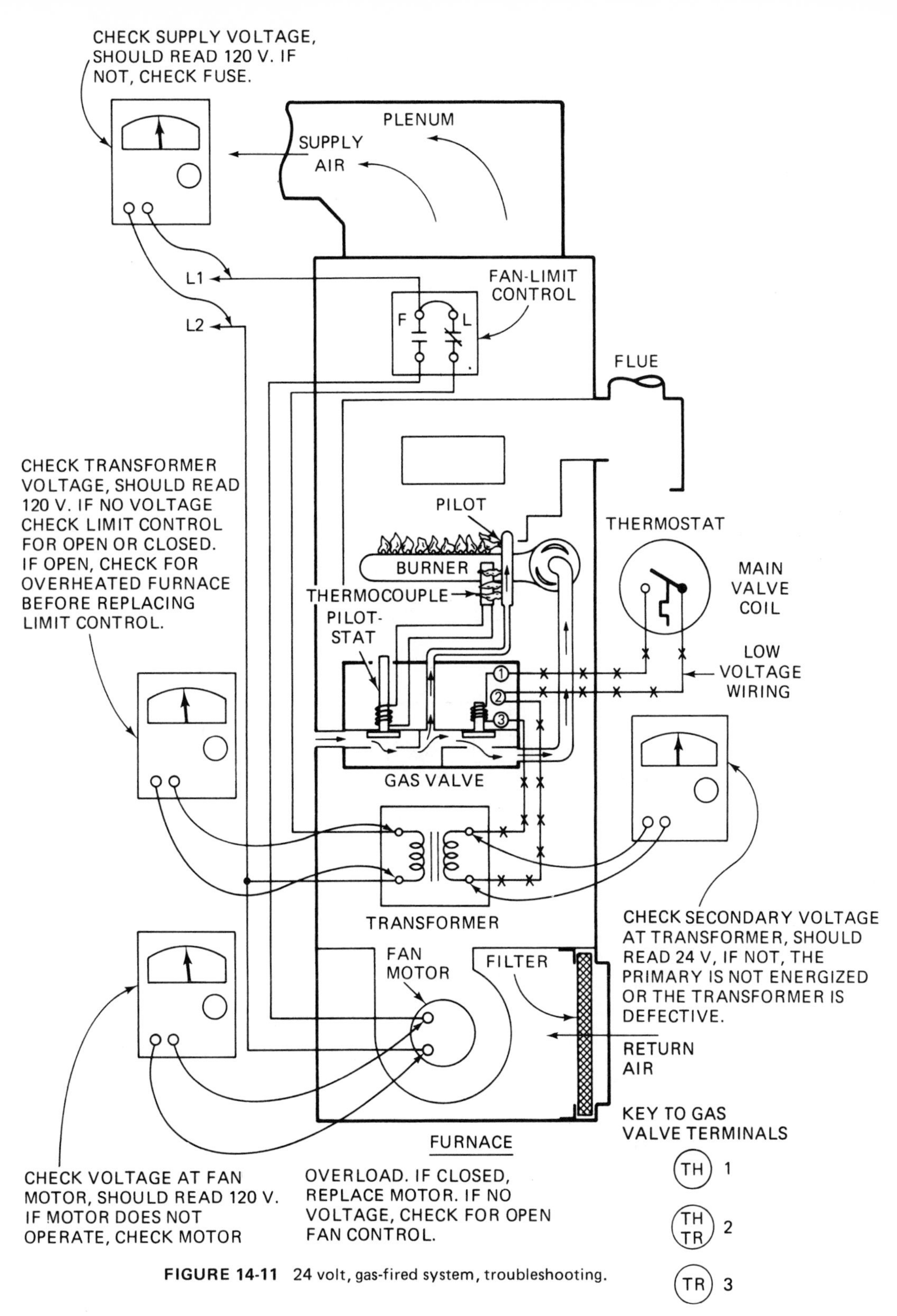

FIGURE 14-11 24 volt, gas-fired system, troubleshooting.

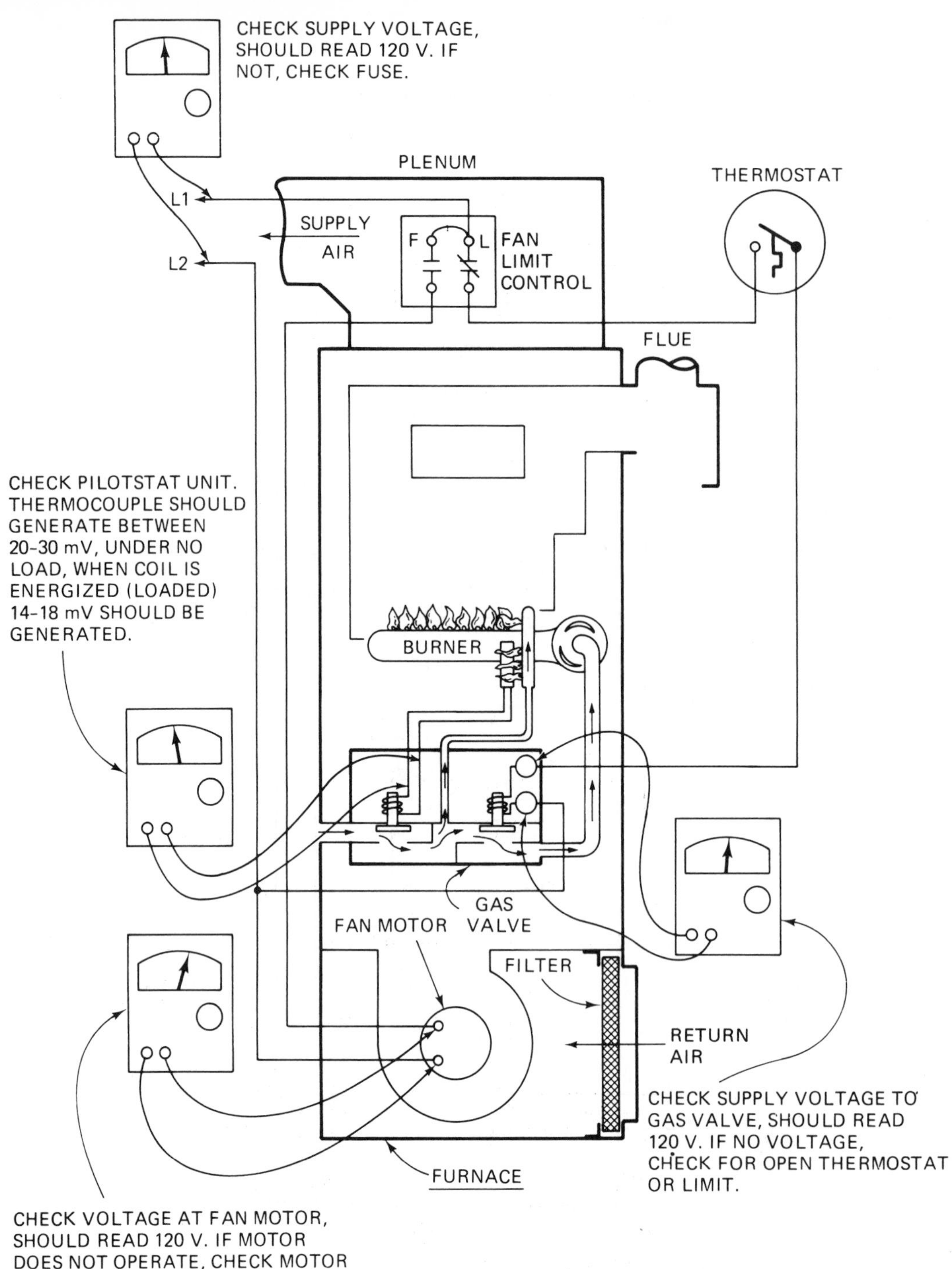

FIGURE 14-12 120 volt, gas-fired system, troubleshooting.

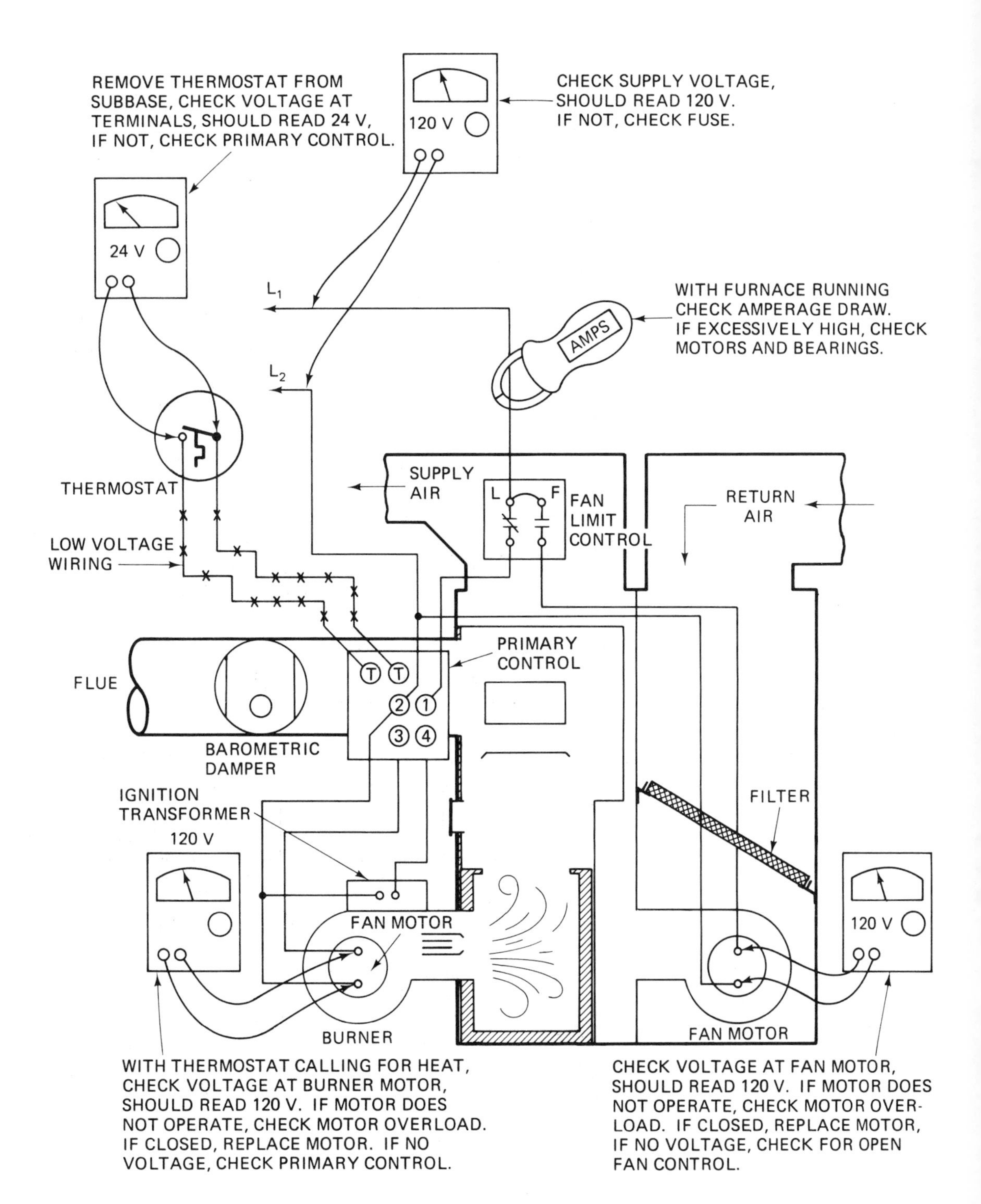

FIGURE 14-13 Oil burner troubleshooting, thermal primary safety control.

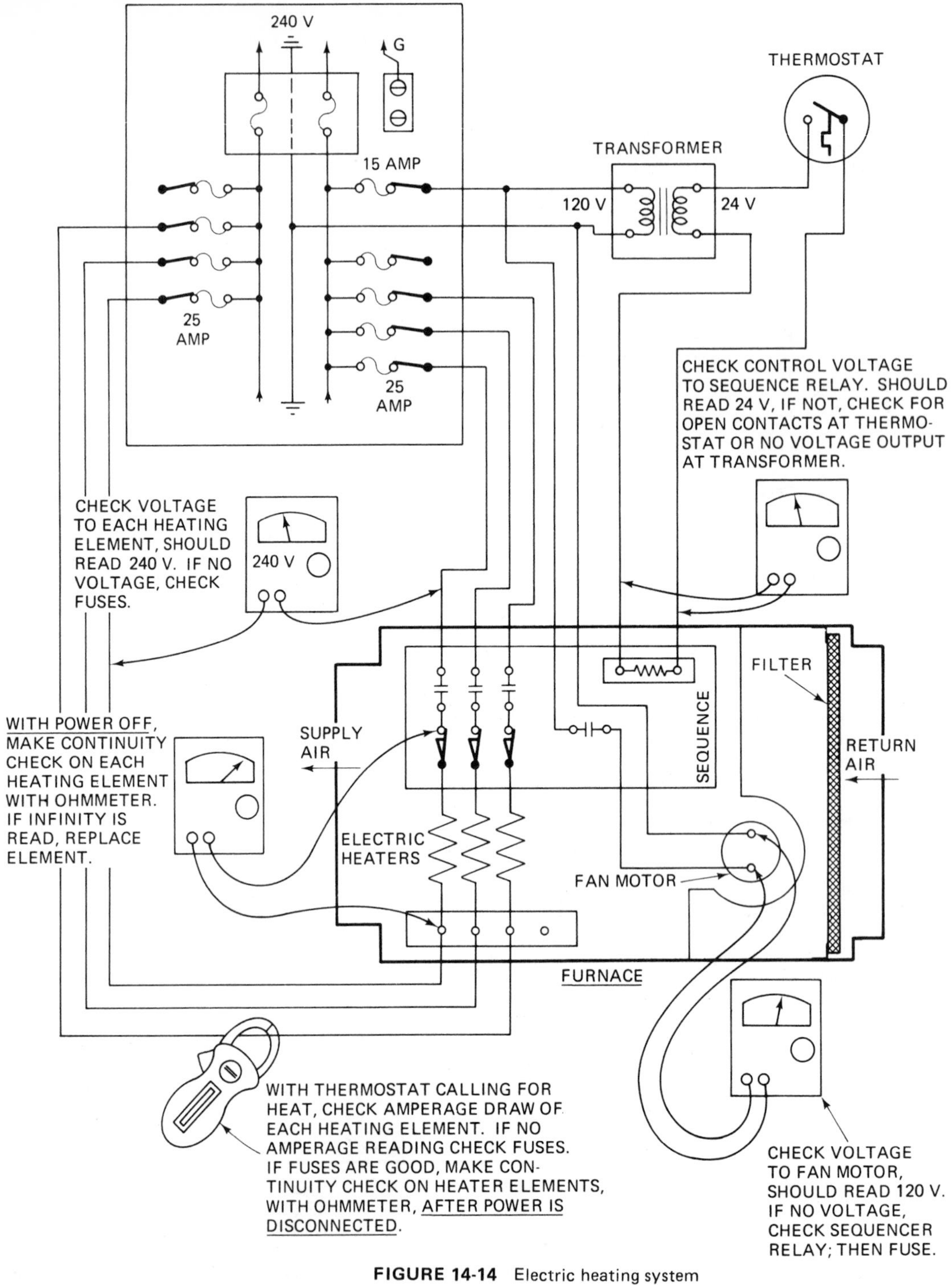

FIGURE 14-14 Electric heating system troubleshooting.

REVIEW QUESTIONS

Select the letter representing your choice of the correct answer.

14-1. Which meter has its own source of power?

(a) Voltmeter

(b) Ammeter

(c) Ohmmeter

(d) Wattmeter

14-2. Which meter is especially useful in testing fuses?

(a) Voltmeter

(b) Ammeter

(c) Ohmmeter

(d) Wattmeter

14-3. Which meter requires calibration each time it is used?

(a) Voltmeter

(b) Ammeter

(c) Ohmmeter

(d) Wattmeter

14-4. Unless a multiplier coil is used, how many wires should be surrounded by the ammeter in measuring current in a circuit?

(a) One

(b) Two

(c) Three

(d) Four

14-5. In measuring millivolts in a thermocouple circuit, what special device is required?

(a) Thermometer

(b) Adapter

(c) Pressure gauge

(d) Galvanometer

14-6. A "no resistance" reading on an ohmmeter means:

(a) Short circuit

(b) Open circuit

(c) Open overload

(d) Bad transformer

14-7. To eliminate the possibility of an incorrect reading of the resistance in a parallel circuit:

(a) Completely remove component tested

(b) Use scale R × 100

(c) Use scale R × 1000

(d) Disconnect one side of the component tested

14-8. Watts in an ac circuit equal:

(a) Volts × amperes × power factor

(b) $(\text{Amperes})^2$ × power factor

(c) $(\text{Volts})^2$ × power factor

(d) Volts × amperes

14-9. If the voltmeter reads voltage when placed across the terminals of a switch, the switch is?

(a) Open

(b) Closed

14-10. In troubleshooting a defective unit, what circuits can be eliminated from the testing procedure?

(a) Power supply circuits

(b) Circuits not causing trouble

(c) Series circuits

(d) Parallel circuits

15

EXTERNAL FURNACE WIRING

OBJECTIVES

After studying this chapter, the student will be able to:

- Evaluate the external electrical wiring of a forced warm air furnace and troubleshoot where necessary

FIELD WIRING

External furnace wiring usually includes the wiring connections to the power supply and the thermostat (Figures 15-1 and 15-2). These connections are also described as *field wiring*, to distinguish them from the wiring done at the factory by the manufacturer. Field wiring is completed by the installation crew or the electrician on the job.

Where the furnace equipment includes accessories such as a humidifier, electronic air cleaner, or cooling, additional field wiring is required.

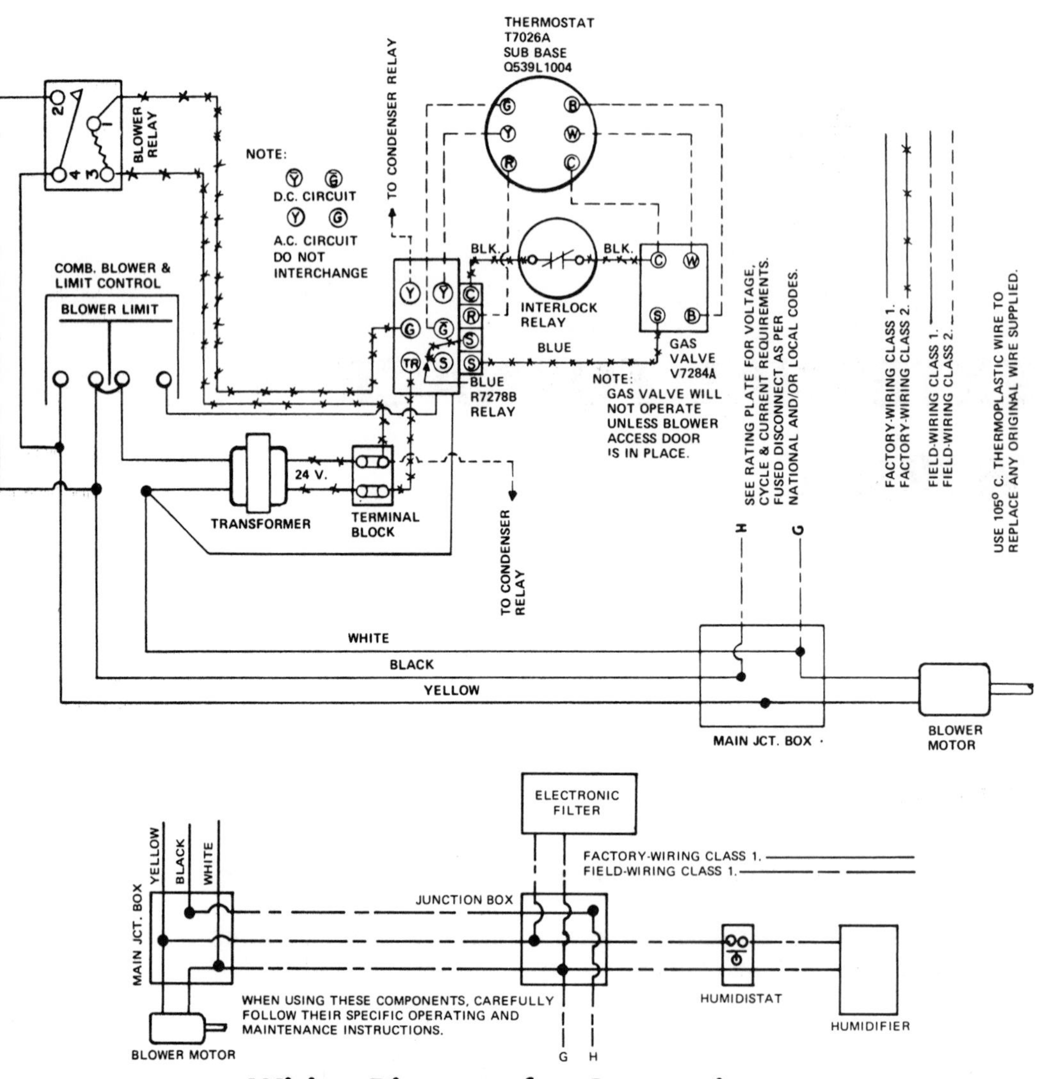

FIGURE 15-1 External gas furnace wiring (Courtesy, Luxaire, Inc.).

Since the performance of the furnace is dependent upon proper field wiring, special attention must be given to this particular part of the installation. Most manufacturers supply detailed instructions for connecting the thermostat and accessories to the furnace. Most field-wiring problems occur because of improper or inadequate connections to the building's power supply.

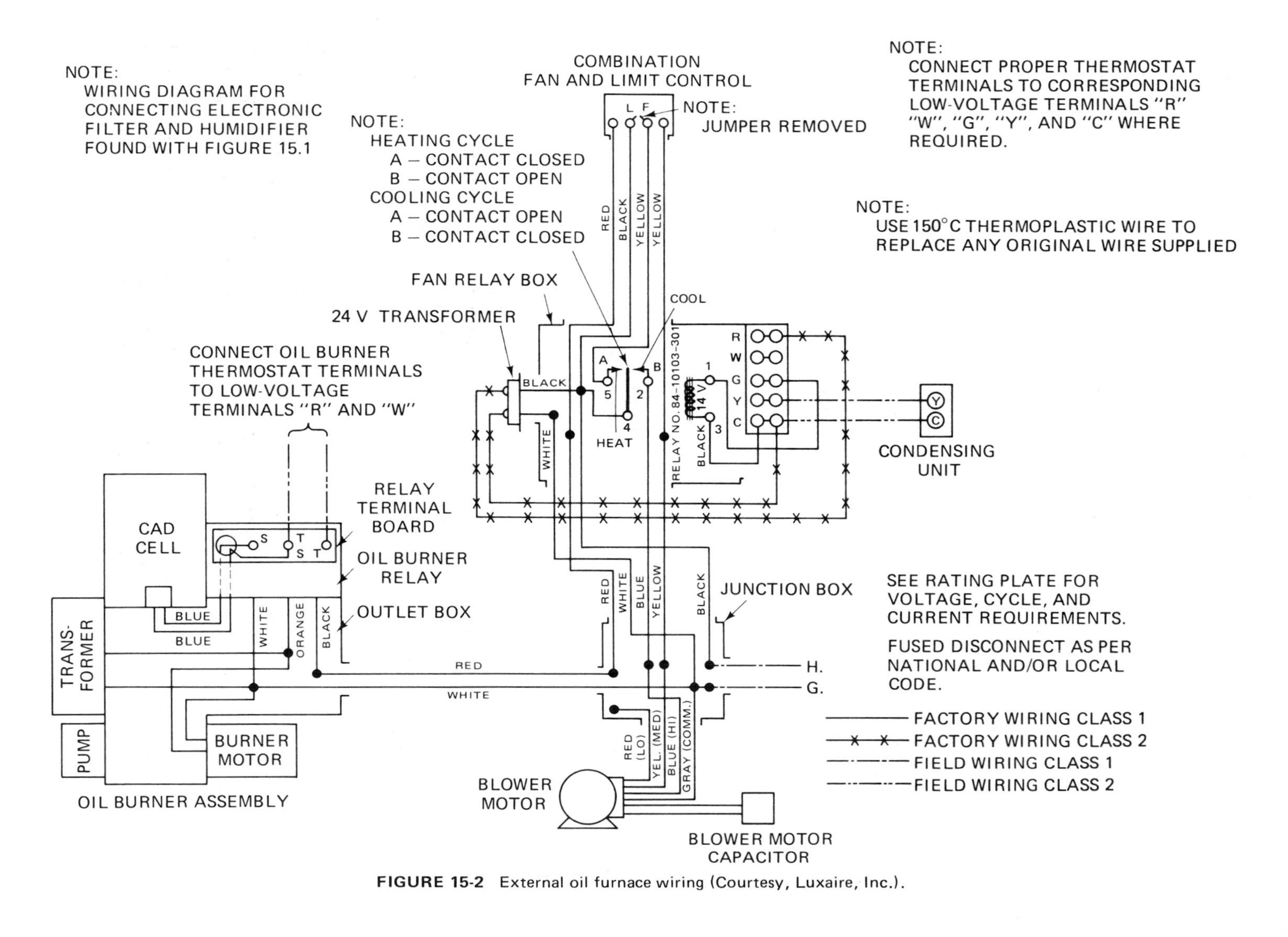

FIGURE 15-2 External oil furnace wiring (Courtesy, Luxaire, Inc.).

If cooling is not one of the accessories, most gas and oil furnaces can operate on a 15-A, 120-V single-phase ac branch circuit. Where electric heating and/or cooling equipment is installed, 240-V single-phase ac power is required. The amperes of service needed depends upon the size of the electric furnace and/or cooling unit.

ANALYZING POWER SUPPLY

A building's electrical system must be analyzed to determine whether or not the existing power supply is adequate. If it is found to be inadequate, additional service needs to be installed. A building's electrical system includes the entire wiring of the building starting from the service entrance. The system can be studied from three standpoints:

1. Main power supply
2. Branch circuits
3. Materials and equipment

All wiring must be installed in accordance with the National Electrical Code and any local codes or regulations that apply.

Main Power Supply

The usual source of the main power supply to a building is an entrance cable, consisting of three wires. On newer services one of these wires is black (hot), one is red (hot), and the third is white (neutral). The size of these entrance wires determines the maximum size of the service panel that can be installed.

THE ENTRANCE CABLE:

The wires in an entrance cable connect the incoming power lines with the building wiring. The cable is connected through the meter socket to the main service panel, as shown in Figure 15-3. Note that the neutral wire at the main panel is connected to a water pipe or other type of approved ground.

When it is necessary to increase the size or capacity of the service, an additional fuse panel or subpanel can be added to the main service panel. However, this is possible only when the entrance wiring is large enough to permit the extra load.

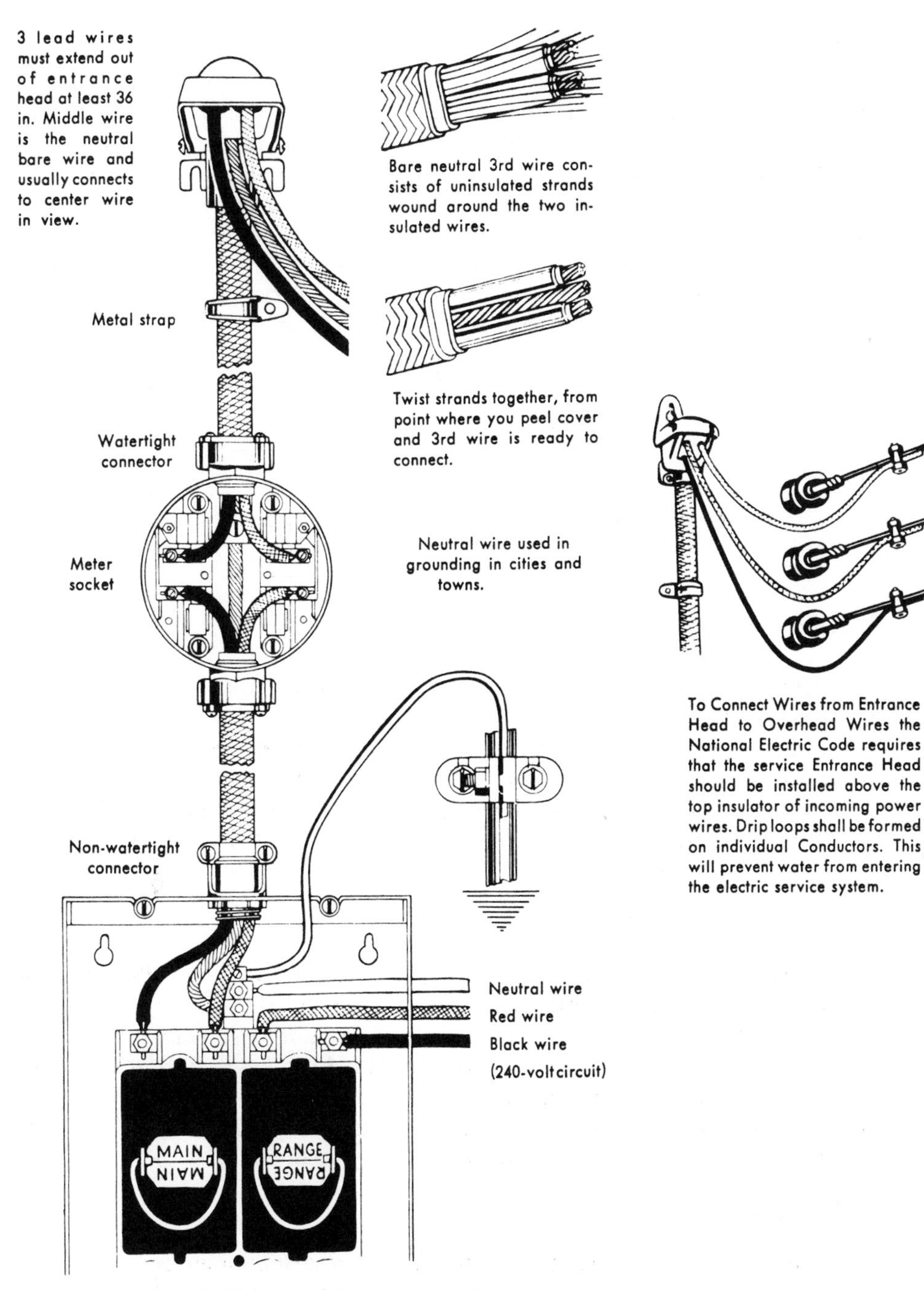

FIGURE 15-3 Entrance cable connections (Courtesy, Sears, Roebuck and Company).

BUILDING'S AMPERE SERVICE:

A building's ampere service is the size or capacity of the power supply available from the main panel for electrical service within the building. The service is usually 60, 100, 150, or 200 A. If a home has an electric range, high-speed drier, and central air conditioning, 150-A service is needed. Homes having electric heating, 200-A service is required. The minimum service for a modern home is 100 A. Many older homes have only 60-A service. The service must always be large enough to supply the needs of the connected loads.

All residential services require a three-wire electric power supply. The size of the wire for various ampere services is as follows:

Copper Wire	*Aluminum & Copper Clad*
· 60-A number 6	number 4
· 100-A number 4	number 2
· 150-A number 1	number 1-2/0
· 200-A number 2/0	number 4/0

Branch Circuits

Branch circuits are the divisions of main power supply that are fused and connected to the loads in a building.

The main power supply is connected to the fuse or circuit breaker box, as shown in Figure 15-4. The two hot lines are fused. The voltage across the two hot lines is usually 208 or 240 V. The neutral line is grounded at the main service panel. The voltage from either of the hot lines to the neutral line is 120 V.

TYPES OF BRANCH CIRCUITS:

Branch circuits can be either 208/240-V (Figure 15-5) or 120-V (Figure 15-6). Branch circuits for 208/240 V have fuses in both hot lines. Branch circuits for 120V have one fuse in the hot line, none in the neutral.

The main fuse box or circuit breaker can have both 208/240-V and 120-V branch circuits. The maximum fuse size on a 120-V branch is 20 A. A 20-A 120-V branch requires No. 12 wire minimum. A 15-A 120-V branch requires No. 14 wire minimum, or as required by local codes.

GROUNDING:

The electrical terms grounding, grounded, and ground may be confusing. *Grounding* is the process of connecting to the earth a wire or other conductor from a motor frame or metal enclosure to a water pipe, buried plate, or

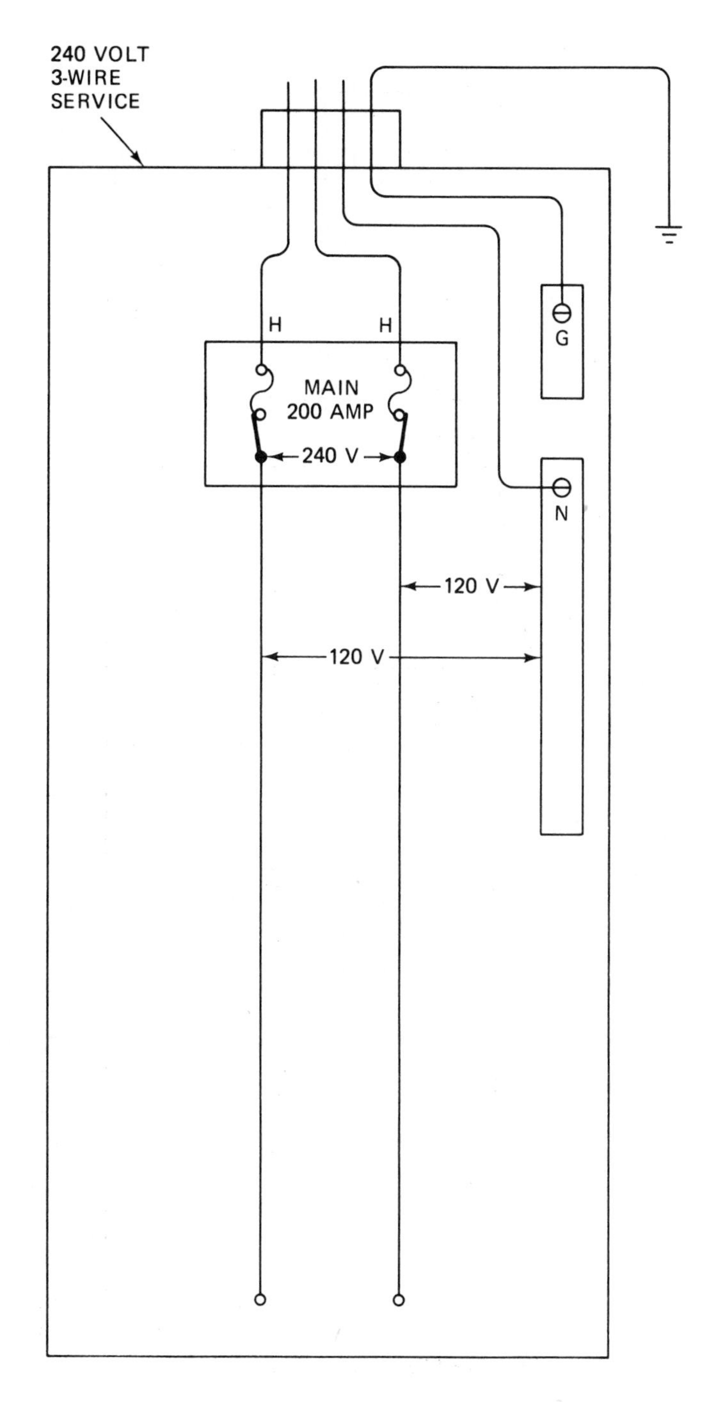

FIGURE 15-4 Main power supply.

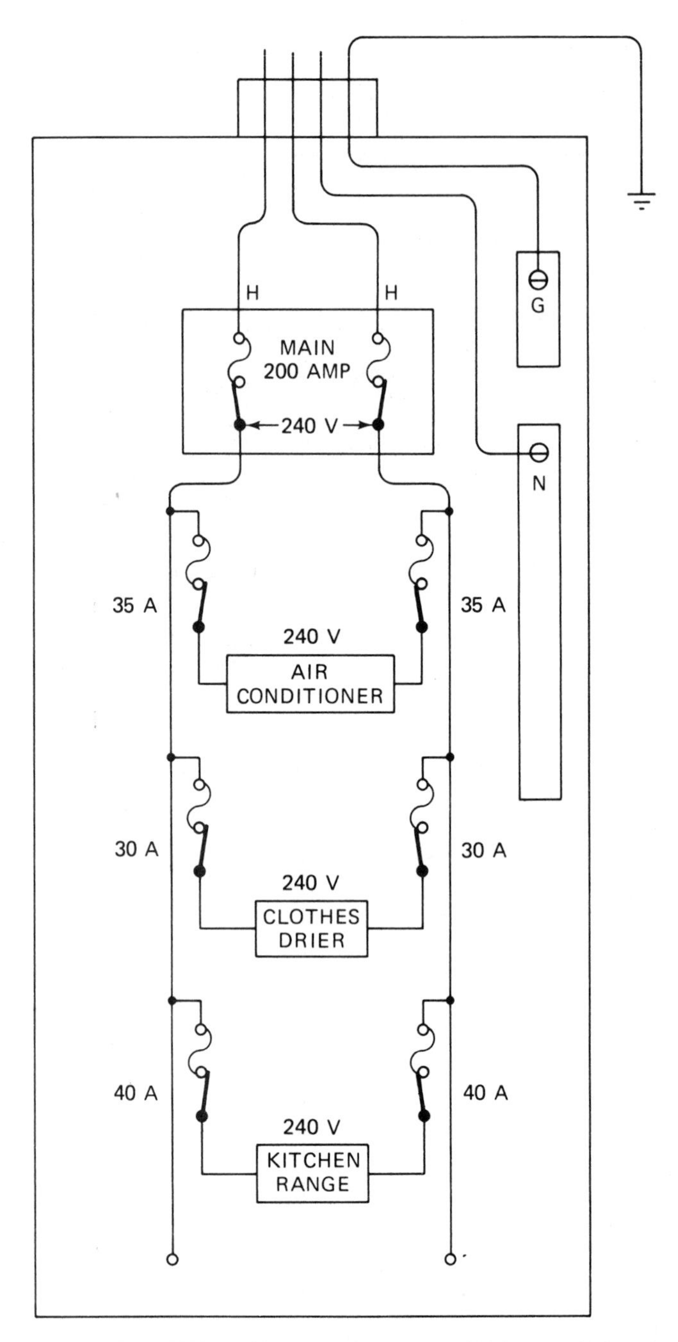

FIGURE 15-5 240-volt branch circuits.

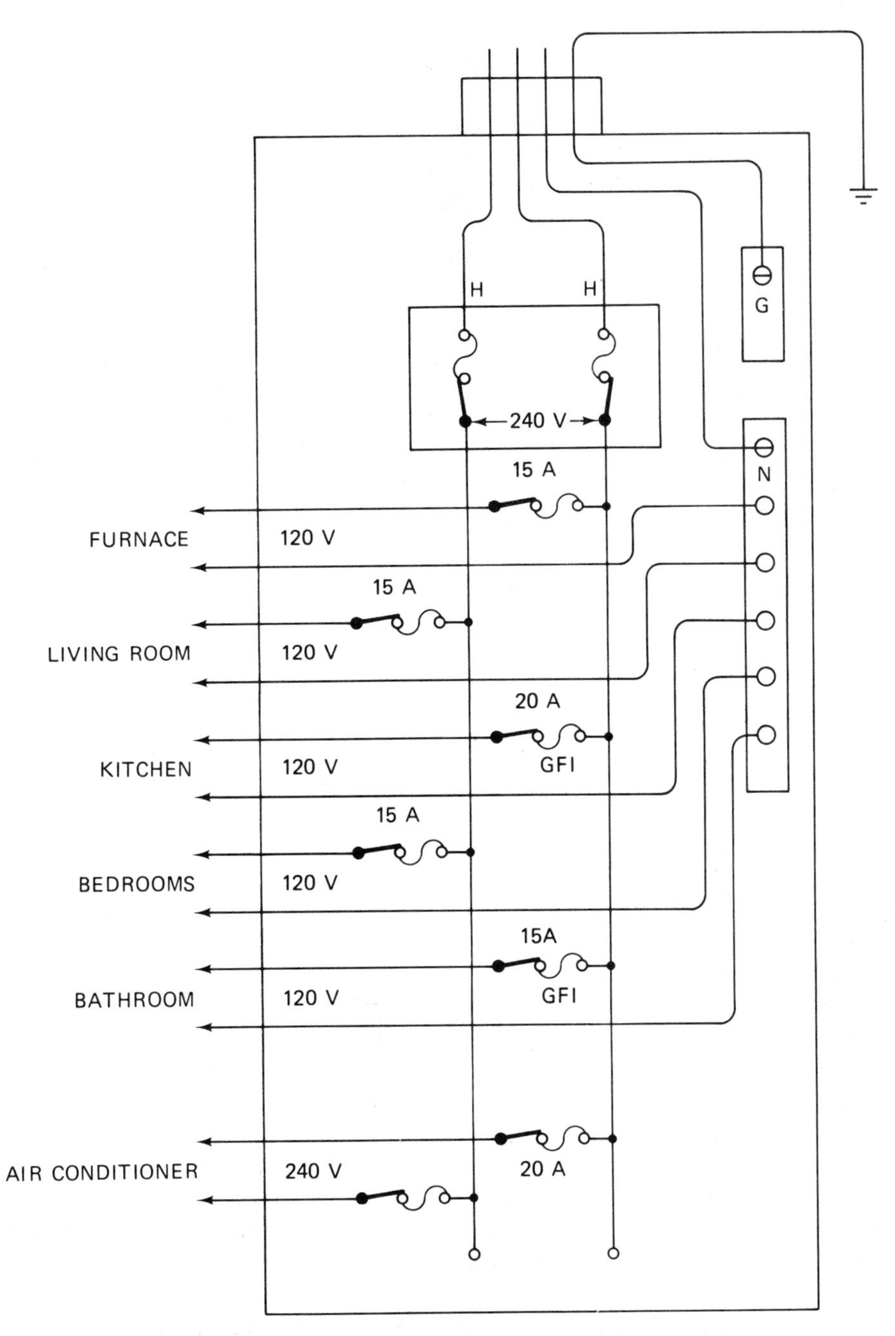

FIGURE 15-6 Branch circuits.

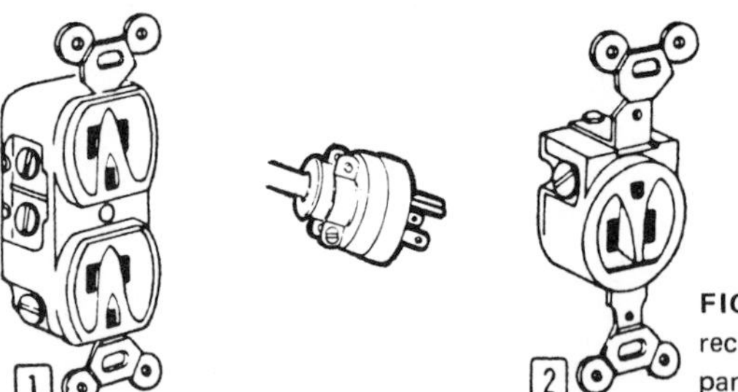

FIGURE 15-7 120-volt three-pronged plug and receptacles (Courtesy, Sears, Roebuck and Company).

other conducting material. Grounding is done chiefly for safety, and all mechanical equipment should be *grounded.* Modern three-pronged plugs and receptacles for 120-V circuits have a grounding wire (Figure 15-7).

A *ground* is the common return circuit in electric equipment whose potential is zero. A ground permits the current to get through or around the insulation to normally exposed metal parts which are hot or "live."

A proper ground protects a user from electrical shock should a short circuit occur.

LOW VOLTAGE:

All load devices are designed to operate at a specified voltage, marked on the equipment. The voltage may be 120, 208, or 240 V. Most electrical equipment will tolerate a variation of voltage from 10% above to 10% below rated (specified) voltage. Thus, a motor rated at 120 V will operate with voltages between 108 and 132 V.

Low or high voltage is usually considered to be any voltage that is not within the tolerated range. Thus, if a 120-V motor is supplied with 100-V power, the voltage would be too low and the motor would probably fail to operate properly.

Low voltage is a much more common problem than high voltage for the reason that all current-carrying wire offers a resistance to flow, which results in a voltage drop. For example, a 50-ft length of No. 14 wire would have a $3\frac{1}{2}$-V drop when carrying 15 A. However, this is not excessive since the resultant voltage is within the 10% allowable limit.

If a circuit is overloaded, the current rises above the rated current-carrying capacity of the wire and the voltage drop can easily exceed the 10% limit established for the load. The low-voltage condition that results from overloaded circuits can cause motors to fail or burn out.

All power supplies should be checked during peak load conditions to determine if proper voltage is being supplied. This should be done at the service entrance, at the furnace disconnect, and at load devices. If there is a problem at the service entrance, the power company should be contacted.

If there is a problem of overloaded circuits in the building, the owner should be notified and an electrician's services secured. If there is a problem in the furnace wiring, a service technician should look for defective wiring and/or check for proper sizes of wires and transformers.

Materials and Equipment

All electrical power devices are important when analyzing a building's electrical system. In addition, certain special equipment is required to complete the external electrical wiring system to the furnace. This equipment includes:

- Main panel
- Power takeoff
- Fused disconnect
- Cable
- Adapter for wire service conversion

MAIN PANEL:

The existing main electrical panel may be suitable or it may need replacing, depending on power requirements for the equipment being installed. On new installations, the building plans should include adequate main panel service.

There are two types of panels: those with fuses (Figure 15-8), and those with circuit breakers for the individual circuits (Figure 15-9). The circuit breaker type of panel has the advantage of being able to simply reset the

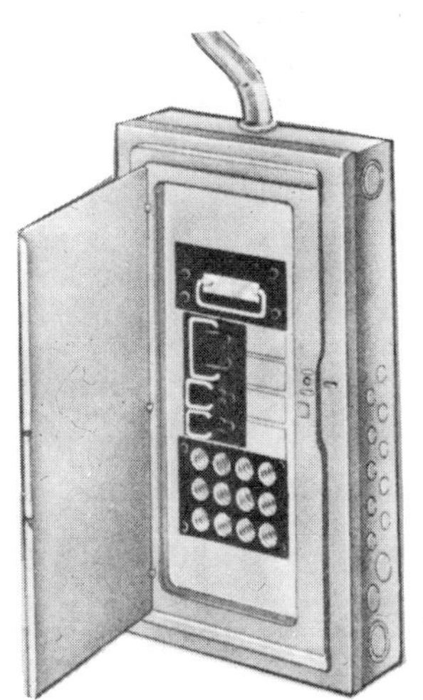

FIGURE 15-8 Fused service panel (Courtesy, Sears, Roebuck and Company).

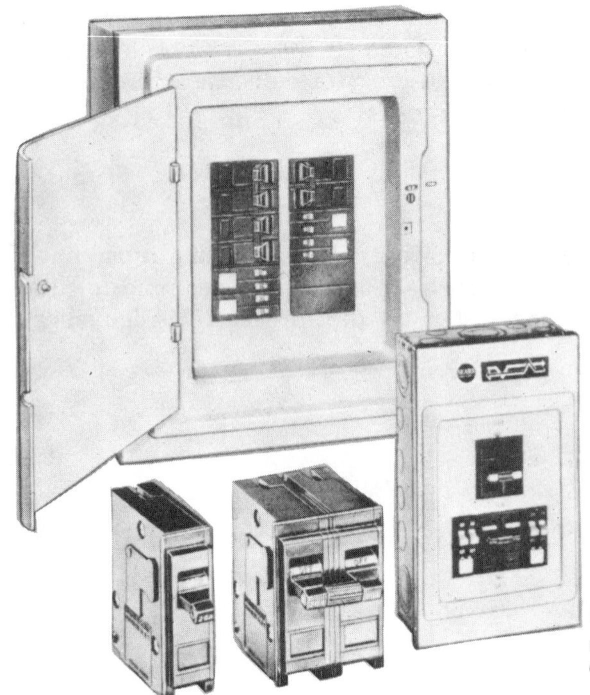

FIGURE 15-9 Circuit breaker service panel (Courtesy, Sears, Roebuck and Company).

breaker rather than having to replace the fuses, making it more convenient to service when an overload occurs. Both types of panels are available with delayed-action tripping to permit motors to draw extra current when starting without shutting down the power supply.

POWER TAKEOFF:

Main panels can have a power takeoff to an additional fuse panel. This arrangement permits adding a circuit where necessary to supply an additional load. A power takeoff using four 15-A circuits is shown in Figure 15-10.

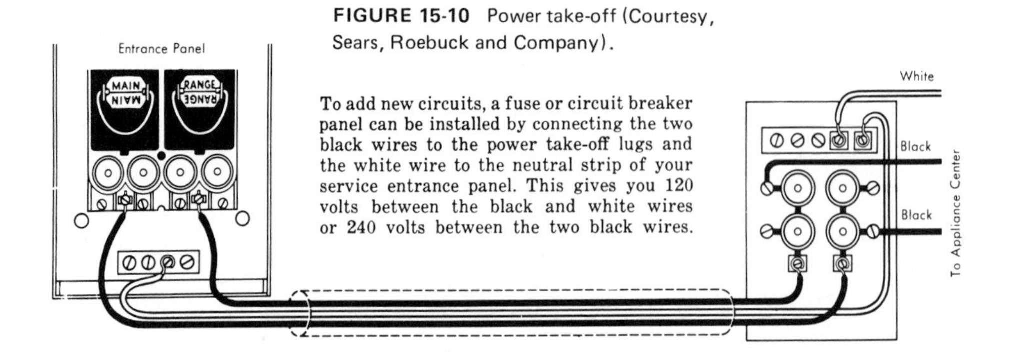

FIGURE 15-10 Power take-off (Courtesy, Sears, Roebuck and Company).

FUSED DISCONNECT:

A fused disconnect is used to disconnect and protect a 120-V ac circuit, and is located within a few feet from the unit it serves. It is sometimes called a "service switch" because it permits a convenient means for disconnecting the power to the furnace.

CABLE:

Cable is a protective shield that encloses electrical wires. In most areas, all line-voltage wiring must be enclosed in approved-type cable with wire connections made in an approved junction box.

There are four types of cable (Figure 15-11):

1. Indoor-type plastic-sheathed
2. Dual-purpose plastic-sheathed
3. Flexible armored
4. Thin-wall and rigid conduit

The dual-purpose type of cable can be used outdoors or indoors without conduit. The method of connecting cable to junction boxes is shown in Figure 15-12 (nonmetallic cable), and Figure 15-13 (thin wall conduit). Grounding is not required when using conduit.

FIGURE 15-11 Types of cable (Sears, Roebuck and Company).

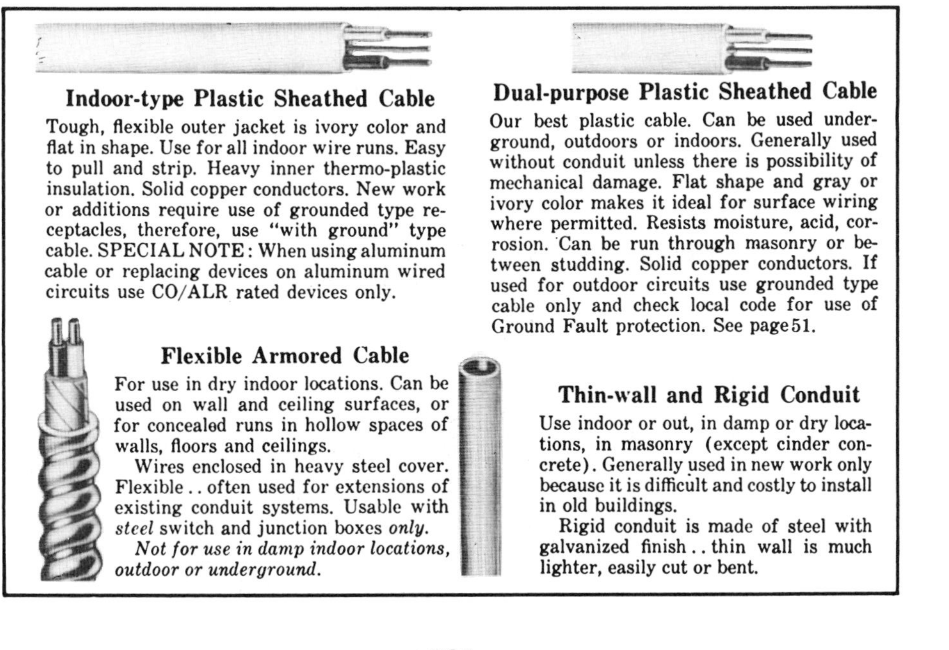

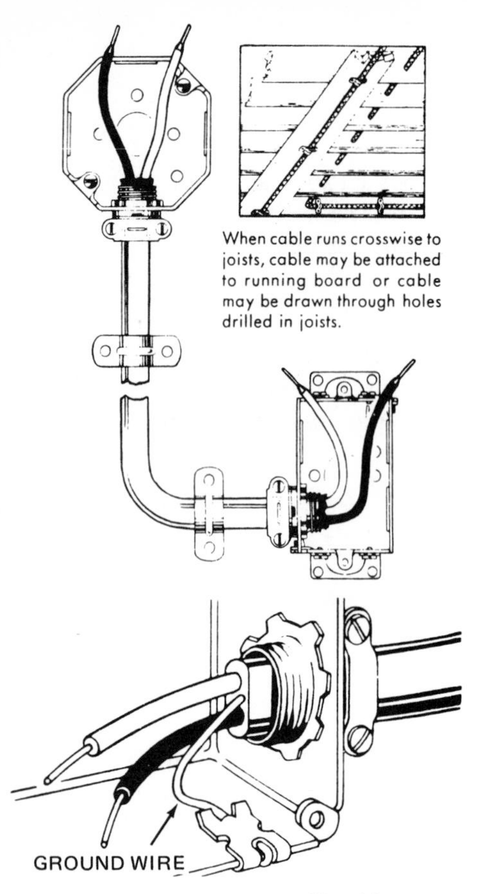

FIGURE 15-12 Nonmetallic cable connectors (Courtesy, Sears, Roebuck and Company).

FIGURE 15-13 Thinwall (metallic) cable connectors (Courtesy, Sears, Roebuck and Company).

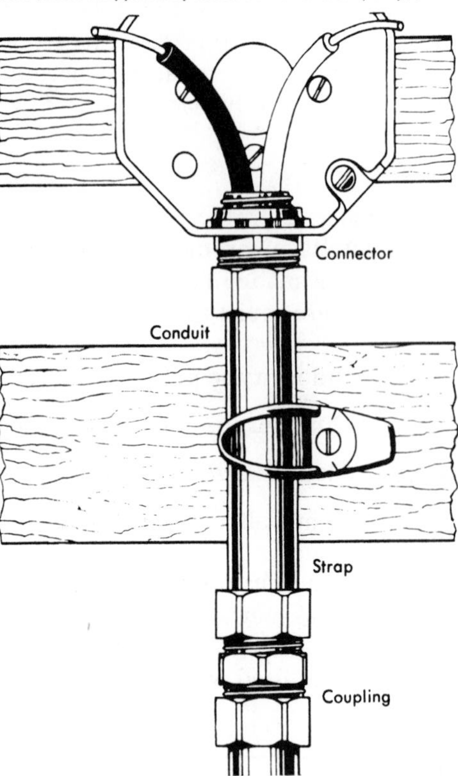

FIGURE 15-14 Adapter for conversion of 2-wire to 3-wire system (Courtesy, Sears, Roebuck and Company).

ADAPTER:

An adapter converts two-wire service to three-wire, 120-V service. After proper grounding, existing systems which include only two wires may be converted to three-wire systems by use of an adapter arrangment, shown in Figure 15-14.

GROUND FAULT CIRCUIT INTERRUPTERS:

Fuses and circuit breakers protect circuits and wire against overloads and short circuits but not against current leakage. Small amounts of leakage can occur without blowing a fuse or tripping a breaker. Under certain conditions the leakage can be hazardous.

A relatively new product adapted for residential use is called the *ground fault circuit interrupter* (GFCI), see Figure 15-15. The GFCI is designed to detect and interrupt the power supply quickly enough to prevent a serious problem. After the problem is corrected the GFCI can be reset and power at that point will be restored.

The National Electrical Code recommends the use of GFCIs on all outdoor circuits. Local codes are especially strong on their use in conjunction with swimming pools that have any electrical connections.

FIGURE 15-15 Ground fault circuit interrupters (Courtesy, Sears, Roebuck and Company).

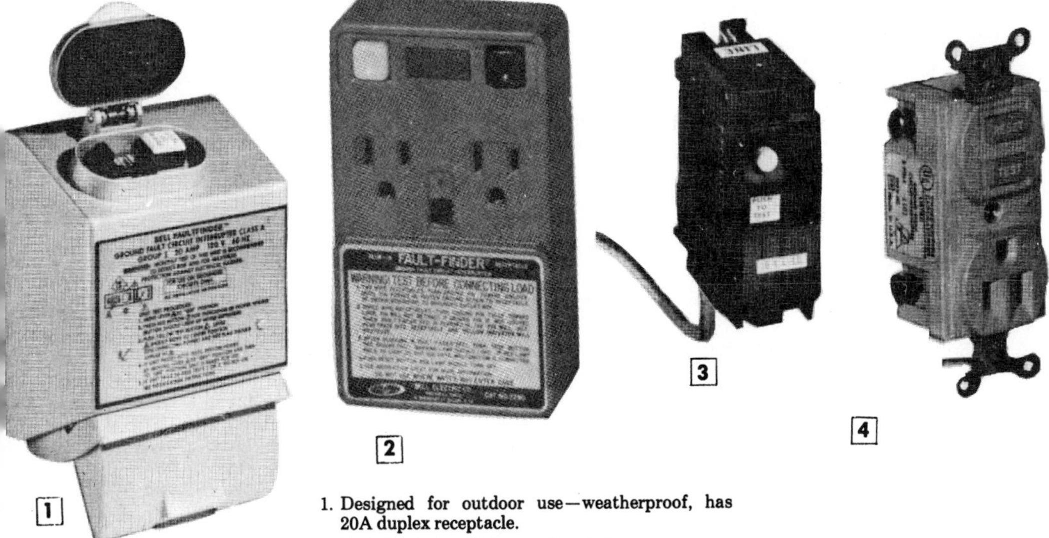

1. Designed for outdoor use—weatherproof, has 20A duplex receptacle.
2. Portable 15A unit designed for indoor use.
3. Breaker type fits only specific panels, check that point carefully. Available in 15 and 20A ratings.
4. Wall Unit receptacle which if wired first into series of receptacles can protect all other outlets on the circuit.

All come with complete wiring instructions.

REVIEW QUESTIONS

Select the letter representing your choice of the correct answer.

15-1. What is the service amperage for most gas furnaces (without cooling)?

(a) 5

(b) 10

(c) 15

(d) 20

15-2. How many wires are used for the main power supply to a building?

(a) Two

(b) Three

(c) Four

(d) Five

15-3. If a residence has electric heating, what amperage service to the building is required?

(a) 600

(b) 100

(c) 150

(d) 200

15-4. How many fuses should be used for a 240-V branch circuit?

(a) One

(b) Two

(c) Three

(d) Four

15-5. What is the minimum voltage that should be used for a motor rated at 120 V ac?

(a) 108

(b) 110

(c) 112

(d) 114

15-6. What is the normal voltage drop using 50 ft of No. 14 wire carrying 15 A?

(a) $1\frac{1}{2}$

(b) $2\frac{1}{2}$

(c) $3\frac{1}{2}$

(d) $4\frac{1}{2}$

15-7. When should power supplies be checked for proper voltage?

(a) During full load

(b) During part load

(c) On startup

(d) With load disconnected

15-8. What is the connection to the main panel called when an additional fuse panel is added?

(a) Disconnect

(b) Conduit

(c) Subpanel

(d) Power takeoff

15-9. Who should be contacted if the voltage is too low coming into the main service panel?

(a) An electrician

(b) The power company

(c) The homeowner

(d) The original contractor

15-10. Metal wire-carrying cable is called:

(a) Indoor-type

(b) Dual-purpose

(c) Romex

(d) Thin wall conduit

16

CONTROLS COMMON TO ALL FORCED AIR FURNACES

OBJECTIVES

After studying this chapter, the student will be able to:

- Identify the common types of electrical control devices used on forced warm air furnaces
- Evaluate the performance of common types of controls in the system

FUNCTION OF CONTROLS

The function of automatic controls is to operate a system or unit in response to some variable condition. Automatic controls are used to turn the various electrical components (load devices) on and off. These devices operate in response to a controller that senses the room, temperature or humidity, thereby activating the equipment to maintain the desired conditions.

Load devices that make up a heating unit are:

- Fuel-burning device or heater
- Fan
- Humidifier
- Electrostatic air cleaner
- Cooling equipment

Each one of these load devices has some type of switch or switches that automatically or manually causes the device to operate. The automatic switches respond to variable conditions through a *sensor*. A sensor is a device that reacts to a change in conditions and is then capable of transmitting a response to one or more switching devices. For example, a thermostat senses the need for heat and switches on the heating unit. When the thermostat senses that enough heat has been supplied, it switches off the heating unit.

COMPONENTS OF CONTROL SYSTEMS

Five elements comprise the control system for a forced warm air heating unit. They are:

1. Power supply
2. Controllers
3. Limit controls
4. Primary controls
5. Accessory controls

THE POWER SUPPLY:

The power supply furnishes the necessary current, at the proper voltage, to operate the various control devices. The fused disconnect and the transformer are parts of the power supply.

CONTROLLERS:

Controllers sense the condition being regulated and perform the necessary switching action on the proper load device. The controller group includes such devices as the thermostat, the humidistat, and the fan control.

LIMIT CONTROLS:

Limit controls shut off the firing device or heater when the maximum safe operating temperature is reached.

PRIMARY CONTROLS:

The gas valve, the flame detector relay for an oil burner, and the sequencer for electric furnaces are types of primary controls. A primary control usually includes some type of safety device. For example, on an oil furnace primary control, the burner will be shut off if a flame is not produced or goes out.

ACCESSORY CONTROLS:

Used to add special features to the control system. One of these is the fan relay, which permits the fan switch on a 24-V thermostat to operate a 120-V fan.

THERMOSTAT

Thermostats control the source of heat to closely maintain the selected temperature in the space being conditioned. A heating system should provide even (consistent) temperatures for the comfort of a building's occupants. It is said that people can sense a change of approximately 1½°F in temperature. A well-regulated system will provide less than 1°F variation.

Sensing Element

A sensing element is usually bimetallic . Thus, the element is composed of two different metals, one is usually copper and the other Invar, bonded together. When heated, copper has a more rapid expansion rate than Invar. When the bimetal is heated, it changes shape. In a heating thermostat, this movement is mechanically connected to a switch that closes on a drop in temperature and opens on a rise in temperature (Figure 16-1).

Bimetallic elements are constructed in various shapes (Figure 16-2). A spiral-wound bimetallic element (Figure 16-3), is compact in construction. For this reason, it is the element used in many thermostats.

Switching Action

Switching action should take place rapidly to prevent arcing, which causes damage to the switch contacts. A magnet is used to provide rapid action. However, the most common type of switching action in use is the

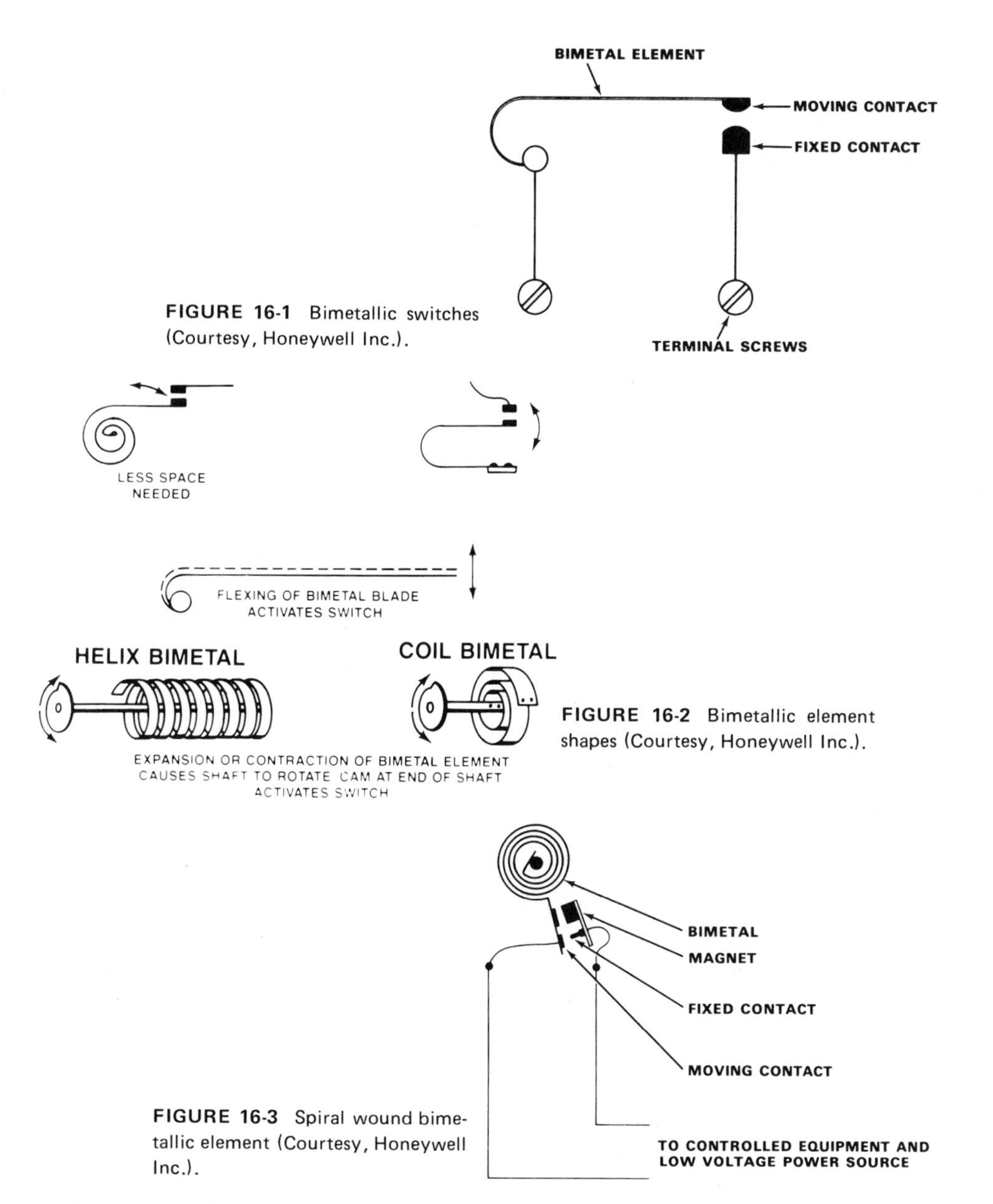

FIGURE 16-1 Bimetallic switches (Courtesy, Honeywell Inc.).

FIGURE 16-2 Bimetallic element shapes (Courtesy, Honeywell Inc.).

FIGURE 16-3 Spiral wound bimetallic element (Courtesy, Honeywell Inc.).

mercury tube arrangement (Figure 16-4). The electrical contacts of the switch are inside the tube together with a globule of mercury. The electrical contacts are located at one end of the tube. When the tube is tipped in one direction, the mercury makes an electrical connection between the contacts and the switch is closed. When the tube is tipped in the opposite direction, the mercury goes to the other end of the tube and the switch is opened.

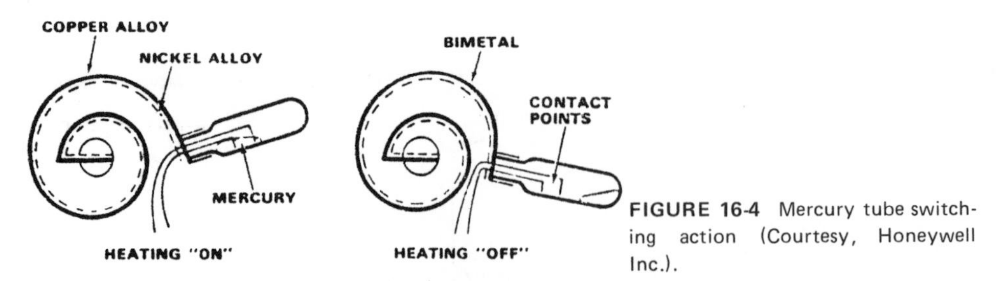

FIGURE 16-4 Mercury tube switching action (Courtesy, Honeywell Inc.).

To produce the required tipping action, the mercury tube is attached to the outside end of the spiral-wound bimetal. Thus, when the temperature drops, the switch closes, and when the temperature rises, the switch opens.

Types of Current

Most residential thermostats are of the low-voltage type (24 V). Low-voltage wires are easier to install between the thermostat and the heating unit. The construction of a low-voltage thermostat provides greater sensitivity to changing temperature conditions than does a line voltage (120-V) thermostat. The materials can be of lighter construction and easier to move, with less chance for arcing.

Some self-generating systems use a millivolt (usually 750-mV) power supply to operate the thermostat circuit. These thermostats are very similar in construction to low-voltage thermostats.

Purpose

Some thermostats are designed for heating only or for cooling only. Other thermostats are designed for a combination of both heating and cooling, (Figure 16-5). With the mercury tube design, a set of contacts is located at one end of the tube for heating and the other end for cooling. The cooling contacts make on a rise in temperature and break on a drop in temperature. The tube with both sets of contacts is attached to a single bimetallic element.

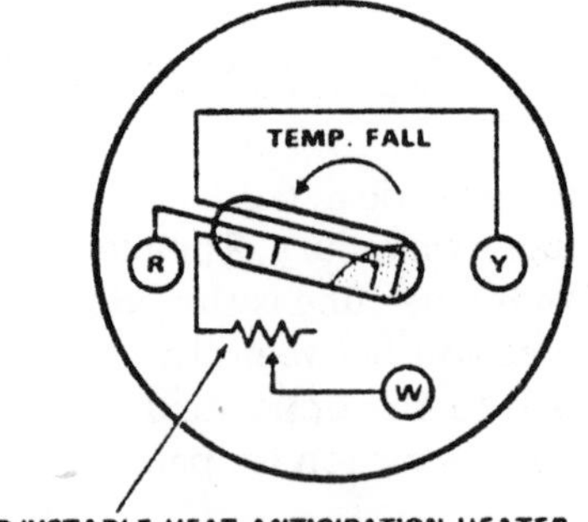

FIGURE 16-5 Heating-cooling thermostat (Courtesy, Honeywell, Inc.).

Anticipator

There is a lag (time delay) between the call for heat by the thermostat and the amount of time it takes for heat to reach a specific area. The differential of the thermostat plus the heat lag of the system could cause a wide variation in room temperature.

There is always a differential (difference) between the temperature at which the thermostat makes (closes) and the temperature at which it breaks (opens). For example, the thermostat may call for heat when the temperature falls to 70° F, and shut the furnace off when the temperature rises to 72° F.

To provide closer control of room temperature, a heat *anticipator* is built into the thermostat (Figure 16-6). A heat anticipator consists of a resistance heater placed in series with the thermostat contacts. On a call for heat, current is supplied to the heater. This action heats the bimetallic element, causing it to respond somewhat ahead of the actual rise in room temperature. By supplying part of the heat with the anticipator, a lesser amount of room heat is required to meet the room thermostat setting. Thus, the furnace shuts off before the actual room temperature reaches the cutout point on the thermostat. Although the furnace continues to supply heat for a short period after the unit shuts off, the actual room temperature does not exceed the setting of the thermostat because the anticipator assists in producing even heating and prevents wide temperature variations.

The heat anticipator is adjustable and should be set at the amperage indicated on the primary control. Small variations from the required setting can be made by the service technician to improve performance on individual jobs.

Decreasing the setting of the anticipator increases the resistance of the anticipator, thus shortening the period that the furnace remains on (shortens the heating cycle). For even heating, the furnace should cycle 8 to 10 times per hour.

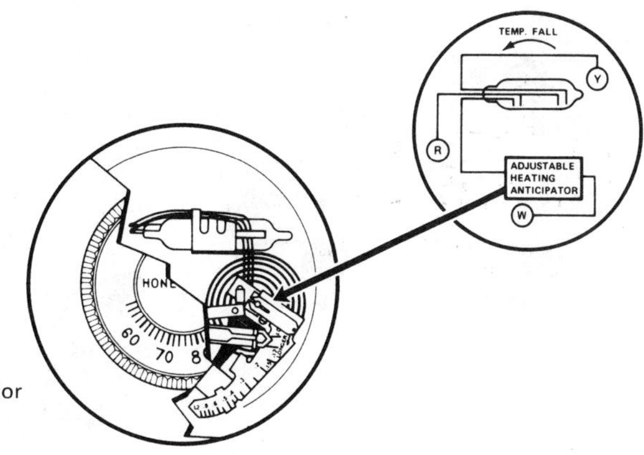

FIGURE 16-6 Heat anticipator (Courtesy, Honeywell, Inc.).

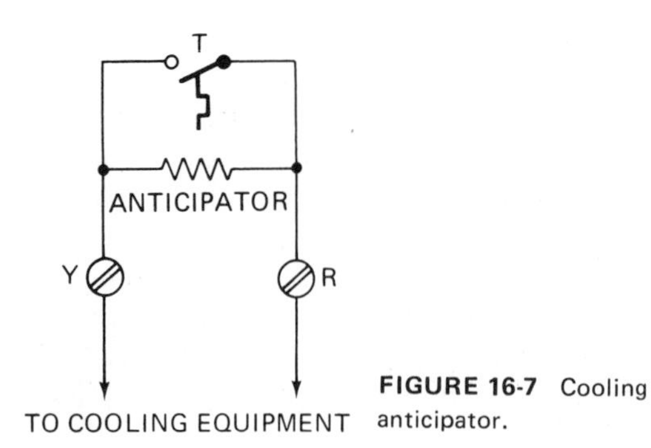

FIGURE 16-7 Cooling anticipator.

Anticipators are also supplied for thermostats that control cooling. A cooling anticipator (Figure 16-7) is placed in parallel with the thermostat contacts. Thus, heat is supplied to the bimetal on the off cycle (when cooling is off) and serves to decrease the length of the shutdown period.

Subbase

Most thermostats have some type of subbase, which serves as a mounting plate (Figure 16-8). A subbase provides a means for leveling and fastening a thermostat to the wall, and contains electrical connections and manual switches. Manual switches offer a homeowner the choice of "heat–off–cool" and fan "auto–on." Switches on a subbase increase the number of functions that can be performed by a thermostat.

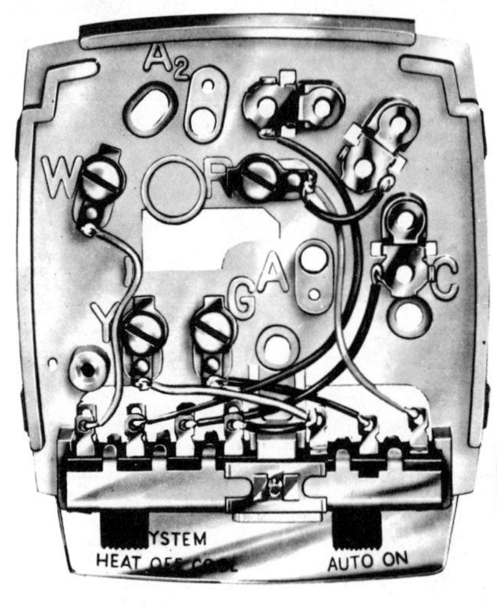

FIGURE 16-8 Thermostat subbases (Courtesy, Robertshaw Controls Company).

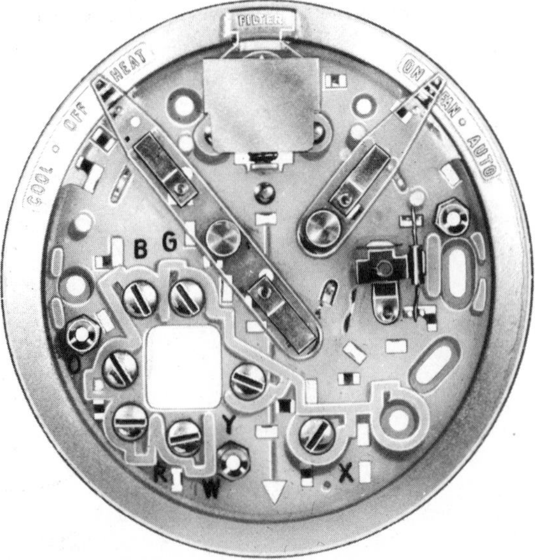

Although a thermostat has only a sensing element and one or two automatic switches, with the addition of subbase manual switches it offers a choice of many switching actions. Some possible switching actions that can be used are described below.

POSITION 1:

One manual switch in "heat" position, the other in fan "auto" position. Heat source operates on a call for heat. Fan is operated by the fan controller.

POSITION 2:

One manual switch in "heat" position, the other in fan "on" position. Heat source operates on a call for heat. Fan runs continuously.

POSITION 3:

One manual switch in "cool" position, the other in fan "auto" position. Cooling equipment and fan operate on a call for cooling.

POSITION 4:

One manual switch in "cool" position, the other in fan "on" position. Cooling equipment operates on a call for cooling. Fan runs continuously.

Letters are used to designate the terminals on a thermostat (Figure 16-9). The two most common designations are by wire color and by function, as shown in the following chart:

	Wire Color	*Function*
Common	R (Red)	V (Common)
Heating	W (White)	H (Heating)
Cooling	Y (Yellow)	C (Cooling)
Fan	G (Green)	F (Fan) or G

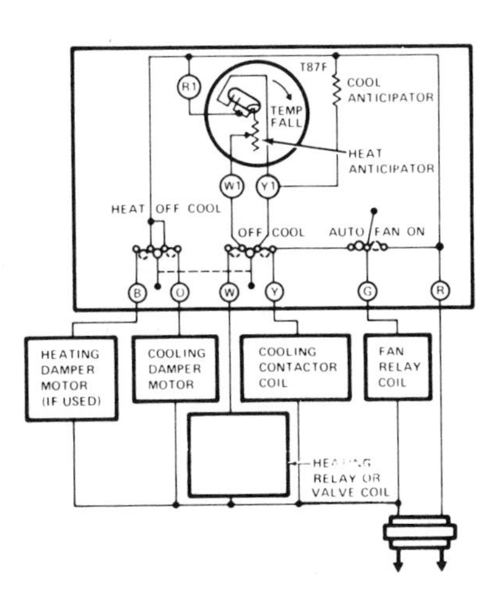

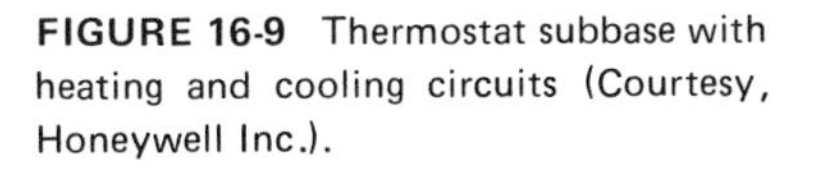

FIGURE 16-9 Thermostat subbase with heating and cooling circuits (Courtesy, Honeywell Inc.).

Using the designated letters in the chart and referring to the manual switch positions shown, the following switching actions are used:

POSITION 1:

R and W make, to call for heat.

POSITION 2:

R and W make to call for heat. R and G make from the manual switch to operate the fan continuously.

POSITION 3:

R and Y make to call for cooling. At the same time, R and G make to operate the fan.

POSITION 4:

R and Y make to call for cooling. R and G make from the manual switch to operate the fan continuously.

Wiring diagrams often show only the terminals of the thermostat and none of the internal circuiting.

Thermostat Installation

The following are some recommended procedures in the installation of a thermostat:

- Install thermostat on an inside wall
- Locate thermostat 5 ft above the floor
- Choose a location in the living room
- Choose an area where natural air movement is good
- Mount the thermostat level
- Adjust the anticipator to agree with the amperage indicated on the primary control
- Seal up any openings in the wall under the mounting plate
- Make all connections secure

HUMIDISTAT

A *humidistat* (Figure 16-10) is a sensing control that measures the amount of humidity in the air and provides switching action for the humidifier. The sensing element is either human hair or nylon ribbon. These materials expand when moist and contract when dry. This movement is used to operate the switching mechanism.

The sensing element for the humidistat can be located either in the return air duct or in the space being conditioned (Figure 16-11). The humidistat is wired in such a way that humidity is added only when the fan is operating.

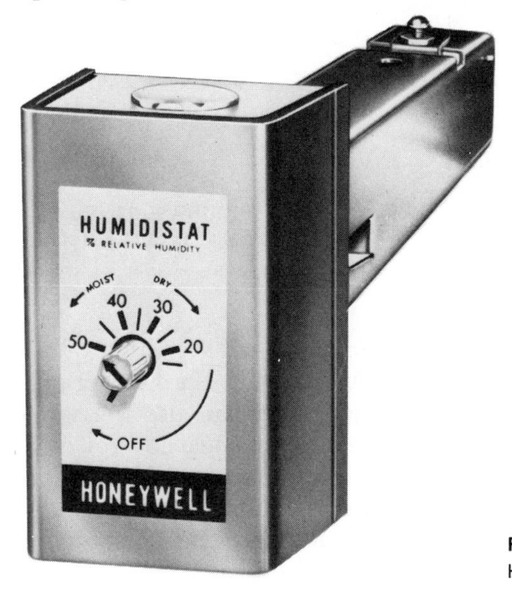

FIGURE 16-10 Humidistat (Courtesy, Honeywell Inc.).

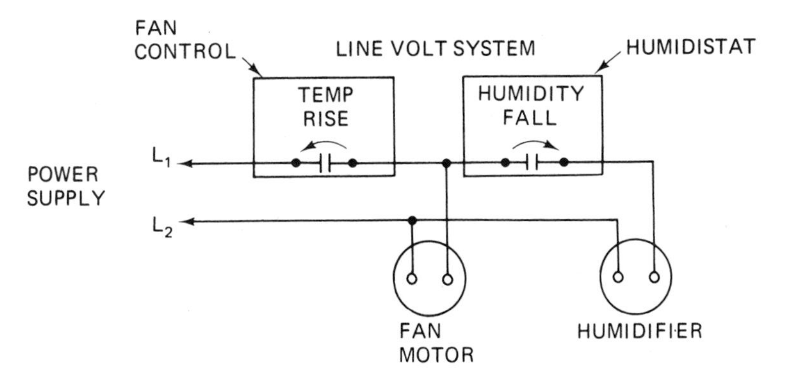

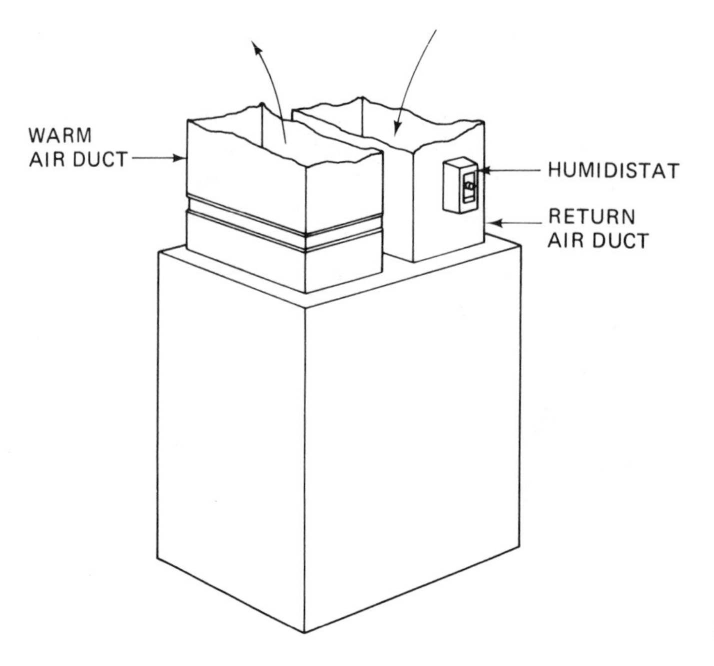

FIGURE 16-11 Duct installation of humidistat.

FAN AND LIMIT CONTROLS

Fan (blower) and limit controls have separate functions in the heating system. However, they are discussed together since both can use the same sensing element and are, therefore, often combined into a single control.

The insertion element (sensor) can be made in a number of forms (Figure 16-12). Some are bimetallic and some are hydraulic. The hydraulic element is filled with liquid which expands when heated, moving a diaphragm connected to a switch.

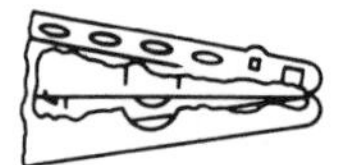

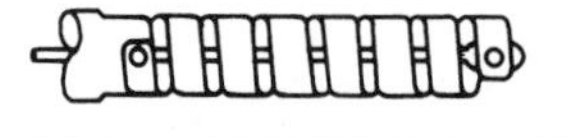

FIGURE 16-12 Types of insertion elements (sensors) (Courtesy, Honeywell Inc.).

The sensing element for these controls is inserted in the warm air plenum of the furnace. It must be located in the moving air stream, where it can quickly sense the warm air temperature rise.

Sensing elements are made in different lengths to fit various applications. When replacing equipment, the original insertion length should always be duplicated.

Fan Control

A fan control (Figure 16-13) senses the air temperature in the furnace plenum and, to prevent discomfort, turns on the fan when the air is sufficiently heated. There are two general types of fan controls:

- Temperature-sensing
- Timed fan start

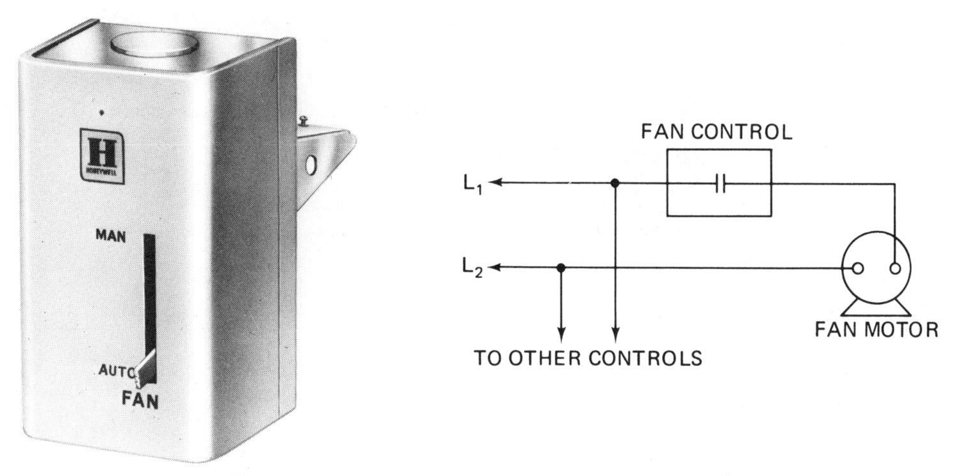

FIGURE 16-13 Fan control (Courtesy, Honeywell Inc.).

TEMPERATURE-SENSING FAN CONTROLS:

These controls depend on the gravity heating action of the furnace to move air across the sensing element. When the air temperature reaches fan "on" temperature (usually about 110° F), the fan starts. When the thermostat is no longer calling for heat, the fan continues to run to move the remaining heat out of the furnace. The fan stops at the fan "off" temperature setting of the fan control (usually about 90° F). Most fan controls are adjustable so that changes in fan operation can be made to fit individual requirements. The fan control circuit is usually line voltage (120 V). The fan control switch is placed in series with the power to the fan motor. Some fan controls have a fixed differential. The fan "on" temperature can be set but the differential (fan "on" minus fan "off" temperature) is set at the factory. The fixed differential is normally 20 to 25° F.

TIMED FAN START CONTROLS:

These controls (Figure 16-14), are used on a downflow or horizontal furnace where gravity air movement over the heat exchanger cannot be depended upon to warm the sensing element. This control has a low-voltage (24-V) resistance heater which is energized when the thermostat calls for heat. The fan starts approximately 60 s after the heat is turned on. Usually, the fan is turned off by a conventional temperature sensing fan control. It is important that the manufacturer's instructions be followed for setting the heat anticipator in the thermostat.

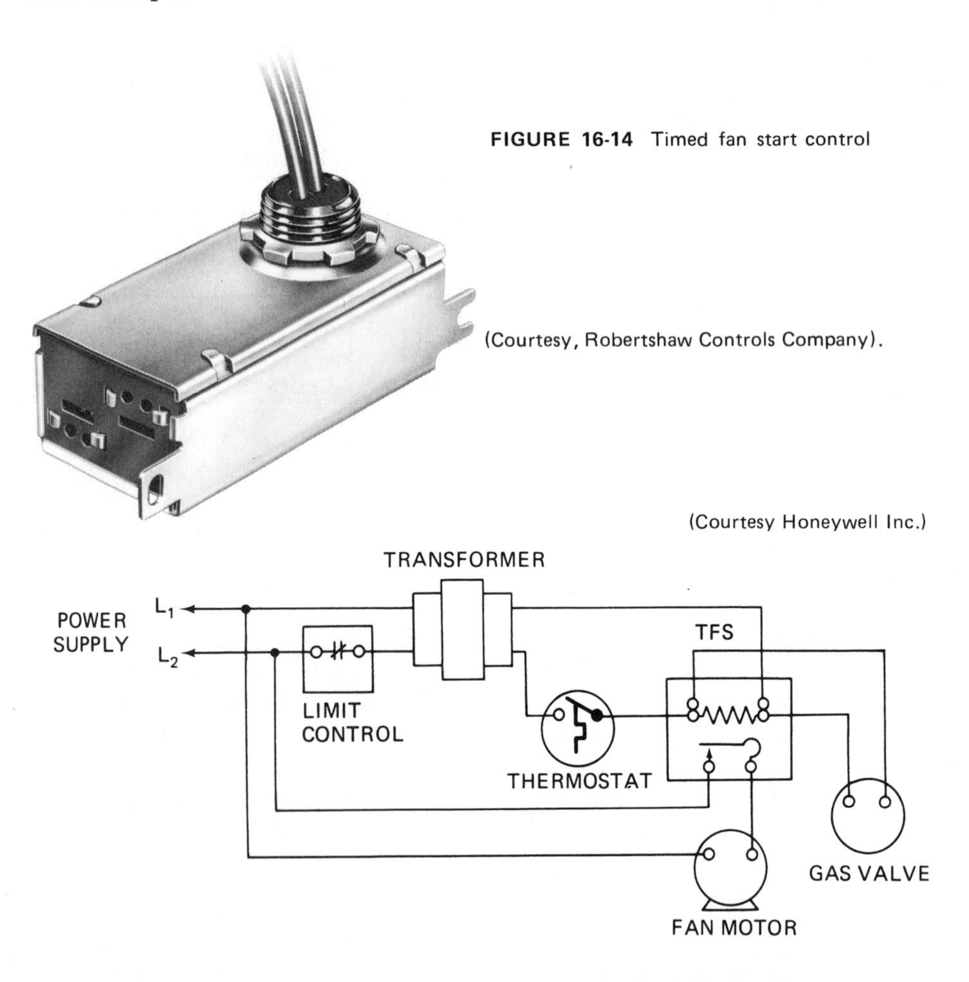

FIGURE 16-14 Timed fan start control (Courtesy, Robertshaw Controls Company).

(Courtesy Honeywell Inc.)

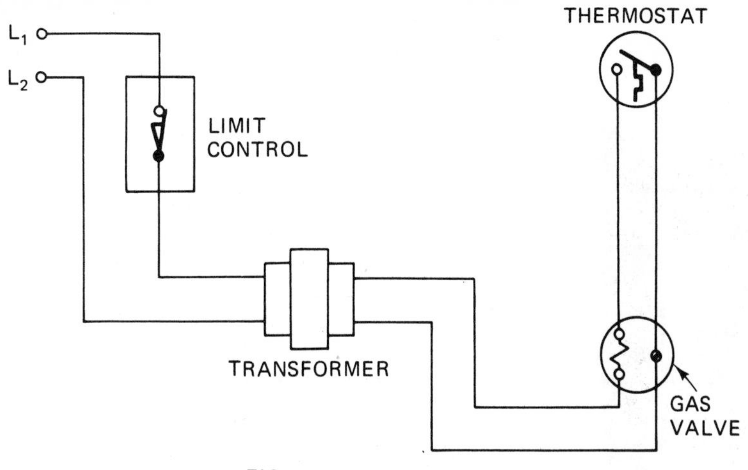

FIGURE 16-15 Limit control.

Limit Control

The *limit control* is a safety device that shuts off the source of heat when the maximum safe operating temperature is reached (Figure 16-15). Like the fan control, it senses furnace plenum temperature or air temperature at the outlet. Warm air furnace limit controls are usually set to cut out at 200°F and automatically cut in at 175°F.

On a gas furnace the line-voltage limit control is placed in series with the transformer. On an oil furnace the line-voltage limit control is placed in series with the primary control. On an electric furnace, the limit controls are placed in series with the heating elements.

On a horizontal or downflow furnace, a secondary limit control is required (Figure 16-16). This control is located above the heating element and is usually set to cut out the source of heat when the sensing temperature reaches 115 to 130°F.

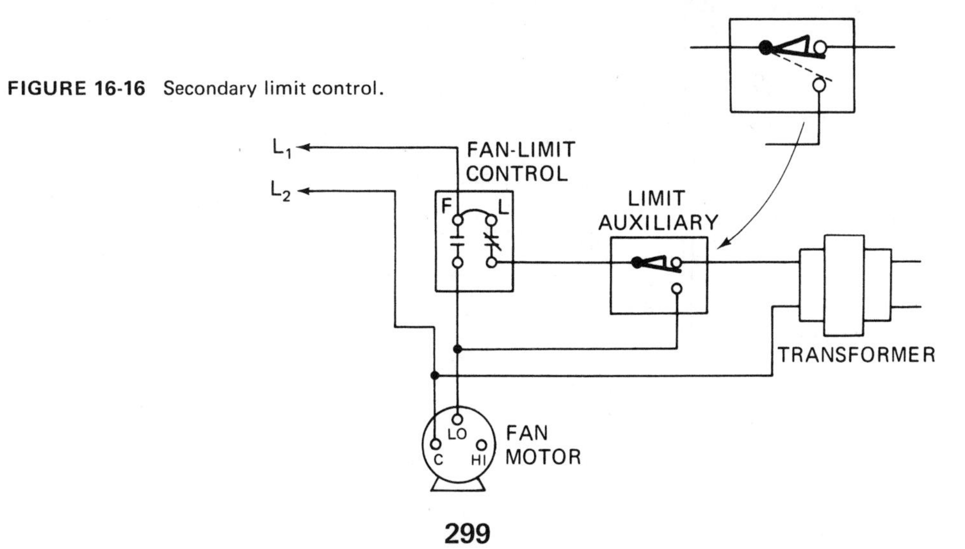

FIGURE 16-16 Secondary limit control.

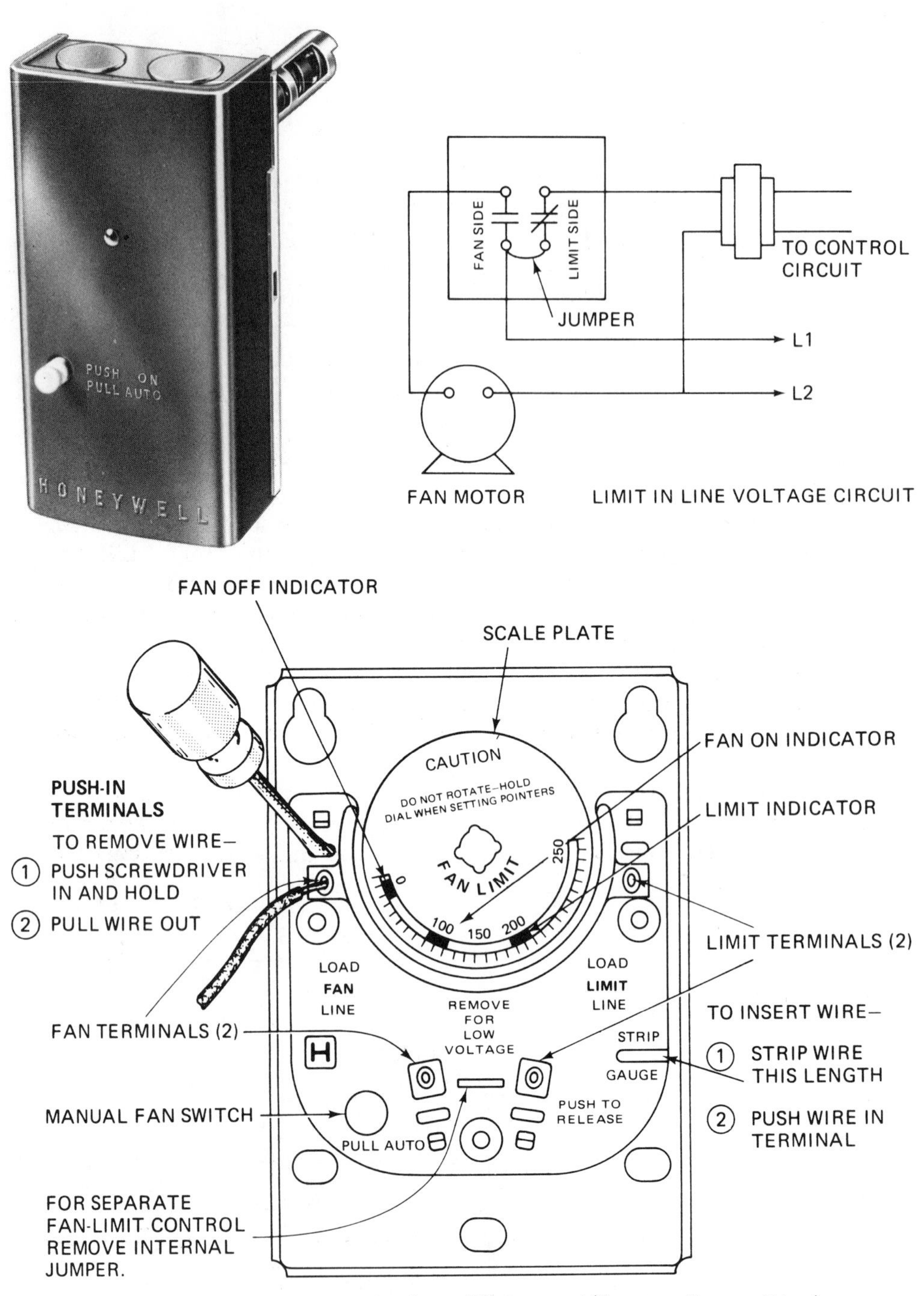

FIGURE 16-17 Combination fan and limit control (Courtesy, Honeywell Inc.).

Where two limit controls are used, they are usually wired in series with each other. Thus, either limit control can turn off the source of heat.

Some secondary limit controls have a manual reset (not automatic). Some have an arrangement for switching on the fan when the limit contacts are opened.

Combination Fan and Limit Control

Since the fan and limit controls both use the same type of sensing element, they are often combined into one control (Figure 16-17).

This is helpful when both the fan and limit controls are of line-voltage type, since this simplifies the wiring. However, the fan and limit controls each have separate switches and separate terminals. The fan control can be wired for line voltage and the limit control for low voltage, if desired.

Two-Speed Fan Control

Two fan speeds can be obtained by adding a fan relay to a standard control system (Figure 16-18). The fan is usually operated at low speed on heating and high speed for ventilation or air conditioning. The relay has a low-voltage coil in series with the G terminal on the thermostat. When the thermostat calls for fan operation, an n.c. switch in the relay opens and an n.o. switch closes. This switches the fan from low to high speed.

FIGURE 16-18 Two-speed fan control (Courtesy, Honeywell Inc.).

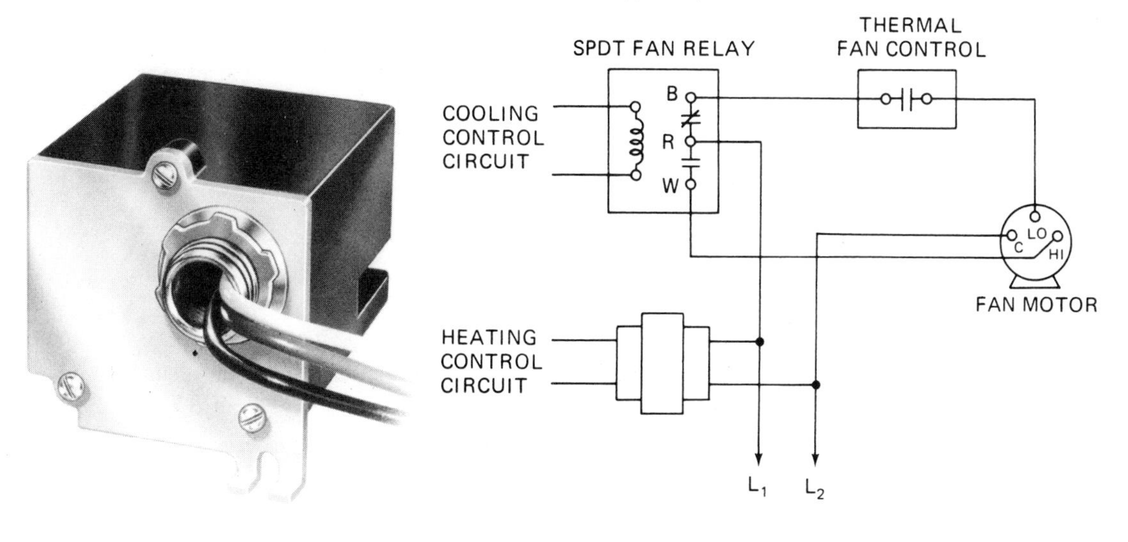

REVIEW QUESTIONS

Select the letter representing your choice of the correct answer(s).

16-1. Which of the following is not a load device?

(a) Fan

(b) Humidifier

(c) Heater

(d) Fan control

16-2. Which basic element of the furnace control system includes the sensor?

(a) Power supply

(b) Gas valve

(c) Limit devices

(d) Fan motor

16-3. What is the smallest variation in °F that people can sense?

(a) ½

(b) 1½

(c) 2½

(d) 4

16-4. A low-voltage control system operates on how many volts?

(a) 24

(b) 30

(c) 120

(d) 240

16-5. On a bimetallic thermostat the metal used in addition to copper is:

(a) Zinc

(b) Chromium

(c) Invar

(d) Iron

16-6. What is the largest number of switches commonly found in a single mercury tube element?

(a) One

(b) Two

(c) Three

(d) Four

16-7. A self-generating system usually uses its own control circuit of how many volts or millivolts?

(a) 30 mV

(b) 750 mV

(c) 24 V

(d) 120 V

16-8. If a thermostat cuts in at 70°F and cuts out at 72°F, the differential is

(a) Zero

(b) 2°F

(c) 70°F

(d) 72°F

16-9. A heating system should cycle how many times per hour?

(a) 1 to 3

(b) 4 to 7

(c) 8 to 10

(d) 11 to 15

16-10. The color code terminals of a thermostat are:

(a) Q, H, C, F

(b) R, G, B, V

(c) T, A, E, H

(d) R, W, Y, G

16-11. A timed fan start delays the operation of the fan approximately how many seconds after the heater is turned on?

(a) 30

(b) 60

(c) 90

(d) 120

16-12. A limit control is usually set to cut out the source of heat at how many °F?

(a) 180

(b) 200

(c) 220

(d) 240

17

GAS FURNACE CONTROLS

OBJECTIVES

After studying this chapter, the student will be able to:

- Identify the various electrical controls and circuits on a gas-fired furnace
- Determine the sequence of operation of the controls
- Service and troubleshoot the gas heating unit control system

USE OF GAS FURNACE CONTROLS

Broadly speaking, gas furnace controls are the electrical and mechanical equipment that manually or automatically operates the unit.

The controls used depend upon:

1. Type of fuel or energy
2. Type of furnace
3. Optional accessories

The controls differ for gas, oil, or electric heat sources. They differ somewhat for the upflow and downflow units. They also become more complex as accessories such as humidifiers, electrostatic filters, and cooling are added.

CIRCUITS

Control circuits for a gas warm air heating unit are:

1. Power supply circuit
2. Fan circuits
3. Pilot circuit
4. Fuel-burning or heater circuit
5. Accessory circuits

Both line-voltage and low-voltage circuits are required for some components. For example, a relay used to start a fan motor may have a low-voltage coil but the switch that starts the fan motor operates on line voltage.

Schematic diagrams are used in the study of circuits. Following is the key to the legends used on schematics described in this chapter.

CAP	Capacitor
CC	Compressor contactor
EAC	Electrostatic air cleaner
EP	Electric pilot
FC	Fan control
FM	Fan motor
FD	Fused disconnect
FR	Fan relay
G	Terminal
GV	Gas valve
H	Humidistat
HS	Humidification system
HU	Humidifier

L Limit

LA Limit auxiliary

L, L_2 Power supply

N Neutral

NC Normally closed

NO Normally open

PPC Pilot power control

SPDT Single-pole double-throw

SPST Single-pole single-throw

TFS Timed fan start

THR Thermocouple

TR Transformer

W Heating terminal

Y Cooling terminal

Power Supply Circuit

The power supply circuit (Figure 17-1) usually consists of a 15-A source of 120-V ac power, a fused disconnect (FD) in the "hot" line, and a transformer (TR) to supply 24-V ac power to the low-voltage controls.

A single source of power is adequate unless a cooling accessory is added, in which case a separate source of 208/240-V ac power is required. Power for cooling requires a double-pole double-throw (DPDT) disconnect with fuses in each "hot" line. Possibly an additional low-voltage transformer would be required, if the one used for heating does not have sufficient capacity.

Fan Circuit

The line-voltage circuit for a single-speed fan consists of a fan control (FC) in series with the fan motor (FM) connected across the power supply (Figure 17-2).

Many heating units have multiple-speed fans. The low speed is used for heating, while the high speed is used for cooling or ventilation.

The operation of the fan with a two-speed motor requires two circuits (Figure 17-3). There is a line voltage circuit to power the fan motor, and a low-voltage circuit to operate the switching circuit.

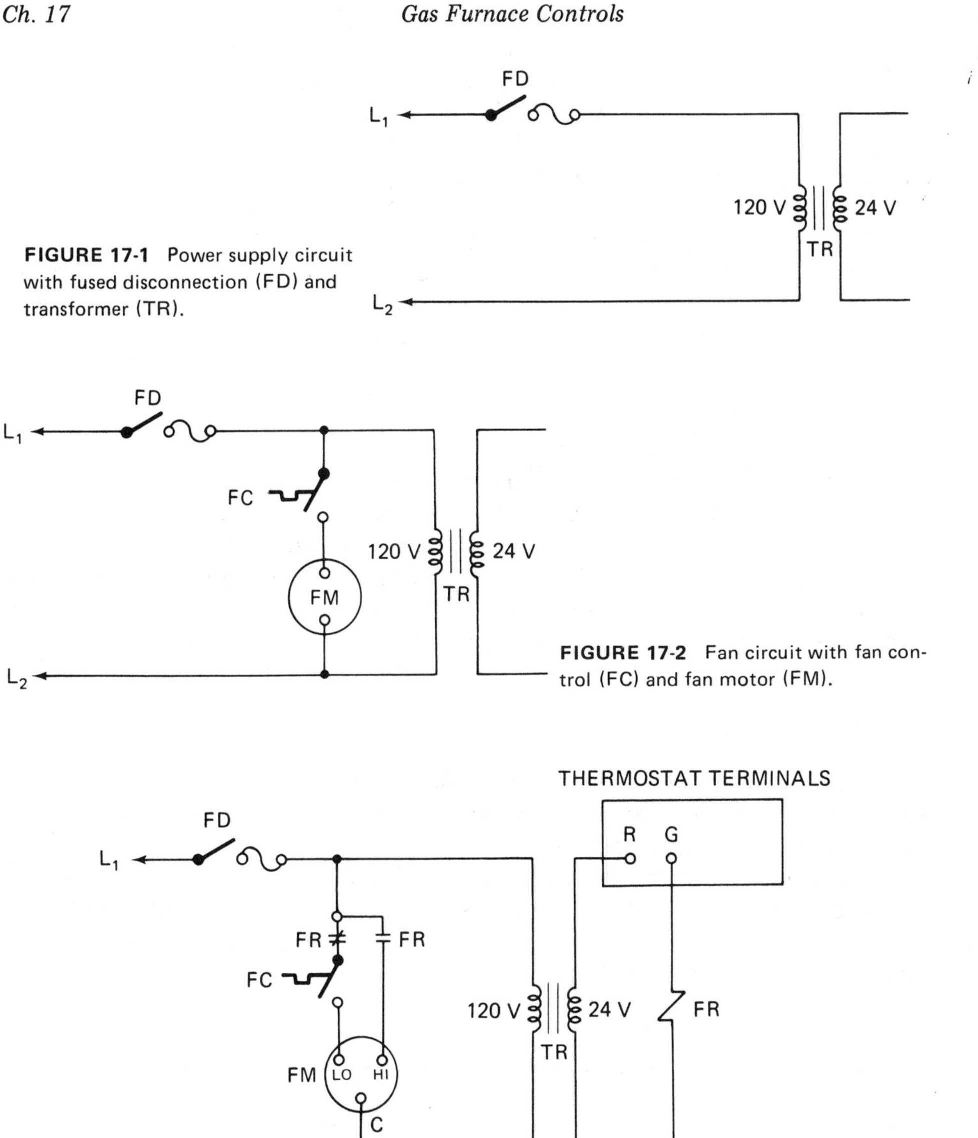

FIGURE 17-1 Power supply circuit with fused disconnection (FD) and transformer (TR).

FIGURE 17-2 Fan circuit with fan control (FC) and fan motor (FM).

FIGURE 17-3 Two-speed fan motor with cooling fan relay (FR).

Under normal operation on heating, the fan motor (FM) is controlled by the fan control (FC). When operating the fan on high speed, the fan switch on the subbase of the thermostat can be moved to the "on" position. This action closes contact between R and G in the thermostat. Current flows through the coil of the cooling fan relay (FR), opening the n.c. switch and closing the n.o. switch operating the fan motor on high speed.

On a downflow or horizontal unit the fan motor can be operated by a timed fan relay (TFS) (Figure 17-4). On these units, with the fan not running and the sensing element of the fan control located in the air outlet, the standard fan control cannot be used to start the fan.

On a call for heating, R and W make in the thermostat. The heater of the timed fan start (TFS) is energized. After about 45 s, the fan motor is operated by the TFS switch. When the thermostat is satisfied and R and W break, the fan motor continues to operate until the leaving air temperature reaches the cut-out temperature setting of the fan control (FC).

Pilot Circuit

The gas safety circuit is powered by a thermocouple (Figure 17-5). When a satisfactory pilot is established the thermocouple generates approximately 30 MV to energize the safety control portion of the combination gas valve. Thus, the gas cannot enter the burner until a satisfactory pilot is established.

Most wiring diagrams omit the pilot circuit, since it is common to all combination gas systems and need not be repeated in each diagram.

Gas Valve Circuit

The gas valve circuit is in a low-voltage circuit (Figure 17-6). When the thermostat calls for heat, R and W make, energizing the gas valve because the limit control is normally closed.

Should excessive temperatures occur (usually 200° F in the plenum), the limit control opens the circuit to deenergize and close the gas valve. The limit control automatically restarts the burner when its cut-in temperature (usually 175° F) is reached.

FIGURE 17-4 Fan control for down-flow furnace using timed fan start (TFS).

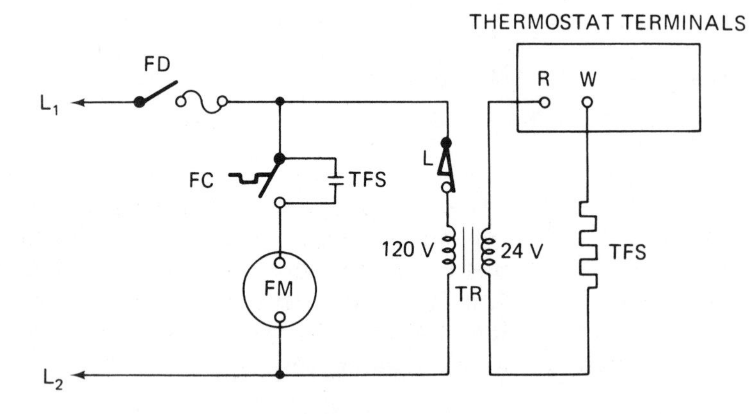

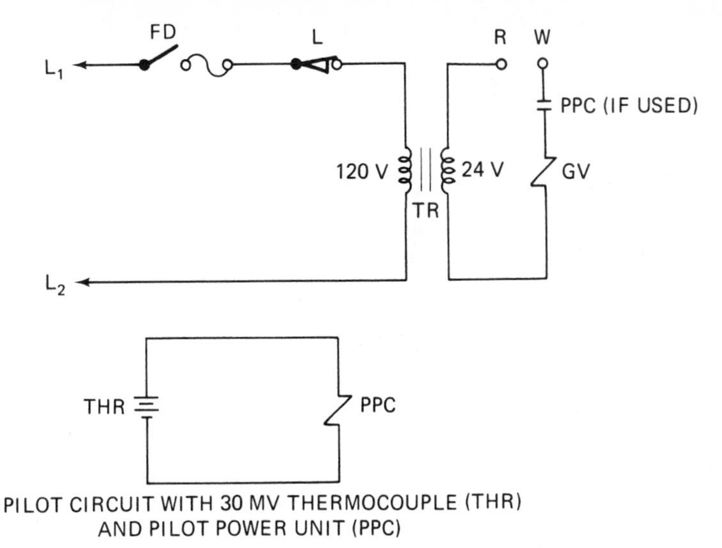

FIGURE 17-5 Pilot circuit with 30mV thermocouple (THR) and pilot power control (PPC).

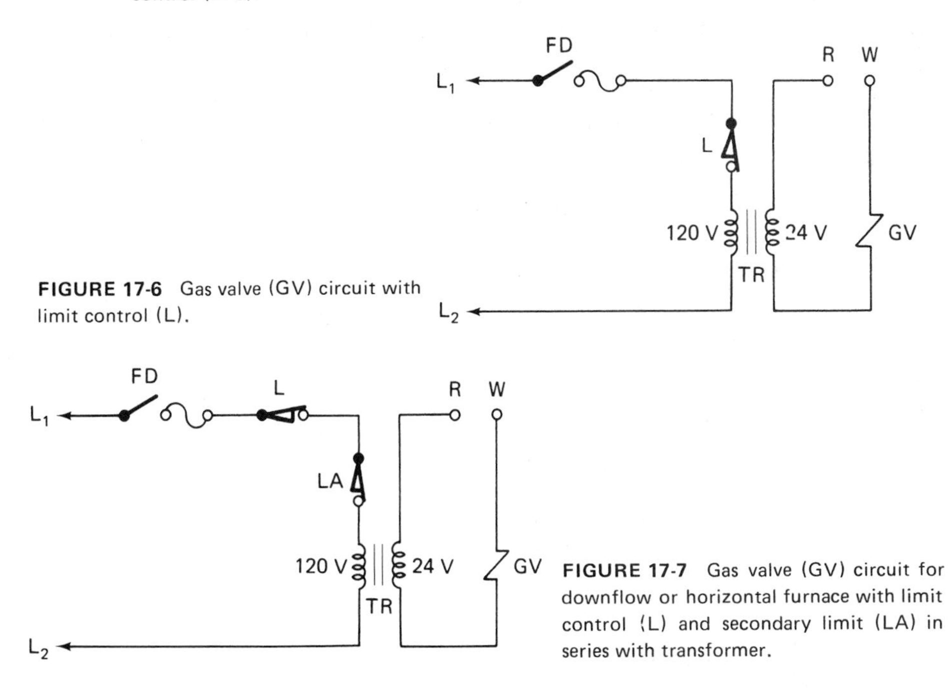

FIGURE 17-6 Gas valve (GV) circuit with limit control (L).

FIGURE 17-7 Gas valve (GV) circuit for downflow or horizontal furnace with limit control (L) and secondary limit (LA) in series with transformer.

On a downflow or horizontal furnace, a limit auxiliary (LA) control is used (Figure 17-7). The LA control is located in a position to sense gravity heat from the heating elements. If the fan fails to start, the LA control will sense excess heat (usually set to cut out at 145° F).

Limit controls are usually line voltage and placed in series with the transformer (Figure 17-8). Some line-voltage limit controls have a separate set of contacts that make when the limit breaks turning on the fan if, for any reason, it is not running.

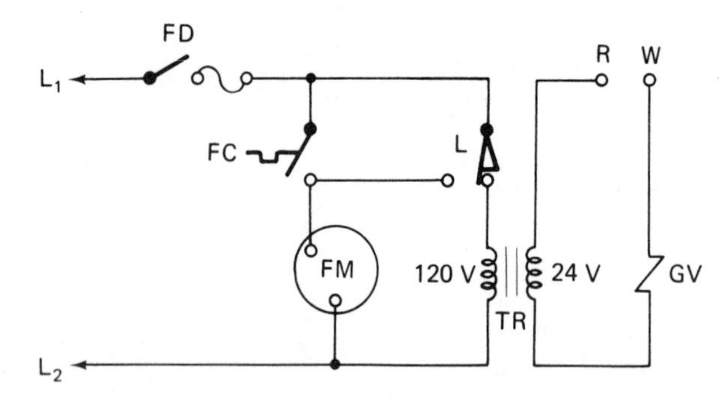

FIGURE 17-8 Single-pole double-throw (SPDT) limit control with set of contacts to operate fan on limit action.

Accessory Circuits

Figure 17-9 shows the schematic of accessory circuits. The humidistat (H) and humidifier (HU) are wired in series and connected to power only when the fan is running. The electrostatic air filter also operates only when the fan is running. These accessories are supplied with line-voltage power.

The cooling contactor (CC) is in series with the Y connection on the thermostat. When R and Y make, the cooling contactor coil is energized, operating the cooling unit.

FIGURE 17-9 Accessory circuits.

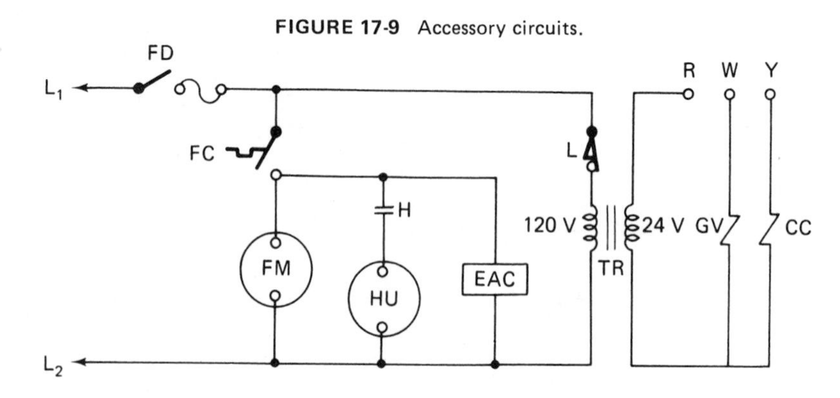

TYPICAL WIRING DIAGRAMS

The foregoing schematic wiring diagrams illustrate the control circuits found on many heating furnaces. A schematic aids a service technician in understanding how the control system operates. However, many manufacturers supply only connection wiring diagrams for their equipment. Therefore, it may be necessary for a technician to construct a schematic in order to separate the circuits for testing and for diagnosing service problems.

Unfortunately, few standards exist which require manufacturers to conform in making connection wiring diagrams. However, if service technicians know the most common controls and their functions, with practice, they can interpret almost any diagram. In some instances when a diagram is not available, a technician can also prepare one to suit the equipment found on the job.

The following is a review of some typical connection diagrams supplied by various manufacturers for their equipment.

Upflow Gas Furnace

The controls in an upflow gas furnace (Figure 17-10) include:

- Thermostat
- Gas valve
- Combination fan and limit control
- Multispeed fan motor with run capacitor
- Transformer
- Fan relay

This system has been prepared for the addition of air conditioning so cooling can be added with very little modification of the control system. A schematic diagram for this unit can be constructed as shown in Figure 17-11.

Downflow Gas Furnace

The controls in a downflow gas furnace (Figure 17-12) include:

- Thermostat
- Gas valve
- Combination fan and limit control
- Auxiliary limit
- Fan motor with run capacitor
- Transformer

A schematic diagram for this unit can be constructed as shown in Figure 17-13.

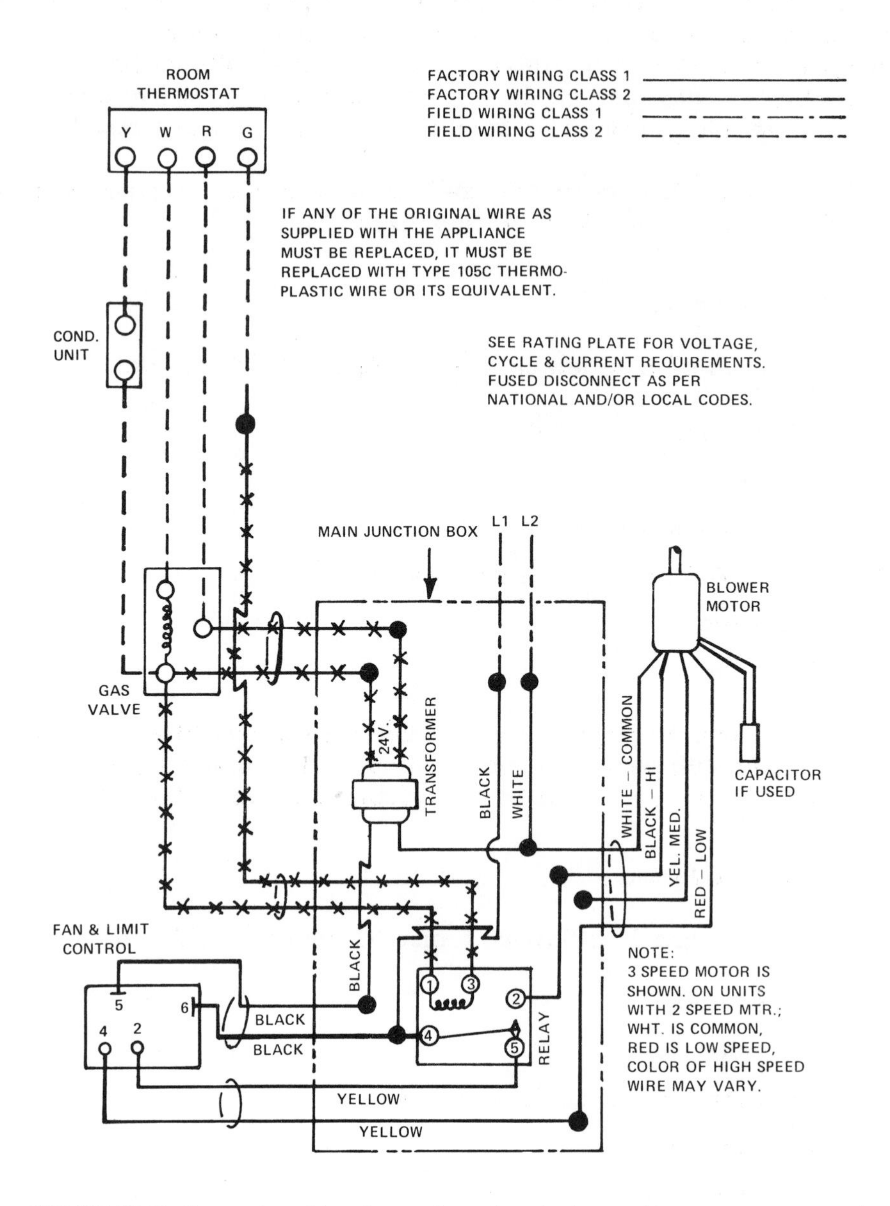

FIGURE 17-10 Connection wiring diagram for upflow gas furnace (Courtesy, Luxaire, Inc.).

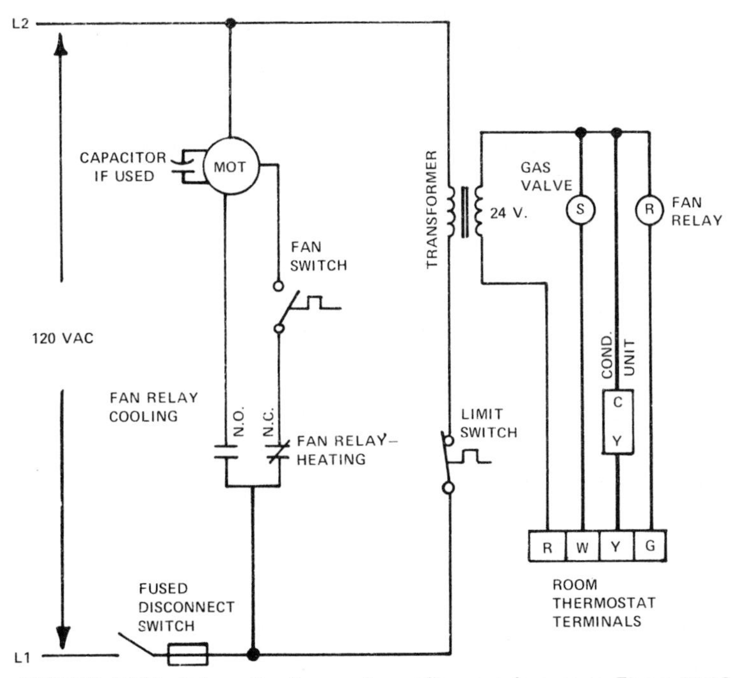

FIGURE 17-11 Schematic diagram for upflow gas furnace in Figure 17-10 (Courtesy, Luxaire, Inc.).

FIGURE 17-12 Connection wiring diagram for downflow gas furnace (Courtesy, Luxaire, Inc.).

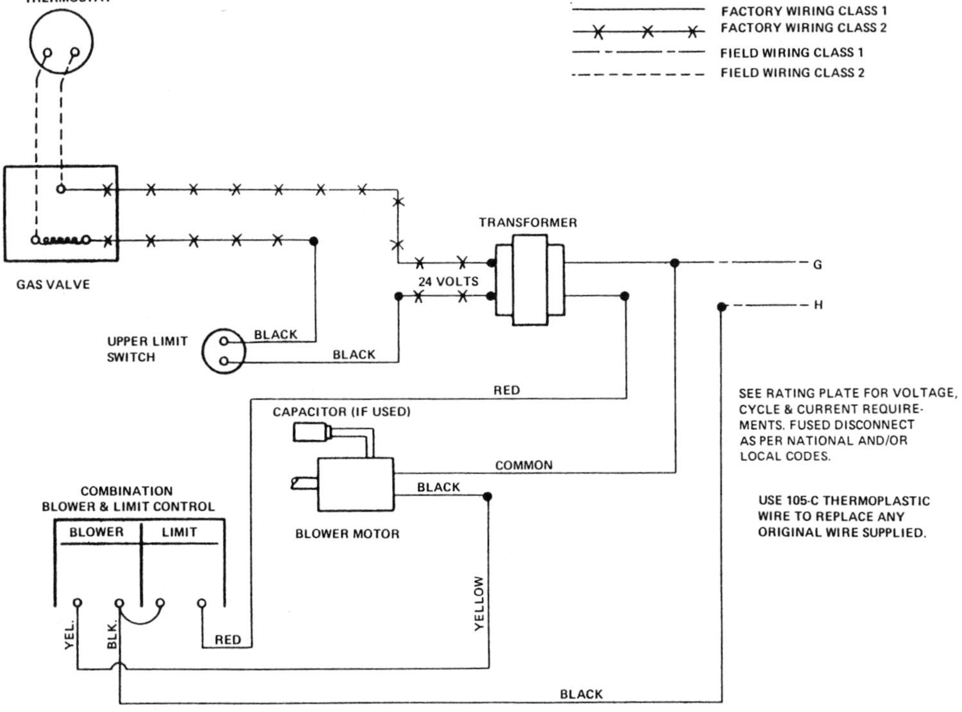

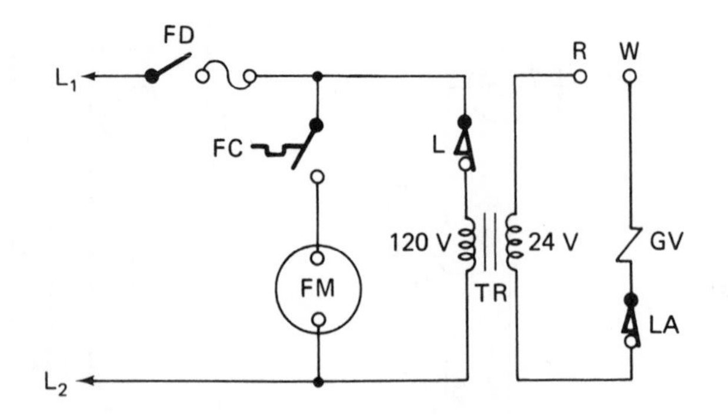

FIGURE 17-13 Schematic diagram for downflow gas furnace shown in Figure 17-11.

Note that Figures 17-12 and 17-13 do not show a timed fan start, which is common to many downflow (counterflow) units. On certain units, the fan control is located in such a position that it senses gravity heat from the heating elements, thus eliminating the need for a timed fan start.

Horizontal Gas Furnace With Cooling Accessory

A connection drawing for a horizontal gas furnace with cooling accessory (Figure 17-14) includes:

- Thermostat
- Gas valve
- Combination fan and limit control
- Auxiliary limit
- Fan motor
- Transformer

A schematic diagram for this unit can be constructed as shown in Figure 17-15.

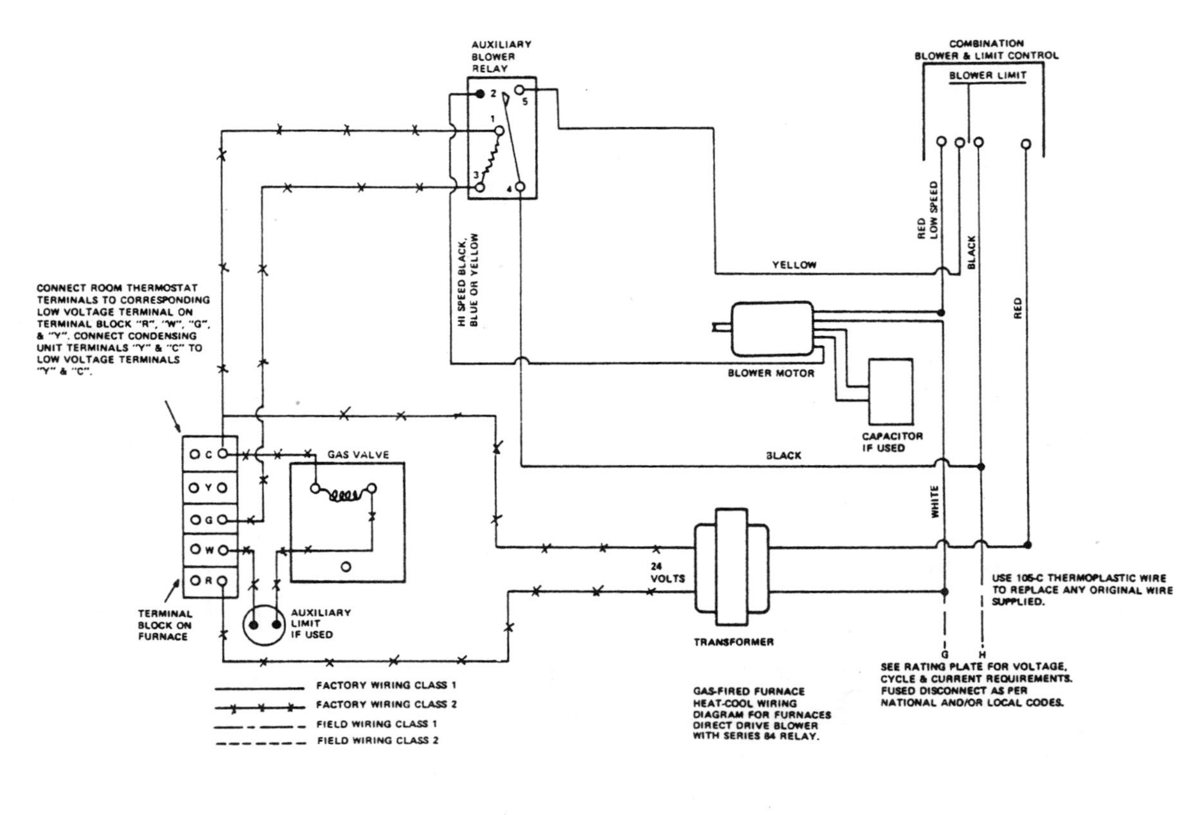

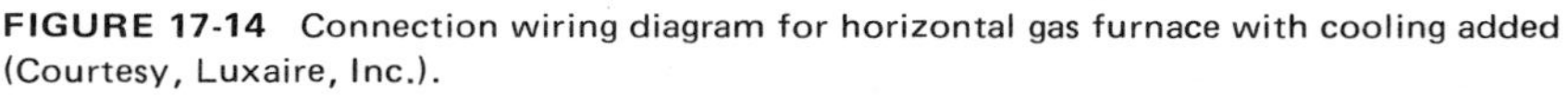

FIGURE 17-14 Connection wiring diagram for horizontal gas furnace with cooling added (Courtesy, Luxaire, Inc.).

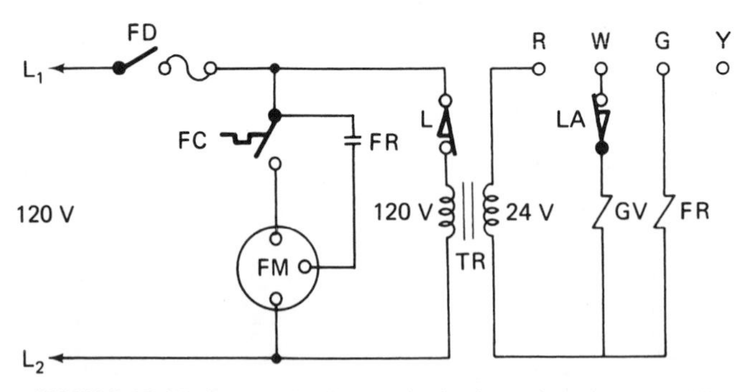

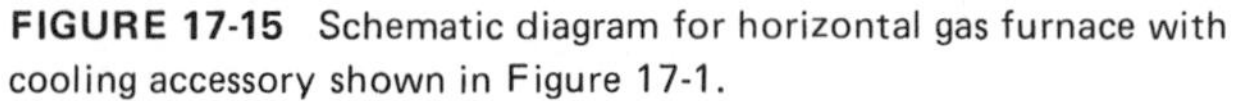
FIGURE 17-15 Schematic diagram for horizontal gas furnace with cooling accessory shown in Figure 17-1.

Upflow Gas Furnace With Printed Circuit Control Center

The controls in an upflow gas furnace with printed circuit control center (Figure 17-16) include:

- Thermostat (subbase shown)
- Gas valve
- Combination fan and limit control
- Fan motor
- Transformer
- Fan relay
- Compressor contactor
- Humidistat
- Humidifier
- Electronic air cleaner
- Electric pilot
- Provision for two-stage gas valve

One special feature of this control system is that the purchaser has the option of using either line-voltage or low-voltage humidification controls and equipment. The schematic diagram (Figure 17-17) shows the use of the line-voltage type. However, the connection drawing shows that an optional plug-in humidifier relay can be used for low-voltage humidification equipment.

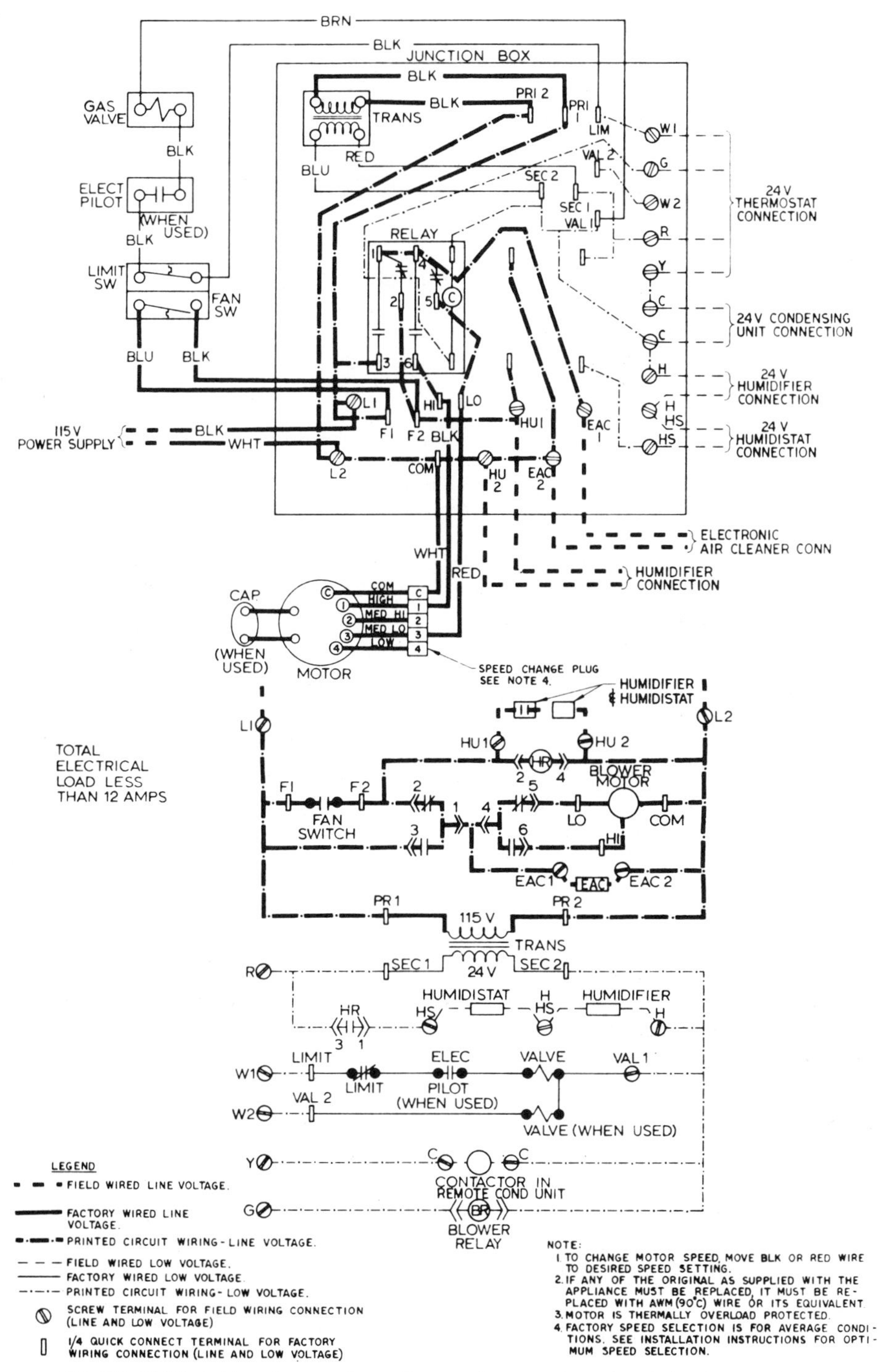

FIGURE 17-16 Wiring diagrams for upflow gas furnace with printed circuit control center (Reproduced by permission of Carrier Corporation, Copyright 1979, Carrier Corporation).

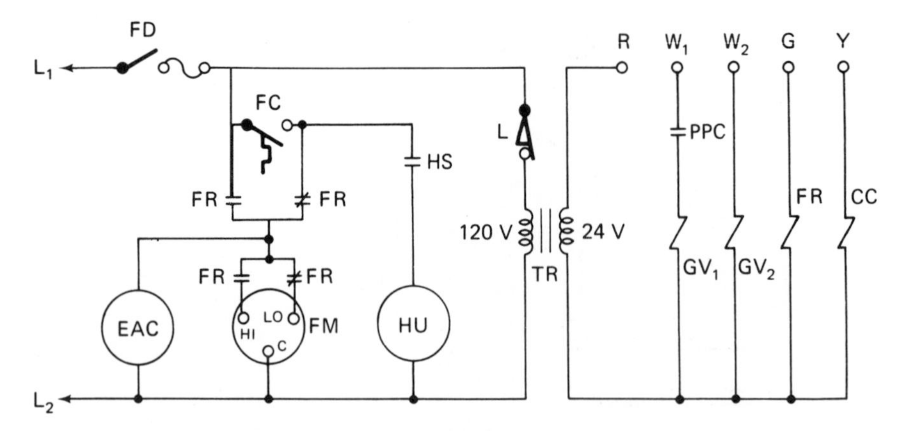

FIGURE 17-17 Schematic diagram for upflow gas furnace with printed circuit control center.

This control system uses a fan relay with two n.o. contacts and two n.c. contacts. This makes possible proper operation of the electrostatic air cleaner and humidification system. The electrostatic air cleaner will be on whenever the fan is running. However, the humidification system will be used only when the fan is running and the unit is heating.

A two-stage gas valve is optional equipment. It requires a two-stage thermostat: R and W_1 are the contacts for the first-stage gas valve (GV_1); R and W_2 are the contacts for the second-stage gas valve (GV_2).

TESTING AND SERVICE

When performing service on a furnace, every effort is made to pinpoint the problem to a specific part or circuit. This avoids extra work in checking over the complete control system to find the problem. For example, if all parts of a furnace operate properly except the fan, the fan circuit can be tested separately to locate the malfunction.

In electrical troubleshooting, it is good practice to "start from power": that is, to start testing where power comes into the unit and continue testing in the problem area until power is either no longer being supplied, or with power available, the load does not operate. When reaching either point the difficulty is located. For example, referring to Figure 17-11, if the complaint is that the fan will not run on high speed, the step-by-step troubleshooting procedure would be as follows:

1. Check power supply at load side of fused disconnect
2. Jumper thermostat terminals R and G to determine if the relay will "pull in"

3. If the relay operates satisfactorily, check power supply at high-speed terminals of fan
4. If power is available at fan, and it still will not run, motor is defective

Use of Electrical Test Meters

Figure 17-18 shows five locations where a test meter can be used on a 24-V gas heating control system. These meters check out the following circuits:

- Power
- Fan
- Pilot
- Gas valve
- Transformer

No accessories are shown. However, if these circuits exist on equipment being serviced, meter readings can be taken at these parts in addition to the ones shown.

Component Testing and Troubleshooting

In troubleshooting, each component (or circuit) has the potential for certain problems that the service technician can check for or test. Some of these potential problems are as follows:

1. *Power supply*
   (a) Switch open
   (b) Blown fuse or tripped breaker
   (c) Low voltages
2. *Thermostat*
   (a) Subbase switch turned off
   (b) Set too low
   (c) Loose connection
   (d) Improper anticipator setting

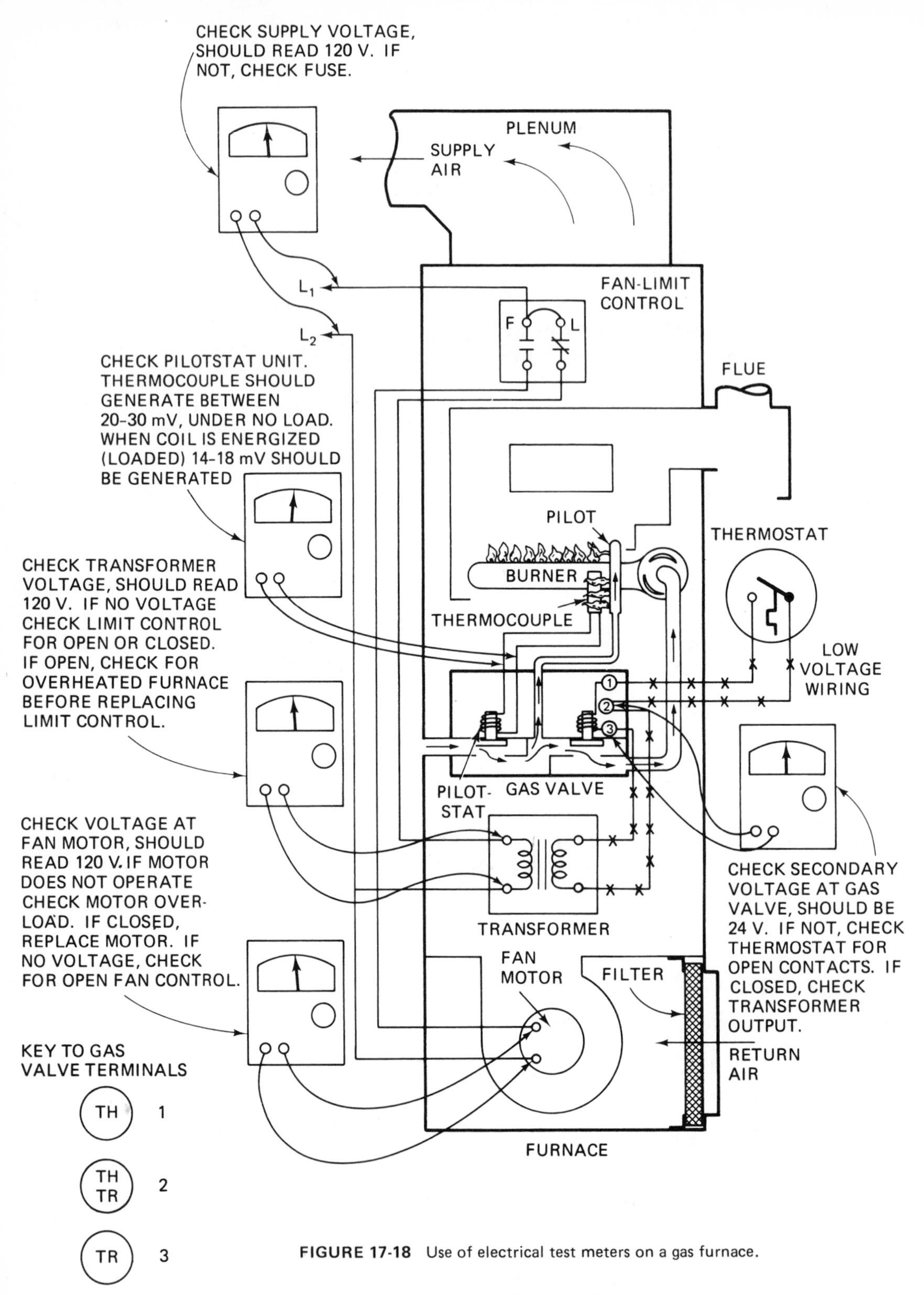

FIGURE 17-18 Use of electrical test meters on a gas furnace.

3. *Pilot*
   (a) Plugged orifice on pilot
   (b) Pilot blowing out due to draft
   (c) Too little or too much gas being supplied to pilot
4. *Thermocouple*
   (a) Loose connection
   (b) Too close or too far from flame
   (c) Defective thermocouple
5. *Transformer*
   (a) Low primary voltage
   (b) Defective transformer
   (c) Blown fuse in 24-V circuit
6. *Limit control*
   (a) Switch open
   (b) Loose connection
   (c) Defective control
7. *Gas valve*
   (a) Defective pilot safety power unit
   (b) Defective main gas valve
8. *Fan control*
   (a) Switch open
   (b) Improper setting
   (c) Defective control

Special Service Techniques

THERMOCOUPLE CIRCUIT:

Millivoltmeter probes can be connected in the thermocouple circuit, as shown in Figure 17-19, using the special adapter. If the meter reading is less than 17 mV, there is insufficient power to operate the pilot safety power unit. This must be corrected by either improving the flame impingement on the thermocouple or by replacing the thermocouple.

ADJUSTING THE ANTICIPATOR CIRCUIT ON A THERMOSTAT:

Normally, the anticipator is set at the amperage shown on the gas valve. However, if this information is not available, the current flow in the gas valve circuit can be measured.

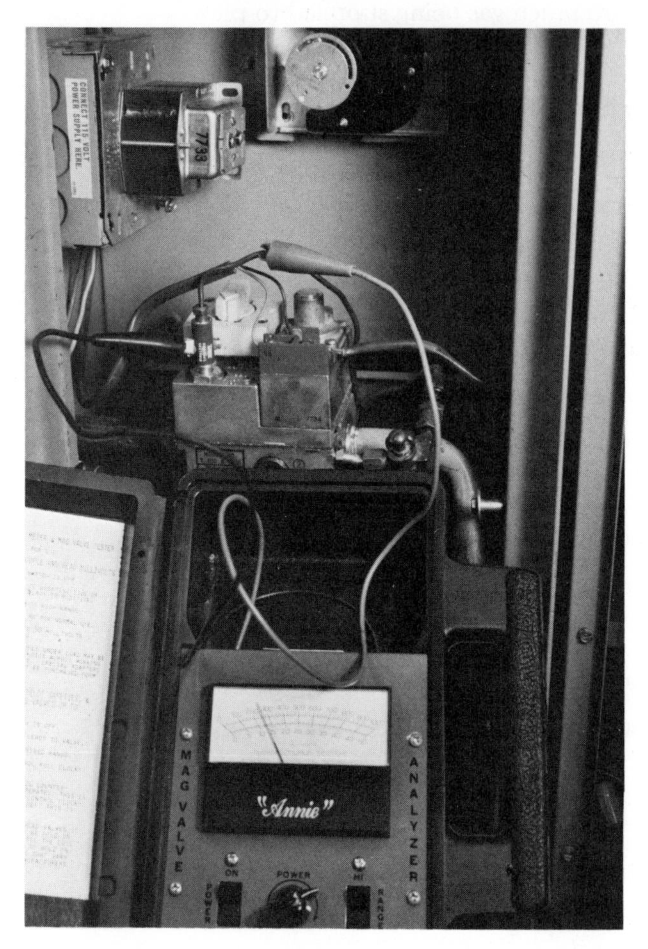

FIGURE 17-19 Testing a thermocouple circuit.

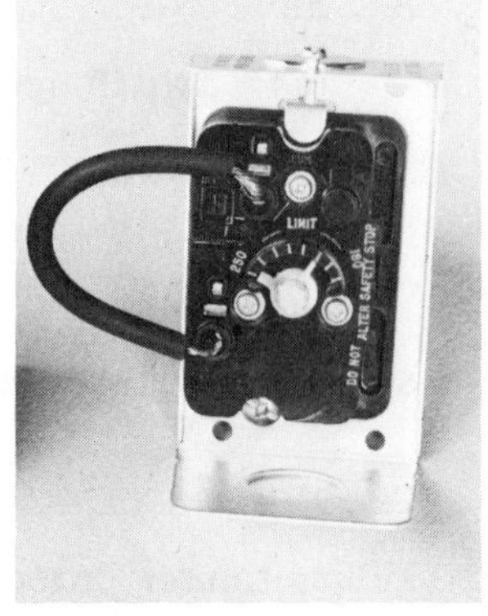

FIGURE 17-20
Jumpering a control switch.

TESTING THE OPERATION OF A CONTROL SWITCH:

Fan-limit and other switches can be tested by *jumpering* the switch contacts as shown in Figure 17-20. Jumpering is the means of making a temporary connection between the terminals for test purposes.

REVIEW QUESTIONS

Select the letter representing your choice of the correct answer.

17-1. What size of branch circuit can usually be recommended by code for a heating furnace?

(a) 10 A

(b) 15 A

(c) 20 A

(d) 25 A

17-2. How many circuits are required to operate a two-speed fan?

(a) One

(b) Two

(c) Three

(d) Four

17-3. On a call for heating, what two terminals on the thermostat close (make)?

(a) R and W

(b) R and Y

(c) R and B

(d) G and W

17-4. With a cut-out temperature of 200°F for the limit control, what would the normal cut-in be?

(a) 150°F

(b) 175°F

(c) 200°F

(d) 225°F

17-5. What is the usual cut-out temperature for the secondary limit control?

(a) 145°F

(b) 160°F

(c) 175°F

(d) 200°F

17-6. When the thermostat calls for cooling, what two terminals on the thermostat close (make)?

(a) R and W

(b) R and Y

(c) R and G

(d) G and W

17-7. What control device activates a timed fan start?

(a) Gas valve

(b) Limit control

(c) Fan control

(d) Thermostat

17-8. What accessory can be added that requires a second power supply?

(a) Humidifier

(b) Cooling

(c) Electrostatic air cleaner

(d) Stoker

17-9. What are the new thermostat terminals required when a two-stage gas valve arrangement is used?

(a) R_1 and R_2

(b) W_1 and W_2

(c) b_1 and b_2

(d) Y_1 and Y_2

17-10. In testing a fan relay, what voltage is usually applied to the relay coil?

(a) 24 V

(b) 120 V

(c) 240 V

(d) 480 V

18

OIL FURNACE CONTROLS

OBJECTIVES

After studying this chapter, the student will be able to:

- Identify the various controls and circuits on an oil-fired furnace
- Determine the sequence of operation of the controls
- Service and troubleshoot the oil heating unit control system

USE OF OIL FURNACE CONTROLS

Many of the oil furnace controls are similar to those of gas furnaces. The main difference is in the control of the fuel-burning equipment. The major components controlled are:

- Oil burner (including ignition)
- Fan
- Accessories

As with gas furnaces, certain variations must be incorporated in the oil furnace control system to comply with various furnace models. A downflow furnace, for example, needs an auxiliary limit control and a timed fan start control, which are not required on an upflow furnace.

CIRCUITS

The control circuits for an oil-burning furnace consist of:

- Power circuit
- Fan circuit
- Ignition circuit
- Oil burner circuit
- Accessory circuits

Primary Control

Several of the control circuits are combined in a single primary control to simplify control construction and field wiring. The primary control is a type of central control assembly that supplies power for the ignition, and oil burner circuits.

The primary control actuates the oil burner and provides a safety device which stops the operation of the burner if the flame fails to ignite or is extinguished for any reason. This safety device prevents any sizable quantity of unburned oil from flowing into the furnace, thus reducing the possibility of an explosion.

Two types of oil primary control sensors are commonly used. They are:

1. Bimetallic sensor
2. Cad cell sensor

BIMETALLIC SENSOR:

The bimetallic sensor type of oil primary control is often called a *stack relay* because it is placed in the flue pipe between the heat exchanger and the barometric damper. It includes a relay which permits the low-voltage (24-V) thermostat to operate the line-voltage (120-V) oil burner. The internal wiring for a bimetallic sensor type of primary control is shown in Figure 18-1.

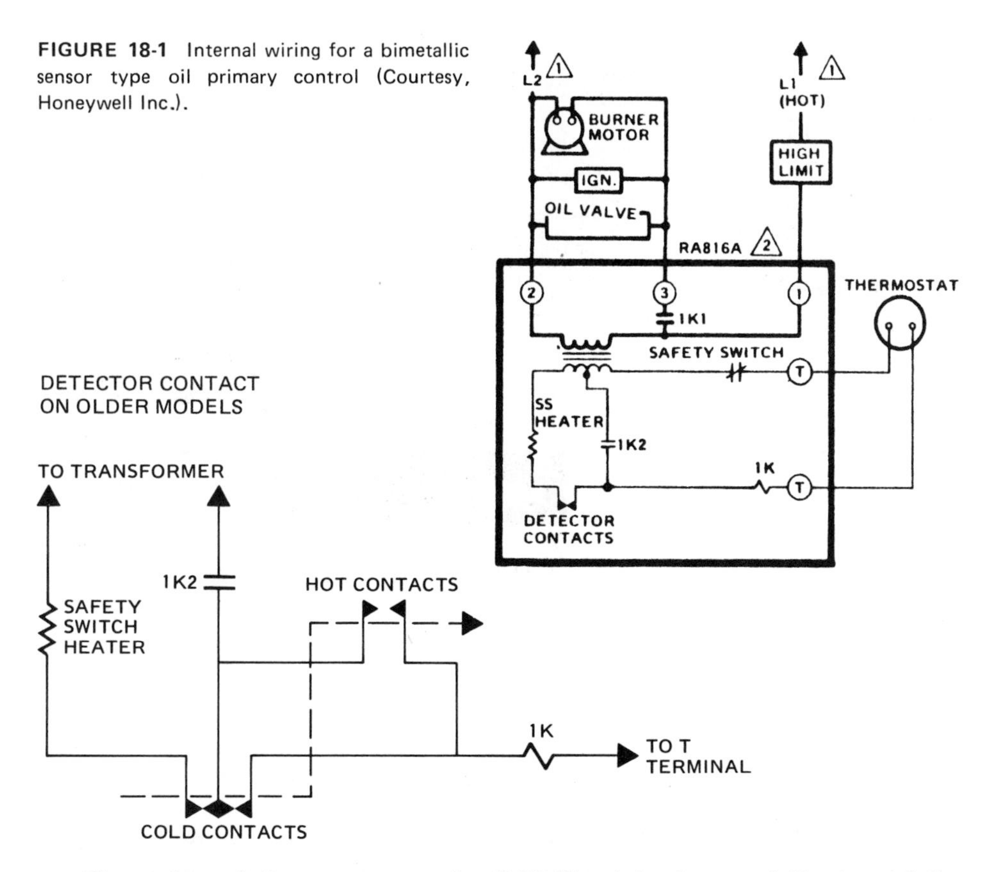

FIGURE 18-1 Internal wiring for a bimetallic sensor type oil primary control (Courtesy, Honeywell Inc.).

The entry of the power supply (120-V ac) is shown at the top of the diagram. Line-voltage power is connected to the oil burner and to the transformer located in the primary control. The balance of the control operates on low voltage.

Figure 18-2 shows the installation of the bimetallic sensor (pyrotherm detector) in the furnace flue. Figure 18-3 shows the mechanical action that takes place when the bimetal expands as it senses heat from the flame. The normal position of the switches in this control are: cold contacts n.c., hot contacts n.o. When the bimetal is heated, the cold contacts break and the hot contacts make.

Referring to Figure 18-1, when the thermostat calls for heat, the cold contacts are made and the safety switch heater is energized. If the flame fails to light and the cold contacts continue closed, the safety switch heater will "warp out" (open) the safety switch. The heaters and switch constitute a type of heat relay. With current flowing through this circuit, about 90 s is required to trip (open) the safety switch. If the burner produces flame and

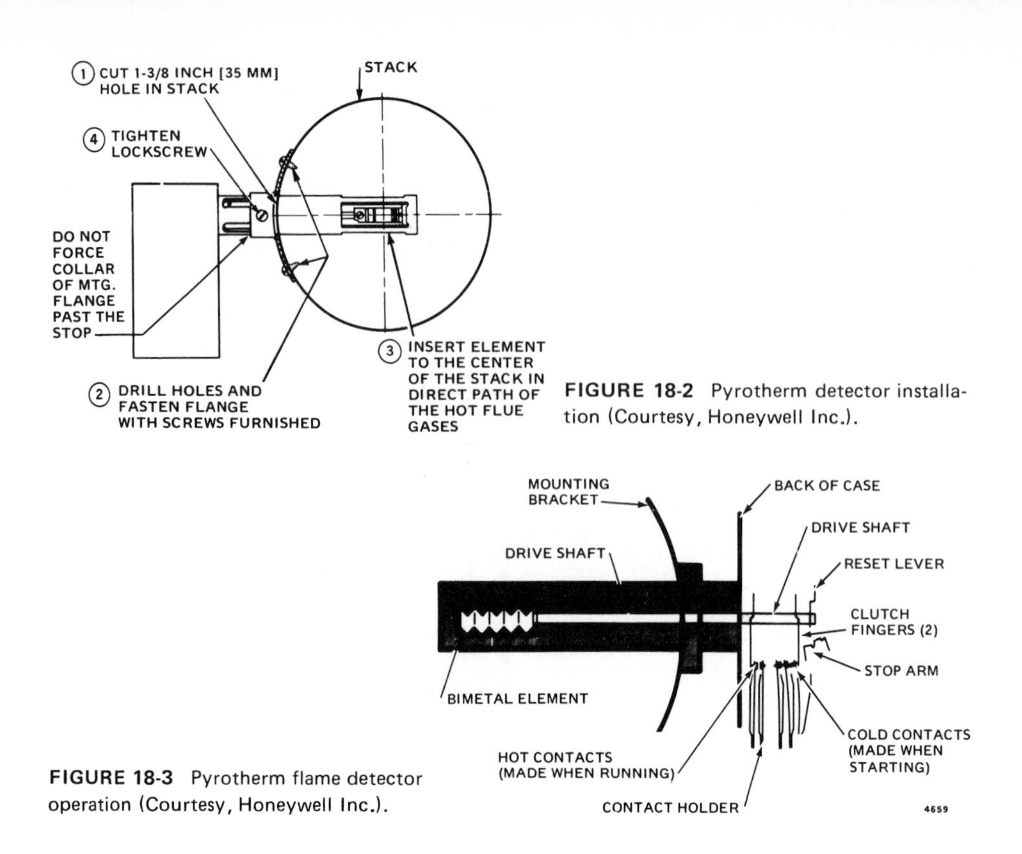

FIGURE 18-2 Pyrotherm detector installation (Courtesy, Honeywell Inc.).

FIGURE 18-3 Pyrotherm flame detector operation (Courtesy, Honeywell Inc.).

heats the bimetallic element the switching occurs, opening the cold contacts and closing the hot contacts. This action causes the current to bypass the safety switch heater and the burner continues to run.

A stack relay type of primary control is shown in Figure 18-4. Line-voltage power is connected across terminals 1 and 2, the oil burner motor is connected across 1 and 3. Constant ignition is also connected across terminals 1 and 3. Intermittent ignition is connected across terminals 1 and 4. The thermostat is connected to terminals T and T (W and B).

Two types of ignition are used in the field: constant and intermittent. With constant ignition the electrodes spark continuously. Intermittent ignition operates only when the burner is started. With constant ignition, both the oil burner motor and the ignition transformer are wired across terminals 1 and 3. Bimetallic-type stack relays are found on many furnaces now in the field. However, the new units are usually equipped with a cad cell type of primary control.

CAD CELL SENSOR:

A cad cell is shown in Figure 18-5 and its location is shown in 18-6. A cad cell sensor type of oil primary control uses cadmium sulfide, a light-sensitive material, to detect the presence of a flame (Figure 18-7). This material has a high electrical resistance (100,000 Ω) when no light reaches it. When it is lighted by the burner flame, its electrical resistance is lowered to 1000 Ω or less.

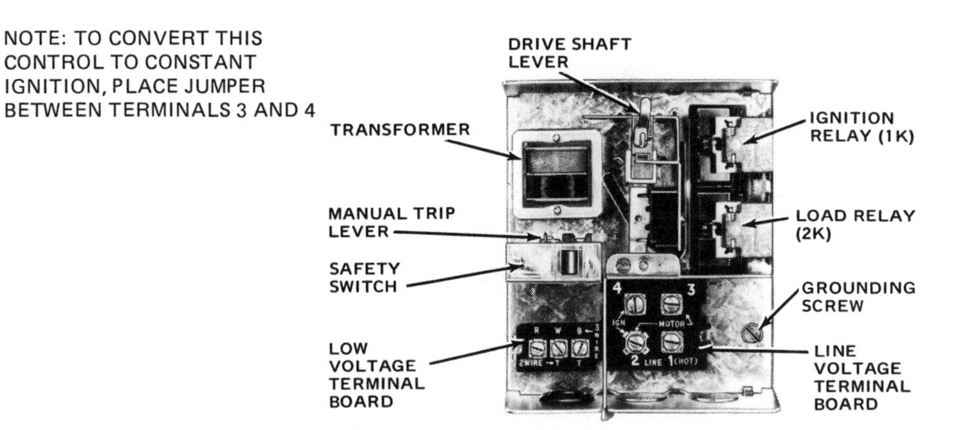

FIGURE 18-4 Typical primary control (pyrotherm type) showing line voltage and low voltage electrical terminals (Courtesy, Honeywell Inc.).

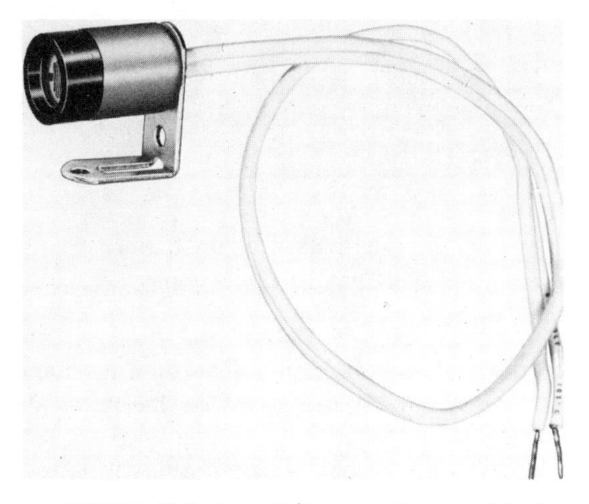

FIGURE 18-5 Cad cell (Courtesy, Honeywell Inc.).

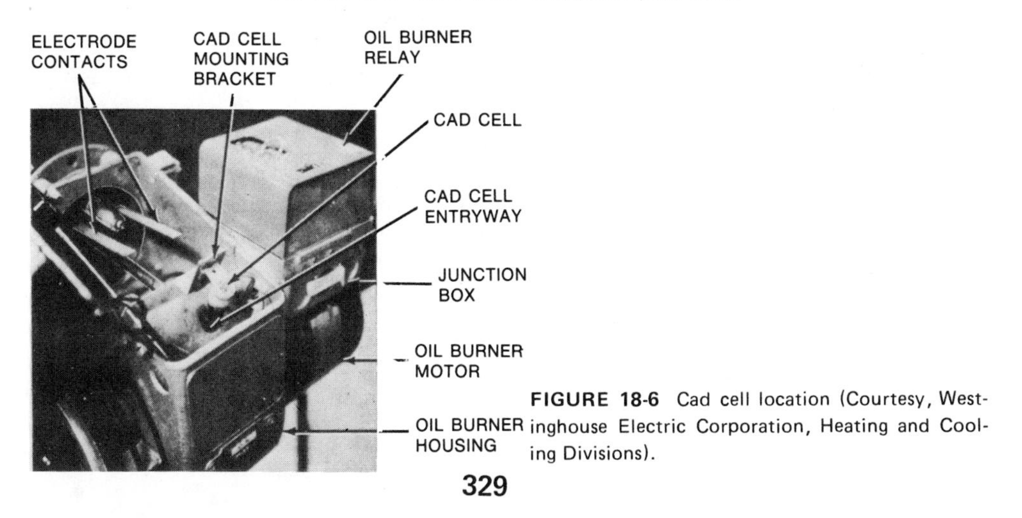

FIGURE 18-6 Cad cell location (Courtesy, Westinghouse Electric Corporation, Heating and Cooling Divisions).

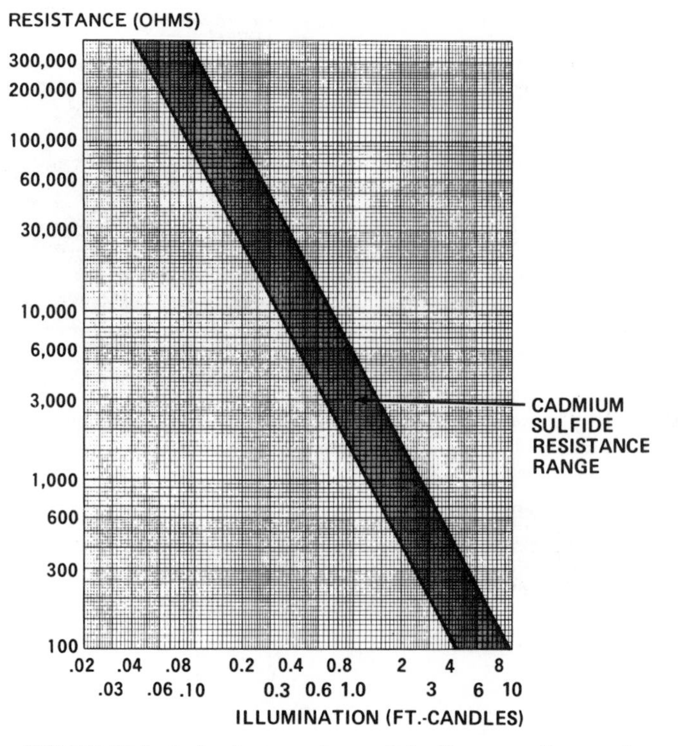

FIGURE 18-7 Cad cell responding to light (Courtesy, Honeywell Inc.).

The ability of the cad cell to change electrical resistance when exposed to different intensities of light makes possible the operation of a safety control circuit.

Figure 18-8 shows the internal wiring connections for the primary cad cell types of primary control as well as the internal circuiting. The power (120-V ac) connection to the burner and primary control transformer are shown on the right side of the diagram. The balance of the control is low voltage (24-V ac). The thermostat and the cad cell are shown connected to the left side of the primary control.

When the thermostat calls for heat, the oil burner starts and the safety switch heater is energized. If the flame fails to ignite, the heater warps out the safety switch in about 30 s, turning off the burner. If the flame is produced, it is sensed by the cad cell. This causes the safety switch heater to be bypassed and the burner stays on.

The cad cell primary control power supply (120-V ac) is wired to the black (hot) and white (neutral) connections. The oil burner motor is wired to the white and orange connections. The thermostat is connected to terminals T and T. The cad cell is connected to terminals S and S (or F and F).

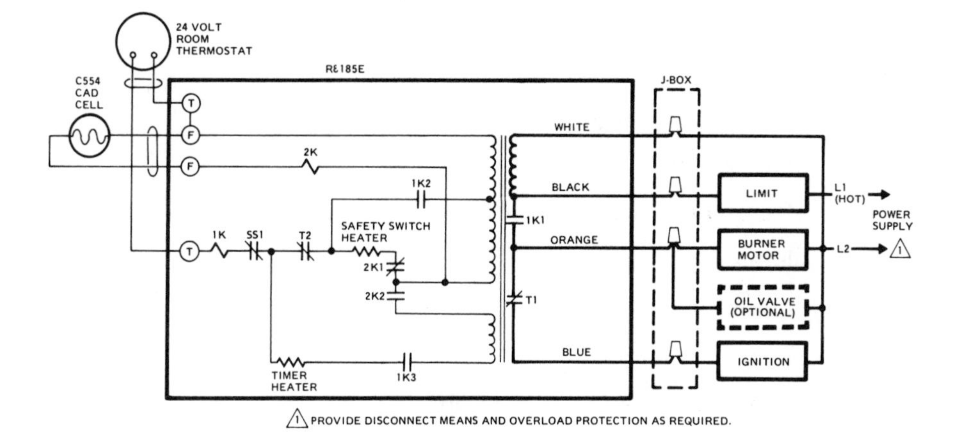

FIGURE 18-8 Wiring for cad cell type oil primary control (Courtesy, Honeywell Inc.).

There are two important options available on the cad cell primary control:

1. Constant or intermittent ignition
2. Variations in motor control arrangement

On constant ignition both the oil burner motor and the ignition transformer are wired across the white and orange connections. On intermittent ignition, the motor is wired across white and orange, and the ignition transformer is wired across white and blue.

The low-voltage connections on the primary control consist of four terminals, as shown in Figure 18-8.

Power Circuit

Following is the key to the schematic diagrams used in the study of the power circuit and the other circuits described in this chapter.

BK	Black-wire color
BL	Blue-wire color
G	Green-wire color
OR	Orange-wire color
R	Red-wire color
W	White-wire color

Y	Yellow-wire color
C	24-V common connection
CC	Compressor relay
FC	Fan control
FM	Fan motor
FR	Fan relay
L	Limit
LA	Limit auxiliary
L_1 L_2	Line 1, line 2 of power supply
S	Cad cell terminal
T	Thermostat terminal
TFS	Timed fan start

Referring to the power circuit shown in Figure 18-9, 120-V ac power is supplied to the primary control. A fused disconnect is placed in the "hot" side of the line. A limit control is wired in series with the hot side of the power supply (L_1) and the black connection on the primary control. The transformer is a part of the primary control.

Fan Circuit

The simplest type of fan circuit consists of a fan control in series with a single-speed fan motor, as shown in Figure 18-10.

Using a multiple-speed fan motor the switch on the subbase of the thermostat changes the fan speed, as shown in Figure 18-11.

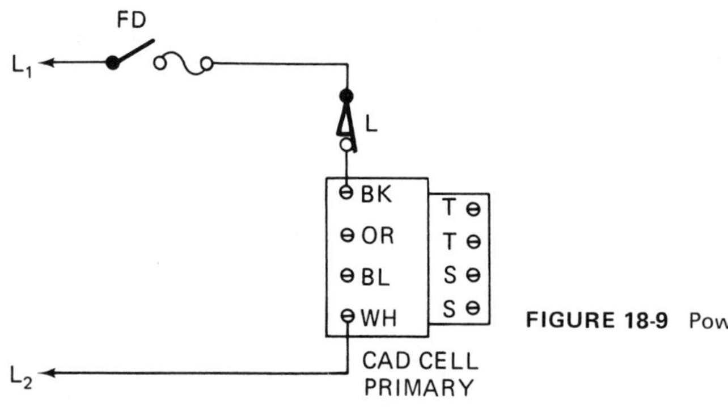

FIGURE 18-9 Power circuit.

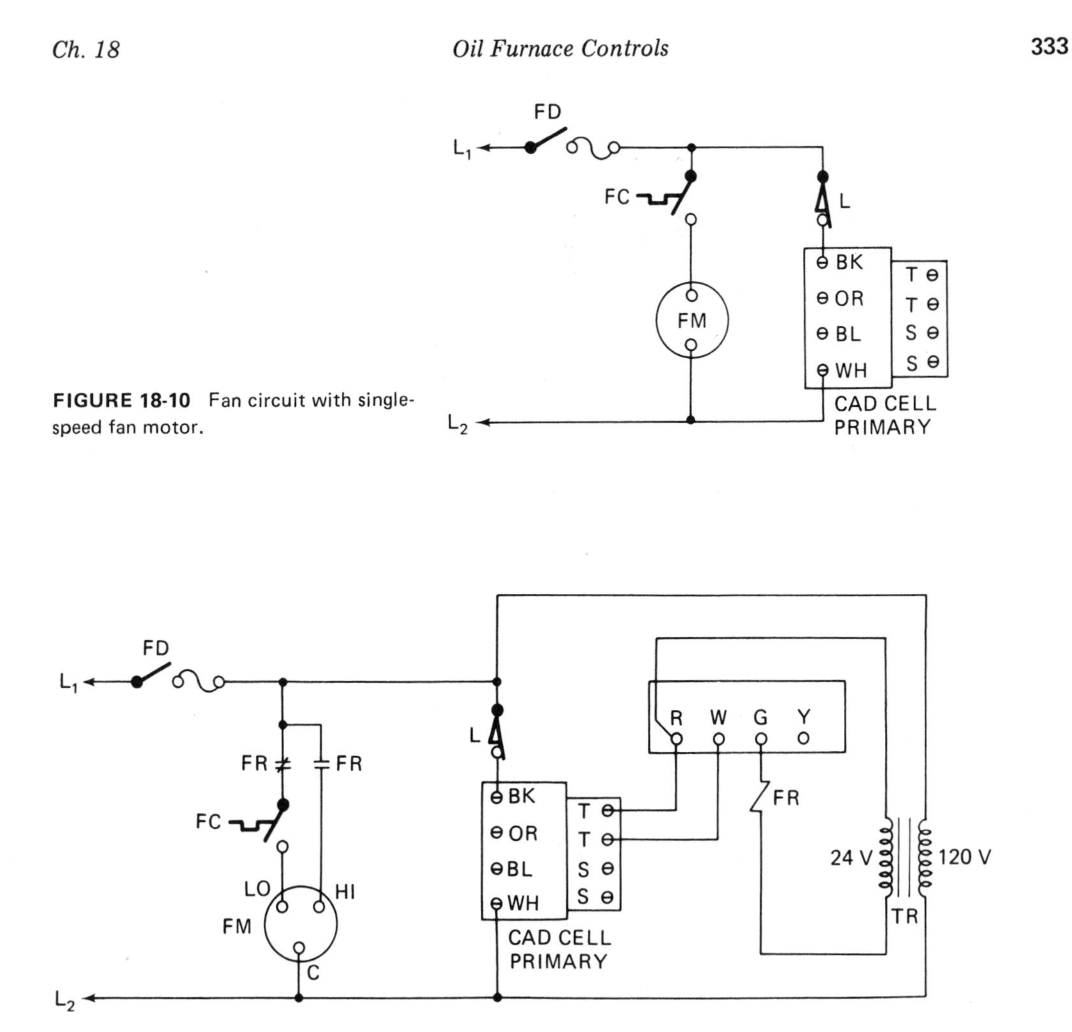

FIGURE 18-10 Fan circuit with single-speed fan motor.

FIGURE 18-11 Fan circuit with multi-speed fan motor.

On downflow and horizontal furnaces, a timed fan start (TFS) is used to start the fan and the fan control (FC) stops it, as shown in Figure 18-12. An auxiliary limit control is used in addition to the regular limit control and is wired in series with the primary control.

Ignition and Oil Burner Circuits

Ignition and oil burner circuits are shown in Figure 18-13. The ignition transformer is wired in parallel with the oil burner motor on a constant ignition system. The burner is wired to the orange and white connections on the primary control.

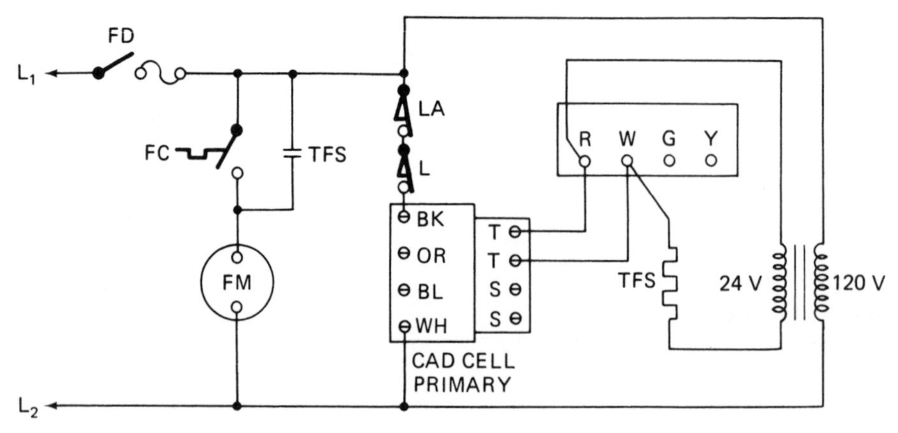

FIGURE 18-12 Fan circuit for downflow and horizontal furnaces.

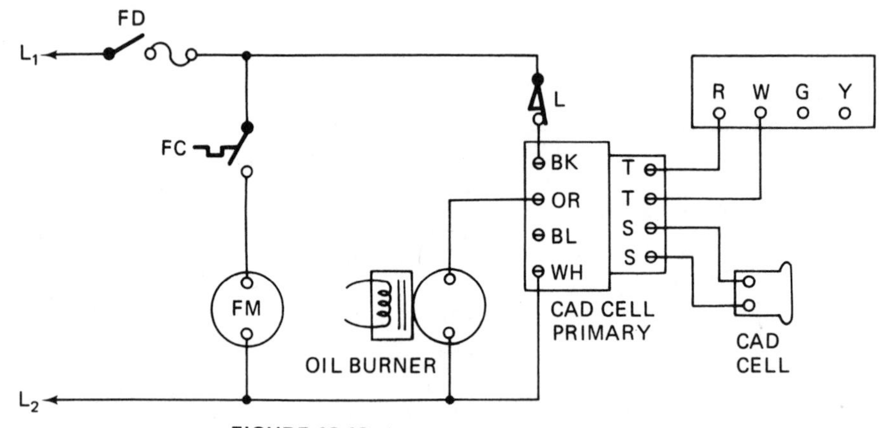

FIGURE 18-13 Ignition and oil burner circuits.

Accessory Circuits

Accessories for oil furnaces can be added in a manner similar to those shown in Chapter 17 for gas furnaces. A separate transformer is usually required for control circuit power when cooling is added. See Figures 18-11 and 18-12.

TYPICAL WIRING DIAGRAMS

Upflow Oil Furnace

Figure 18-14 shows a typical connection diagram for an upflow oil furnace. Note that on this unit most of the wiring connections are brought to a common junction box. Since both the fan and limit controls are of line-

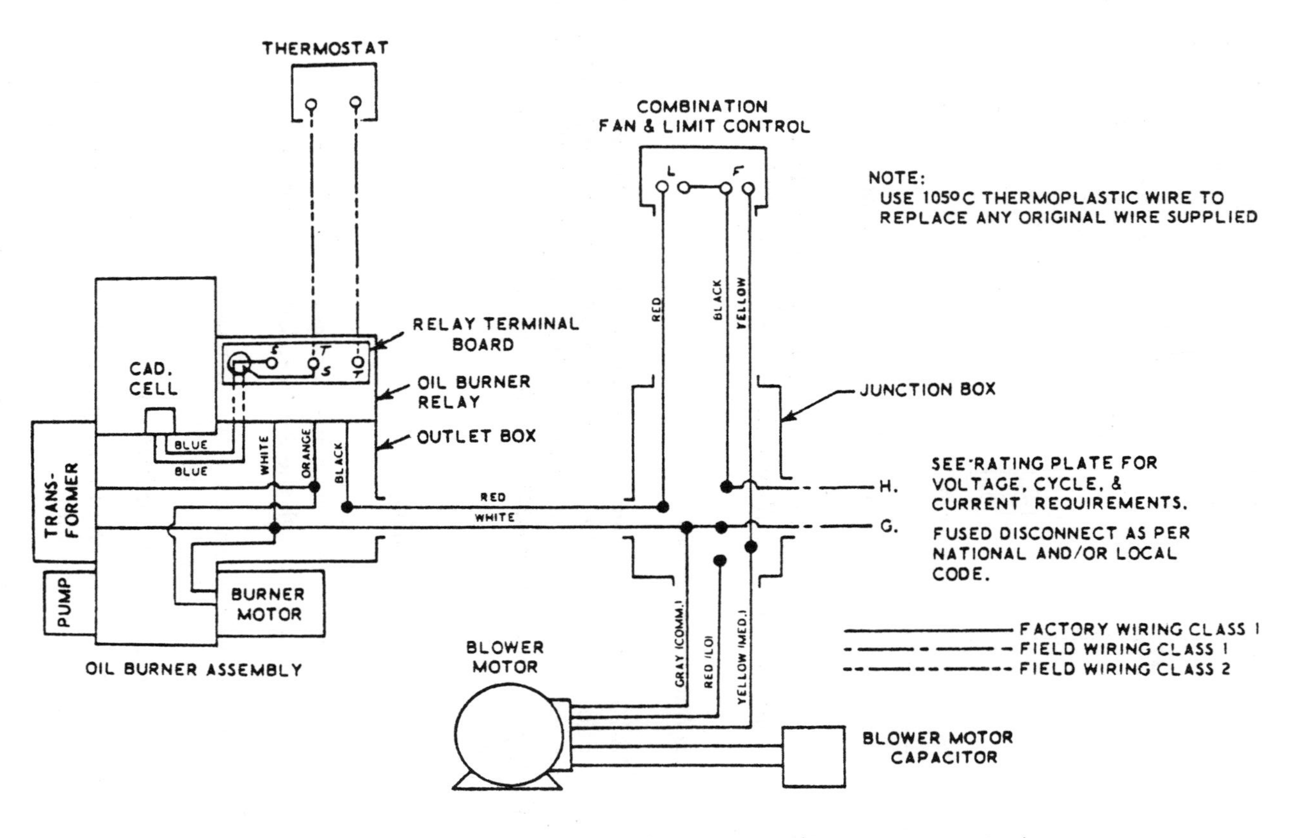

FIGURE 18-14 Connection diagram for upflow oil furnace (Courtesy, Luxaire, Inc.).

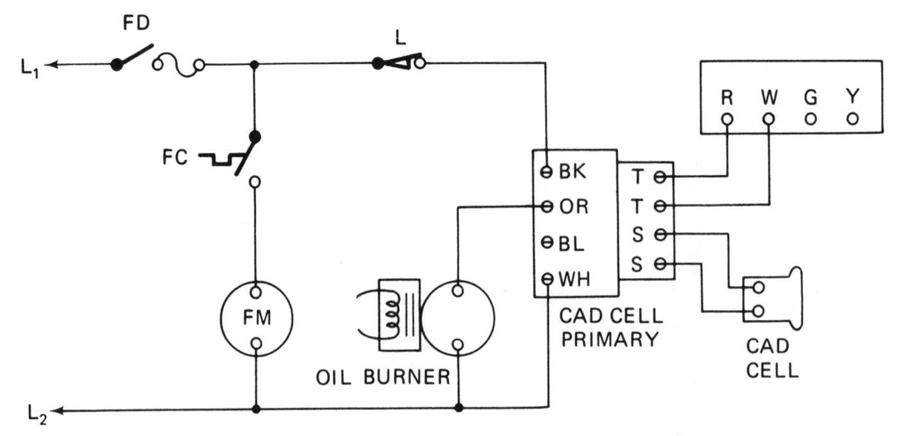

FIGURE 18-15 Schematic diagram for upflow oil furnace.

voltage type, a jumper can be used between common terminals. The schematic wiring diagram for the upflow unit is shown in Figure 18-15.

Oil Furnace With Cooling Added

Figure 18-16 shows a connection diagram for an oil furnace with cooling accessories added. All wiring connections are made either in the primary control junction box or in a separate junction box that includes the indoor fan relay. The schematic wiring diagram for the oil furnace with cooling added is shown in Figure 18-17.

TROUBLESHOOTING AND SERVICE

In troubleshooting an oil-fired furnace, attention should be given to the electrical and fluid flows through the furnace, for the following reasons:

- Electrical power is required to operate the control system, ignite the fuel, and energize the loads
- Fuel oil is required for combustion
- Air is required for combustion and to convey heat from the furnace to the spaces being heated

Electricity, oil, and air each has a circuit or path of movement through the furnace. Each must be supplied in the proper place, in the proper quantity, and at the proper time.

Troubleshooting the fuel oil and air supply have been covered in other chapters. Therefore, the concern here is chiefly electrical power and control.

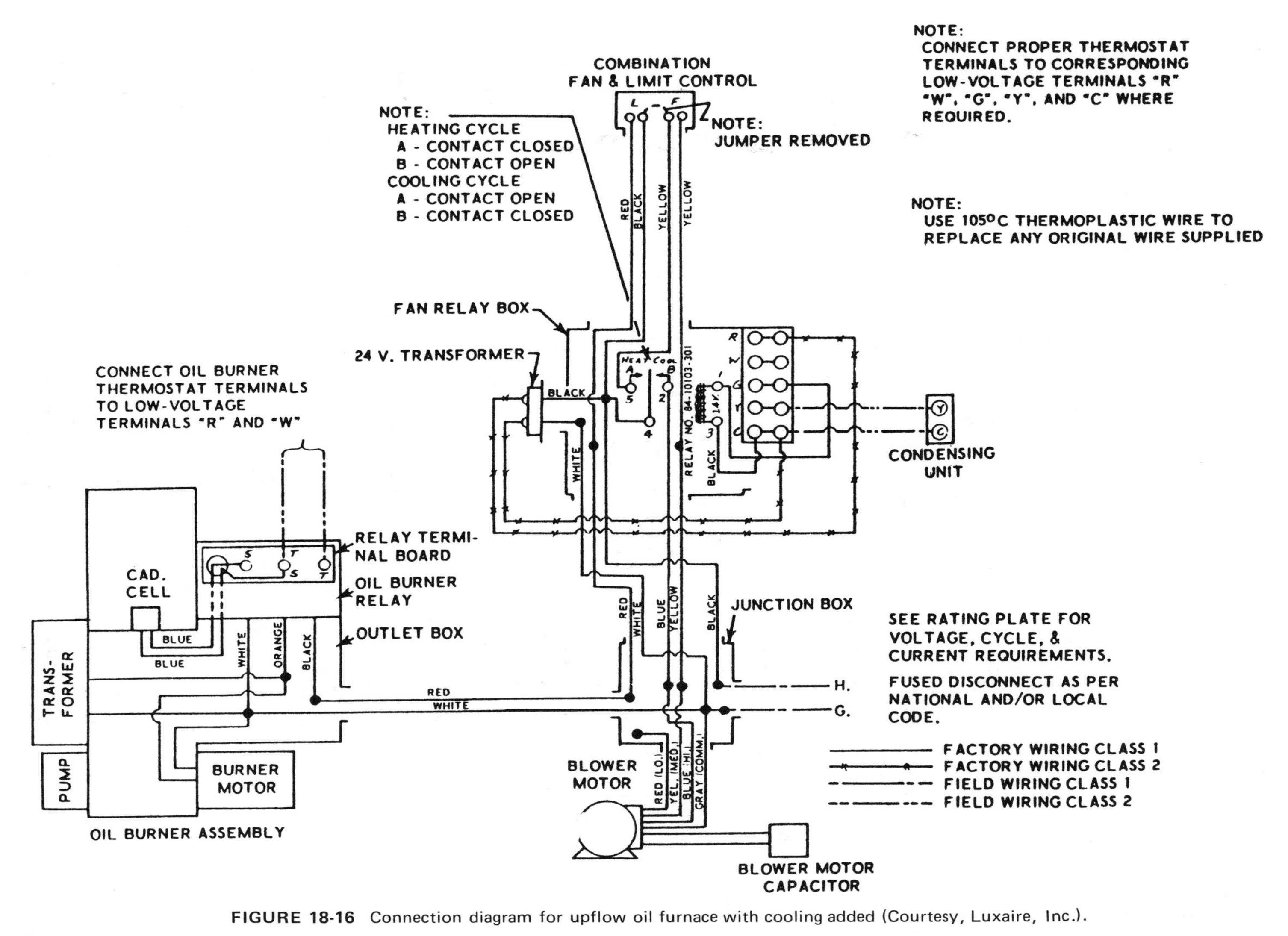

FIGURE 18-16 Connection diagram for upflow oil furnace with cooling added (Courtesy, Luxaire, Inc.).

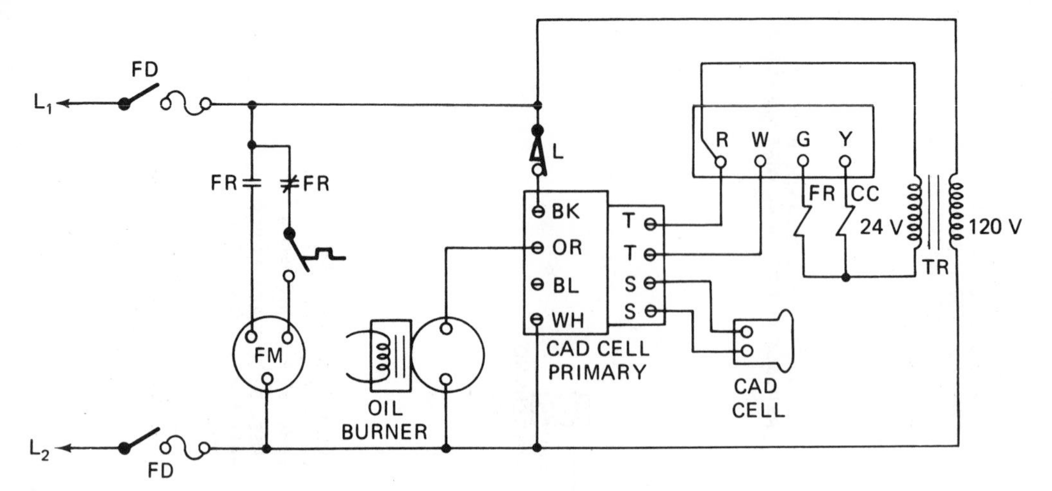

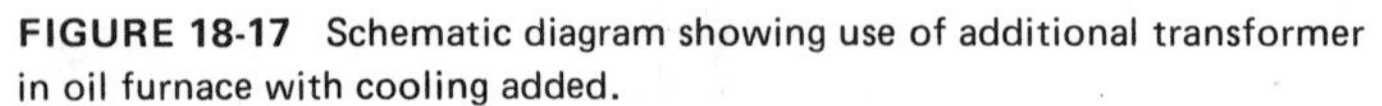
FIGURE 18-17 Schematic diagram showing use of additional transformer in oil furnace with cooling added.

Use of Test Meters

Figure 18-18 shows the use of various test meters in measuring current in the electrical system of the furnace. The technique in electrical troubleshooting is to "start from power" and follow the availability of power through the control system to the load. If the power supply is stopped for any reason along the proper path, the reason for the stoppage should be determined, and corrected. If power in the proper quantity is supplied to the load and it does not operate, the load device is at fault.

Seven locations where meter readings are taken along the path of the electrical current are shown in Figure 18-18. Locations to be checked are:

1. Voltage supply to the unit
2. Voltage across the limit control, to determine if this control switch is properly closed
3. Voltage across the thermostat, to be certain that it is calling for heat
4. Voltage at the oil burner motor and ignition transformer, to be certain that 120 V ac is being supplied
5. Voltage across the fan control, to determine that it is calling for fan operation
6. Voltage at the fan motor, to be certain that 120 V ac is being supplied
7. Amperage to the fan motor, to measure the running amperes. This should agree with the data on the motor nameplate

TROUBLESHOOTING OIL BURNER SYSTEMS

NOTE: When checking to determine if any switch is open or closed: Check voltage across switch with power on; zero voltage indicates switch is closed; if any voltage is indicated, switch is open.

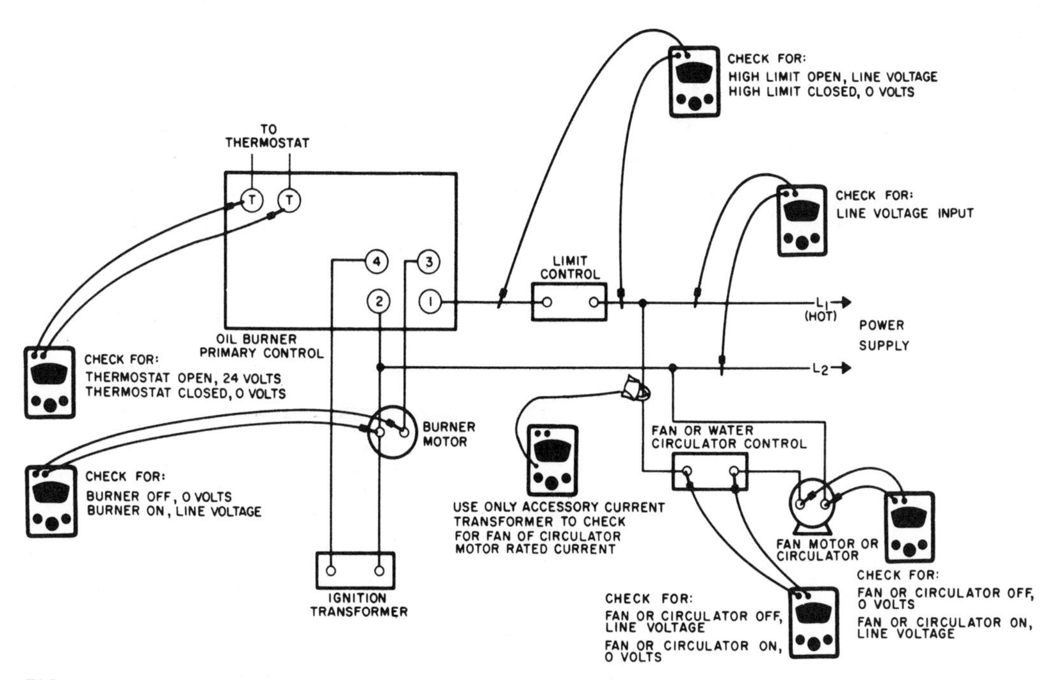

FIGURE 18-18 The use of various electrical meters in troubleshooting oil burner systems (Courtesy, Honeywell Inc.).

In meter readings 2, 3, and 5, a voltmeter is used to test a switch. If the switch is open, a voltage shows on the meter. If the switch is closed, the voltmeter reads zero.

Testing Components

The most complex element in an oil-fired furnace control system is the primary control. Therefore, it is this element which has the greatest potential need for service. Specific testing procedures are necessary to determine if the primary control is functioning properly.

PRIMARY CONTROL WITH BIMETALLIC SENSOR:

For the service technician's protection, it is important that the "hot" line of the power supply be brought to the number 1 terminal. In accordance with good practice, the switching action of this control should be in the hot line.

If the thermostat is calling for heat, 120-V ac power must be available at terminals 1 and 3, on a constant ignition system. However, if intermittent ignition is used, 120-V ac power must also be available at terminals 1 and 4. The thermostat terminals can be jumpered to simulate the thermostat calling for heat.

The control should be tested without supplying heat to the bimetallic element. When this is done, the safety switch should shut off the power to the burner in about 90 s. If heat is supplied to the bimetallic element, the burner should continue to run.

Sometimes these controls get "out of step." That is, the cold contacts do not remake when the bimetal cools down. This should be corrected by following the restepping procedure shown in Figure 18-19.

To test the contact terminals, jumper across the cold contacts, as shown in Figure 18-20. If the bimetal is cold and contacts cannot be restepped, replace the primary control.

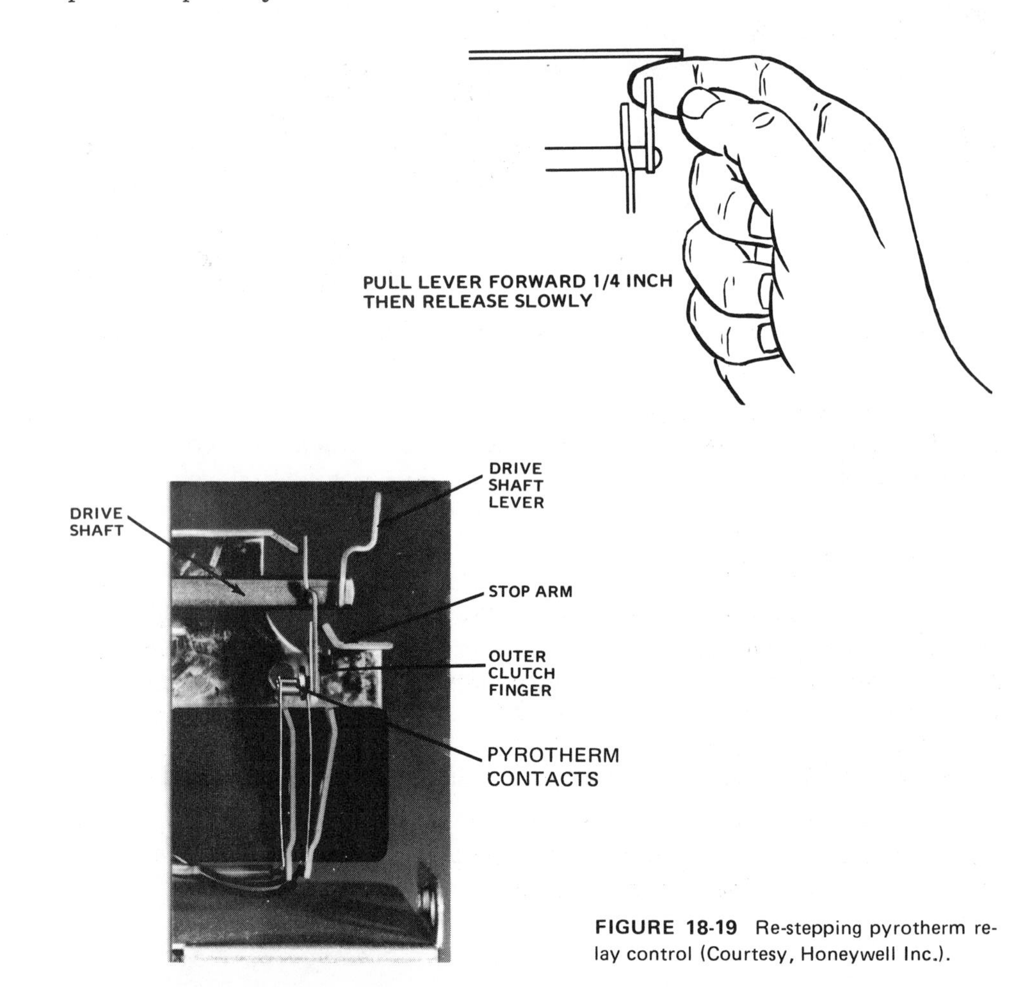

FIGURE 18-19 Re-stepping pyrotherm relay control (Courtesy, Honeywell Inc.).

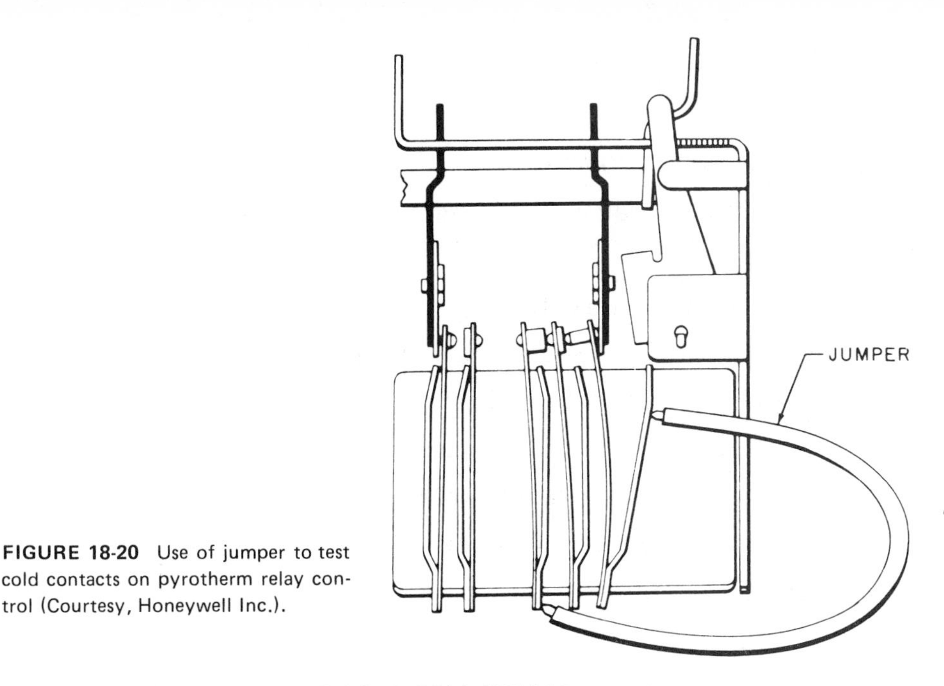

FIGURE 18-20 Use of jumper to test cold contacts on pyrotherm relay control (Courtesy, Honeywell Inc.).

PRIMARY CONTROL WITH CAD CELL SENSOR:

The cad cell primary control has one feature that is not present in the bimetallic sensor. If the cad cell senses stray light (light not from the flame) before the thermostat calls for heat, the burner cannot be started by the thermostat. This is a protective arrangment. Therefore, when testing the operation of the cad cell primary, block off all light that might reach the cad cell.

The cad cell itself can be tested with a suitable ohmmeter. The ohmmeter must be capable of reading up to 100,000 Ω. In the absence of light, the cad cell should have a resistance of about 100,000 Ω. In the presence of light, its resistance should not exceed 1500 Ω.

To test the primary control, start the burner and within 30 s jump the cad cell terminals with a 1000-1500-Ω resistor. If the burner continues to run, the cad cell primary is operating correctly. In actual operation, a poor or inadequate oil burner flame can be a problem. The safety arrangement on a cad cell primary control usually stops the burner in about 30 s if the cad cell does not sense an adequate flame.

Other Servicing Techniques

FUSES AND OVERLOADS:

Most oil-burning furnaces have a manual reset overload on the oil burner motor and a reset-type safety switch on the primary control. After an overload, these devices must be manually reset. The tripping of an overload switch is an indicator of a problem that must be found and corrected. Resetting the overload switches may start the equipment but unless the problem is resolved, nuisance "tripout" will continue to recur.

BURNER SERVICE ROUTINE:

Regardless of the complaint, much time can often be saved at first by checking the following routine conditions:

- Does the tank contain fuel
- Is power being delivered to the building
- Are all hand-operated switches closed
- Are all hand valves in the oil supply lines open
- Are all limit controls in their normal (closed) position
- Is the thermostat calling for heat
- Are all overload switches closed

LOW VOLTAGE:

Low voltage can cause such problems as low oil pump speed, motor burnouts from overload, and burners that fail to operate because relays do not pull in.

The power supply voltage should be checked while the greatest power usage is occurring. Low voltage should never be overlooked as a source of service problems. Where the service to the building is at fault, the local power company should be contacted. If the problem is within the building, the services of an electrician may be required to solve the problem.

REVIEW QUESTIONS

Select the letter representing your choice of the correct answer.

18-1. On the stack relay type of oil primary control, are the cold contacts normally open or normally closed?

(a) Normally open

(b) Normally closed

18-2. What is the light-sensitive material used on a cad cell?

(a) Cadmium nitrate

(b) Copper sulfate

(c) Sodium chloride

(d) Cadmium sulfate

18-3. What is the resistance of a cad cell in the absence of light?

(a) 5000 Ω

(b) 50,000 Ω

(c) 100,000 Ω

(d) 150,000 Ω

18-4. To which terminals on the primary control is the cad cell usually connected?

(a) T and S

(b) S and S

(c) R and W

(d) W and R

18-5. How many low-voltage terminals on a cad cell primary control?

(a) Two

(b) Four

(c) Six

(d) Eight

18-6. On a cad cell primary control the "hot" side of the power supply is wired to what connection?

(a) Black

(b) Orange

(c) Blue

(d) White

18-7. On the oil burner stack relay, the power supply is wired to which terminals?

(a) 0 and 1

(b) 1 and 2

(c) 2 and 3

(d) 3 and 4

18-8. In checking a switch with a voltmeter, a reading of zero indicates:

(a) A closed switch

(b) An open switch

(c) A short

(d) A ground

18-9. Using a stack relay, if the flame fails on start-up the burner will go on safety in about:

(a) 30 s

(b) 60 s

(c) 90 s

(d) 120 s

18-10. When the cad cell senses light, what is its usual resistance?

(a) 3,500 Ω or less

(b) 1,500 Ω or less

(c) 500 Ω or less

(d) 0 Ω

19

ELECTRIC HEATING

OBJECTIVES

After studying this chapter, the student will be able to:

- Identify the various types of controls and circuits used on an electric furnace
- Determine the sequence of operation of the controls
- Service and troubleshoot the electric furnace control system

CONVERSION OF ELECTRICITY TO HEAT

An electric heating furnace converts energy in the form of electricity to heat. The conversion takes place in resistance heaters.

Electric furnaces differ from gas or oil furnaces in that no heat exchanger is required. Return air from the space being heated passes directly over the resistance heaters and into the supply air plenum.

The amount of heat supplied depends upon the number and size of the resistance heaters used. The conversion of electricity to heat (the heat equivalent for 1 W of electrical power) takes place in accordance with the following formula:

$$1 \text{ W} = 3.415 \text{ Btu}$$

ELECTRIC FURNACE COMPONENTS

Figure 19-1 shows an electric forced warm air furnace (upflow model). Return air is brought into the blower compartment through the filters, then passed over the heating elements and sent out into the distribution system.

An electric furnace requires no flue. All of the heat produced is used in heating the building. Input is equal to output. Thus, it operates at 100% efficiency.

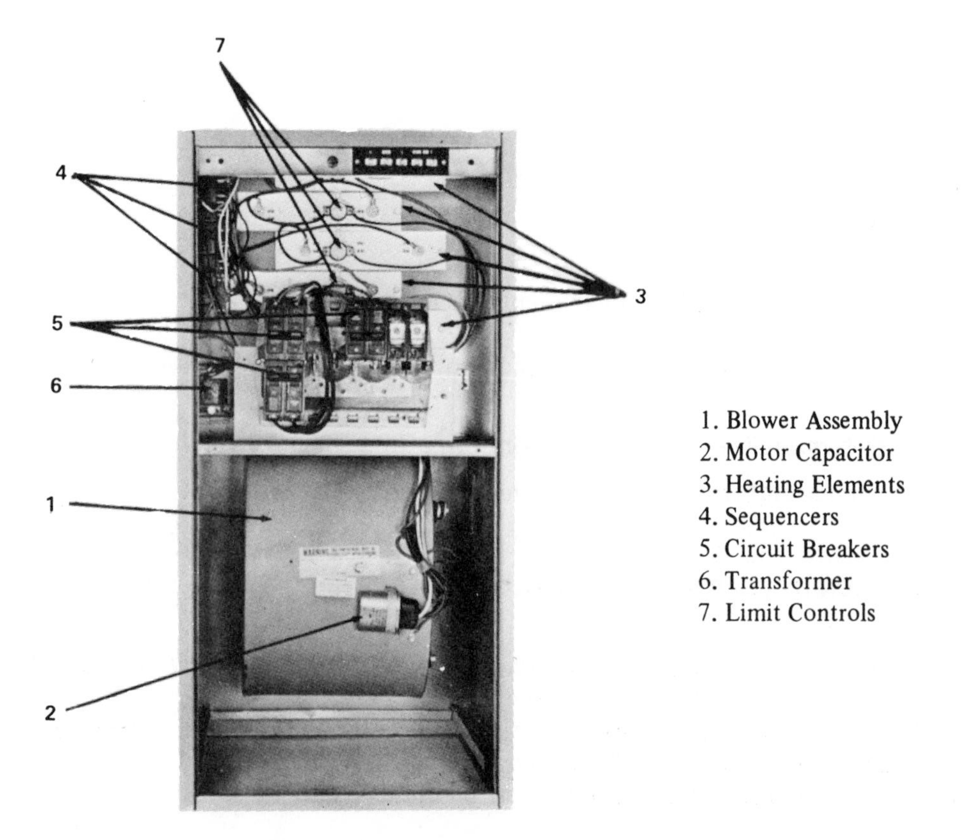

FIGURE 19-1 Electric forced warm air furnace, upflow model (Courtesy, Luxaire, Inc.).

Major components, excluding the controls, are:

- Heating elements
- Blower and motor assembly
- Furnace enclosure
- Accessories, such as filters, humidifier, and cooling (optional)

Following is the key to the schematic diagrams used in this chapter.

AC	Auxiliary contacts
C	24-V common connection
FC	Fan Control
FD	Fused disconnect
FL	Fuse link
FM	Fan motor
FU	Fuse
HR	Heat relay
L	Limit control
L_1 L_2	Line 1, line 2 of power supply
OT	Outdoor thermostat
R	Red wire, thermostat connection
SQ	Sequencer
TFS	Timed fan start
TR	Transformer
W	White wire, thermostat connection

Heating Elements

The heating elements are made of Nichrome, a metal consisting chiefly of nickel and chromium. Heating elements are rated in kilowatts of electrical power consumed. (One kilowatt equals 1000 watts.)

A typical heating element assembly is shown in Figure 19-2. The Nichrome wire is supported by insulating material placed in such a way as to provide a minimum amount of air resistance and a maximum amount of contact between the air and the resistance heaters.

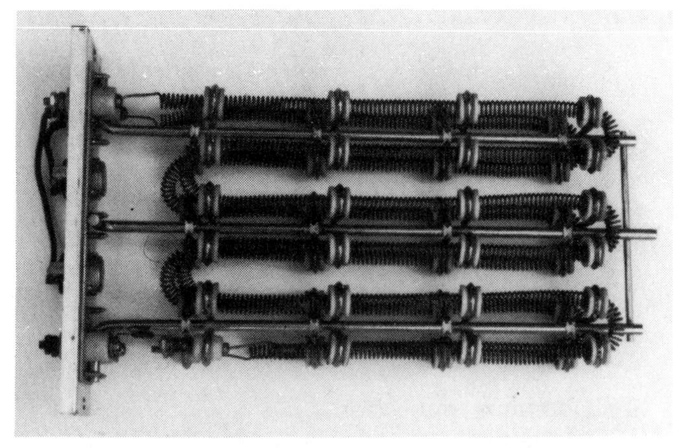

FIGURE 19-2 Typical heating element.

Each element assembly contains a thermal fuse and a safety limit switch (Figure 19-3). The thermal fuse and safety limit switch are shown in schematic form in Figure 19-4. The limit switch is usually set to open at 160° F and close when the temperature drops to 125° F. The thermal fuse, a backup for the safety limit switch, is set to open at a temperature slightly higher than the limit switch.

In addition to the safety controls in the element assembly, each assembly is fused where it connects to the power supply. Thus, triple safety protection is provided.

FIGURE 19-3 Heating element assembly showing thermal fuse and safety limit switch (Courtesy, Luxaire, Inc.).

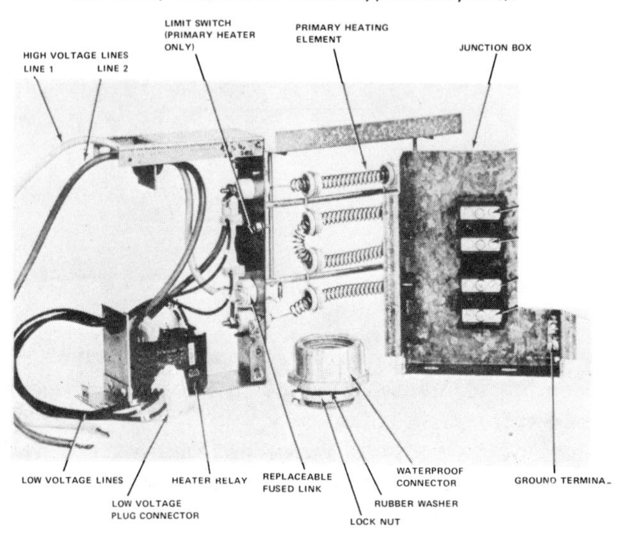

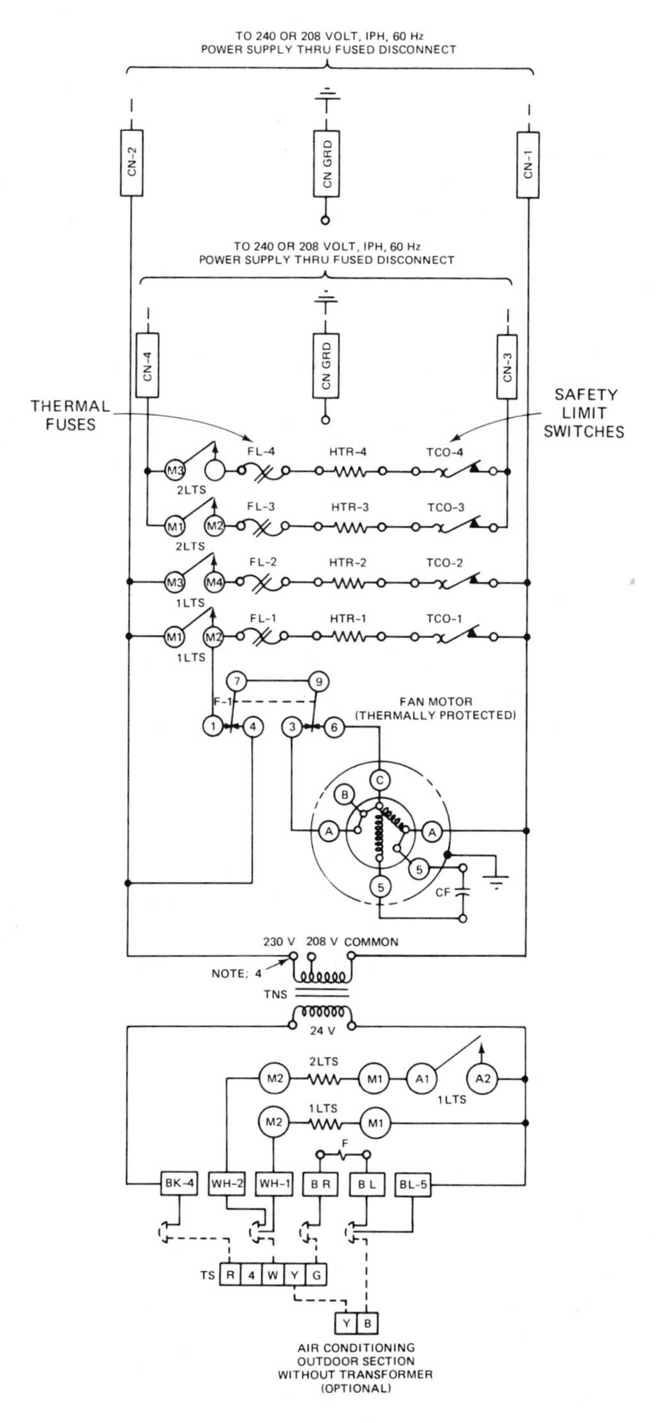

FIGURE 19-4 Thermal fuse and safety limit switch shown in schematic diagram (Courtesy, General Electric Company).

Electric furnaces have various individual numbers and sizes of heating element assemblies, depending on the total capacity of the equipment. Elements are placed on the line (power supply) in stages so as not to overload the electrical system on startup. In most units, the minimum size of an element energized or deenergized at one time is 5 kW (17,075 Btuh).

Fan and Motor Assembly

The fan and motor assembly is similar to that used on a gas or oil furnace. Fans may be either direct-drive or belt-driven. There is a trend toward the use of multispeed direct-drive fans since they facilitate adjustment of the air flow by changing the blower speed. This is usually necessary when cooling is added, since larger air quantities and, consequently, higher speeds are required.

Furnace Enclosure

The exterior of the casing is similar to that of a gas or oil furnace but without the flue pipe connection. The interior is designed to permit the air to flow over the heating elements. The section supporting the heating elements is usually insulated from the exterior casing by an air space.

Accessories

Filters, humidifiers, and cooling are added to an electric heating furnace in a manner similar to gas and oil furnaces. Electric heating, therefore, can provide all of the related climate control features provided by other types of fuel.

POWER SUPPLY

The power supply for an electric furnace is 208/240-V, single-phase, 60-H_z This power is supplied by three wires: two are "hot" and one is ground. Fused disconnects are placed in the hot lines leading to the furnace. The National Electric Code limits the amount of service in a single circuit to 48 A. Therefore, if the running current exceeds this amount, additional circuits and fused disconnects must be provided.

The fuses for 48-A service must not exceed 60 A. This is a National Electric Code requirement, specifying that fuses should not exceed 125% of full load amperes ($48 \times 1.25 = 60$).

All wiring must be enclosed in conduit with proper connectors. Since 280/240-V is considerably more dangerous than lower voltages, every possible protection must be provided. The National Electric Code also requires that the furnace be grounded. The ground wire in the power supply is provided for this purpose.

CONTROL SYSTEM

Since the control system for an electric furnace includes more electrical parts than are in a gas or oil furnace, the wiring is more involved. It is extremely important to use a schematic diagram to assist in troubleshooting. If a schematic is not available from the manufacturer, one should be drawn by the service technician.

The control system consists of the following electrical circuits:

- Power circuit
- Fan circuit
- Heating element circuits
- Control circuits operated from the thermostat

Power Circuit

The power supply consists of one or more 208/240-V ac sources, directed through fused disconnects to the load circuits, including the 208/240-V/24-V transformer. The load lines are both "hot," as shown in Figure 19-5.

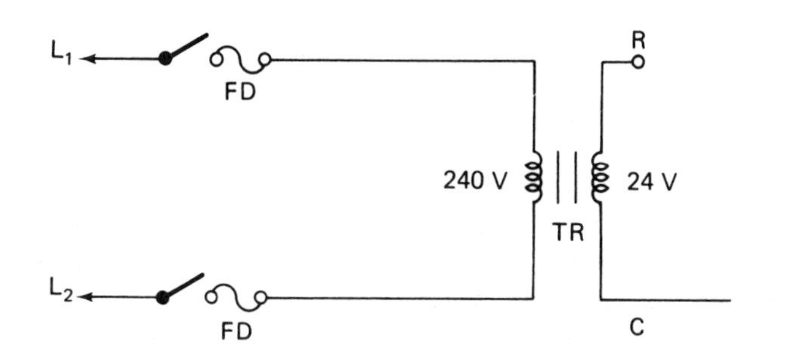

FIGURE 19-5 Power circuit showing "hot" load lines.

Fan Circuit

The fan circuits on an electric furnace are similar to those used with gas or oil except for the following changes:

1. The fan motor is usually 208/240-V ac
2. A timed fan start is used to start the fan. This is usually a part of the sequencer (an electrical device for staging the loads).
3. The fan starts from the timed fan start but is stopped by the regular thermal fan control

Details of the fan circuit arrangement are shown in Figure 19-6. When the thermostat calls for heating, R and W in the thermostat are made, energizing the low voltage heater in the sequencer. After approximately 45 s, the fan operates. Just as soon as the temperature leaving the furnace rises to the fan control cut-in temperature, this control makes. However, it has no effect on the fan since it is already running.

Approximately 45 s after the thermostat is satisfied, the timed fan start switch opens. However, it has no effect on the fan since the fan control is made. When the leaving air temperature drops to the cut-out setting of the fan control, the fan stops.

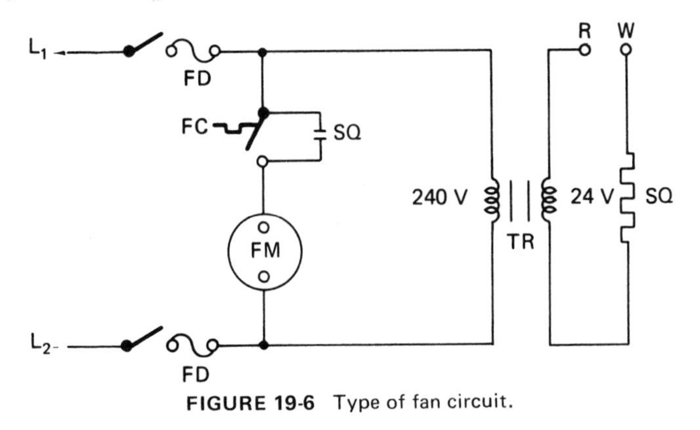

FIGURE 19-6 Type of fan circuit.

Heating Element Circuits

The heating element assemblies are connected in parallel to the power supply. These assemblies are turned on by the sequencer or heat relays so that the load is gradually placed on the line. Each heating element has a sequencer switch wired in series with the element to start its operation. Figure 19-7 shows the fan and the heating elements operated by a sequencer.

When the thermostat calls for heating, R and W make, energizing the low-voltage heater (SQ) in the sequencer. In approximately 45 s, the switch to the fan (SQ_1) and the first heating element (SQ_2) are made simultane-

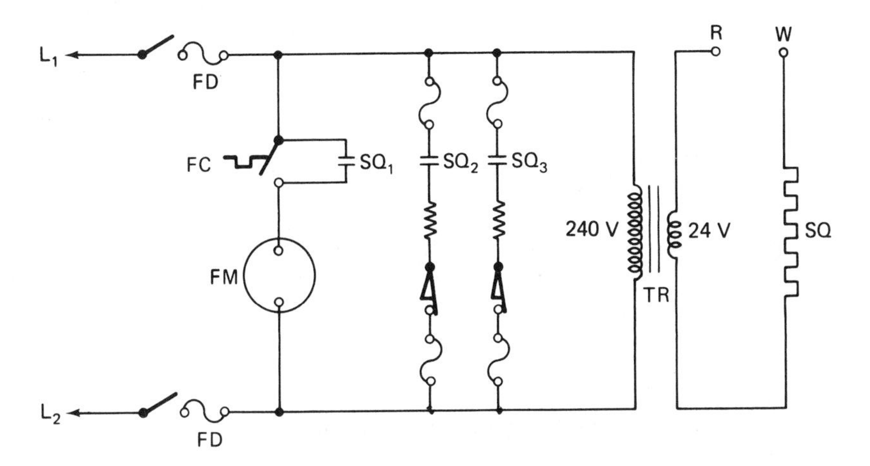

FIGURE 19-7 Fan and heating elements operated by a sequencer.

ously. Approximately 30 s later, the switch to the second heater element (SQ_3) is made. If additional heating elements are used, there is a comparable 30-s delay before each succeeding element is operated.

A two-stage thermostat is used on some systems. This thermostat has two mercury bulb switches, one for each stage. The first stage is set to make at a temperature a few degrees higher than the second stage. The first stage controls the fan and some of the heating elements. The second stage controls the balance of the heating elements.

Referring to Figure 19-8, when the first stage of the thermostat calls for heat R and W_1 make, energizing the heater SQ in the sequencer. This action switches on the first element and the fan, then the second element will be energized in 30 s. If the room temperature continues to drop to the setting of the second-stage of the thermostat, R and W_2 are made, energizing the heat relay heater (HR). In approximately 45 s the switch to the third heating element (HR_1) will be made, operating the final stage of heating.

FIGURE 19-8 Thermostat assembly in two-stage heating.

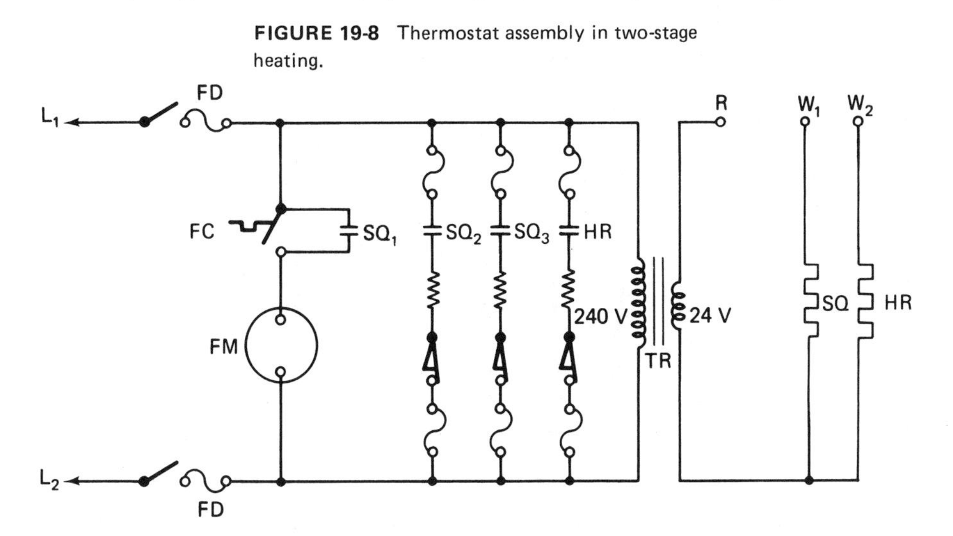

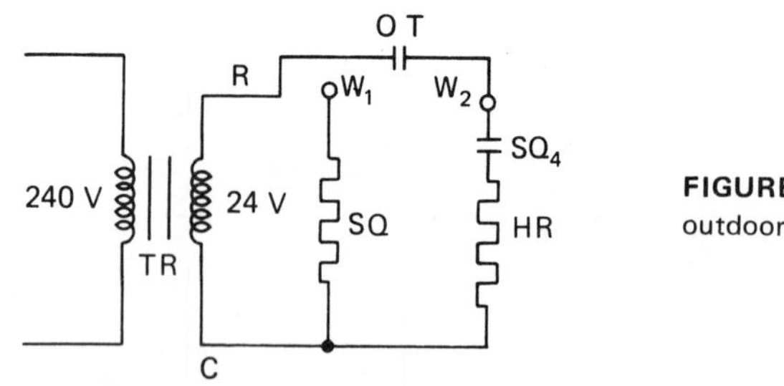

FIGURE 19-9 Use of outdoor thermostat.

The second stage thermostat can also be an outdoor thermostat rather than a part of the first-stage thermostat assembly. The low-voltage control circuit for this type of installation is shown in Figure 19-9.

The first stage (standard) thermostat operates the first-stage heating in the usual manner. A switch on the sequencer (SQ_4), an auxiliary switch, is also made in stage 1 to permit the second stage to operate. When the outside temperature reaches the setting of the outdoor thermostat (OT), the second-stage heating is operated in a manner similar to that shown in Figure 19-8.

Separate heat relays for each load can be used as an alternative to the use of the sequencer for staging the loads on an electric heating furnace. These relays function in a manner similar to the sequencer, as shown in Figure 19-10.

When R and W make, the heater on heat relay 1 (HR_1) is energized along with the fan relay (FR). In this arrangement the fan starts immediately and in about 30 s heat relay 1 switch (HR_1) closes, operating the first heater element. At the same time, the other HR_1 switch in the 24-V circuit closes, energizing the heater on heat relay 2. In approximately 30 s the two switches

FIGURE 19-10 Use of heat relays (single-stage thermostat).

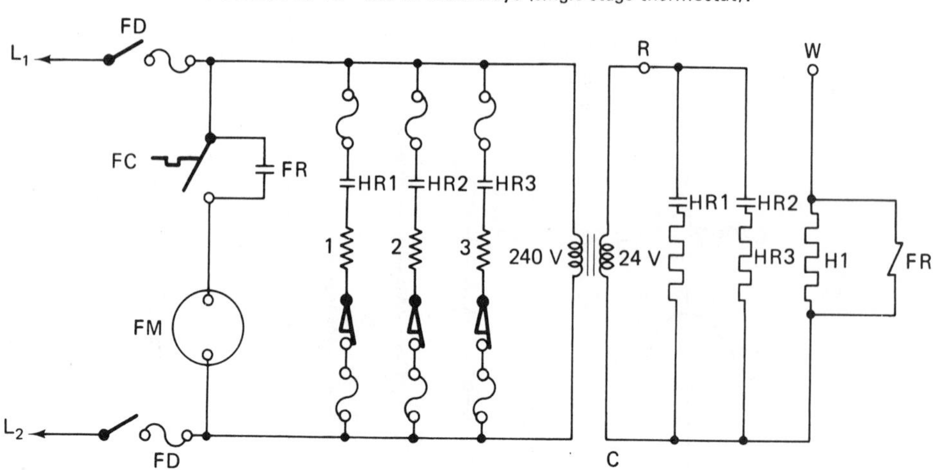

(HR_2) in heat relay 2 close. One operates heating element 2 and the other energizes the heater in heat relay 3. In approximately 30 s the switch (HR_3) closes, operating heating element 3. Thus, the fan and all heating elements are placed on the line in sequence.

USES OF SEQUENCERS AND HEAT RELAYS

A typical example of the use of sequencers and heat relays to stage the loads on an electric furnace is shown in Figures 19-11 and 19-12.

The furnace shown in Figure 19-11 has four heating elements. This unit uses a single-stage thermostat. When R and W are made in the thermostat, the fan immediately starts on low speed. At the same time, the heater in the sequencer is energized and the heater elements are operated in stages at approximately 30-s intervals. The fan can be turned on manually to high speed at any time by making R and G in the thermostat circuit. Sequencer details are shown in Figure 19-12.

SERVICE AND TROUBLESHOOTING

The methods for checking at the thermostat and the fan motor of an electric furnace are similar to those used for gas and oil furnaces. The external wiring of an electric furnace has a fused disconnect (or disconnects) similar to a gas furnace but uses a higher voltage and more power. The method of checking the external wiring is similar to that used for gas and oil furnaces, with additional allowances for the larger power supply.

In troubleshooting, it is good practice to use the nature of the complaint as a key to the area in which service is required. For example, if the complaint of a homeowner is that the fan will not run, troubleshooting should be confined to this area until the problem is found and corrected.

If the complaint is more general, such as "no heat," then a thorough and systematic check of the electrical system must be made. This requires an electrical check, starting where power is available and tracing through the electrical system to determine where it is no longer available. Switches must be checked to be sure that they are in their proper positions. Loads must be checked to be certain that they operate when supplied with the proper power.

A schematic wiring diagram is extremely helpful in determining the sequence in which power travels through the unit and for checking the correct operation of switches and loads.

(Abbreviations and Symbols — see back page.)

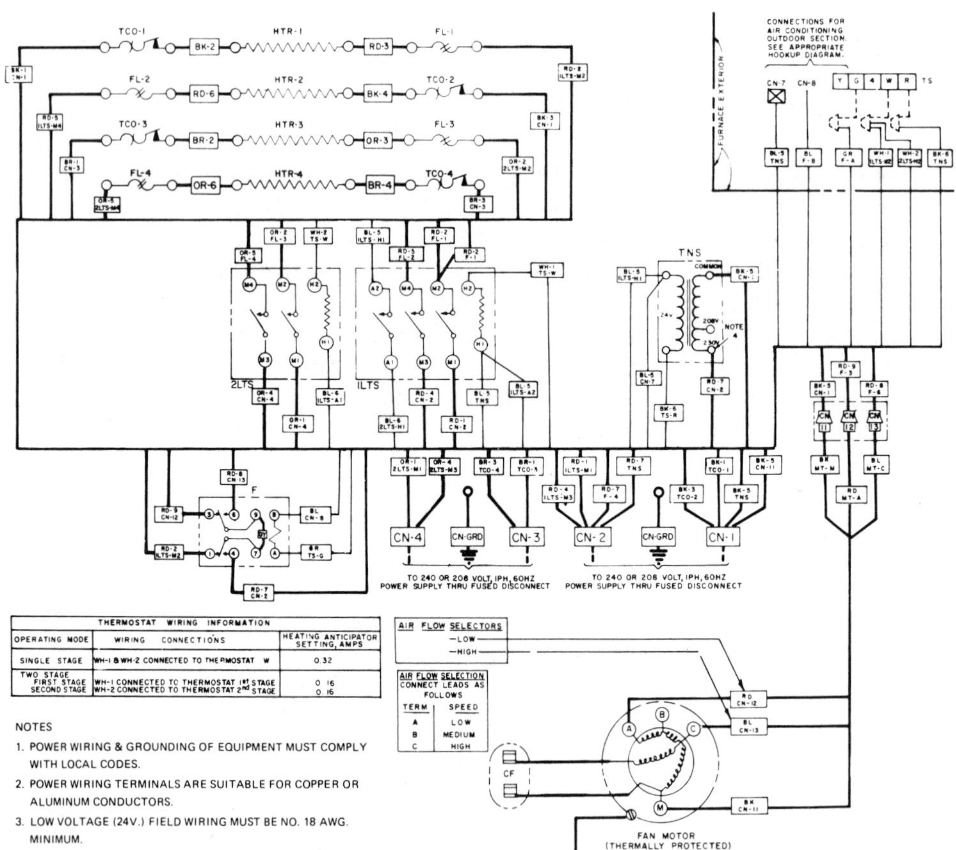

THERMOSTAT WIRING INFORMATION		
OPERATING MODE	WIRING CONNECTIONS	HEATING ANTICIPATOR SETTING, AMPS
SINGLE STAGE	WH-1 & WH-2 CONNECTED TO THERMOSTAT W	0.32
TWO STAGE FIRST STAGE SECOND STAGE	WH-1 CONNECTED TO THERMOSTAT 1st STAGE WH-2 CONNECTED TO THERMOSTAT 2nd STAGE	0.16 0.16

NOTES

1. POWER WIRING & GROUNDING OF EQUIPMENT MUST COMPLY WITH LOCAL CODES.
2. POWER WIRING TERMINALS ARE SUITABLE FOR COPPER OR ALUMINUM CONDUCTORS.
3. LOW VOLTAGE (24V.) FIELD WIRING MUST BE NO. 18 AWG. MINIMUM.
4. WHEN FURNACE IS CONNECTED TO A 208 VOLT POWER SUPPLY MOVE THE RD–7 LEAD TO THE 208 VOLT TRANSFORMER TAP.

FIGURE 19-11 Typical example of the use of sequencers, connection diagram (Courtesy, General Electric Company).

1. When room thermostat calls for heating, 24 volts is applied to time delay heater, terminals H1 and H2.
2. 30 to 90 seconds after heater is energized contacts M1 and M2 close, 10 seconds later M3-M4 close and then 10 seconds later A1-A2 closes.
3. Closing these contacts completes a 240 volt circuit to the first two blanks of heaters.
4. Sequencers with the auxilliary contacts A1-A2 are used to energize a second relay for additional heater banks.

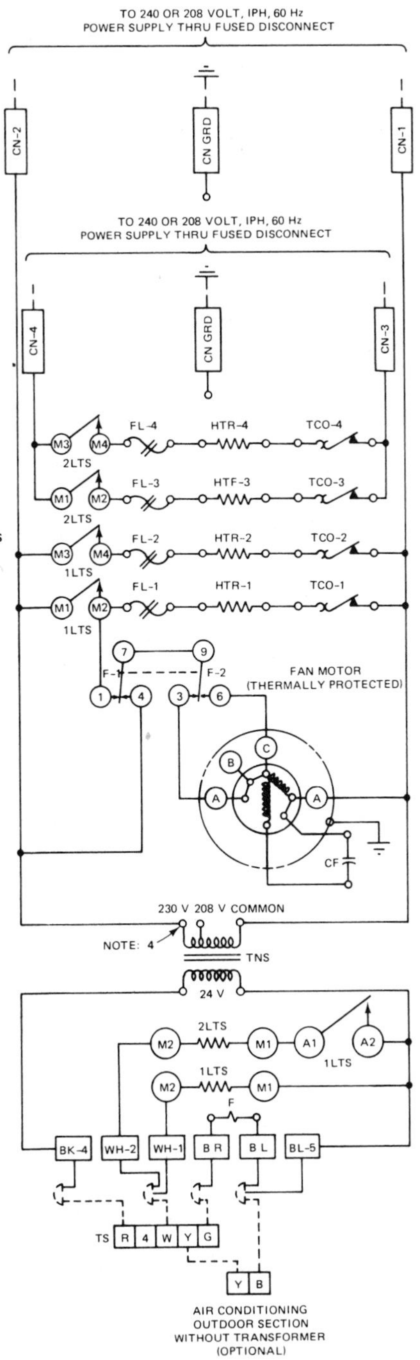

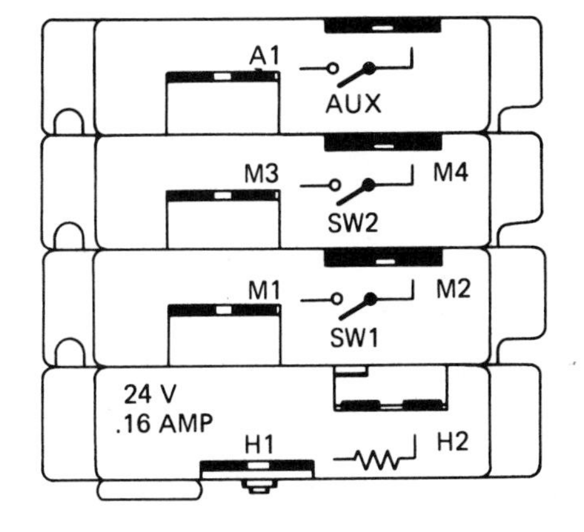

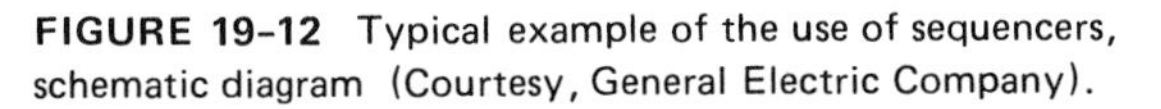

FIGURE 19-12 Typical example of the use of sequencers, schematic diagram (Courtesy, General Electric Company).

SYSTEMATIC CHECK

Systematic troubleshooting of an electrical furnace includes the following checklist:

1. Thermostat
2. Power
3. Transformer
4. Heating elements
5. Fan relay (if used)
6. Sequencer or heat relays

Techniques involved in checking various components include:

1. Mechanical check
2. Checking power with voltmeter
3. Checking switches with
   (a) Voltmeter
   (b) Ohmmeter
   (c) Jumper
4. Checking loads with
   (a) Voltmeter
   (b) Ammeter

Mechanical Check

1. Visually check for broken wires or cracked insulation
2. Check to see that terminal screws are tight
3. Check for frayed wires that may place power in incorrect locations
4. Check to see that controls are secure and that the thermostat is level

Checking Power With Voltmeter

It is essential that proper power be supplied to the equipment and to load devices in the furnace. Manufacturers' data on the nameplate indicates the proper voltage. Most equipment can be operated within a range of 10%

above or below the rated voltage. If power is not available within these limits, the equipment will not operate properly.

Checking Switches

(a) **Voltmeter:** This test can be used only when power is "on" and the voltmeter leads are placed across the two terminals of the switch. When the switch is closed, the voltmeter reads "0." When the switch is open, the voltmeter reads the voltage in the circuit

(b) **Ohmmeter:** *This test is used only when the power is off.* The leads of the ohmmeter are placed across the two terminals of the switch. The switch must be disconnected from the circuit. A "0" reading indicates that the switch is closed. An "∞" (infinity) reading indicates the switch is open.

(c) **Jumper:** This test is used with power on. If a jumper across the two terminals of the switch operates the load, the switch is open.

Checking Loads

(a) **Voltmeter:** Test with power on. Determine if proper voltage is available at the load. With proper voltage the load should operate.

(b) **Ammeter:** This test is performed with power on.

The ammeter test is used to check the current used by the total furnace or any one of its load components (Figure 19-13). The meter readings are compared with the data on the nameplate. This test offers an excellent means of checking the current through the sequencer to be certain that the elements are being staged "on" at the proper time. The jaws of the clamp-on ammeter are placed around one of the main power supply lines.

FIGURE 19-13 Ammeter used to read current draw of electric furnace.

REVIEW QUESTIONS

Select the letter representing your choice of the correct answer.

19-1. The heat equivalent for 1 W of electrical power is:

(a) 4.315 Btu

(b) 3.415 Btu

(c) 5.341 Btu

(d) 1.534 Btu

19-2. The efficiency of an electric furnace is:

(a) 75%

(b) 80%

(c) 90%

(d) 100%

19-3. The cut-out point on the limit control of an electric furnace is:

(a) 140°F

(b) 150°F

(c) 160°F

(d) 200°F

19-4. The maximum-size heater element on an electric furnace is usually:

(a) 3 kW

(b) 4 kW

(c) 5 kW

(d) 6 kW

19-5. The power supply for an electric furnace is:

(a) 24-V ac

(b) 120-V ac

(c) 240-V ac

(d) 480-V ac

19-6. The control circuit transformer on an electric heating system is:

(a) 240 V/24 V ac

(b) 120 V/24 V ac

(c) 24 V/440 V ac

(d) 24 V/120 V dc

19-7. Are the heating elements connected in series or in parallel with the power supply?

(a) Parallel

(b) Series

19-8. The staging of heating elements on an electric furnace is accomplished by use of a:

(a) Change in voltage

(b) Change in amperage

(c) Fan control

(d) Sequencer

19-9. An outdoor thermostat is used in place of:

(a) A sequencer

(b) The second stage of a room thermostat

(c) A two-stage fan control

(d) End switches on a damper motor

19-10. The instrument usually used to check the current flow through an electric furnace is the:

(a) Millivolt scale on a VOM meter

(b) Ohmmeter

(c) Voltmeter

(d) Clamp-on ammeter

20

EQUIPMENT ADJUSTMENT, PREVENTIVE MAINTENANCE, AND CUSTOMER RELATIONS

OBJECTIVES

After studying this chapter, the student will be able to:

- Recommend a planned maintenance program for a heating system
- Keep a service record
- Instruct the owner on use of the heating system

EFFICIENCY STANDARDS

To assure continued good performance and customer satisfaction with a warm air heating installation, proper attention must be given to:

- Equipment adjustment
- Preventive maintenance
- Customer relations

Efficient operation of equipment depends upon the care it receives, so it is important that a customer be made aware of the continuing maintenance practices.

EQUIPMENT ADJUSTMENT

Certain components require adjustment on the job after the installation has been completed or after a period of use. These adjustments are made on the:

1. Burner
2. Fan motor
3. Air balancing
4. Thermostat anticipator
5. Fan control

Details for performing many of these adjustments have been given in previous chapters. The student should refer to these for instructions to regulate individual components. This chapter views the system as a whole, concentrating on the effect of the performance of the parts on total comfort conditions.

Burner Adjustment

Two important questions regarding burner adjustment are:

1. Is the firing rate correct?
2. Is the fuel being burned efficiently?

All furnaces have the Btuh input and bonnet output stamped on the nameplate. On a gas furnace the output is rated at 80% of the input. On an oil-burning furnace the input is based on the nozzle size, with the output rated at 80% of the input. On an electric furnace the input is equal to the output.

In all cases the firing rate must be checked in relation to the input. For example, on a gas furnace with an input rating of 100,000 Btuh and using 1000 Btu/ft^3 gas, the input should be 100 ft^3 of gas per hour. In testing, if the input is found to be incorrect, an adjustment must be made on the gas pressure regulator or the size of the burner orifices.

The efficiency of a gas or oil unit can be checked by using the combustion test instruments described in Chapter 10.

Fan Motor Adjustment

The easiest way to determine the volume of air handled by the furnace is by measuring the temperature rise through the furnace. The temperature rise plus the output rating of the furnace provides the information necessary for calculating the cfm handled by the fan:

$$\text{cfm} = \frac{\text{output Btuh}}{\text{temperature rise} \times 1.08}$$

If the air volume handled by the fan does not comply with the manufacturer's requirement, the fan speed must be adjusted. On a direct-drive fan, it is necessary to select the proper motor taps. On a belt-driven fan, it is necessary to adjust the variable-pitch motor pulley.

The air filters must be clean when testing any system for air volume (Figure 20-1). A dirty filter creates an unnecessary restriction, thereby reducing air volume.

When cooling is added to an existing furnace, the evaporator coil adds an additional pressure drop in the distribution system. In most cases, cooling also requires more air than heating. If the furnace does not already have a multispeed motor for the fan, it is advisable to install one. For heating, the fan speed is usually slower than that used for cooling. A combination fan relay and transformer (Figure 20-2) can be installed. This control supplies additional power for the compressor contactor and provides a means of automatically changing the speed of the fan motor from heating to cooling.

FIGURE 20-1 Replacing air filters (Courtesy, American Air Filter Company).

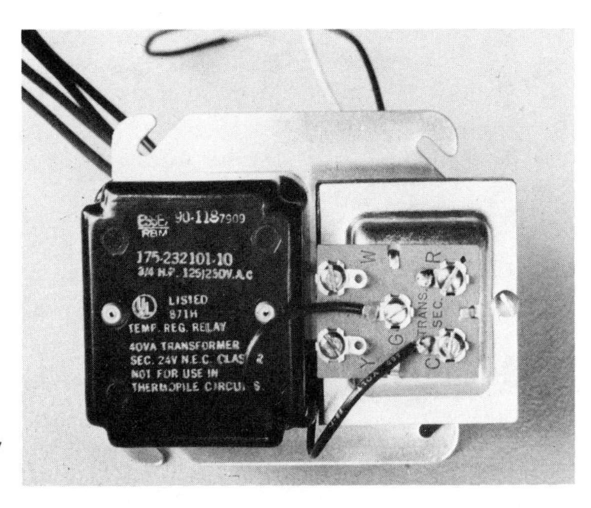

FIGURE 20-2 Combination fan relay and transformer.

Air Balancing

Air balancing is performed by adjusting the branch duct dampers to produce uniform temperatures throughout the building (Figure 20-3). The procedure is as follows:

1. Place a thermometer in each room
2. Open all supply and return air duct dampers
3. Open all grilles and register dampers
4. Set thermostat to call for heat
5. Adjust dampers while furnace is running to produce uniform temperatures in all rooms

The balancing should be done during weather cold enough to permit continuous operation of the heating equipment for a substantial period of time. If continuous fan operation is normally used with the furnace cycling at the rate of 8 to 10 times per hour, this condition provides good operation for balancing.

Thermostat Anticipator Adjustment

The thermostat anticipator is adjusted to control the length of the operating cycle. Raising the amperage setting of the anticipator lengthens the cycle. Lowering the amperage setting of the anticipator shortens the cycle.

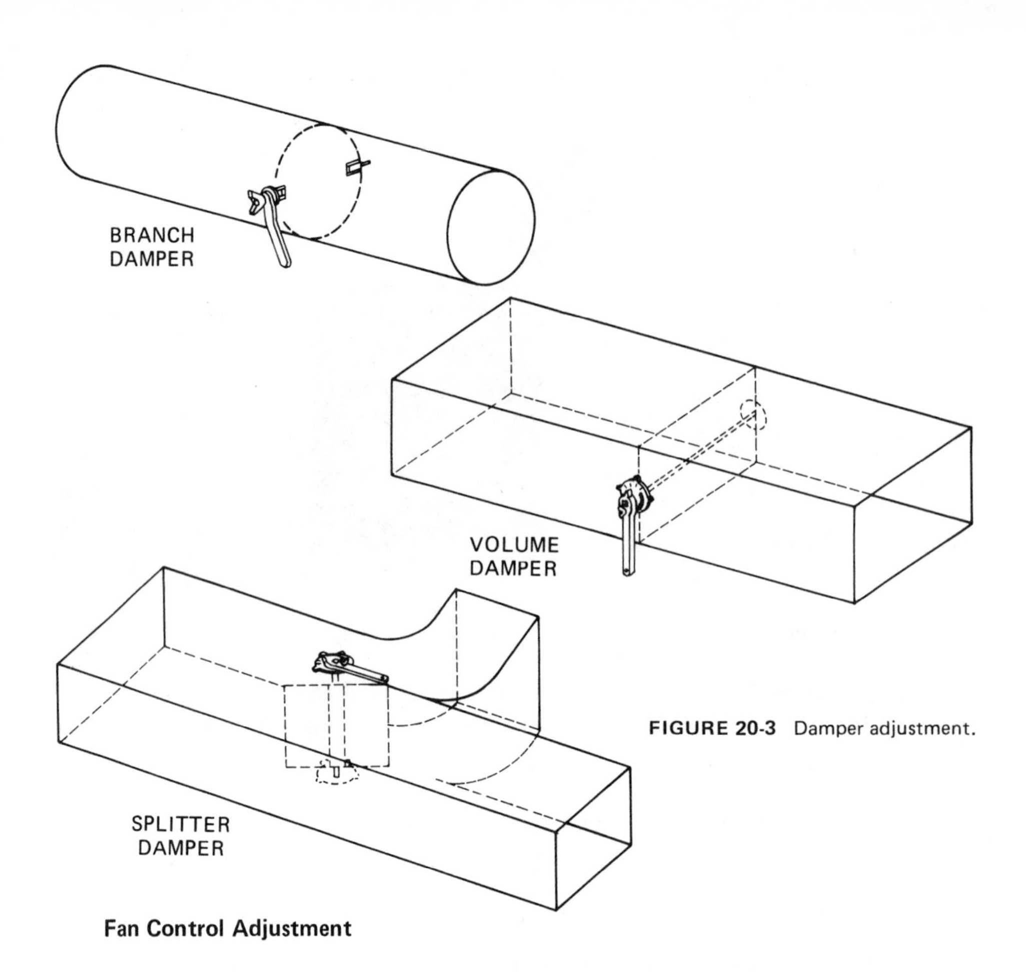

FIGURE 20-3 Damper adjustment.

Fan Control Adjustment

Where continuous fan action is not used, the fan control on an upflow furnace starts the fan when the bonnet temperature reaches its cut-in point (Figure 20-4). If the fan is delivering air that is too cool for comfort, the cut-in point of the fan control can be raised. It is advantageous to have the longest possible fan-on time and still not blow cold air. The differential of the fan control (cut-in temperature minus cut-out temperature) should be great enough to prevent short cycling of the fan.

PREVENTIVE MAINTENANCE

Preventive maintenance is the performing of service on equipment before a breakdown occurs. The furnace should receive periodic preventive maintenance for continuous good performance. Most of this work can be done during seasons when the equipment is normally shut down. Two important items are:

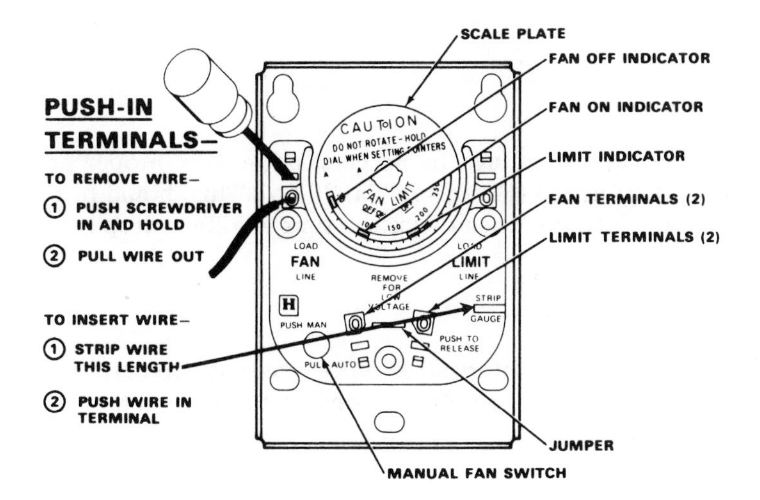

FIGURE 20-4 Fan control adjustment (Courtesy, Honeywell Inc.).

- Motors should be oiled twice a year
- Air filters should be inspected periodically

Major points where maintenance is required include the:

1. Thermostat
2. Power supply
3. Air filters
4. Fan motor assembly
5. Firing device
6. Wiring and controls
7. Heat exchanger and flue passages
8. Accessories

Thermostat

Check to be certain that the thermostat is secure and level and that all wiring connections are tight. Carefully remove any dust or lint that may have collected on the moving parts.

Power Supply

Check the supply voltage to be certain that the proper power is being delivered to the furnace (Figures 20-5a, 20-5b). This should be done during a period of heavy loading, since this is the time a voltage drop can occur.

Air Filters

Clean or replace air filters as often as necessary to permit the proper air flow. Replacement filters should be similar to those removed in material, thickness, and size.

Fan Motor Assembly

Clean out the blades of the fan wheel. Check the belt (if used) for wear and tightness. Oil the fan motor. If there is too much end play on the shaft, move the thrust collar closer to the bearing.

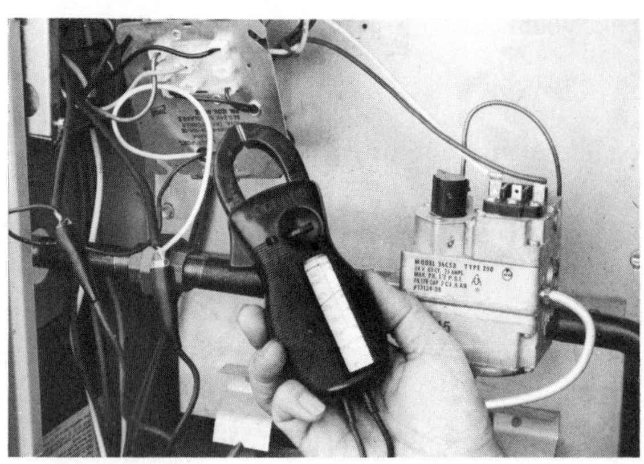

FIGURE 20-5(a) Checking the power supply on a gas-fired furnace.

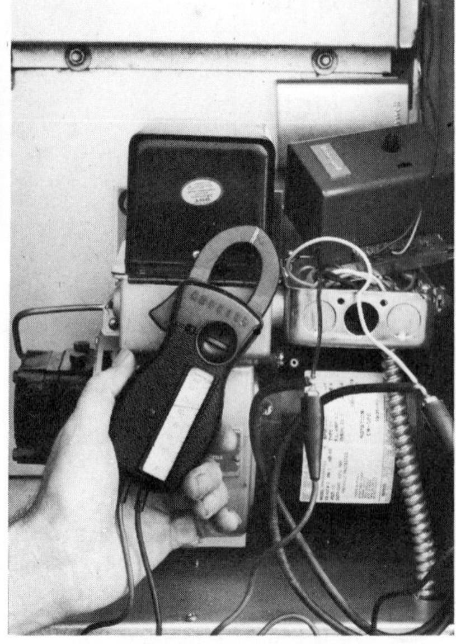

FIGURE 20-5(b) Checking the power supply on an oil-fired furnace.

Firing Device

GAS FURNACE:

Observe the flame. It should be a soft blue color without yellow tips. Adjust primary air if necessary. If the flame will not clean up when adjusted, remove the burners and clean them both inside and outside.

Observe the pilot flame. It should be a soft blue color. Clean and adjust if necessary.

Measure the manifold gas pressure (Figure 20-6). It should be approximately 3 to 4 in. of water column. Adjust if necessary.

Check the thermocouple. The thermocouple, under no load, should generate between 18–30 mV ac. When under load it should generate at least 7 mV dc.

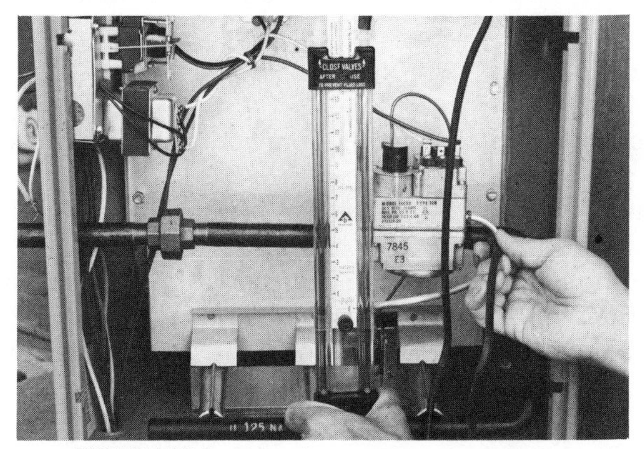

FIGURE 20-6 Measuring manifold gas pressure.

OIL BURNER:

Clean the burner and run a complete combustion test. Replace the oil burner nozzle if necessary. Adjust the air volume for maximum efficiency. Replace the oil line strainer. Clean the bimetallic element on the stack relay (if used).

Observe the flame after burner is restarted. Flame should be centered in the combustion chamber.

Check the oil pressure (Figure 20-7). It should be 100 psig during operation and 85 psig immediately after shutdown.

ELECTRIC FURNACE:

Set thermostat to call for heat. Check by reading the incoming power amperage to be certain that all heater elements are operating.

FIGURE 20-7 Checking the oil pressure (Courtesy, Sundstrand Hydraulics).

Wiring and Controls

Check to be sure that all wiring connections are tight and that wiring insulation is in good condition.

Using a thermometer, check the cut-in and cut-out points of the fan control.

With a thermometer check the cut-out point of the limit control by shutting down the fan. Bonnet temperature must not exceed 200° F.

Operate switches in their various positions to be certain that the equipment responds properly.

Heat Exchanger and Flue Passages

Clean heat exchanger and flue passages if necessary (Figure 20-8). Check furnace flue pipe for wear and corrosion. Replace flue pipe if necessary.

FIGURE 20-8 Cleaning heat exchanger.

Accessories

Clean and service accessories in accordance with manufacturer's instructions. Humidifiers may require cleaning more than once a year depending on water conditions.

The condenser surface on condensing units used for cooling must be kept clean. Condenser fans usually require lubrication.

CUSTOMER RELATIONS

Good practice in customer relations has many benefits. Continuing good relationships with customers increases their awareness of the responsibility they have toward properly maintaining the equipment. Satisfied customers usually mean repeat business and new customers. Here are some of the important areas in this relationship:

- Instructions in equipment care
- Service company communication
- Warranty
- Preventive maintenance contracts
- Confidence

Instructions in Equipment Care

At the time of the original installation, the customer should be fully informed about the care and operation of the equipment. Original instructions should be available and posted so that any future owner will also be informed (Figure 20-9).

Service Company Communication

The name and telephone number of the service company should be displayed in a prominent location near the installation. A service log should be attached to the furnace and kept up to date by any technician servicing the unit.

Warranty

The customer should be fully informed of the terms of the warranty. A sample equipment warranty is shown in Figure 20-10. A new installation usually carries a 1-year warranty against defective materials and workman-

HOME OWNER INSTRUCTIONS

Your heating unit is a valuable piece of equipment, designed and manufactured by the most modern methods. Proper care of this unit should result in many years of service and comfort. An annual check-up of the heating unit by a competent service man is recommended.

So that you may understand the operating parts of you equipment better, the following subjects have been itemized:

1. **WARNING: These units are manufactured for use with a specific type of fuel. Natural gas units must not be converted to liquefied petroleum gas.**

2. **INPUT:** The correct heat output of the furnance is regulated by the burner orifices and the gas pressure. The proper orifices are furnished but the gas pressure regulator must be adjusted by the contractor or Gas Company. This is a "one-time" adjustment.

3. **CYCLING PILOT BURNER:** This unit has a Natural Gas Non-100% shut off system and is supplied with a Pilot Relight control designed to automatically light the pilot burner each time the thermostat "calls" for heat.

If the pilot burner (or main burners) should fail to light, a "No-Heat" condition will result, that will produce the following condition:

—the safety control will prevent gas from reaching the main burners. Unit will not return to normal heating cycle until abnormal conditions (that caused failure of pilot burner to light) are corrected. Contact Heating Contractor or Gas Company for service to insure that proper operating conditions are restored.

4. **MAIN BURNER:** These burners do not normally require frequent servicing. Lint may cause some burners to burn improperly and necessitate cleaning. The presence of a yellow flame, or delayed ignition, is a warning that service is required.

5. **COMBINATION AUTOMATIC GAS VALVE:** This valve is controlled by the thermostat, the limit switch and the cycling pilot relight system. If the automatic gas valve fails to shut-off the gas to the main burners, once the thermostat is "satisfied", close the main manual gas valve to unit and call the heating contractor or gas company.

6. **THERMOSTAT:** Thermostats may have mercury tube contacts or they may have open, snap acting contacts. Those having mercury tube contacts must be installed level for proper performance. The open contact style type are not dependent upon the degree of level. Lint accumulation may affect the degree of accuracy of control of either type of thermostats. An annual cleaning is recommended.

Heat generated by devices other than the furnance may interfere with thermostat performance. Therefore, lamps, radios, television sets, etc. should not be placed near the thermostat.

7. **LIMIT CONTROLS:** These controls are so wired that they cause the main automatic gas valve to close should the air temperature rise above their settings. These controls normally do not require any service. Failure of the blower motor to run, broken fan belt, dirty filters and closed registers will result in the limit control and/or auxillary limit control operating to close the gas valve.

8. **FAN CONTROL:** This control turns the blower "on" and "off" at predetermined temperatures. Normally no service is required.

9. **BLOWERS:** Belt driven blower bearings are permanently lubricated and with normal usage do not require servicing. (See motors for direct-drive bearings.) The air volume delivered by the blower is regulated by the distribution system and blower speed. The adjustments made by the installing contractor should be adequate.

10. **MOTORS:** Lubrication of the motor should be 15 to 30 drops of SAE #20 oil annually.

11. **BELT:** Belt should be checked for wear when filters are checked. A spare belt on hand may avoid a period without heat.

12. **FILTERS:** This unit must be equipped with an efficient air filter to remove dirt and lint from the air before it is heated and delivered to the living space. Unit should never be operated without filters in place for any length of time. Dirt and lint entering the motor will shorten its life.

    When filters become dirt laden, insufficient air will be delivered by the blower, heating costs will increase and ability of unit to heat the building will be hampered. Inability to heat, increasingly hotter air temperature from registers and cycling of the burner from the limit control are symptoms of dirty filters.

    If filters are dirty, replace then with the same size and type. If permanent filters are used, wash, and reinstall. Keeping replacement filters on hand is advisable when disposable filters are used.

13. **REGISTERS:** Warm air and return air registers must be open when furnance is in operation. Never place carpets, rugs or low furniture over registers.

14. **FUSES:** If neither the gas valve nor the blower will operate, check for burned out or loose fuse.

FIGURE 20-9 Furnace operating instructions (Courtesy, Westinghouse Electric Corporation, Heating and Cooling Divisions).

Heating Equipment Warranty

The Manufacturer warrants to the original user of the heating product identified hereon that such product has been manufactured in accordance with its standard manufacturing practices. The Manufacturer further warrants that the product is fit for the ordinary purpose or purposes specified in its catalog when installed, operated, serviced and used in accordance with (i) all specifications and instructions attached to or accompanying the product or set forth in the Manufacturer's catalog, and (ii) standards equivalent to those set forth by the National Environmental Systems Contractors Association. Except as stated herein, MANUFACTURER MAKES NO OTHER WARRANTY, EXPRESS OR IMPLIED, INCLUDING MERCHANTABILITY.

The Manufacturer shall, at its sole option, repair or replace (or make an appropriate allowance or credit for the repair or replacement of) any component part thereof (other than filters) which shall prove to the satisfaction of the Manufacturer to have been defective in material or workmanship at the time of shipment from its factory and which is returned to Manufacturer, through an authorized wholesaler, freight prepaid, to Manufacturer's factory. (With the exception of heat exchangers which shall be held by the wholesaler for inspection by the manufacturer's representative) within (i) IN THE CASE OF HEAT EXCHANGERS ONLY, the period specified (see table below) with respect to the appropriate unit as measured from the earlier of: (a) the date of installation, or (b) a date six (6) months from the date of manufacture, and (ii) IN THE CASE OF ALL OTHER COMPONENT PARTS, a period of one year measured from the earlier of (a) the date of installation, or (b) a date six (6) months from the date of manufacture. Any claim not so made within the applicable period shall conclusively be deemed waived by the user. The Manufacturer's responsibility hereunder shall be conditioned upon inspection of the product and, if authorized by the Manufacturer, return of the product or component part to Manufacturer without expense to the Manufacturer.

THIS WARRANTY SHALL NOT APPLY TO ANY HEAT EXCHANGER WHICH UPON INSPECTION IS DETERMINED TO HAVE BEEN OPERATED IN A CORROSIVE ATMOSPHERE.

No field charges for labor or other expenses incurred in the removal, repair or replacement of the product or any component claimed to be defective shall be paid by the Manufacturer nor shall Manufacturer be liable for any expense incurred by the user in order to remedy any defect in the product. Manufacturer shall not be liable for any consequential, special or contingent damage or expense, arising directly or indirectly from any defect in the product or from the use thereof. THE REMEDIES SET FORTH HEREIN SHALL BE THE EXCLUSIVE REMEDIES AVAILABLE TO THE USER AND ARE IN LIEU OF ALL OTHER REMEDIES.

Service under the warranty covering this heating unit is the responsibility of the dealer or contractor who installs the equipment. In the event service under this warranty is required, the owner should request such service directly from the dealer or contractor from whom he purchased the unit, under the provisions of any applicable service agreement.

Furnace heat exchangers have extended warranty protection as follows:

MODEL (S)	PERIOD
BTS, BTL, HAS, HBS, HTL, WAS, WTL and NTL	20 Years
NAS	10 Years
KAS (stainless steel element)	10 Years
HWC (heating element only)	10 Years
SA, SAC (Heating element only)	5 Years
SBC, SAM	5 Years
L13 Industrial Oil	5 Years
MBE, EWC (heating element only)	1 Year
KAS (aluminized element)	1 Year

FIGURE 20-10 Sample equipment warranty (Magic Chef Heating & Cooling Division, Columbus, Ohio).

COMBUSTION SERVICE RECORD

HIGH PRESSURE GUN-TYPE BURNERS

For Use with BACHARACH Combustion Testing Instruments

Order No. 4244 Date 2-20-79
Taken by Smith

Condition Reported

- [ ] No Fire
- [x] Insufficient Heat
- [x] Excessive Oil Consumption
- [ ] Odor
- [ ] Burner Ignites, then Goes Out
- [ ] Noise
- [ ] Burner puffs . . . [ ] On Start; [ ] On Stop

OTHER

When Service Wanted

DATE 2-21-79 TIME Before Noon

- [ ] PHONE FOR APPOINTMENT

Job Assigned to:

NAME Ryan DATE 2-21-79

Job Completed:

DATE 2-21-79 TIME 10:25 AM

BY Tom Ryan
Signature of Service Man

Owner John Doe
Street 426 Maple Drive
City Centerville Phone CE-5481-J
If not home get key 428 Maple Drive
Occupant Same
Street
City Phone
Work Authorized [x] by Owner [ ] by Occupant John Doe
Signature of person authorizing work

I—Preparing for Combustion Test

1. Open main burner switch.
2. Inspect and clean out accumulated oil in combustion chamber.
3. Advance thermostat. (5-10° F.)
4. Close remote control burner switch.
5. Make ¼" diameter hole in flue pipe and overfire (for BACHARACH test instruments).
6. Insert TEMPOINT thermometer (200-1000° F. range). through ¼" diameter hole in flue pipe.
7. Open inspection port or door.
8. Adjust flame mirror.
9. Close main burner switch. (Starting burner.)

II—Combustion Test Procedure and Inspection Data

STEP	Observe—and mark with √		TEST NO. 1	2	3	4
1	**FLAME IGNITION**	Instant	✓	✓		
		Delayed				
		Doesn't Ignite				
2	**FLAME COLOR** If flame shows two colors check both	Orange		✓		
		Yellow	✓	✓		
		White	✓			
		Sparks				
3	**FLAME SHAPE**	Uniform	✓			
		Lop-sided				
4	**FLAME IMPINGEMENT**	At bottom				
		At sides				
		At rear				
5	**ODOR** Near burner; observation door, draft regulator	None	✓			
		Slight				
		Heavy				
6	**NOISE** Mark (x) when Excessive Mark (√) if Moderate	Rattle				
		Hum				
		Pulsation				
		Start				
		Running				
		After fire				
7	**SOOT DEPOSIT** Mark (x) if Heavy; (√) if Light; (o) if None	Flue				
		Comb. Chamber				
		Furnace/Boiler				

STEP	Observe—and write in data	TEST NO. 1	2	3	4
8	(Close observation door) **OVERFIRE DRAFT** in inches Water	.030	.020		
9	**TEMPOINT READING FLUE GAS TEMP. °F.** When constant temperature is reached	710	610		

STEP	Observe—and write in data		TEST NO. 1	2	3	4
10	**BASEMENT AIR TEMP. °F.**		60	60		
11	**NET STACK TEMP. °F.** Subtract basement temp. (step 10) from flue gas temp. (step 9)		650	550		
12	**FLUE DRAFT** in inches water, (use same hole used for stack temp. test)		.035	.025		
13	**FYRITE READING % CO_2** (use same hole used for stack temp. test)		5½	9½		
14	**TRUE-SPOT SMOKE READING** (use same hole used for stack temp. test)		½	1		
15	**FIRE EFFICIENCY FINDER % COMBUSTION EFFICIENCY**		60¼	76½		
16	(Open Main Burner Switch) **FLAME CUT-OFF (Seconds)** Estimate time required in seconds for flame to disappear after burner stops		2	2		
17	(Close Main Burner Switch) **OIL PRESSURE (psi)** Measured with oil gauge installed on pump					
18	**FEED LINE SUCTION (inches)** Measured with vacuum gauge installed in feed line					
19	(Open Main Burner Switch Remove Nozzle Assembly) **NOZZLE** (service if necessary, then reinstall)	Size—Gph				
		Type—S/H				
		Spray Angle				
20	**COMBUSTION CHAMBER SIZE**	Depth "				
		Length "				
		Width "				
		Area sq. in.				

III—Adjustments and Repairs

Make adjustments, install replacements, and tune-up as required. Indicate, in spaces provided below, work done before repeating the tests listed under "II".

Write in "A" for "Adjust"; "C" for "Clean", "R" for "Replace". Mark "√" for other work, and describe it on the back of this sheet, if necessary.

WORK PERFORMED	BEFORE TEST NO. 2	3	4
BURNER AIR SHUTTER	A		
SEAL AIR LEAKS	✓		
BURNER AIR BLOWER	C		
TURBULATOR			
AIR CONE			
BAROMETRIC DAMPER	A		
BURNER IGNITION—SAFETY CONTROL			
LIMIT CONTROL			
ELECTRODES			
ELECTRODE CABLE			
TRANSFORMER			
AIR FILTERS	R		
NOZZLE			
NOZZLE STRAINER			
PUMP STRAINER			
PUMP			
OIL FILTER	R		
OIL PRESSURE			
PUMP CUT OFF			
COMBUSTION CHAMBER			
BURNER POSITION			
BELT-COUPLING			
OIL LINE			
CHIMNEY REPAIRS			
FURNACE/BOILER CLEANED			

IV—Final Inspection

(a) Repeat the combustion check-ups listed under "II", and enter data in proper spaces.

(b) Check each of the following for proper setting, operation, or condition.

- [x] MAIN BURNER SWITCH
- [x] THERMOSTAT
- [x] BLOWER CONTROL
- [x] LIMIT CONTROL
- [x] PUMP CONTROL
- [x] LUBRICATION
- [ ] LOW WATER CUT OFF
- [x] OIL LEAKS
- [x] CIRCULATING-AIR FAN
- [x] AIR FILTERS

CONDITION OF FUEL OIL good
FLAME FAILURE CUT OFF TIME 120 SEC.
IGNITION CUT OFF TIME 45 SEC.

FIGURE 20-11 Sample combustion service record for oil-fired heating units (Courtesy, Bacharach Instrument Company).

ship. Heat exchangers often have an extended warranty period. The details of the contractor's warranty should be clearly stated to the customer in writing.

Preventive Maintenance Contracts

All heating equipment should have a routine maintenance check once a year. Most customers prefer to have this service performed during the shutdown period. It is good practice to include this yearly service in a maintenance contract which also arranges for in-between service calls. The contract should also state which items are excluded from servicing, such as the changing of air filters. This is considered a customer's responsibility.

It is particularly important that oil-burning equipment service be included on a regular maintenance contract. A sample check sheet for oil-fired heating units is shown in Figure 20-11. Service once a year on these units is absolutely essential to continued good performance.

Confidence

To build confidence in a relationship with the customer, a service technician must be courteous, reliable, honest, clean, and efficient. By exhibiting these qualities, a technician benefits personally, and is also shown to be the representative of a reputable service company.

REVIEW QUESTIONS

Select the letter representing your choice of the correct answer.

20-1. Which of the following components does not require adjustment?

(a) Thermostat

(b) Fan control

(c) Heat exchanger

(d) Gas burner

20-2. Which of the following has a direct relation to the firing rate?

(a) Input

(b) Fan

(c) Thermostat

(d) Transformer

20-3. The easiest method of determining the air quantity the furnace is handling is by checking the:

(a) Anemometer

(b) Velometer

(c) Pilot tube

(d) Temperature rise

20-4. What is the relation of air volume used for cooling compared to heating?

(a) More

(b) Less

(c) The same

(d) No relation

20-5. In balancing a job, where are the thermometers located?

(a) In the duct work

(b) In the heat exchanger

(c) In each room

(d) In the furnace

20-6. How often should a furnace cycle per hour to maintain even temperatures?

(a) 3 to 5

(b) 8 to 10

(c) 9 to 12

(d) 14 to 20

20-7. How many times a year should motors be oiled?

(a) Once

(b) Twice

(c) Three times

(d) Four times

20-8. What is the color of a properly adjusted gas flame?

(a) Yellow

(b) Blue with yellow tip

(c) Green

(d) Soft light-blue

20-9. What is the cutoff pressure of an oil burner?

(a) 50 lb

(b) 70 lb

(c) 85 lb

(d) 100 lb

20-10. For what maximum period of time do some manufacturers warrant their heat exchangers?

(a) 1 year

(b) 2 years

(c) 5 years

(d) 20 years

21

ENERGY CONSERVATION

OBJECTIVES

After studying this chapter, the student will be able to:

- Evaluate a system from an energy use standpoint
- Recommend changes or adjustments to improve energy consumption

HEATING INDUSTRY ENERGY NEEDS

The heating industry uses approximately 18% of the total energy consumed in the United States, see Figure 21-1. The heating business is therefore greatly affected by changes in the energy supply and changes in energy costs.

United States' energy needs are increasing, as shown in Figure 21-2. The supply is limited, based on the estimated number of years availability. The costs of energy are rising. It is therefore important that known supplies be conserved and that new sources of energy be found. Some new sources are available in unlimited quantities, such as solar energy.

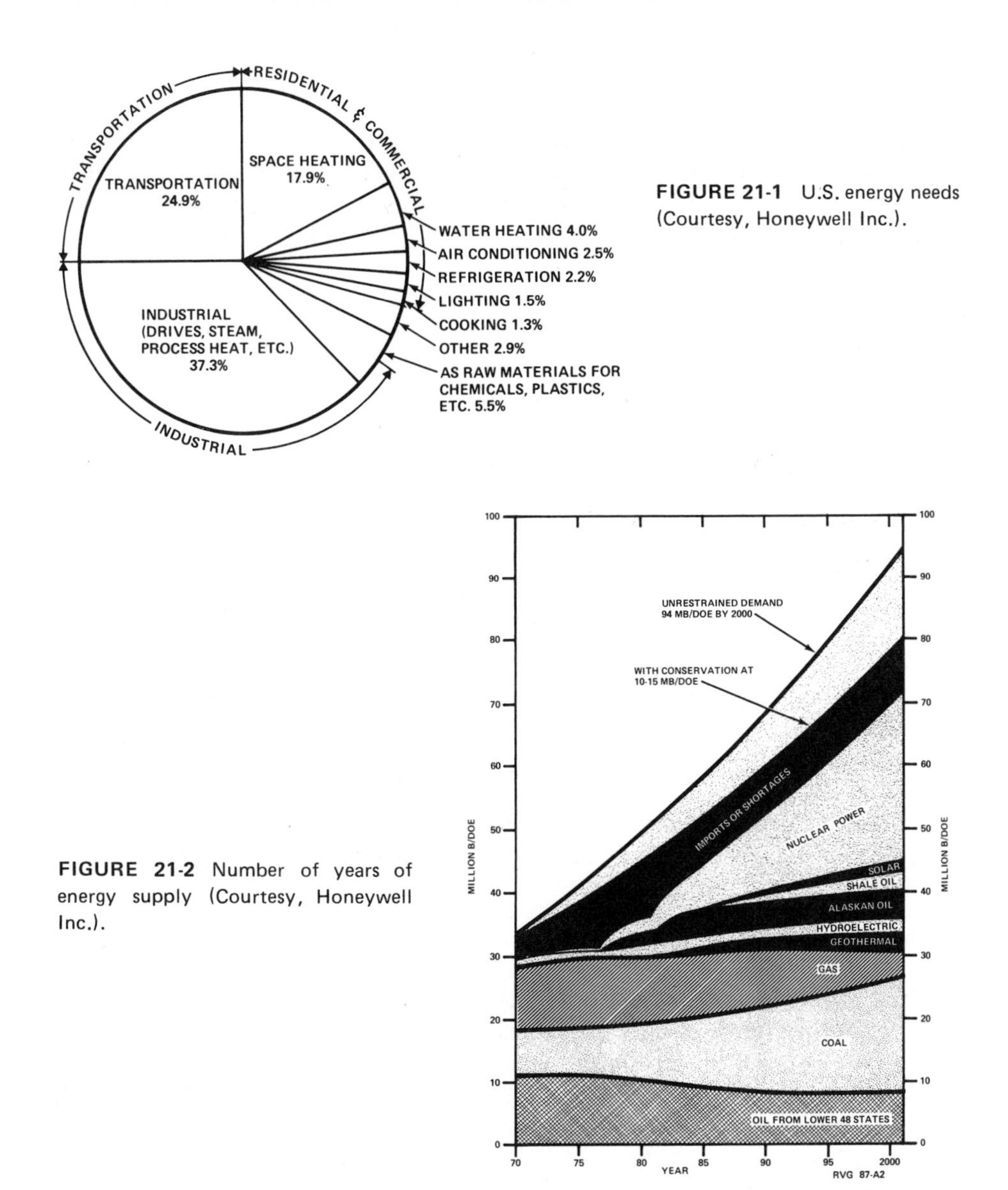

FIGURE 21-1 U.S. energy needs (Courtesy, Honeywell Inc.).

FIGURE 21-2 Number of years of energy supply (Courtesy, Honeywell Inc.).

WAYS TO CONSERVE ENERGY

The steps that need to be taken to improve our energy usage are:

1. Eliminate as much waste as possible. A rough estimate indicates that approximately 50% of our present usage is wasted
2. Develop new sources of supply

Much of this text is concerned with the general principles of warm air heating. These principles are also useful when making the alterations needed to conserve energy. A number of changes usually occur when energy conservation methods are applied:

1. The cost of materials and equipment is increased
2. Fuel usage is decreased
3. The comfort level may be altered

Some ways that fuel can be saved are:

1. *Improve insulation* of the structure and reduce air leakage
2. *Size the heating system* to fit the actual load requirements
3. *Alter the heating unit* or the distribution system to improve its efficiency
4. *Operate the system to reduce fuel consumption*, particularly when the load requirements are low
5. *Provide improved maintenance* of the system
6. *Install solar equipment* or multiple fuel burning equipment, such as wood-oil-gas or wood-coal-gas, to utilize available energy sources

Improve Insulation

The insulating value of various types of materials is often given in terms of an *R factor.* R represents the thermal resistance of the material or materials. To compare the insulating value of various substances, the R values are usually given in terms of an inch thickness of each material. The R values for various types of building materials, including insulation, are listed in Figure 21-3.

Certain standards are recommended for the R value of the exposed areas of a building.

Material or surface	*R*
Exterior surface resistance	0.17
Bevel-lapped siding, $\frac{1}{2}$ by 8 in.	0.81
Fiberboard insulating sheathing, $\frac{3}{4}$ in.	2.10
Insulation, mineral wool batts, $3\frac{1}{2}$ in.	11.00
$\frac{3}{8}$-in. gypsum lath and $\frac{3}{8}$-in. plaster	0.42
Interior surface resistance	0.68
Total 12 factor	15.18

MATERIAL	THICKNESS (Inches)	R VALUE
Air Film and Spaces:		
Air space, bounded by ordinary materials	¾ or more	.91
Air space, bounded by aluminum foil	¾ or more	2.17
Exterior surface resistance	—	.17
Interior surface resistance	—	.68
Masonry:		
Sand and gravel concrete block	8	1.11
	12	1.28
Lightweight concrete block	8	2.00
	12	2.13
Face brick	4	.44
Concrete cast in place	8	.64
Building Materials — General:		
Wood sheathing or subfloor	3/4	1.00
Fiber board insulating sheathing	3/4	2.10
Plywood	5/8	.79
	1/2	.63
	3/8	.47
Bevel-lapped siding	1/2 x 8	.81
	3/4 x 10	1.05
Vertical tongue and groove board	3/4	1.00
Drop siding	3/4	.94
Asbestos board	1/4	.13
3/8" gypsum lath and 3/8" plaster	3/4	.42
Gypsum board (sheet rock)	3/8	.32
Interior plywood panel	1/4	.31
Building paper	—	.06
Vapor barrier	—	.00
Wood shingles	—	.87
Asphalt shingles	—	.44
Linoleum	—	.08
Carpet with fiber pad	—	2.08
Hardwood floor	—	.71
Insulation Materials (mineral wool, glass wool, wood wool):		
Blanket or batts	1	3.70
	3 1/2	11.00
	6	19.00
Loose fill	1	3.33
Rigid insulation board (sheathing)	3/4	2.10
Windows and Doors:		
Single window	—	approx. 1.00
Double window	—	approx. 2.00
Exterior door	—	approx. 2.00

FIGURE 21-3 Insulating valve of common materials (Courtesy, U.S. Department of Commerce, National Bureau of Standards).

In most northern climates it is good practice to provide construction of walls with an R factor of 15.0 or greater, and ceilings with a factor of 25.0 or greater. Windows should be constructed with double glass or single glass with a storm window.

It is necessary in figuring heating loads to convert to a *U factor* (the heat flow in Btuh). The U factor can be determined by dividing 1 by the R factor. The U factor is an overall heat-transfer coefficient. For example, if the R factor for an exposed wall is 15.18 the heat-transfer coefficient would be determined as follows:

$$U = \frac{1}{R} = \frac{1}{15.18} = 0.07\ \text{Btu/ft}^2/^\circ\text{F}$$

The U factor is calculated in Btu per square foot of surface per degree difference in temperature. With this information, the effectiveness of adding insulation can be evaluated in terms of heat savings. Figure 21-4 shows the distribution of heat loss through the various components of a conventionally insulated house. It points to the areas that could receive additional attention in terms of insulation, caulking, and weatherstripping. See Figure 21-5.

To protect the insulation against condensation a vapor barrier is recommended, with variations related to weather conditions as shown in Figure 21-6. An excellent example of energy efficient wall construction is shown in Figure 21-7. Note the position of the polyethylene vapor barrier and the use of sheathing with a high R factor.

FIGURE 21-4 Heat loss through a typical conventionally insulated home (Courtesy of © 1978 The Dow Chemical Company).

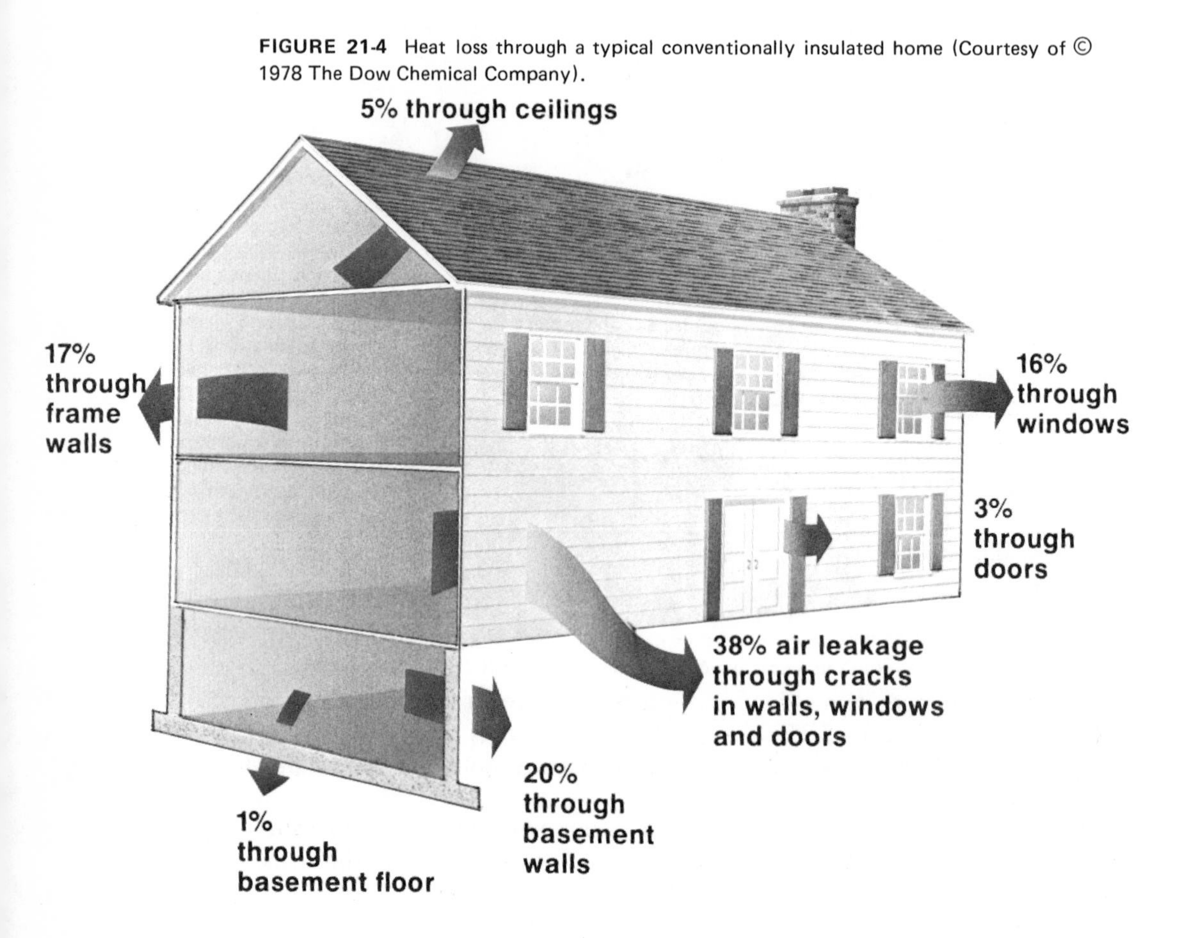

[1] This home features fiberboard sheathing, insulated doors, dual glazed windows, R-19 ceiling insulation and R-11 wall insulation. Varying the size and shape of the house and its window area will, of course, alter its heat-loss distribution. The heat losses are determined in accordance with standard methods recommended by the American Society of Heating, Refrigeration and Air Conditioning Engineers, Inc. (ASHRAE). These calculations are presented in their 1977 Handbook of Fundamentals, the authoritative text for energy transfer in buildings.

TYPE	COMMENTS	APPLICATION
BATTS/BLANKETS Preformed glass fiber or rock wool with or without vapor barrier backing	Fire resistant, moisture resistant, easy to handle for do-it-yourself installation, least expensive and most commonly available	Unfinished attic floor, rafters, underside of floors, between studs
FOAMED IN PLACE Plastic installed as a foam under pressure. Hardens to form insulation 1.) Urethane 2.) Ureaformaldehyde 3.) Polystyrene	1.) Has highest R-value. If ignited, burns explosively and emits toxic fumes. Should be covered with ½″ gypsum wallboard to assure fire safety 2.) Fire resistant, high R-value, first choice of many experts. Requires installation by reliable, experienced contractor (as all foams do) 3.) Lacks fire resistance as does urethane and has lower R-value than urethane	Finished frame walls, floors, ceilings
RIGID BOARD 1.) Extruded polystyrene bead 2.) Extruded polystyrene 3.) Urethane 4.) Glass fiber	All have high R-values for relatively small thickness 1.2.3.) Are not fire resistant, require installation by contractor with ½″ gypsum board to insure fire safety 3.) Is its own vapor barrier; however, when in contact with liquid water, it should have a skin to prevent degrading 1.4.) Require addition of vapor barrier 2.) Is its own barrier	Basement walls, new construction frame walls, commonly used as an outer sheathing between siding and studs
LOOSE FILL (POURED-IN) 1.) Glass fiber 2.) Rock wool 3.) Treated cellulosic fiber	All easy to install, require vapor barrier bought and applied separately. Vapor barrier may be impossible to install in existing walls 1.2.) Fire resistant, moisture resistant 3.) Check label to make sure material meets federal specifications for fire and moisture resistance and R-value	Unfinished attic floor uninsulated existing walls
LOOSE FILL (BLOWN-IN) 1.) Glass fiber 2.) Rock wool 3.) Treated cellulosic fiber	All require vapor barrier bought separately, all require space to be filled completely. Vapor barrier may be impossible to install in existing walls 1.,2.) Fire resistant, moisture resistant. 3.) Fills up spaces most consistently. When blown into closed spaces, has slightly higher R-value, check label for fire and moisture resistance and R-value.	Unfinished attic floor, finished attic floor, finished frame walls, underside of floors.

FIGURE 21-5 Insulation materials chart (Courtesy, U.S. Dept. of Commerce, National Bureau of Standards).

TYPE	COMMENTS	APPLICATION
CAULKING COMPOUNDS 1.) Oil or resin base 2.) Latex, butyl, polyvinyl base 3.) Elastomeric base Silicones, polysulfides, polyurethanes	1.) Lowest cost, least durable—replacement time approximately 2 years. 2.) Medium priced, more durable —look for guarantees on time of durability. 3.) Most expensive, most durable. Urethanes at $2–$3/cartridge. Recommended by some experts as a best buy. Note: Lead base caulk is not recommended because it is toxic.	At stationary joints, exterior window and door frames; whenever different materials or parts of building meet.
WEATHERSTRIPPING 1.) Felt or foam strip 2.) Rolled vinyl-with or without metal backing 3.) Thin spring metal 4.) Interlocking metal channels	1.) Inexpensive, easy to install, not very durable. 2.) Medium priced, easy to install, durable, visible when installed. 3.) More expensive, somewhat difficult to install, very durable, invisible when installed. 4.) Most expensive, difficult to install, durable, excellent weather seal.	At moving joints, perimeter of exterior doors, inside of window sashes.

FIGURE 21-5 (cont.)

FIGURE 21-6 Water vapor control requirements (Courtesy, Owens-Corning Fiberglas Corporation).

The following recommendations for vapor barrier and vapor relief strip installation have been developed through on-going research programs for exterior sheathed frame wall systems with cavity filled insulations.

Vapor Barriers:

Moisture buildup in the cavity can be a problem. However, a properly installed 6 mil polyethylene interior vapor barrier greatly reduces the potential of moisture accumulation. When properly installed, it provides an excellent perm rating and has the mechanical strength to resist many on-site punctures. The following installation procedures are recommended:

- The polyethylene must cover the entire height of the wall, including the bottom and top wall plates. (Overlap at floor and ceiling are advised.)
- Horizontal joints should be avoided. Vertical joints should overlap at wall studs.
- The polyethylene should be cut tight to penetrating elements.
- Vapor barrier penetrations in high humidity areas such as bath and laundry rooms should be avoided.
- The vapor barrier should be inspected and necessary repairs made prior to installing the interior wall finish.

Vapor Control Strips:

If moisture passage into the cavity wall area does occur, vapor relief strips have been designed to supplement the interior vapor barrier providing moisture relief with no significant effect on thermal performance. Vapor relief strips should be used in colder climates or wherever additional insurance against moisture accumulation in cavities is desired. Corrugated plastic vapor control strips are available in various widths and are easily installed.

Based on the climatic conditions in your area, the following are the recommendations for vapor barriers and vapor control strip usage:

Area I: (4000 winter degree days or less) use:

A properly installed 6 mil polyethylene vapor barrier on the interior side of the wall or foil back gypsum wallboard.

Area II: (Above 4000 winter degree days) use:

A properly installed 6 mil polyethylene vapor barrier on the interior side of the wall.

Area III: (8000 winter degree days or more)

Vapor relief strips should be used to supplement a properly installed 6 mil polyethylene interior vapor barrier.

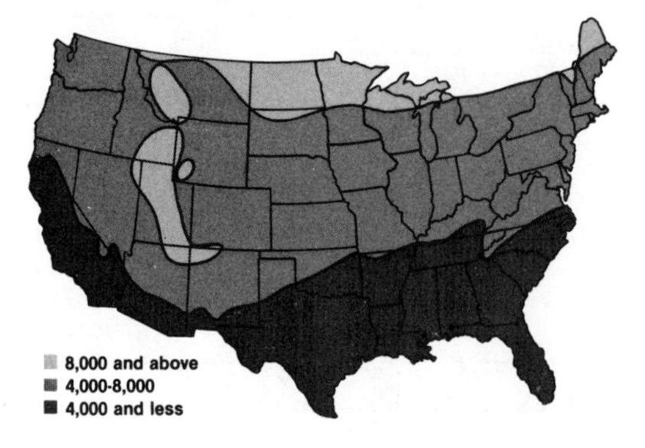

FIGURE 21-7 An example of energy efficient wall construction (Courtesy, Owens-Corning Fiberglas Corporation).

Triple-glazed windows with built-in thermal barrier reduces the heat loss through the glass and the framing, as shown in Figure 21-8. One window manufacturer's rating chart is shown in Figure 21-9. Note that it includes single-glazed, insulating glass, double-glazed, and triple-glazed units. The total unit U and R values are given for each, as is the percent relative humidity when condensation will appear on the innermost surface. The inside glass surface temperature is also shown. All of these factors are based on an outside temperature of 0° F, an inside temperature of 70° F, no inside air movement and uniform heating conditions.

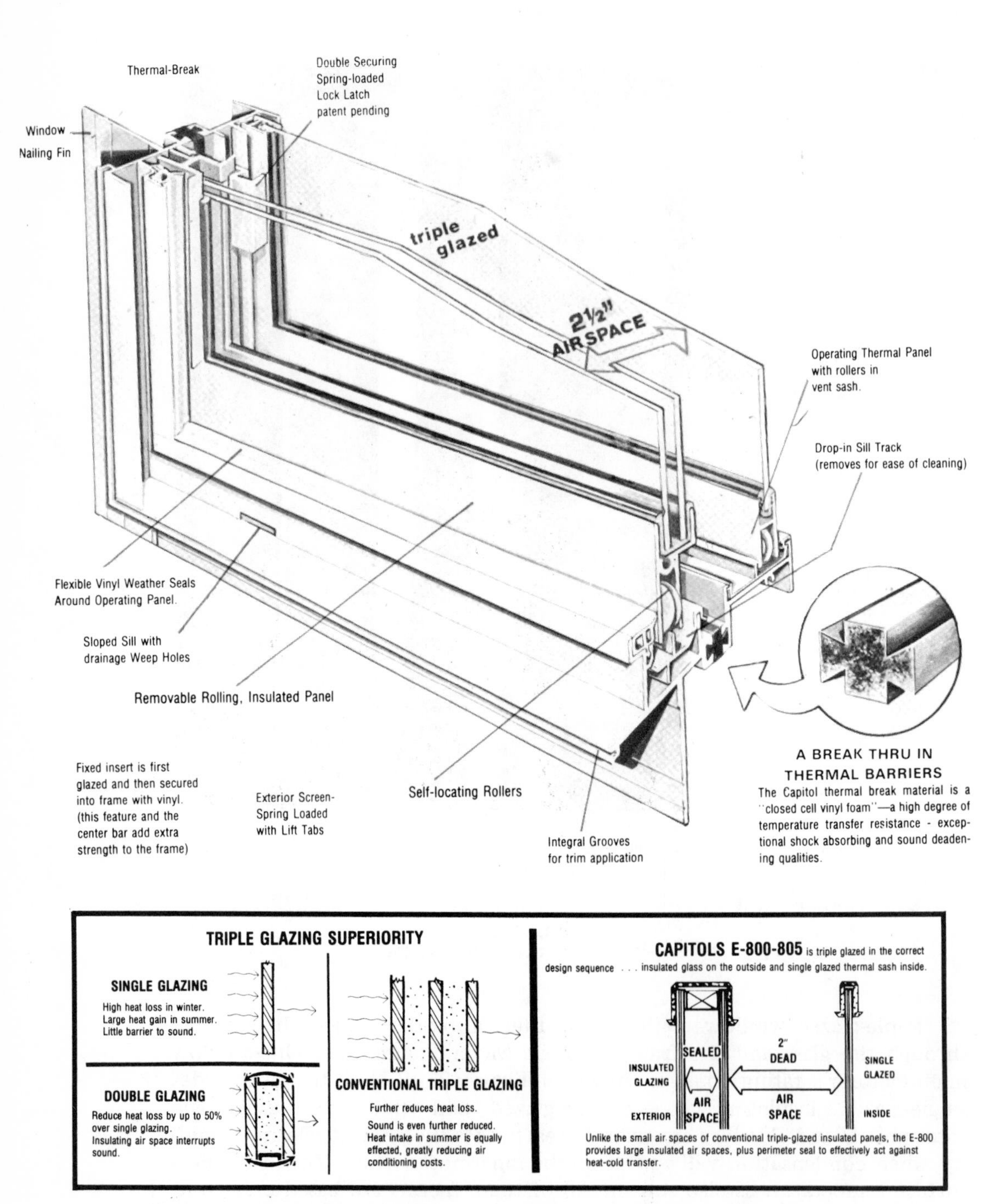

FIGURE 21-8 Triple-glazed, thermal-break aluminum window (Courtesy, Capitol Products Corporation, a subsidiary of Ethyl Corporation).

Andersen® Units	Total Unit "U" Value	Total Unit "R" Value	% Relative Humidity When Condensation Appears On Innermost Surface	Inside Glass Surface Temperature	Type of Glazing
SINGLE-GLAZED					
Wood Gliding Door, Prefinished Wood Gliding Door, Perma-Shield® Gliding Door	1.10	.91	12%	14°F	Single-Pane Safety Glass
Primed Casement, Prefinished Basement/Utility	1.04	.96	12%	14°F	Single-Pane Glass
INSULATING GLASS					
Wood Gliding Door, Prefinished Wood Gliding Door, Perma-Shield Gliding Door	.58	1.72	35%	41°F	Double-Pane Safety Insulating Glass
Primed Casement Picture Wdws. #135, 144, W126, Perma-Shield Casement Picture Window #CP35	.55	1.82	36%	41°F	Double-Pane Insulating Glass
Perma-Shield Casement, Awning, Gliding, Narroline®; Primed Casement	.52	1.92	37%	42°F	Double-Pane Insulating Glass
Primed Casement Picture Windows #145, W144, W135, W145, W136, W146	.47	2.13	43%	46°F	Double-Pane Insulating Glass
DOUBLE-GLAZED					
Primed Casement, Prefinished Basement/Utility	.48	2.08	41%	45°F	Single-Pane Glass With Outside or Inside Storm Panel
TRIPLE-GLAZED					
Perma-Shield Narroline Perma-Shield Gliding Window	.33	3.03	53%	52°F	Dbl.-Pane Insul. Glass With Combination – 2" Air Space
Perma-Shield Casement, Awning	.32	3.12	55%	53°F	Dbl.-Pane Insul. Glass With Outside Storm Panel — ½" Air Space
Primed Casement	.32	3.12	55%	53°F	Dbl.-Pane Insul. Glass With Outside Storm Panel – ⅝" Air Space

The above figures were calculated by using the following:
Outside Temperature — 0 Degrees F
Outside Wind Velocity —15 MPH
Inside Room Temperature —70 Degrees F
No Air Movement Inside & Uniform Heating Conditions

NOTE: Total unit "U" and "R" values equal: Glass only "U" and "R" values, times correction factors as per American Society of Heating and Ventilation Engineers Manual.

FIGURE 21-9 Insulating values for various types of windows (Courtesy, Andersen Corporation, Bayport, MN 55003).

Energy-efficient doors are available. These are doors that incorporate either polystyrene or polyurethane insulation and a thermal barrier to also reduce heat loss through the framing (Figure 21-10). One door manufacturer's rating chart gives the overall R factor for various types of door construction (Figure 21-11). With the use of a full polyurethane core, the R factor is 15.15, which is almost equal to the wall factor of 15.18 calculated earlier in this chapter. With such a door a storm door is virtually not needed. In fact, a storm door might be counterproductive, because of the tendency to leave the primary door open at times to avoid the inconvenience of opening and closing two doors.

Knowledge about insulation, caulking, weatherstripping, vapor barriers, and energy-efficient windows and doors does not necessarily motivate homeowners to make changes. Recognizing this, the U.S. government has worked out a simple procedure for determining what is "best" for each home, based on its geographical location and the utility costs. It is suggested that this step-by-step procedure be applied. An example and worksheets are provided.

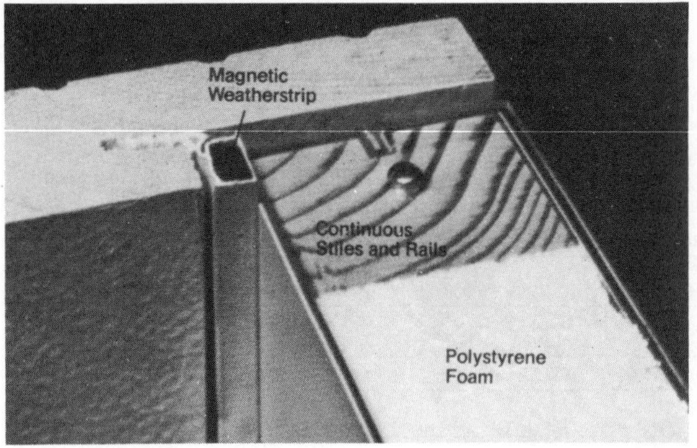

FIGURE 21-10 An example of energy efficient door construction (Courtesy, Pease Company, Ever-Strait Division).

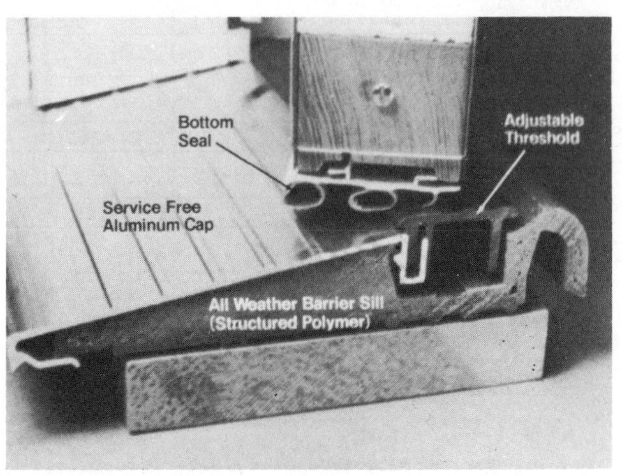

FIGURE 21-11 "R" factors for various types of exterior doors (Courtesy, Pease Company, Ever- Strait Division).

	R factor
Solid core wood door	2.90*
Stile and rail wood door	2.79*
Hollow core wood door	2.18*
Steel door with polystyrene core	7.14
Steel door with ¾" urethane "honeycomb" core	8.42
Storm door (aluminum)	1.84**
THERMA-TRU door (with a full polyurethane core door)	15.15

**R-factor reference according to ASHRAE (American Society of Heating, Refrigeration and Air Conditioning Engineers) figures.*
***Based on engineering calculations.*

FIGURING YOUR ENERGY CONSERVATION BUDGET

To find the "best combination" of energy conservation measures for your climate and fuel prices, use the tables on the following pages. This best combination gives you the largest, long run net savings on your heating and cooling costs for your investment. By comparing this best combination with what already exists in your house, you can figure out how much more needs to be added to bring your house up to the recommended levels.

The recommended improvements apply to most houses to the extent they can be installed without structurally modifying the house. Recommended improvements are based on sample costs given in Table 7.

Follow the steps outlined below and fill in the information for your house on Worksheet A. We have filled in the information for a typical house located in Indianapolis, Indiana.

Locate your city on the Heating Zone Map below. (Our house is located in Heating Zone III.)

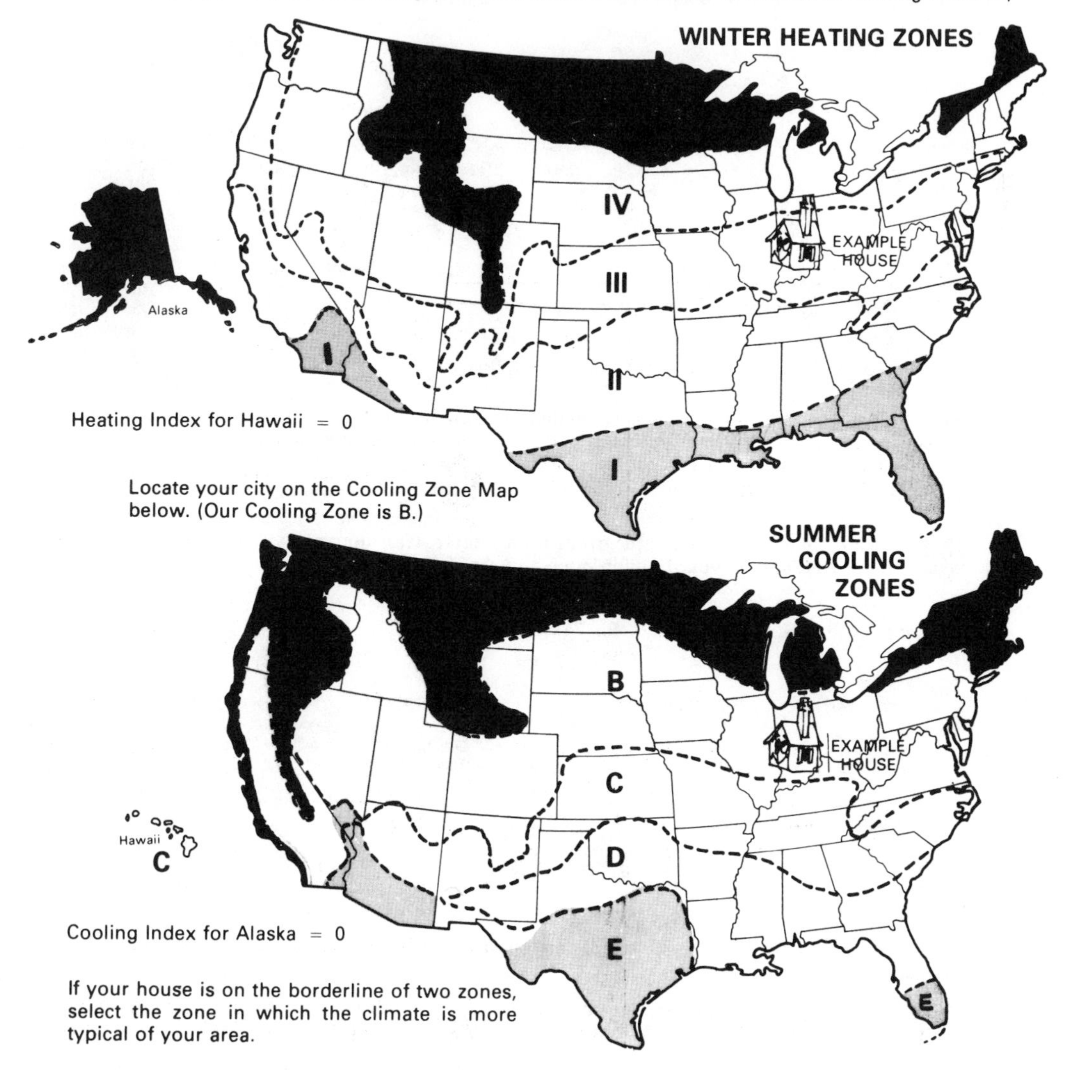

(Courtesy, U.S. Dept. of Commerce, National Bureau of Standard).

3 Our house currently uses fuel oil at a cost of 34¢ a gallon to heat. It uses electricity at 4¢ a kilowatt hour to cool. Obtain your unit heating and cooling costs from the utility companies as follows: Tell your company how many therms (for gas) or kilowatt hours (for electricity) you use in a typical winter month and summer month (if you have air conditioning). The number of therms or kilowatt hours is on your monthly fuel bill. Ask for the cost of the last therm or kilowatt hour used, including all taxes, surcharges, and fuel adjustments. For oil heating, the unit fuel cost is simply your average cost per gallon plus taxes, surcharges, and fuel adjustments.

4 Locate your Heating Index from Table 1 by finding the number at the intersection of your Heating Zone row and heating fuel cost column (to the nearest cost shown). (Our house has a Heating Index of 20.)

If your house is air conditioned, or you plan to add air conditioning, find your Cooling Indexes from steps 5 and 6. If your house is not air conditioned and it is not planned, your Cooling Indexes are zero.

TABLE 1 HEATING INDEX

Type of fuel:	Cost per unit*									
Gas (therm)	9¢	12¢	15¢	18¢	24¢	30¢	36¢	54¢	72¢	90¢
Oil (gallon)	13¢	17¢	21¢	25¢	34¢	42¢	50¢	75¢	$1.00	$1.25
Electric (kWh)				1¢	1.3¢	1.6¢	2¢	3¢	4¢	5¢
Heat pump (kWh)	1¢	1.3¢	1.7¢	2¢	2.6¢	3.3¢	4¢	6¢	8¢	10¢
HEATING ZONE I	2	2	3	3	4	5	6	9	12	15
II	5	6	8	9	12	15	18	27	36	45
III	8	10	13	15	20	25	30	45	60	75
IV	11	14	18	21	28	35	42	63	84	105
V	14	18	23	27	36	45	54	81	108	135
VI	22	28	36	42	56	70	84	126	168	210

Note: In Tables 1-3, if your fuel costs fall midway between two fuel costs listed, you can interpolate. For example, if our fuel oil costs were 38¢ a gallon, our Heating Index would be 22.5.

5 Locate your Cooling Index for Attics from Table 2 by finding your Cooling Zone and cooling cost to the nearest cost shown. (Our house has a Cooling Index for Attics of 5.)

6 Locate your Cooling Index for Walls from Table 3 by finding your Cooling Zone and cooling cost to the nearest cost shown in the table. (Our house has a Cooling Index for Walls of 2.)

TABLE 2 COOLING INDEX FOR ATTICS

Type of air conditioner:	Cost per unit*						
Gas (therm)	9¢	12¢	15¢	18¢	24¢	30¢	36¢
Electric (kWh)	1.5¢	2¢	2.5¢	3¢	4¢	5¢	6¢
COOLING ZONE A	0	0	0	0	0	0	0
B	2	2	3	4	5	6	7
C	3	5	6	7	9	11	13
D	5	6	8	9	12	15	18
E	7	9	11	14	18	23	27

TABLE 3 COOLING INDEX FOR WALLS

Type of air conditioner:	Cost per unit*						
Gas (therm)	9¢	12¢	15¢	18¢	24¢	30¢	36¢
Electric (kWh)	1.5¢	2¢	2.5¢	3¢	4¢	5¢	6¢
COOLING ZONE A	0	0	0	0	0	0	0
B	1	1	2	2	2	3	4
C	2	2	3	4	5	6	7
D	3	3	4	5	7	8	10
E	4	5	6	8	10	13	15

*Cost of last unit used (for heating and cooling purposes) including all taxes, surcharges, and fuel adjustments.

Find the sum of your Heating Index and Cooling Index for Attics. (Our sum is 25.)

8 Find the sum of your Heating Index and Cooling Index for Walls. (Our sum is 22.)

(Courtesy, U.S. Dept. of Commerce, National Bureau of Standard).

Energy savings result from decreasing the heat flow through the exterior shell of the building. The resistance, or "R," value of insulation is the measure of its ability to decrease heat flow. Two different kinds of insulation may have the same thickness, but the one with the higher R value will perform better. For that reason, our recommendations are listed in terms of R values with the approximate corresponding thickness. R values for different thicknesses of insulation are generally made available by the manufacturers.

9 Find the resistance value of insulation recommended for your attic and around attic ducts from Table 4. (For our house the recommended resistance value is R-30 for attic floors and R-16 for ducts.)

TABLE 4 ATTIC FLOOR INSULATION AND ATTIC DUCT INSULATION

INDEX Heating Index Plus Cooling Index for Attics	ATTIC INSULATION R-Value	Approximate Thickness: Mineral Fiber Batt/Blanket	Approximate Thickness: Mineral Fiber Loose-Fill**	Approximate Thickness: Cellulose Loose-Fill**	DUCT INSULATION* R-Value	DUCT INSULATION* Approximate Thickness
1-3	R-0	0"	0"	0"	R-8	2"
4-9	R-11	4"	4-6"	2- 4"	R-8	2"
10-15	R-19	6"	8-10"	4- 6"	R-8	2"
16-27	R-30***	10"	13-15"	7- 9"	R-16	4"
28-35	R-33	11"	14-16"	8-10"	R-16	4"
36-45	R-38	12"	17-19"	9-11"	R-24	6"
46-60	R-44	14"	19-21"	11-13"	R-24	6"
61-85	R-49	16"	22-24"	12-14"	R-32	8"
86-105	R-57	18"	25-27"	14-16"	R-32	8"
106-130	R-60	19"	27-29"	15-17"	R-32	8"
131—	R-66	21"	29-31"	17-19"	R-40	10"

* Use Heating Index only if ducts are not used for air conditioning. ** High levels of loose-fill insulation may not be feasible in many attics. ***Assumes that joists are covered; otherwise use R-22.

10 Find the recommended level of insulation for floors over unheated areas from Table 5. (Our house should have R-19.) Using Table 5, check to see whether storm doors are economical for your home. Storm doors listed as optional may be economical if the doorway is heavily used during the heating season.

TABLE 5 INSULATION UNDER FLOORS AND STORM DOORS

INDEX Heating Index Only	INSULATION UNDER FLOORS* R-Value	INSULATION UNDER FLOORS* Mineral Fiber Batt Thickness	STORM DOORS
0-7	0**	0"**	None
8-15	11**	4"**	None
16-30	19	6"	Optional
31-65	22	7"	Optional
66—	22	7"	On all doors

* If your furnace and hot water heater are located in an otherwise unheated basement, cut your Heating Index in half to find the level of floor insulation.
**In Zone I and II R-11 insulation is usually economical under floors over open crawlspaces and over garages; in Zone I insulation is not usually economical if crawlspace is closed off.

11 Find the recommended level of insulation for your walls and ducts in unheated areas from Table 6. (Our house should have full-wall insulation if none existed previously and R-16 insulation around ducts.) Table 6 also shows the minimum economical storm window size in square feet for triple-track storm windows. (Our house should have storm windows on all windows 9 square feet in size or larger where storm windows can be used.)

TABLE 6 WALL INSULATION, DUCT INSULATION, AND STORM WINDOWS

INDEX Heating Index Plus Cooling Index for Walls	WALL INSULATION (blown-in)	INSULATION AROUND DUCTS IN CRAWLSPACES AND IN OTHER UNHEATED AREAS (EXCEPT ATTICS)* Resistance and Approximate Thickness	STORM WINDOWS (Triple-Track) Minimum Economical Window Size
0-10	None	R-8 (2")	none
11-12	Full-Wall Insulation Approximately R-14	R-8 (2")	20 sq. ft.
13-15		R-8 (2")	15 sq. ft.
16-19		R-16(4")	12 sq. ft.
20-28		R-16(4")	9 sq. ft.
29-35		R-16(4")	6 sq. ft.
36-45		R-24(6")	4 sq. ft.
46-65		R-24(6")	All windows**
66—		R-32(8")	All windows**

* Use Heating Index only if ducts are not used for air conditioning. **Windows too small for triple-track windows can be fitted with one-piece windows.

Weather stripping and caulking.

Regardless of where you live or your cost of energy, it is almost always economical to install weather stripping on the inside around doors and windows where possible and to caulk on the outside around doors and window frames—if you do it yourself. This is especially true for windows and doors which have noticeable drafts.

(Courtesy, U.S. Dept. of Commerce, National Bureau of Standard).

YOU NOW KNOW your best combination of energy conservation improvements. Of course, the size of your investment depends on your existing insulation and the size of your house.

In addition, some of the recommended improvements in this booklet are not appropriate for all houses. For instance, insulation cannot be added under floors in houses built on concrete slabs. In such cases, the other recommended improvements should still be added to the extent indicated in this booklet. Similarly, R-30 insulation may be recommended for your attic although only R-19 may fit at the eaves or in areas where the attic is floored. In this case, you should still put R-30 insulation wherever it fits.

Use Worksheet B and Table 7 (or your own cost information) to calculate how much you need to add to reach your best combination and how much this will cost. We have provided this information on Worksheet B for our example house. Our house only has R-11 attic insulation, some wall insulation, and R-8 attic duct insulation to begin with. To reach our best combination, the improvements would cost about $1200.

WORKSHEET A

EXAMPLE:		YOUR CALCULATIONS:	
Climate:		Climate:	
Heating Zone	III	Heating Zone	
Cooling Zone	B	Cooling Zone	
Fuel Costs:		Fuel Costs:	
Heating Energy	Oil	Heating Energy	
Cost per Unit	34¢/gal.	Cost per Unit	
Cooling Energy	Electric	Cooling Energy	
Cost per Unit	4¢/KWH	Cost per Unit	
Indexes:		Indexes:	
Heating	20	Heating	
Cooling (Attic)	5	Cooling (Attic)	
Cooling (Wall)	2	Cooling (Wall)	
Heating + Cooling (Attic)	25	Heating + Cooling (Attic)	
Heating + Cooling (Wall)	22	Heating + Cooling (Wall)	

BEST COMBINATION			BEST COMBINATION	
Attic Insulation (Batt)	R-30 (10 inches)	FROM TABLE 4	Attic Insulation	
Duct Insulation (in attics)	R-16 (4 inches)	FROM TABLE 4	Duct Insulation (in attics)	
Insulation Under Floors	R-19 (6 inches)	FROM TABLE 5	Insulation Under Floors	
Storm Doors	optional	FROM TABLE 5	Storm Doors	
Wall Insulation (blown-in)	full-wall R-14 (3½ inches)	FROM TABLE 6	Wall Insulation (blown-in)	
Duct Insulation (in unheated crawl-spaces, etc.)	R-16 (4 inches)	FROM TABLE 6	Duct Insulation (in unheated crawlspaces, etc.)	
Storm Windows (minimum size)	9 sq. ft.	FROM TABLE 6	Storm Windows (minimum size)	
Weather strip and caulk windows and door frames	all		Weather strip and caulk windows and door frames	

(Courtesy, U.S. Dept. of Commerce, National Bureau of Standard).

WORKSHEET B

OUR EXAMPLE:

ATTIC INSULATION

1. Attic area (sq. ft.)	1200	
2. Recommended level	R-30 (10")	
3. Existing level	R-11 (4")	
4. Add	R-19 (6")	
5. Cost/sq. ft.	$.25	
6. Total cost (1×5)		$300

WALL INSULATION (BLOWN-IN)

1. Wall area (sq. ft.)	900	
2. Recommended level	full-wall	
3. Existing level	some	
4. Add	0	
5. Cost/sq. ft.	$.60	
6. Total cost (1×5)		0

FLOOR INSULATION

1. Floor area (sq. ft.)	1200	
2. Recommended level	R-19 (6")	
3. Existing level	0"	
4. Add	R-19 (6")	
5. Cost/sq. ft.	$.30	
6. Total cost (1×5)		$360

DUCT INSULATION (ATTIC)

1. Length (ft.)	30'	
2. Perimeter (ft.)	2'	
3. Area (1×2×1.5)*	90 sq. ft.	
4. Recommended level	R-16 (4")	
5. Existing level	R-8 (2")	
6. Add	R-8 (2")	
7. Cost/sq. ft.	$.30	
8. Total cost (3×7)		$27

DUCT INSULATION (OTHER AREAS)

1. Length (ft.)	30'	
2. Perimeter (ft.)	2'	
3. Area (1×2×1.5)*	90 sq. ft.	
4. Recommended level	R-16 (4")	
5. Existing level	0"	
6. Add	R-16 (4")	
7. Cost/sq. ft.	$.50	
8. Total cost (3×7)		$45

STORM WINDOWS (over 9 sq. ft.)

size (sq. ft.)	number	cost each	sub-total
20	2	$35	$ 70
15	4	30	120
12	3	30	90
9	2	30	60

Total cost $340

STORM DOORS

1. Doors Needed	1 (Optional)	
2. Cost per door	$75	
3. Total cost		$75

WEATHER STRIPPING (MATERIALS ONLY)

1. Linear feet	200	
2. Cost per foot	$.10	
3. Total cost		$20

CAULKING (MATERIALS ONLY)

1. Variable costs	$20-50	
2. Estimated cost		$33

Total cost of all improvements $1200

YOUR ESTIMATES:

ATTIC INSULATION

1. Attic area (sq. ft.) ______
2. Recommended level ______
3. Existing level ______
4. Add ______
5. Cost/sq. ft. ______
6. Total cost (1×5) ______

WALL INSULATION (BLOWN-IN)

1. Wall area (sq. ft.) ______
2. Recommended level ______
3. Existing level ______
4. Add ______
5. Cost/sq. ft. ______
6. Total cost (1×5) ______

FLOOR INSULATION

1. Floor area (sq. ft.) ______
2. Recommended level ______
3. Existing level ______
4. Add ______
5. Cost/sq. ft. ______
6. Total cost (1×5) ______

DUCT INSULATION (ATTIC)

1. Length (ft.) ______
2. Perimeter (ft.) ______
3. Area (1×2×1.5)* ______
4. Recommended level ______
5. Existing level ______
6. Add ______
7. Cost/sq. ft. ______
8. Total cost (3×7) ______

DUCT INSULATION (OTHER AREAS)

1. Length (ft.) ______
2. Perimeter (ft.) ______
3. Area (1×2×1.5)* ______
4. Recommended level ______
5. Existing level ______
6. Add ______
7. Cost/sq. ft. ______
8. Total cost (3×7) ______

STORM WINDOWS

size (sq. ft.)	number	cost each	sub-total

Total cost ______

STORM DOORS

1. Doors needed ______
2. Cost per door ______
3. Total cost ______

WEATHER STRIPPING (MATERIALS ONLY)

1. Linear feet ______
2. Cost per foot ______
3. Total cost ______

CAULKING (MATERIALS ONLY)

1. Variable costs ______
2. Estimated cost ______

Total cost of all improvements ______

*1.5 is an adjustment factor for increased width of insulation needed to fit around duct.

(Courtesy, U.S. Dept. of Commerce, National Bureau of Standard).

TABLE 7 SAMPLE IMPROVEMENT COSTS

These sample costs were used in estimating the best combination of energy conservation improvements for the various climates and fuel prices covered in this booklet. They include an allowance for commercial installation, except in the case of weather stripping and caulking which is considered to be a do-it-yourself project. While these costs are typical of 1975 prices, there may be considerable variation among specific materials, geographic locations, and suppliers. It usually is worth your time to obtain several estimates for materials and installation before making any purchase. Many of these items can be purchased at substantial discounts if you watch the advertised sales. Considerable savings may be made by installing these yourself, where possible.

ATTIC INSULATION
(ALL MATERIALS)

Installed cost per square foot of attic:

R-11	= 15¢	R-44	= 57¢
R-19	= 25¢	R-49	= 64¢
R-22	= 29¢	R-57	= 74¢
R-30	= 39¢	R-60	= 78¢
R-33	= 43¢	R-66	= 86¢

WALL INSULATION
(ALL MATERIALS)

Installed cost = 60¢ per square foot of net wall area*

FLOOR INSULATION
(MINERAL FIBER BATT)

Installed cost:

R-11 = 20¢
R-19 = 30¢
R-22 = 34¢

DUCT INSULATION
(MINERAL FIBER BLANKET)

Installed cost per square foot of material:

R-8	= 30¢	R-32	= 90¢
R-16	= 50¢	R-40	= $1.10
R-24	= 70¢		

STORM WINDOWS
(TRIPLE-TRACK, CUSTOM-MADE AND INSTALLED**)

Up to 100 united inches (height + width) = $30.00
Greater than 100 united inches = $30.00 + $.60 per united inch greater than 100"

STORM DOORS
(CUSTOM-FITTED AND INSTALLED**)

All sizes = $75.00

WEATHER STRIPPING AND CAULKING

Prices vary according to material used. Use the most durable materials available.

* Price includes allowance for painting inside surface of exterior walls with water vapor-resistant paint.

**Prices may be considerably less for stock sizes, homeowner-installed.

If you find that the costs of any of the improvements to your house are substantially different from the sample costs in Table 7, you can easily compensate for the difference.

Take the Index Number appropriate for the improvement in question, multiply this by our sample cost, and divide the result by your cost. This will give you an Adjusted Index Number with which you can find the best level of investment for that particular improvement.

$$\frac{\text{Original Index} \times \text{Our Cost}}{\text{Your Cost}} = \text{Adjusted Index}$$

EXAMPLE

For our example house, we might find that we can get good quality storm windows for $20 apiece instead of our $30 estimate. Our Index Number for storm windows was 22. Our new Adjusted Index Number for storm windows would be: $\frac{22 \times \$30}{\$20} = 33$

Using our Adjusted Index Number of 33 we find that storm windows are economical on all windows 6 square feet in size or larger, instead of 9 square feet in size. In other words, if your costs are substantially less than ours, you will want to go beyond the recommended level.

Similarly, if R-30 insulation in the attic costs 65¢ per square foot instead of our 39¢ price, the Index Number of 25 for attic insulation would be adjusted to $\frac{25 \times 39¢}{65¢} = 15$

From Table 4 we find that R-19 insulation is now recommended instead of R-30. In other words, if your costs are substantially greater than ours, you may want to use a little less than the recommended level.

(Courtesy, U.S. Dept. of Commerce, National Bureau of Standard).

Fitting the Heating System to the Load

In new construction, the heating unit can be accurately sized to fit the load requirements. But what can be done to improve the efficiency of a system where the original furnace is too large or where insulation has been added to the structure, reducing the load below the original requirement?

Studies indicate that the average efficiency of residential heating installations is between 50 and 65%, whereas it should be higher. This is partially due to the inefficient operation of a furnace that is too large for the job. Two modifications that may be made to improve performance are:

1. Reduce the orifices in the gas burners to reduce input
2. Use flue restrictors to hold the hot gases in the furnace longer

Both of these modifications must be made by an expert technician and in strict accordance with local code and American Gas Association (AGA) safety restrictions.

However, the safest method of providing the proper match between furnace and load is to change the furnace size, to match the load.

Altering the Heating Unit

Other than the modifications mentioned above, some of the changes that have been used to improve efficiency are:

INSTALLATION OF AN AUTOMATIC VENT DAMPER:

This device (Figures 21-12 and 21-13) automatically closes the vent on a gas furnace when the unit shuts down to prevent loss of heat through the chimney during the period the heating element is cooling down.

INSTALLATION OF A CYCLE-PILOT:

This device (Figure 21-14) lights the pilot by electric ignition when the thermostat calls for heat. After the thermostat is satisfied, the pilot goes out to save fuel until the next startup.

USE OF A TIME-ON/TEMPERATURE-OFF FAN CONTROL:

This device conserves heat by operating the circulator fan longer, using the maximum amount of heat for the building heating rather than permitting heat to be wasted by going up the chimney.

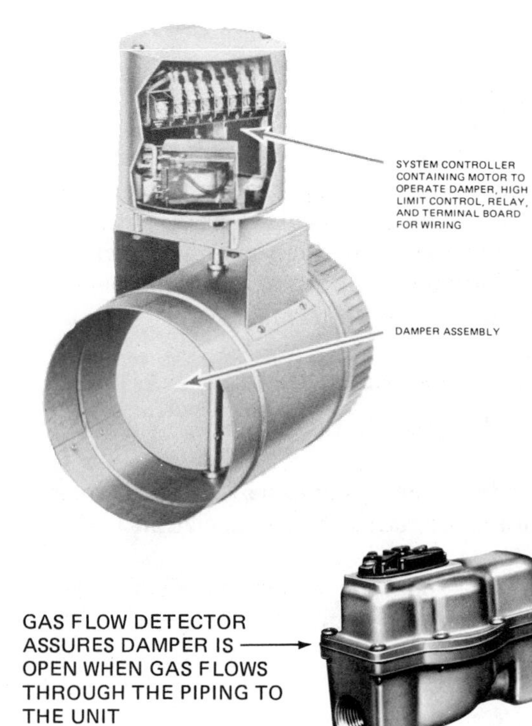

FIGURE 21-12 Automatic flue damper, with gas flow detector.

FIGURE 21-13 Automatic flue damper installed on energy efficient furnace (Courtesy, Michigan Furnace Company).

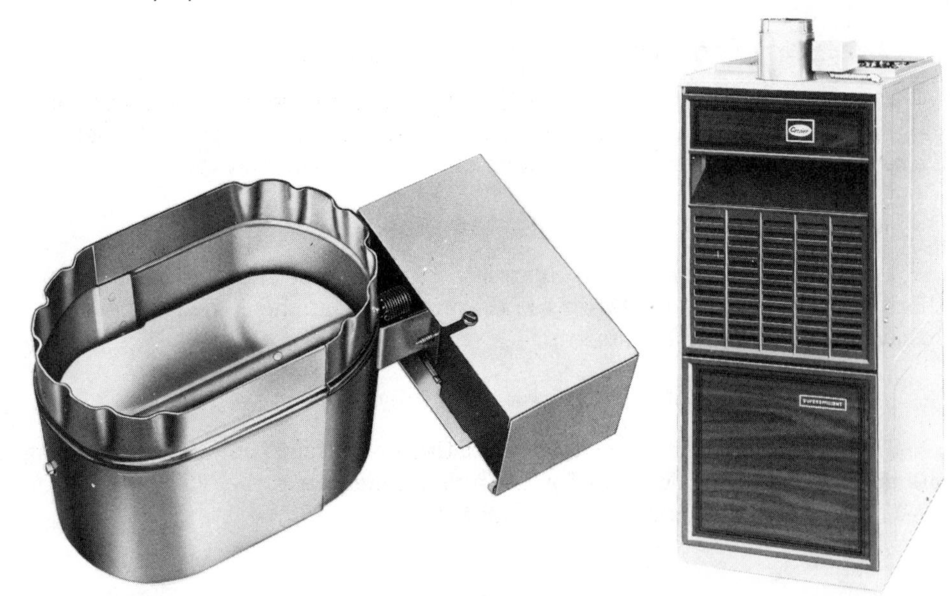

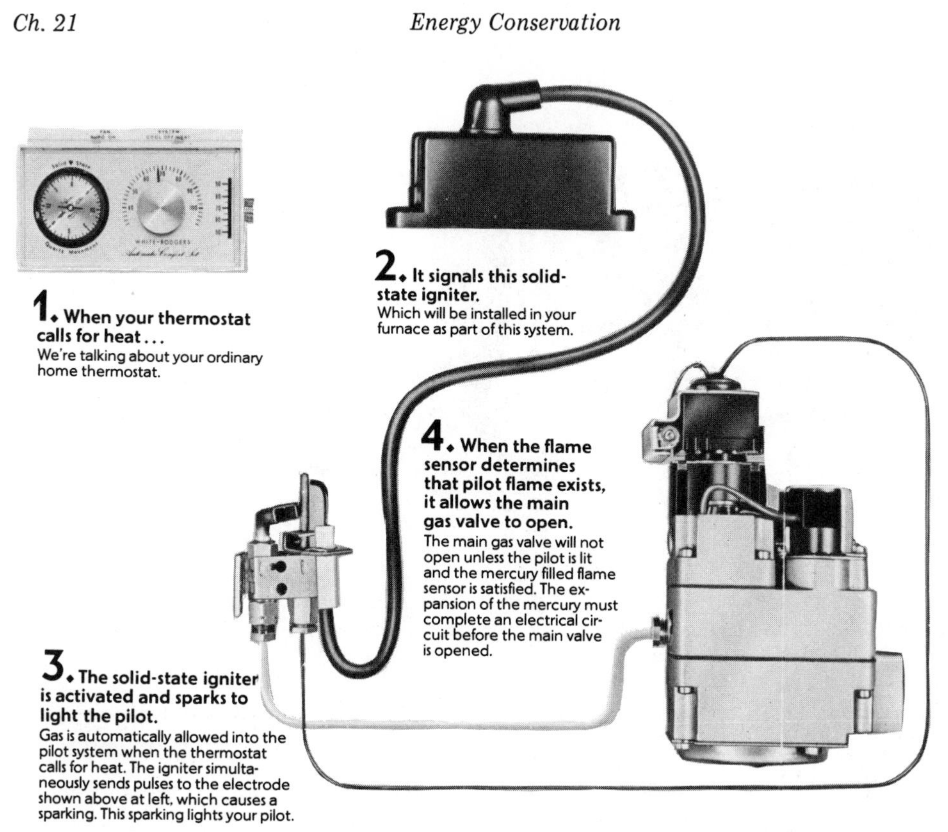

FIGURE 21-14 The operation of the cycle-pilot (Courtesy, White-Rodgers Div., Emerson Electric Company).

USE OF A TWO-STAGE GAS VALVE:

This type of valve permits a low fire for light loads, thus providing for more continuous operation of the heating unit. Even heating produces more efficient operation.

System Modifications

The use of an automatic temperature setback is one example of a system modification that can be made (Figures 21-15, 21-16). A 5 to 12% saving in fuel can be realized by lowering the temperature in the space 5° for 8 h at night. A saving of 9 to 16% can be realized by lowering the space temperature 10° for 8 h at night. A table showing night setback savings for various parts of the country is shown in Figure 21-17.

Setback periods of less than 8 h are not recommended, since they may increase fuel consumption.

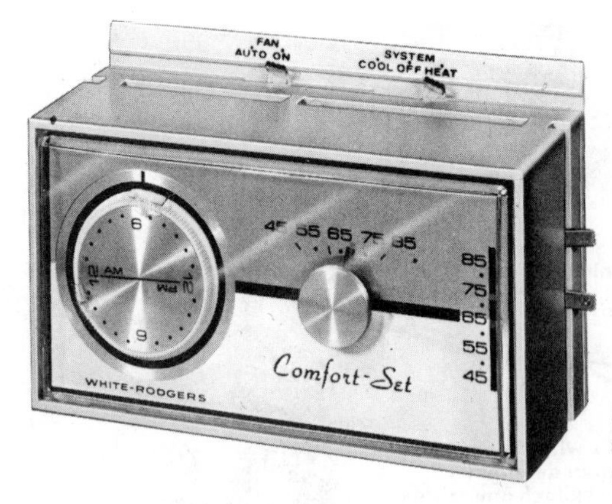

FIGURE 21-15 Night setback thermostat (Courtesy, White-Rodgers Div., Emerson Electric Company).

FIGURE 21-16 Internal components of night setback thermostat (Courtesy, White-Rodgers Div., Emerson Electric Company).

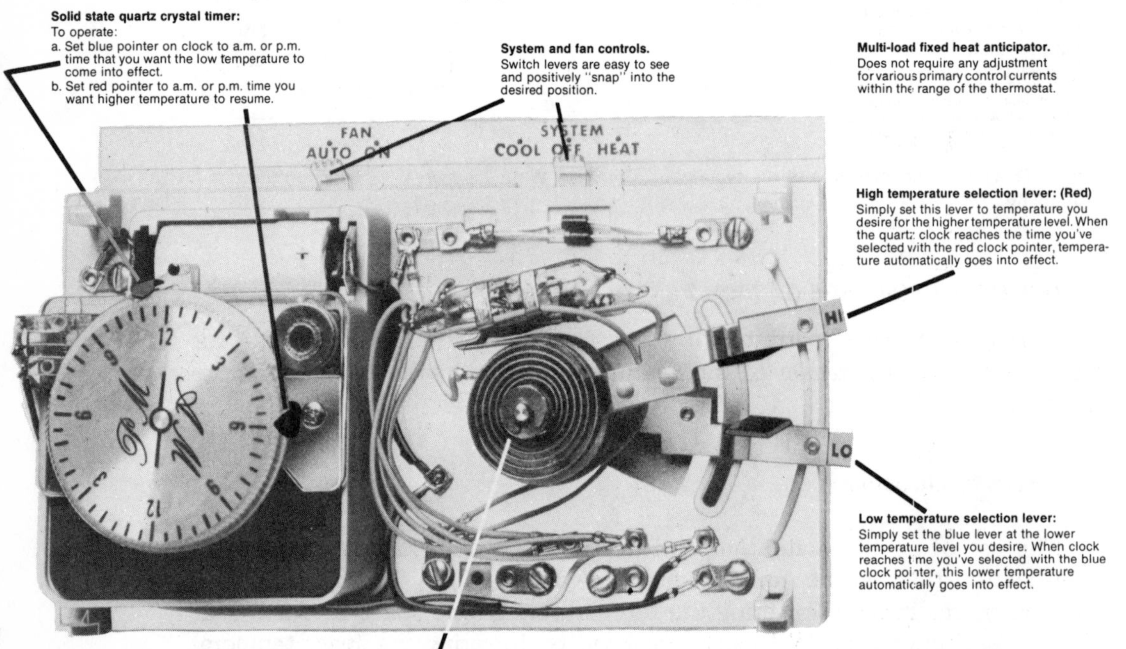

CITY	5° SETBACK	10° SETBACK	CITY	5° SETBACK	10° SETBACK
Atlanta	11	15	Minneapolis	5	9
Boston	7	11	New York City	8	12
Buffalo	6	10	Omaha	7	11
Chicago	7	11	Philadelphia	8	12
Cincinnati	8	12	Pittsburgh	7	11
Cleveland	8	12	Portland	9	13
Dallas	11	15	Salt Lake City	7	11
Denver	7	11	San Francisco	10	14
Des Moines	7	11	St. Louis	8	12
Detroit	7	11	Seattle	8	12
Kansas City	8	12	Washington, D.C.	9	13
Los Angeles	12	16			
Louisville	9	13			
Milwaukee	6	10			

These percentages of fuel saving were calculated on a 75-degree temperature setting during the day — and a night set-back time period of eight hours, from 10 p.m. to 6 a.m.

FIGURE 21-17 Percent fuel savings with night setback (Courtesy, Honeywell Inc.).

Improved Maintenance

Firing devices should be cleaned and serviced on a regular basis. Throw-away filters should be replaced when dirty. Motors should be oiled. Regular maintenance is important on any heating system and, if neglected, the efficiency and reliability will decrease. Where efficiency can be easily measured, as it can on an oil-fired installation, this should be done at least once a year to assure good performance.

Distribution systems should be kept clean. Thermostats should be kept in calibration. Duct work should be balanced to provide an even distribution of heat.

The thermostat location should be changed or the thermostat replaced if it does not provide adequate temperature control.

Proper maintenance is essentially good housekeeping practice which results in improved performance.

Many new types of energy-saving devices are becoming available on the market because of the urgent need for saving fuel. Anyone involved in the heating business should carefully evaluate these innovations, since fuel should be saved by all practical means.

Heating Units with Energy-Saving Devices

The sequence of operation for an intermittent electric pilot lighter is shown in Figure 21-14. A detailed description of the sequence of operation for such a furnace is shown in Figure 21-18.

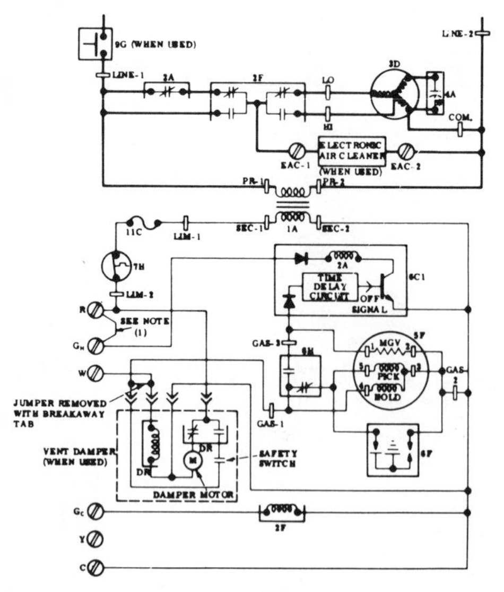

LEGEND

1A-Transformer 115/25
2A-Heat Relay (SPST-N.C.)
2F-Cool Relay (DPDT)
3D-Blower Motor
4A-Run Capacitor
5E-Gas Valve
5F-Gas Valve
6C1-Printed-Circuit Board
6F-Pilot Igniter
6H-Safety Pilot (Flame Sensing)
7H-Limit Switch (SPST-N.C.)
11C-Fusible Link

SEQUENCE OF OPERATION

Gas and electrical supplies must be turned on at the furnace.

NOTE: When power is applied to heat relay coil 2A in the control circuit, the normally closed contacts in the supply circuit will open.

Gas Valve

When the thermostat "calls for heat," the control circuit is closed between terminals R and W. Power from transformer 1A through fusible link 11C and limit switch 7H energizes the pilot valve portion of automatic gas valve 5F and pilot igniter 6F. The pilot valve opens, permitting gas flow to the pilot burner where it is ignited.

The pilot valve portion of automatic gas valve 5F is a solenoid consisting of a "pick" and a "hold" coil. Both the "pick" and the "hold" coils must be energized to open the valve, but only the "hold" coil must be energized to **keep** it open.

When the pilot flame is established, pilot 6H switches its contacts in approximately 40 to 60 seconds, energizing the main valve portion of gas valve 5F and deenergizing pilot igniter 6F and the "pick" coil of the pilot solenoid portion in gas valve 5F.

The main valve portion of gas valve 5F is heat motor operated; therefore, after approximately 10 seconds, this portion of the valve opens, permitting gas flow to the main burners, where the gas is ignited by pilot 6H.

Blower Circuit

With power through the solid-state time-delay circuit on printed-circuit board 6C1 and heat relay 2A, blower motor 3D is energized on heating speed approximately 75 seconds after gas valve 5E has been energized (or the pilot flame has been proven in the case of BDP 646 Gas Valve 5F.)

Limit Control

If the furnace overheats for any reason, limit control 7H switches, breaking the circuit to automatic gas valve 5E or 5F. The gas valve closes immediately, stopping gas flow to the main burners and the pilot. In addition, blower motor 3D continues to operate because heat relay 2A is deenergized to cool down the furnace.

Fusible link 11C is provided in the transformer 1A secondary circuit as protection from overheating conditions in the vestibule area of the furnace. Should this condition exist, the fuse opens and deenergizes gas valve 5E or 5F and heat relay 2A, stopping the gas flow to the burners and starting blower motor 3D.

When the thermostat is satisfied, the circuit between R and W is broken, deenergizing automatic gas valve 5E or 5F, pilot 6H (when used), and the solid-state time-delay circuit on printed-circuit board 6C1. The gas flow stops immediately to the pilot and main burners with the Gas Valve.

After approximately 75 seconds, heat relay 2A is energized and blower motor 3D stops.

Vent Damper (when used)

When the thermostat "calls for heat," the control circuit is closed between terminals R and W. Power from transformer 1A energizes the damper motor relay coil, causing the normally closed relay contacts to open, deenergizing the damper motor and causing the spring-loaded damper to open. When the automatic vent damper is open, the circuit is completed to automatic gas valve 5E or 5F. The sequence from this point on is the same as that for *heating*.

When the thermostat is satisfied, the circuit between R and W is broken, deenergizing the damper motor relay, and causing it to switch its contacts. The damper motor starts and closes the damper.

Printed-Circuit Control Center

Each furnace features a printed-circuit control center. This will aid the installer and serviceman when installing and servicing the unit. A low-voltage terminal board is marked for easy connection of field wiring.

FIGURE 21-18 Schematic wiring diagram and sequence of operation for gas furnace with automatic flue damper and intermittent pilot (Reproduced by permission of Carrier Corporation, Copyright 1979, Carrier Corporation).

REVIEW QUESTIONS

Select the letter representing your choice of the correct answer.

21-1. Of the total energy consumed, what is the approximate percentage used for heating?

(a) 8

(b) 15

(c) 18

(d) 25

21-2. What proven source of energy is most plentiful?

(a) Gas

(b) Coal

(c) Oil

(d) Electricity

21-3. What is the approximate percentage of energy being wasted?

(a) 15

(b) 30

(c) 50

(d) 75

21-4. If in Detroit a 5° night setback in temperature will produce a 7% saving; a 10° setback will produce what percent saving?

(a) 7%

(b) 11%

(c) 14%

(d) 20%

21-5. What does R represent?

(a) Thermal resistance

(b) Thermal conductance

(c) 1/K

(d) U per inch

21-6. The biggest factor in the thermal resistance of a wall is:

(a) Sheathing

(b) Siding

(c) Insulation

(d) Wind resistance

21-7. A ceiling should have an R factor equal to or greater than:

(a) 10.0

(b) 15.0

(c) 20.0

(d) 25.0

21-8. An R factor of 16.35 is equal to a U factor of:

(a) 2.60

(b) 1.60

(c) 0.60

(d) 0.06

21-9. The efficiency of a heating system in an average existing home is:

(a) 10 to 25%

(b) 30 to 45%

(c) 50 to 65%

(d) 70 to 85%

21-10. How long should night temperature be maintained to produce a reasonable saving?

(a) 8 h or more

(b) 4 h or less

(c) 10 h

(d) 12 h

22

SOLAR HEATING

OBJECTIVES

After studying this unit, the student will be able to

- Describe the common types of active solar heating systems
- Evaluate the selection and performance of solar heating system components

HEAT FROM THE SUN

One of the ways to conserve energy and to lower operating costs is to install solar heating. Figure 22-1 shows a home, located in St. Mary's, Maryland, that has a solar heating system. Three hundred square feet of Grumman Sunstream Model 200 solar collectors have been designed to supply 75% of the total space heating and hot water needs. This house contains 2,000 square feet of habitable floor area. In most sections of the country, using a flat-plate collector, heat from the sun can be collected at a rate of 200 to 300 Btuh per square foot. A flat-plate collector is a permanent surface, directed toward the sun, which has the capability of absorbing a large amount of the sun's energy in the form of heat.

FIGURE 22-1 Typical solar applications (Courtesy, LOF Solar Energy Systems, Inc.).

(Courtesy, Gruman Energy Systems, Inc.).

CURRENT LEGISLATION

Federal legislation providing an income tax credit for money spent on qualifying energy conservation expenditures was made retroactive back to April 20, 1977. The maximum allowance is $300, based on a 15% credit on the first $2000 of expenditures. Items eligible for credit include:

1. Insulation in ceilings, walls, floors, roofs, or on water heaters
2. Exterior storm or thermal windows or doors

3. Caulking or weather stripping
4. Furnace replacement burners which reduce the amount of fuel used
5. A device that makes flue openings for the heating system more efficient
6. An electrical or mechanical furnace ignition system that replaces a gas pilot light
7. An automatic energy-saving setback thermostat
8. A meter that displays the cost of energy usage

Federal income tax credit is also provided for the installation of renewable energy source systems up to maximum of $2200. This is based on 30% of the first $2000 and 20% of the next $8000 of expense. Approved items include active and passive solar systems, wind, and geothermal devices.

USES OF SOLAR HEATING

Solar heating is used principally by a homeowner for

- Space heating
- Domestic hot water heating
- Heating swimming pool water

From an economic standpoint, domestic hot water solar heating has the greatest return for the average homeowner since hot water is needed year-round. This type of heating permits utilization of the summer solar radiation, which is greater than that available in winter for most parts of the country. Figure 22-2 shows a solar closed system domestic water heating package. This particular package consists of a series of components, hooked up and ready to be connected to collectors and water tank. The collectors and tank are not part of the package, so must be ordered separately. Any good closed system collectors of ample capacity can be used.

Swimming pool water solar heating also utilizes the summer heat to advantage. A simple hookup is shown in Figure 22-3.

Since space heating is the area of greatest interest to the reader of this text, the remainder of this unit describes various types of solar heating systems and their applications.

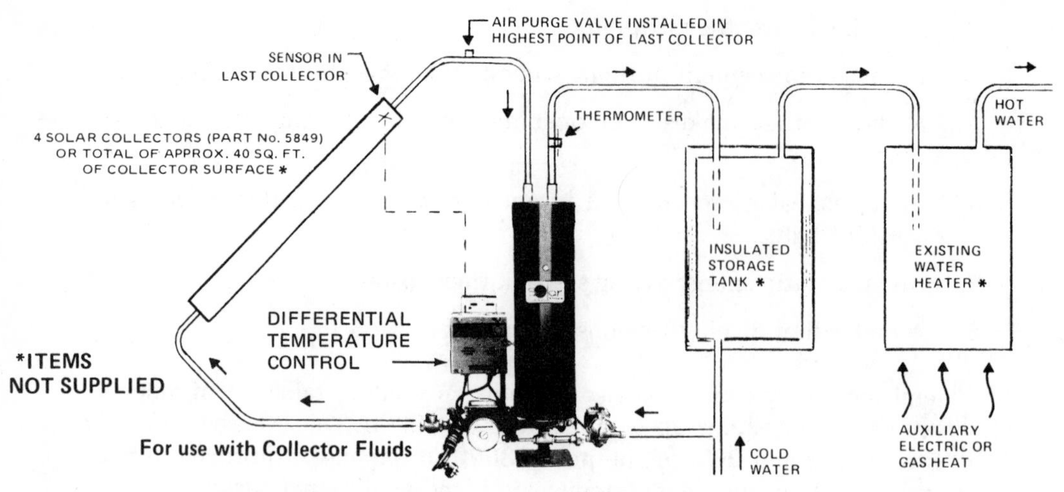

FIGURE 22-2 Solar domestic water heating package (Courtesy, Solar Research, Div. of Refrigeration Research, Inc.).

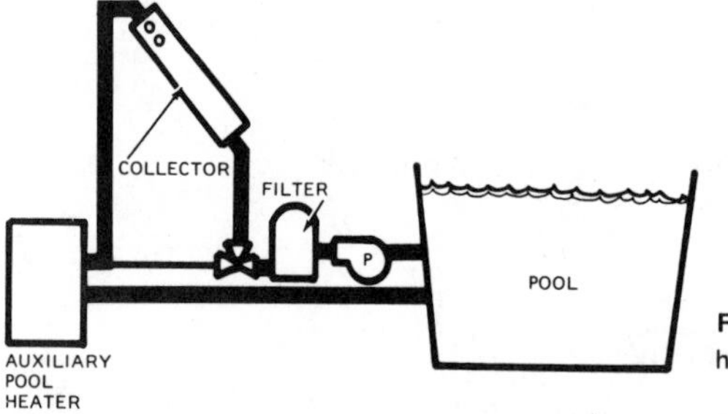

FIGURE 22-3 Solar swimming pool heater (Courtesy, Honeywell, Inc.).

SPACE HEATING

The feasibility of installing solar space heating depends on such factors as

- The size and type of house
- The geographic location
- The portion of the heating load to be supplied by solar heating
- Whether the house is existing or is of new construction
- Possible tax advantages in making the installation

Solar Heating Supply

In most portions of this country it is not feasible to supply 100% solar heating because in the coldest periods of the winter the available heat from the sun is at a minimum. Storage requirements could not be economically justified.

Therefore, in many areas the most practical application of solar heating is to supply 50 to 70% of the annual heating load requirements. A conventional heating system must be installed to handle the full maximum heating load even though solar heat is installed. This is necessary because the greatest heating load may occur when the solar supply is not available. For this reason, an added solar supplement does not reduce the size of a conventional heating system.

Types of Systems

There are two principal types of systems:

- Air system
- Liquid system

In air systems the heat in the collectors is absorbed by air. Thus, heated air is used either directly to heat the space or to heat a rock pile to store excess heat for requirements when the sun is not shining. Designs for an air-type solar system are shown in Figures 22-4 and 22-5.

In liquid systems the heat in the collector is absorbed by a liquid, usually water. A design for a liquid-type solar heater is shown in Figure 22-6. Heated water is used directly or through a heat exchanger to heat the space. A heat exchanger is a transfer unit, such as a coil, where heat in one medium is used to heat a second medium. For example, hot water passing through a coil may be used to heat air, as shown in Figure 22-7.

In the liquid system, excess heat that is not immediately required for space heating is used to heat water in a tank. Thus, heat can be stored for future use.

Collecting Energy

A flat-plate liquid collector uses a surface, usually black, to absorb the sun's energy (Figure 22-8). This surface can be constructed of metal or plastic, depending on the type of collector. On a water-type collector, liquid-carrying tubes are bonded to the surface to absorb the heat. The collector is provided with a frame to support one or two layers of glass or clear plastic.

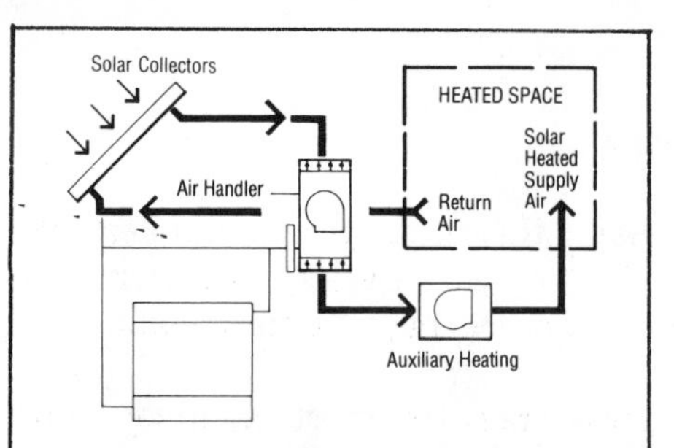

HEATING FROM COLLECTOR Air, the circulating heat transfer medium is drawn through the collector where it is normally heated to about 120-150°F. When the space requires heat, the solar heated air is drawn through the air handling unit in which motorized dampers are automatically opened to direct the hot air to the space. The air then returns to the collector where it is again heated and the cycle repeats itself.

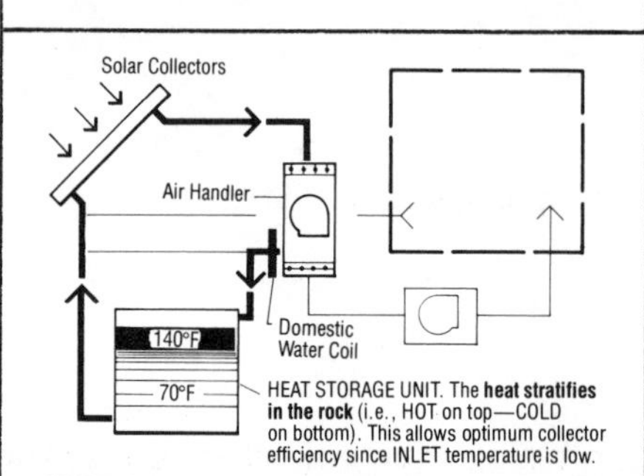

STORING HEAT When the space temperature is satisfied the automatic control system diverts the air into the heat storage unit where the heat is absorbed by the pebble bed. Domestic water is simultaneously heated. The air returns to the collector where it is heated and the cycle repeats.

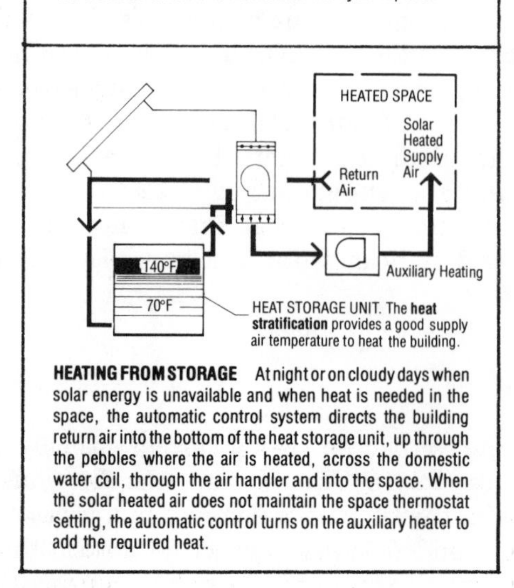

HEATING FROM STORAGE At night or on cloudy days when solar energy is unavailable and when heat is needed in the space, the automatic control system directs the building return air into the bottom of the heat storage unit, up through the pebbles where the air is heated, across the domestic water coil, through the air handler and into the space. When the solar heated air does not maintain the space thermostat setting, the automatic control turns on the auxiliary heater to add the required heat.

FIGURE 22-4 Designs for air-type solar systems (Courtesy, Solaron Corporation).

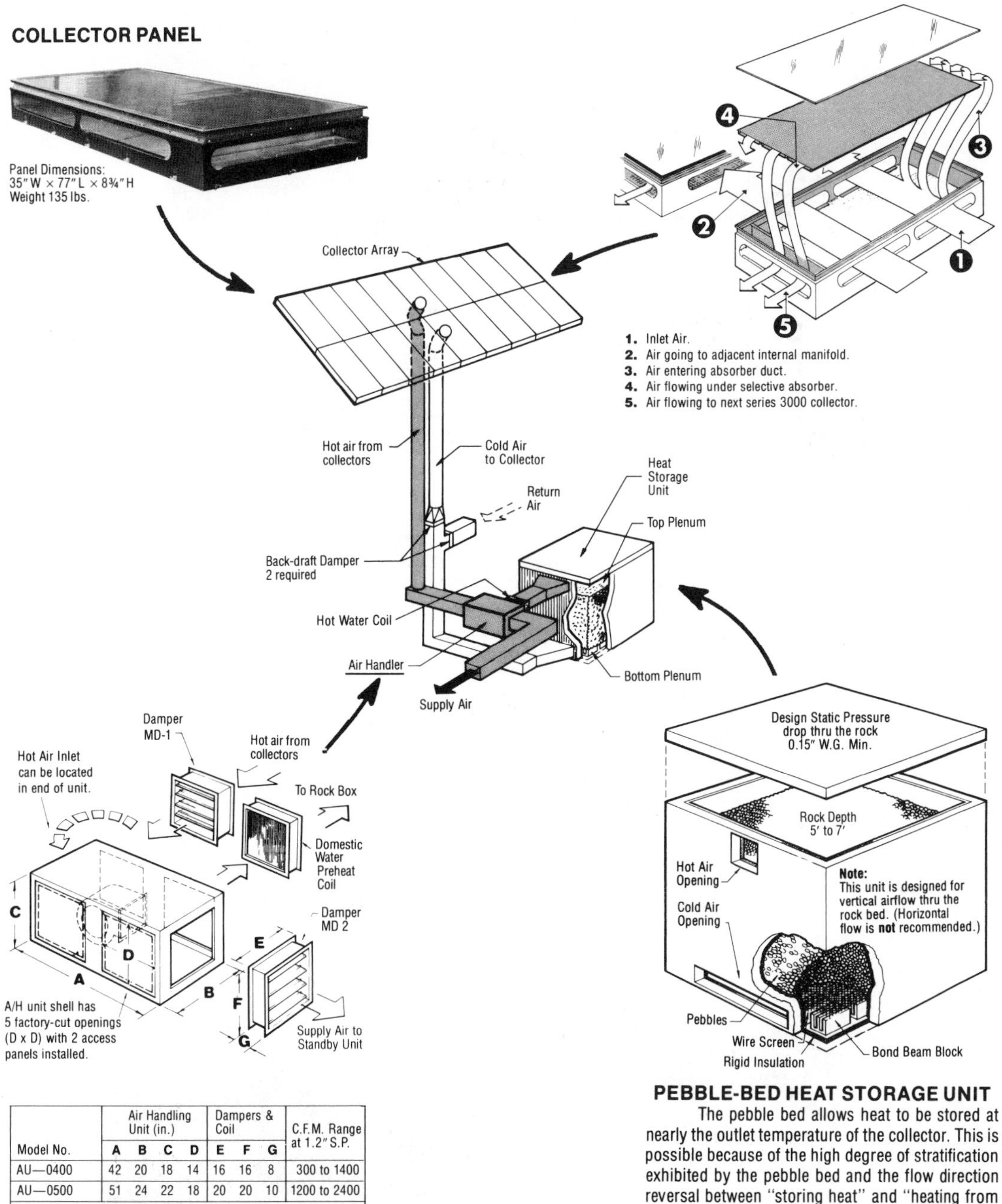

Model No.	Air Handling Unit (in.)				Dampers & Coil			C.F.M. Range at 1.2″ S.P.
	A	B	C	D	E	F	G	
AU—0400	42	20	18	14	16	16	8	300 to 1400
AU—0500	51	24	22	18	20	20	10	1200 to 2400
Larger, custom built air handlers are also available.								

AIR HANDLING UNIT

Solaron provides standard factory pre-assembled air handling units including blowers and motor driven dampers. A separate backdraft damper pair is furnished for mounting in the duct system.

PEBBLE-BED HEAT STORAGE UNIT

The pebble bed allows heat to be stored at nearly the outlet temperature of the collector. This is possible because of the high degree of stratification exhibited by the pebble bed and the flow direction reversal between "storing heat" and "heating from storage."

When "storing heat," the high temperature air from the collector outlet enters the top of the pebble bed where it gives up its heat to the pebbles and returns to the collector as cool air. This allows the collector to operate at the highest possible efficiency.

FIGURE 22-5 Solar air heating system and components (Courtesy, Solaron Corporation).

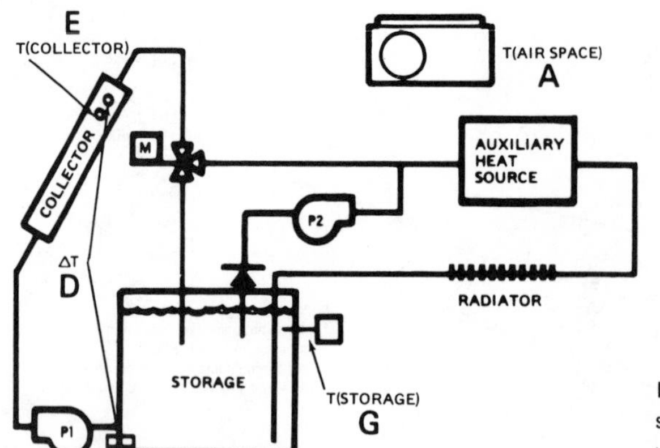

FIGURE 22-6 Design for liquid-type solar heating systems (Courtesy, Honeywell, Inc.).

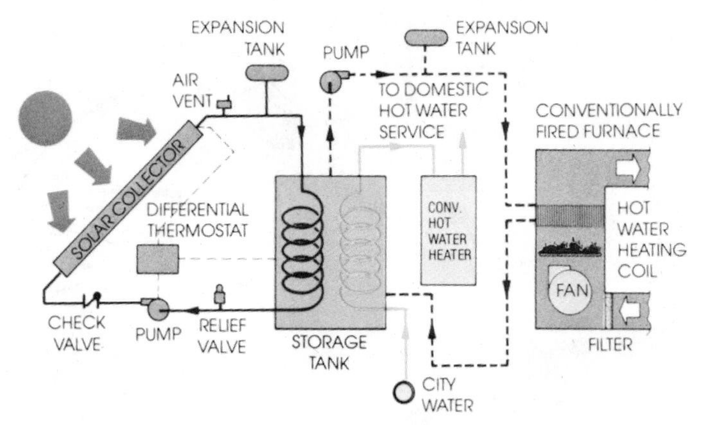

FIGURE 22-7 Solar heated hot water used to heat air in conventionally-fired furnace.

FIGURE 22-8 Cross-section of flat-plate liquid-type solar collector (Courtesy, LOF Solar Energy System, Inc.).

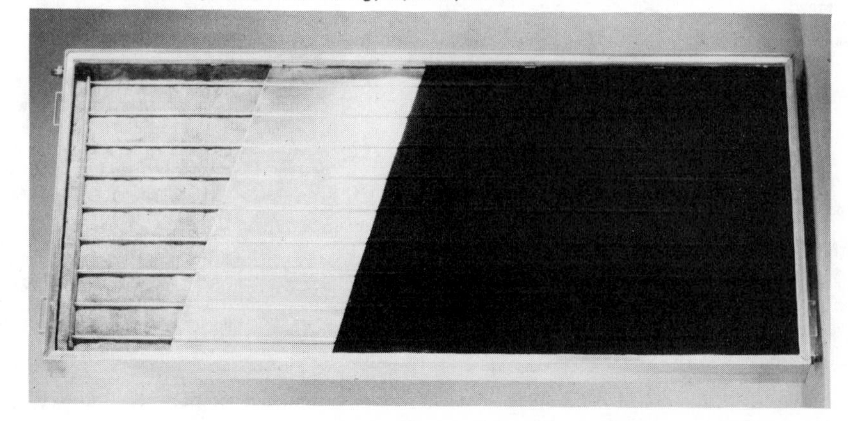

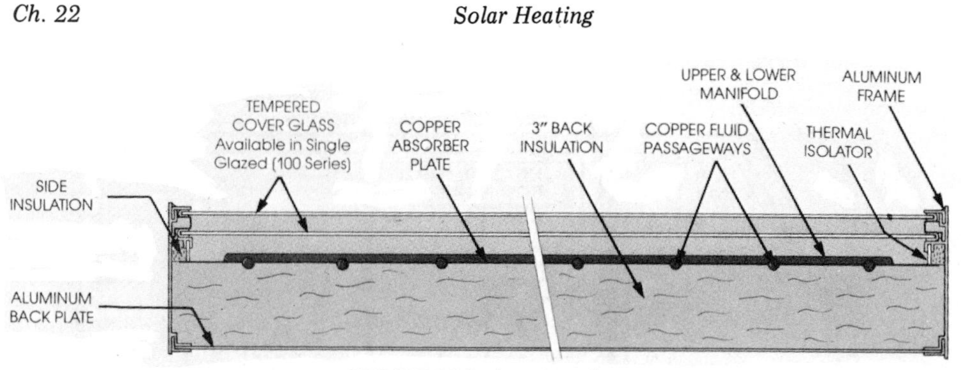

FIGURE 22-8 (continued)

These layers are placed above the collector plate to prevent heat from leaving the collector surface. The back of the plate is insulated to reduce losses.

An air-type collector is shown in Figure 22-5. On a collector of this type, air is passed over a finned absorber surface. Baffle plates (separators) are used to direct a series of paths across the face of the collector. Other air-type collectors provide a space behind the collector surface, between the metal and the insulation, for the passage of air.

When the collector surface is heated by the sun's radiation, the absorber's temperature is greater than the temperature of the area around it. Heat is lost by conduction, radiation, and convection. This loss becomes greater as the temperature difference between the collector surface and the surrounding air becomes greater. Thus, the quality of insulation provided by the "dead air space" between the glass and the insulation material behind and around the collector plate becomes extremely important in collecting heat during cold weather.

Transferring Heat

Heat is collected by using either air or a liquid to convey the energy from the place of collection to the space to be heated or to storage (Figure 22-9). To force the flow of the medium (air or water) to the appropriate area, a liquid system uses pumps and an air-type system uses fans.

Storage of Heat

In a solar heating system heat is collected during the period when the sun is shining. If the supply of heat is in excess of that required to heat the space, the balance is transferred to storage. In a liquid-type system an insulated tank of water is used for storage. In an air system an insulated container filled with rocks is used (Figure 22-5). (Water or rocks are used because of their high heat storage capacity and/or low cost.)

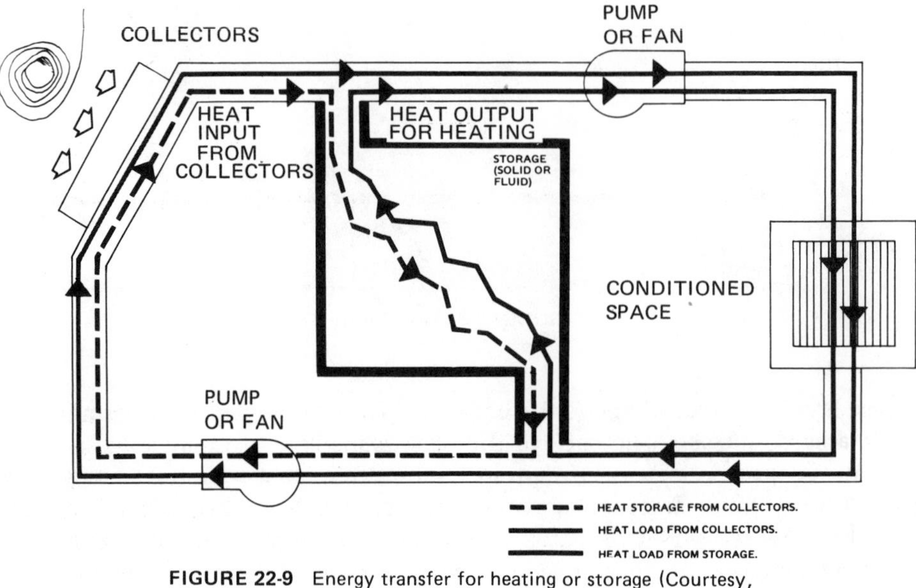

FIGURE 22-9 Energy transfer for heating or storage (Courtesy, Honeywell, Inc.).

The following formula is used to calculate the heat-holding capacity of a storage unit:

$$Q = W \times \mathrm{SH} \times \mathrm{TD}$$

where: Q = quantity of heat, Btu

W = weight of material, lb

SH = specific heat, in Btu/lb/°F

TD = temperature difference, °F above the minimum use temperature

For example, if the minimum use temperature for water is 100°F and water can be obtained from the collector up to a temperature of 180°F, the heat stored in 1000 gal of water (8.33 lb/gal) would be

$$Q = W \times \mathrm{SH} \times \mathrm{TD}$$

$$Q = 1000 \text{ gal} \times 8.33 \text{ lb/gal} \times 1.0 \text{ Btu/lb/}^\circ\text{F} \times (180^\circ - 100^\circ)$$

$$Q = 1000 \times 8.33 \times 80$$

$$Q = 666{,}400 \text{ Btu}$$

To obtain an equal supply of heat from a rock pile with a minimum use temperature of 75° F (specific heat of rock 0.2 Btu/lb/° F):

$$Q = W \times \text{SH} \times \text{TD}$$

$$666{,}400 = W \times 0.2 \times (180° - 75°)$$

$$666{,}400 = \text{W} \times 0.2 \times 105°$$

$$W = \frac{666{,}400}{0.2 \times 105°}$$

$$= 31{,}733 \text{ lb or approximately 16 tons}$$

Placing of Collectors

In locating flat-plate collectors, preference is given to tilting the surface to take greatest advantage of the winter sun position. The best position for most sections of this country is to face the collector toward the south on an unobstructed path of the sun's rays at an angle 10 degrees greater than the latitude. Thus, if the latitude is 40°, the angle of the collector is 50° from the horizontal. Any angle greater than this is more advantageous in the middle of the winter, but not as effective in the fall and spring (see Figure 22-10).

FIGURE 22-10 Collector angle (Courtesy, Honeywell, Inc.).

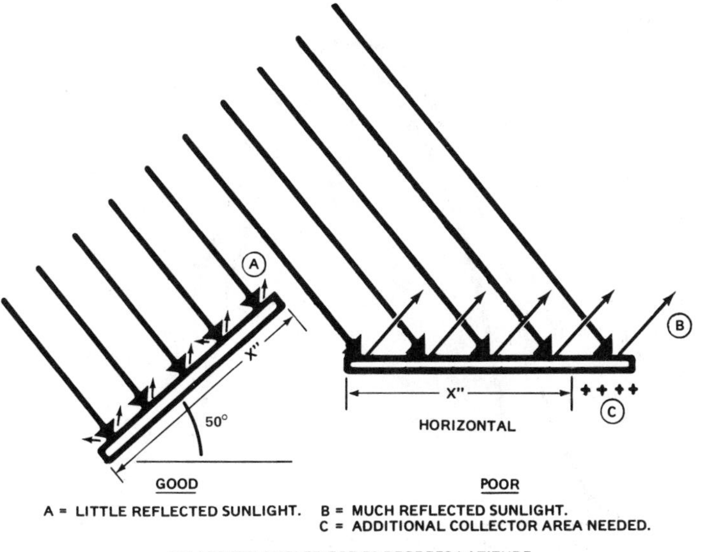

Sizing the Collector

Factors to be taken into consideration when determining the size of the collector surface include the:

- Average insolation (solar radiation) available per month for a given location
- Collector's efficiency
- Heating load of the structure
- Percentage of total heating to be handled by solar installation
- Limitations of the storage unit

Collectors should be sized in accordance with good design practice. Manufacturers of collectors can provide information relating to the application and use of their products.

Insolation is the amount of solar radiation striking a surface on the earth for a given period, usually expressed in Btu per square foot of surface per day (Figure 22-11). Average daily insolation figures are available for many areas.

FIGURE 22-11 Solar insolation (Courtesy, Honeywell, Inc.).

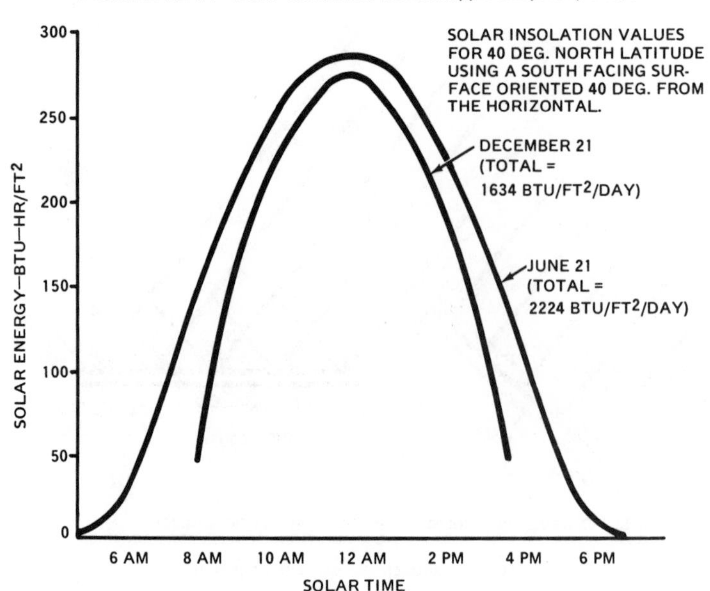

SIZING THE STORAGE UNIT

It is usually considered good practice to size the storage unit to provide the capacity equal to 1 day's heating load in January. This provides at least 3 days' requirement for October and about 2 days' requirement for November. Storage of this size would be used on a system capable of supplying approximately one-half the total requirements for the total season with solar heating.

For example, if the requirements for an average day in January are 400,000 Btu, the amount of water storage would be

$$\text{gal of water} = \frac{400{,}000 \text{ Btu}}{\text{SH} \times \text{TD} \times 8.33 \text{ lb/gal}}$$

$$= \frac{400{,}000 \text{ Btu}}{1.0 \times (180^\circ - 100^\circ) \times 8.33} = \frac{400{,}000}{666}$$

$$= 600 \text{ gal}$$

$$\text{tons of rock} = \frac{400{,}000 \text{ Btu}}{\text{SH} \times \text{TD} \times 2000 \text{ lb/ton}}$$

$$= \frac{400{,}000 \text{ Btu}}{0.2 \text{ Btu/lb/}^\circ\text{F} \times (180^\circ - 75^\circ) \times 2000 \text{ lb/ton}} = \frac{400{,}000}{42{,}000}$$

$$= 9.52 \text{ tons}$$

NOTE: This is based on heating the storage unit to 180° F, using water storage heat down to 100° F, and air storage heat down to 75° F.

Freeze Protection for Liquid Systems

Some protection must be provided on liquid systems to prevent damage caused by water freezing in the collector or piping when the liquid is not heated by the sun. Two practical ways to accomplish this include:

1. Self-draining arrangement. When its temperature gets near freezing the water drains out of the collector
2. The use of an anti-freeze solution such as ethylene glycol in the collector and a heat exchanger circuit, instead of water (Figure 22-12).

Solar heating technology, while still in its infancy, improves each year as people become increasingly aware of the vast resources potentially available in solar energy.

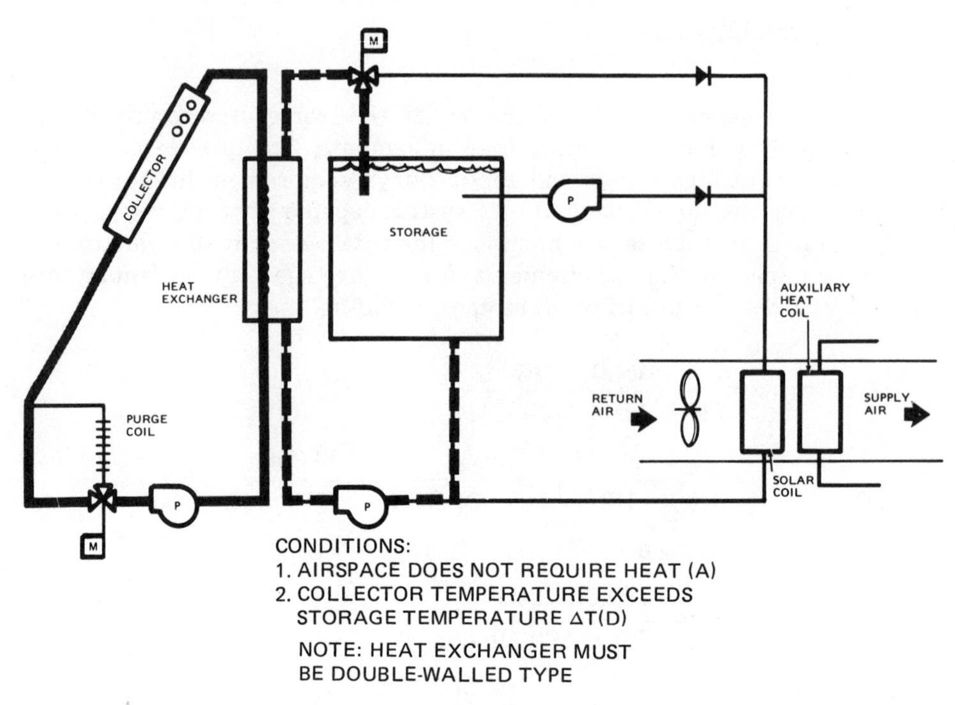

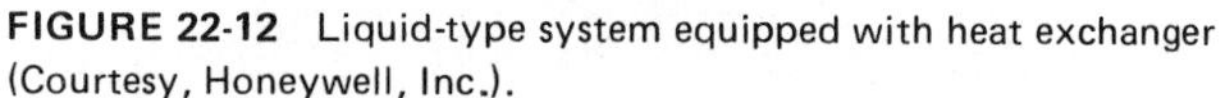

FIGURE 22-12 Liquid-type system equipped with heat exchanger (Courtesy, Honeywell, Inc.).

Scientific experts envision the use of many other possible forms of energy in the future, such as geo-thermal, nuclear fusion, fuel from plants, and power stations orbiting the earth or afloat at sea.

Any of these visions could become a reality in the future. For the present, solar heating, coupled with energy conservation, continues to be one of the very promising methods of solving the energy shortages.

REVIEW QUESTIONS

Select the letter representing your choice of the correct answer

22-1. What is the maximum Btuh per square foot of surface absorbed by many flat-plate collectors?

(a) 150

(b) 200

(c) 250

(d) 300

22-2. What is the most feasible use of solar heating from an energy-conservation standpoint in residential application?

(a) Domestic hot water

(b) Swimming pool heating

(c) Space heating

22-3. How many factors should be considered in determining the feasibility of a residential solar space heating system?

(a) One

(b) Two

(c) Three

(d) Four

(e) Five

22-4. What percentage of the total heating load is usually practical for the solar heating supplement?

(a) 10 to 30

(b) 30 to 50

(c) 50 to 70

(d) 70 to 90

22-5. Air and liquid are the two most practical methods of transferring heat from the sun to residential space heating.

(a) True

(b) False

22-6. What is done with excess solar heat not used immediately for space heating?

(a) It is stored

(b) It is wasted

22-7. Which occupies more space for the same number of Btu stored, water storage or rock storage?

(a) Water

(b) Rock

22-8. What is the specific heat of rock?

(a) 0.2

(b) 0.4

(c) 0.6

(d) 0.8

22-9. What solution can be used in a liquid system to prevent freeze-up?

(a) Mercury

(b) Salt

(c) Alcohol

(d) Ethylene glycol

22-10. What is the best angle (from the horizontal) for a residential heating flat-plate collector?

(a) Latitude + 10°

(b) Latitude - 10°

(c) Latitude + 20°

(d) Latitude - 20°

APPENDICES

INDEX FOR APPENDIX SERVICE INFORMATION

Electric Furnace

Gas Furnace

Oil Furnace

Systems

Miscellaneous

RESIDENTIAL HEATING SYSTEM TROUBLE ANALYSIS CHART

COMPLAINT	AREA OF TROUBLE	POSSIBLE CAUSES	CORRECTIVE ACTION
No heat	Power Supply	1. Tripped overload. 2. Open switch. 3. Bad transformer. 4. Bad connection. 5. Line voltage fluctuating or too low.	1. Determine cause – replace fuse or reset breaker. 2. Correct. 3. Replace. 4. Correct. 5. Inform power company.
	Thermostat	Refer to Thermostat Trouble Analysis Chart	
	Limit	1. System out on limit. 2. Manual reset limit tripped. 3. Dirty or pitted contacts. 4. Poor connection. 5. Defective.	1. Determine cause. 2. Reset, correct cause. 3. Clean or replace. 4. Correct. 5. Replace.
	Fuel delivery and pilot system (gas)	1a. Bad main or pilot gas valve. 1b. Bad thermocouple, thermopile, or power unit. 2. Fuel line blocked or low pressure.	1. Refer to the gas handbook for complete troubleshooting procedures for gas control systems. 2. Check manual valves, check for obstructions, check street supply, refer to equipment manufacturer's instructions.
	Fuel delivery	1. Failure in primary control. 2. No oil being delivered (primary control okay). 3. No ignition (primary control okay).	1. Refer to the oil handbook for complete troubleshooting, procedures for oil control systems. 2. Check for empty tank, blocked fuel line, bad oil valve or pump clogged nozzle; refer to equipment manufacturer's instructions. 3. Check connections, transformer, electrode spacing, insulators; refer to equipment manufacturer's instructions.
Not enough heat	Thermostat	Refer to Thermostat Trouble Analysis Chart	
	Limit	1. Set too low. 2. Improperly located.	1. Raise setting. 2. Relocate if necessary; consult equipment manufacturer.

(Courtesy, Honeywell, Inc.).

RESIDENTIAL HEATING SYSTEM TROUBLE ANALYSIS CHART

COMPLAINT	AREA OF TROUBLE	POSSIBLE CAUSES	CORRECTIVE ACTION
Low Input	Low Input (gas or oil)	1. Undersized furnace or boiler.	1. Replace if condition extreme.
		2. Poor burner adjustment.	2. Refer to manufacturer's instructions.
		3. Sooted heat exchanger.	3. Determine cause (poor draft, burner adjustment, etc.) and correct.
		4. Poor draft.	4. Clean system; check chimney size and design.
		5. Low boiler water.	5. Add make up water.
		6. Distribution system improperly balanced.	6. Correct
		7. Failure of fan or circulator.	7. Repair according to manufacturer's instructions.
		8. Low fan output	8. Adjust pulley for higher speed.
		9. Restricted air flow.	9. Check filters; check obstructions in ducts or in front of registers.
	Low input (gas)	1. Improperly adjusted pressure regulator.	1. Adjust regulator to proper manifold pressure.
		2. Improperly sized orifice—too small.	2. Replace with manufacturer's recommended size.
		3. Malfunctioning regulator.	3. Replace regulator.
		4. Clogged burner orifice.	4. Clean or replace burner orifice.
		5. Malfunctioning gas gas valve.	5. Repair or replace.
		6. Dirt in valve restricting gas flow.	6. Clean valve and seat.
		7. Manual valve partially closed.	7. Move to full open position.
		8. Low supply pressure from gas main.	8. Call gas company.
	Low input (oil)	1. Clogged oil filter or fuel line.	1. Replace filter and/or clean fuel line.
		2. Clogged, faulty, or improperly sized nozzle.	2. Clean or replace nozzle.
		3. Defective or improperly adjusted pump.	3. Replace or readjust pump pressure.
Too much heat, or overshoot	Thermostat	Refer to Thermostat Trouble Analysis Chart.	

(Courtesy, Honeywell, Inc.).

THERMOSTAT TROUBLE ANALYSIS CHART (Courtesy, Honeywell, Inc.).

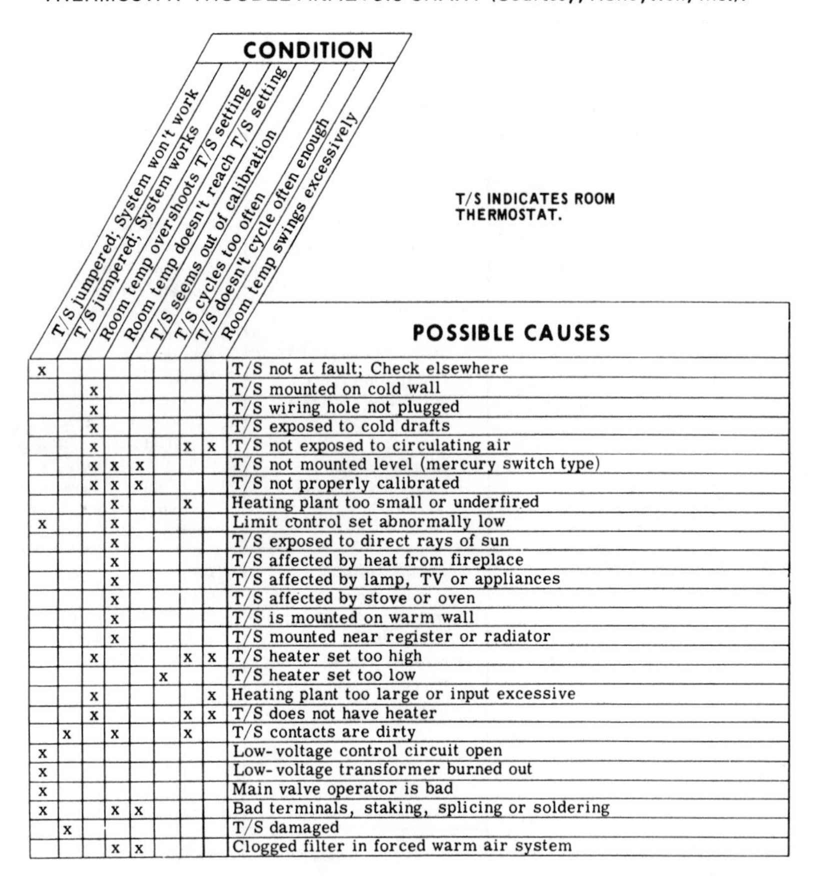

T/S INDICATES ROOM THERMOSTAT.

T/S jumpered; System won't work	T/S jumpered; System works	Room temp overshoots T/S setting	Room temp doesn't reach T/S setting	T/S seems out of calibration	T/S cycles too often	T/S doesn't cycle often enough	Room temp swings excessively	POSSIBLE CAUSES
x								T/S not at fault; Check elsewhere
		x						T/S mounted on cold wall
		x						T/S wiring hole not plugged
		x						T/S exposed to cold drafts
		x				x	x	T/S not exposed to circulating air
		x	x	x				T/S not mounted level (mercury switch type)
		x	x	x				T/S not properly calibrated
			x			x		Heating plant too small or underfired
x			x					Limit control set abnormally low
			x					T/S exposed to direct rays of sun
			x					T/S affected by heat from fireplace
			x					T/S affected by lamp, TV or appliances
			x					T/S affected by stove or oven
			x					T/S is mounted on warm wall
			x					T/S mounted near register or radiator
		x				x	x	T/S heater set too high
					x			T/S heater set too low
		x					x	Heating plant too large or input excessive
		x				x	x	T/S does not have heater
	x		x			x		T/S contacts are dirty
x								Low-voltage control circuit open
x								Low-voltage transformer burned out
x								Main valve operator is bad
x			x	x				Bad terminals, staking, splicing or soldering
	x							T/S damaged
			x	x				Clogged filter in forced warm air system

THERMOSTATS WITH ADJUSTABLE HEAT ANTICIPATOR SCALE ADJUSTMENT

Adjust heater to match current rating of primary control; this rating is usually stamped on the control nameplate. Move the indicator on the scale to correspond with this rating, and the heater will be properly adjusted for optimum comfort with most types of heating systems.

A slightly higher setting to obtain longer "burner-on" times (and thus fewer cycles per hour) may be desirable on some systems such as a one-pipe steam system. Example—If "burner-on" time is too short, proceed as follows:

If the nominal heater setting is 0.4, adjust to 0.45 setting and check system operation, adjust to 0.5 setting and recheck, etc., until the desired "burner-on" time is obtained. If the nominal heater setting is 0.2, adjust to 0.225 to achieve the desired burner-on time.

If the room temperature overshoots the thermostat setting excessively, decreasing the burner-on time may result in more constant temperature. To accomplish this, adjust the heater setting from the nominal 0.4 down to 0.35, or from the nominal 0.2 down to 0.18, and recheck operation of the system.

HEATING SERVICE GUIDE (Courtesy, Robertshaw Controls Company).

A. FORCED AIR FURNACE CHECK LIST

I BURNERS GAS-FIRED

CONDITIONS	POSSIBLE CAUSES	POSSIBLE CURES
FLAME TOO LARGE	1. PRESSURE REG. SET TOO HIGH. 2. DEFECTIVE REGULATOR 3. BURNER ORIFICE TOO LARGE.	1. RESET, USING MANOMETER 2. REPLACE 3. REPLACE WITH CORRECT SIZE.
NOISY FLAME	1. TOO MUCH PRIMARY AIR. 2. NOISY PILOT 3. BURR IN ORIFICE	1. ADJUST AIR SHUTTERS 2. REDUCE PILOT GAS 3. REMOVE BURR OR REPLACE ORIFICE
YELLOW TIP FLAME	1. TOO LITTLE PRIMARY AIR. 2. CLOGGED BURNER PORTS 3. MISALIGNED ORIFICES 4. CLOGGED DRAFT HOOD	1. ADJUST AIR SHUTTERS 2. CLEAN PORTS 3. REALIGN 4. CLEAN
FLOATING FLAME	1. BLOCKED VENTING 2. INSUFFICIENT PRIMARY AIR	1. CLEAN 2. INCREASE PRIMARY AIR SUPPLY
DELAYED IGNITION	1. IMPROPER PILOT LOCATION. 2. PILOT FLAME TOO SMALL. 3. BURNER PORTS CLOGGED NEAR PILOT 4. LOW PRESSURE	1. REPOSITION PILOT. 2. CHECK ORIFICE, CLEAN, INCREASE PILOT GAS. 3. CLEAN PORTS 4. ADJUST PRESSURE REGULATOR.
FAILURE TO IGNITE	1. MAIN GAS OFF 2. BURNED OUT FUSE 3. LIMIT SWITCH DEFECTIVE 4. POOR ELECTRICAL CONNECTIONS. 5. DEFECT GAS VALVE. 6. DEFECTIVE THERMOSTAT.	1. OPEN MANUAL VALVE. 2. REPLACE 3. REPLACE 4. CHECK, CLEAN AND TIGHTEN. 5. REPLACE 6. REPLACE
BURNER WON'T TURNOFF	1. POOR THERMOSTAT LOCATION. 2. DEFECTIVE THERMOSTAT. 3. LIMIT SWITCH MALADJUSTED. 4. SHORT CIRCUIT 5. DEFECTIVE OR STICKING AUTOMATIC VALVE	1. RELOCATE 2. CHECK CALIBRATION. CHECK SWITCH AND CONTACTS. REPLACE. 3. REPLACE 4. CHECK OPERATION AT VALVE. LOOK FOR SHORT AND CORRECT. 5. CLEAN OR REPLACE.
RAPID BURNER CYCLING	1. CLOGGED FILTERS. 2. EXCESSIVE ANTICIPATION. 3. LIMIT SETTING TOO LOW. 4. POOR THERMOSTAT LOCATION.	1. CLEAN OR REPLACE. 2. ADJUST THERMOSTAT ANTICIPATOR FOR LONGER CYCLES. 3. READJUST OR REPLACE LIMIT. 4. RELOCATE.
RAPID FAN CYCLING	1. FAN SWITCH DIFF. TOO LOW. 2. BLOWER SPEED TOO HIGH.	1. READJUST OR REPLACE. 2. READJUST TO LOWER SPEED.
BLOWER WON'T STOP	1. MANUAL FAN "ON". 2. FAN SWITCH DEFECTIVE. 3. SHORTS	1. SWITCH TO AUTOMATIC. 2. REPLACE 3. CHECK WIRING AND CORRECT.

II MOTOR AND BLOWER

CONDITION•	POSSIBLE CAUSES	POSSIBLE CURES
NOISY	1. FAN BLADES LOOSE. 2. BELT TENSION IMPROPER. 3. PULLEYS OUT OF ALIGNMENT. 4. BEARINGS DRY. 5. DEFECTIVE BELT. 6. BELT RUBBING.	1. REPLACE OR TIGHTEN. 2. READJUST (USUALLY 1 INCH SLACK) 3. REALIGN 4. LUBRICATE 5. REPLACE 6. REPOSITION

HEATING SERVICE GUIDE

III LIMIT – FAN – THERMOSTAT – AUTOMATIC PILOT – AUTOMATIC VALVE

CONDITIONS	POSSIBLE CAUSES	POSSIBLE CURES
BURNER WON'T TURN ON	1. PILOT FLAME TOO LARGE OR TOO SMALL. 2. DIRT IN PILOT ORIFICE 3. TOO MUCH DRAFT. 4. DEFECTIVE AUTOMATIC PILOT VALVE. 5. DEFECTIVE THERMOCOUPLE. 6. IMPROPER THERMOCOUPLE POSITION. 7. DEFECTIVE WIRING. 8. DEFECTIVE THERMOSTAT 9. DEFECTIVE AUTOMATIC VALVE.	1. READJUST 2. CLEAN 3. SHIELD PILOT. 4. REPLACE 5. REPLACE 6. PROPERLY POSITION THERMOPILE 7. CHECK CONNECTIONS. TIGHTEN, REPAIR SHORTS. 8. CHECK FOR SWITCH CLOSURE. REPAIR OR REPLACE. 9. REPLACE
BLOWER WON'T RUN	1. POWER NOT ON. 2. FAN CONTROL ADJUSTMENT TOO HIGH. 3. LOOSE WIRING 4. DEFECTIVE MOTOR OVERLOAD, PROTECTOR OR MOTOR.	1. CHECK POWER SWITCH. CHECK FUSES–REPLACE IF NECESSARY. 2. READJUST OR REPLACE. 3. CHECK AND TIGHTEN 4. REPLACE MOTOR
	POOR HEATING RESULTS	
NOT ENOUGH HEAT	1. THERMOSTAT SET TOO LOW. 2. LAMP, T.V., ETC. CLOSE TO THERMOSTAT 3. THERMOSTAT IMPROPERLY LOCATED. 4. DIRTY AIR FILTER. 5. THERMOSTAT OUT OF CALIBRATION. 6. LIMIT SET TOO LOW. 7. FAN SPEED TOO LOW.	1. RAISE SETTING 2. MOVE HEAT SOURCE AWAY FROM THERMOSTAT 3. RELOCATE THERMOSTAT. 4. CLEAN OR REPLACE. 5. RECALIBRATE OR REPLACE. 6. RESET OR REPLACE 7. CHECK MOTOR AND FAN BELT. TIGHTEN IF TOO LOOSE.
TOO MUCH HEAT	1. THERMOSTAT SET TOO HIGH. 2. THERMOSTAT OUT OF CALIBRATION. 3. SHORT IN WIRING. 4. VALVE STICKS OPEN. 5. THERMOSTAT IN DRAFT OR ON COLD WALL. 6. BY-PASS OPEN.	1. LOWER SETTING. 2. RECALIBRATE OR REPLACE. 3. REPAIR SHORT. 4. REPLACE VALVE. 5. RELOCATE THERMOSTAT TO SENSE AVERAGE TEMPERATURE. 6. CLOSE BY-PASS.

B. THERMOPILE SYSTEMS

CONDITIONS	POSSIBLE CAUSES	POSSIBLE CURES
MAIN BURNER DOES NOT COME ON UPON CLOSURE OF THE THERMOSTAT SWITCH	1. LIMIT OR THERMOSTAT NOT MAKING GOOD CONTACT (VALVE OPENS WHEN THERMOSTAT AND OR LIMIT JUMPERED) 2. LOOSE OR POOR CONNECTIONS 3. THERMOPILE OUTPUT TOO LOW 4. DAMAGED VALVE	1. CLEAN CONTACTS OR REPLACE THERMOSTAT AND OR LIMIT SWITCH 2. CLEAN AND TIGHTEN ALL CONNECTIONS CHECK FOR BROKEN WIRES 3. ADJUST PILOT FLAME. CHECK OUTPUT. REPLACE THERMOPILE IF NECESSARY 4. REPLACE VALVE
MAIN BURNER DOES NOT SHUT OFF WITH THERMOSTAT WIRES DISCONNECTED	1. DIRT ON MAIN VALVE FACE 2. IMPROPER MOUNTING 3. DAMAGED VALVE	1. CLEAN VALVE FACE OR REPLACE VALVE 2. REMOUNT IN PROPER ATTITUDE 3. REPLACE VALVE

GENERAL CHECK LIST FOR INSTALLATION AND MAINTENANCE
(Courtesy, Robertshaw Controls Company).

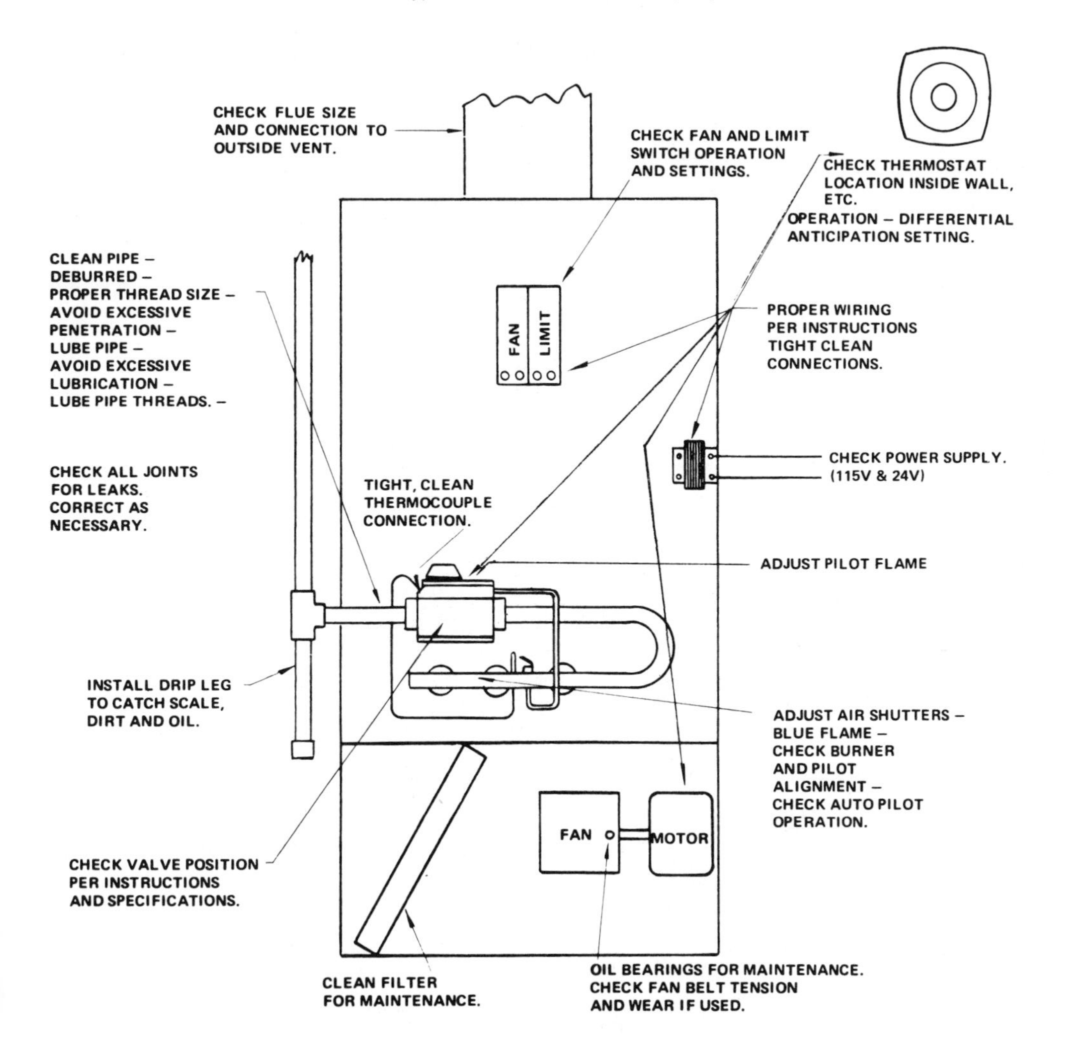

1. How U.S. and metric units relate

Quantity	Current U.S. unit	Conversion Factor*	Metric unit	SI Symbol
Length	inch	25.400	millimeter	mm
	foot	0.305	meter	m
	yard	0.914	meter	m
	mile	1.609	kilometer	km
Area	square inch	645.160	square millimeter	mm^2
	square foot	0.093	square meter	m^2
	square yard	0.836	square meter	m^2
	square mile	2.590	square kilometer	km^2
	acre	0.405	hectare (10,000 m^2)	ha
Weight	ounce	28.350	gram	g
	pound	0.454	kilogram	kg
	ton (2,000 pounds)	0.907	metric ton	t
Volume	fluid ounce	29.574	milliliter	mL
	pint	0.473	liter	L
	quart	0.946	liter	L
	gallon	3.785	liter	L
	gallon per square foot	40.746	liter per square meter	L/m^2
	cubic foot	0.028	cubic meter	m^3
	cubic yard	0.765	cubic meter	m^3
	barrel (petroleum)	0.159	cubic meter	m^3
Force	pound-force	4.448	newton	N
Pressure and Stress	psi (pounds per square inch)	6.895	kilopascal	kPa
	psf (pounds per square foot)	0.048	kilopascal	kPa
	atmosphere (14.7 psi)	101.325	kilopascal	kPa
	ton per square foot square foot	95.76	kilopascal	kPa
Electric Current	ampere +	no conversion	ampere	A
Light	lumen +	no conversion	lumen	Lm
	candela +	no conversion	candela	cd
	footcandle	10.764	lux	$lx(lm/m^2)$
Heat, work or energy	foot pound	1.356	joule	J
	kilowatt hour	3.600	megajoule	MJ
	BTU	1.055	kilojoule	kJ
	BTUs per square foot	11.357	kilojoules per square meter	kJ/m^2
Power	foot pound per second	1.356	watt	W
	BTU/hour	0.293	watt	W
	horse power	0.746	kilowatt	kW
	tons (refrigeration)	3.517	kilowatt	kW
	watts per square foot	10.76	watts per square meter	W/m^2
Heat factors	U value (heat transfer co-efficient)	5.678	metric U value	$W/(m^2 \cdot °C)$
	k value (thermal conductivity)	1.731	metric k value	$W/(m \cdot °C)$
	R value (thermal resistance)	0.176	metric R value	$m^2 \cdot °C/W$
Temperature	degree Fahrenheit	+ +	degree Celsius	°C
	degree Fahrenheit	+ + +	kelvin	K
Cost	¢ per running foot	3.281	¢ per running meter	¢/m
	$ per square foot	10.764	$ per square meter	$\$/m^2$
	¢ per kilowatt hour	0.278	¢ per megajoule	¢/MJ

Conversion factors have been rounded to the third decimal place. To convert from U.S. customary to metric, multiply by this factor; to convert from metric to U.S. customary, divide by this factor.

\+ *Already in common use*

\+ + *To convert °F to °C subtract 32 and divide by 1.8; to convert °C to °F multiply by 1.8 and add 32.*

\+ + + *To convert °F to K add 460 and divide by 1.8.*

(Courtesy, Building Design & Construction Magazine).

METRIC REFERENCES*

Width and length
1 millimeter = thickness of a dime
6 millimeters = diameter of a pencil
10 millimeters = width of man's little fingernail
75 millimeters = length of stick of gum
100 millimeters = long cigaret
190 millimeters = wood pencil length
300 millimeters = standard manila folder width
900 millimeters = width of an ordinary door
2,000 millimeters = length of a mattress for a queen size bed
21 meters = width of an Olympic swimming pool
50 meters = length of an Olympic swimming pool
90 meters = length of a football field

Height
450 millimeters = an ordinary chair seat
1 meter = tennis net at center plus the diameter of a tennis ball
2 meters = most household doorways
50 meters = Niagara Falls
170 meters = Washington Monument
190 meters = St. Louis Arch
445 meters = Sears Tower, Chicago
630 meters = tallest U.S. structure (TV Tower, Blanchard, S.D.)

Area
3 square meters = mattress for queen size bed
150 square meters = living space of average house built in 1977
260 square meters = doubles tennis court
435 square meters = basketball court
750 square meters = baseball infield
1,000 square meters = Olympic swimming pool
4,500 square meters = football field
53,000 square meters = New Orleans Superdome
200,000 square meters = gross floor space, Empire State Building, New York City.
380,000 square meters = Merchandise Mart, Chicago
580,000 square meters = Pentagon
340 hectares = Central Park, New York City

Weight
1 gram = paper clip
2 grams = dime
45 grams = golf ball
175 grams = apple
450 grams = standard loaf of bread
1.7 kilograms = telephone
7 kilograms = man's bowling ball
18 kilograms = IBM Selectric typewriter
50 kilograms = jockey
100 kilograms = Muhammed Ali
1 metric ton = 4-door Chevette car
12 metric tons = empty Greyhound bus
95,000 kilograms = fully loaded Boeing 727-200c plane

Speed
20 to 30 kilometers per hour = moderate breeze
65 to 75 kilometers per hour = gale
90 kilometers per hour = open road speed limit for cars
1,200 kilometers per hour = speed of sound (sea level)
370 kilometers per hour = peak wind recorded in U.S.

Volume
250 milliliters = coffee cup
50 liters = gas tank capacity of 4-door Chevette
80 liters = gas tank capacity of 4-door Impala and Caprice
360 cubic meters = average house built in 1977

U values-W/(m^2·°C)
0.3 metric U value = modern refrigerator wall
0.6 metric U value = 1 inch of stucco plus 4 inches of heavy weight concrete plus 2 inches of insulation
0.9 metric U value = standard picnic cooler (1 inch of styrofoam)
1.8 metric U value = 5/8 inch plaster, 4 inches of cement block and 5/8 inch plaster
2.8 metric U value = standard storm window, summer
5.7 metric U value = standard window glass, summer

Pressure, Stress, Force
1 pascal = downward pressure exerted by a dollar bill on a table
100 kilopascals = one atmosphere
180 kilopascals = standard car tire pressure
1,400 kilopascals = pressure exerted by a standard high heel by an average woman
7 megapascals = typical unit compressive strength of concrete block
20 megapascals = commonly used low-strength poured concrete
55 megapascals = average compressive strength of brick
100 megapascals = shear strength for the cross sectional area of structural steel
250 megapascals = A36 structural steel minimum yield strength
200 megapascals = yield strength of typical low-carbon steel

Heat
1 kilojoule = 1 wood kitchen match burned

Temperature
-30 to 0°C = winter temperature range in Chicago
0 to +10°C = cold
10 to 20°C = cool
20°C = room temperature
20 to 30°C = comfortable
37°C = normal body temperature
30 to 40°C = hot

Cost per square meter
$97 = typical new office space in Chicago in January, 1978

**Dimensions are rounded to within 5%*

(Courtesy, Building Design & Construction Magazine).

(Courtesy, Robertshaw Controls Company).

CONVERSION FACTORS

MULTIPLY	BY	TO OBTAIN
Atmospheres (Std.) 760 MM of Mercury at 32°F.	14.696	Lbs./sq. inch
Atmospheres	76.0	Cms. of mercury
Atmospheres	29.92	In. of mercury
Atmospheres	33.90	Feet of water
Atmospheres	1.0333	Kgs./sq.cm.
Atmospheres	14.70	Lbs./sq. inch
Atmospheres	1.058	Tons/sq. ft.
Brit. Therm. Units	0.2520	Kilogram-calories
Brit. Therm. Units	777.5	Foot-lbs.
Brit. Therm. Units	0.000393	Horse-power-hrs.
Brit. Therm. Units	0.293	Watt-hrs.
BTU/min.	12.96	Foot-lbs./sec.
BTU/min.	0.02356	Horse-power
BTU/min.	0.01757	Kilowatts
BTU/min.	17.57	Watts
Calorie	0.003968	BTU
Centimeters	0.3937	Inches
Centimeters	0.03280	Feet
Centimeters	0.01	Meters
Centimeters	10	Millimeters
Centmtrs. of Merc.	0.01316	Atmospheres
Centimtrs. of merc.	0.4461	Feet of water
Centimtrs. of merc.	136.0	Kgs./sq. meter
Centimtrs. of merc.	27.85	Lbs./sq. ft.
Centimtrs. of merc.	0.1934	Lbs./sq. inch
Cubic feet	2.832×10^4	Cubic cms.
Cubic feet	1728	Cubic inches
Cubic feet	0.02832	Cubic meters
Cubic feet	0.03704	Cubic yards
Cubic feet	7.48052	Gallons U.S.
Cubic feet/minute	472.0	Cubic cms./sec.
Cubic feet/minute	0.1247	Gallons/sec.
Cubic foot water	62.4	Pounds @ 60°F.
Feet	30.48	Centimeters
Feet	12	Inches
Feet	0.3048	Meters
Feet	1/3	Yards
Feet of water	0.02950	Atmospheres
Feet of water	0.8826	Inches of mercury
Feet of water	0.03048	Kgs./sq. cm.
Feet of water	62.43	Lbs./sq. ft.
Feet of water	0.4335	Lbs./sq. inch
Feet/min.	0.5080	Centimeters/sec.
Feet/min.	0.01667	Feet/sec.
Feet/min.	0.01829	Kilometers/hr.
Feet/min.	0.3048	Meters/min.
Feet/min.	0.01136	Miles/hr.
Foot-pounds	0.001286	BTU
Gallons	3785	Cu. centimeters
Gallons	0.1337	Cubic feet
Gallons	231	Cubic inches
Gallons	128	Fluid ounces
Gallons	3.785	Liters
Gallons water	8.35	Lbs. water @ 60°F.
Horse-power	42.44	BTU/min.
Horse-power	33,000	Foot-lbs./min.
Horse-power	550	Foot-lbs./sec.
Horse-power	0.7457	Kilowatts
Horse-power	745.7	Watts
Horse-power (boiler)	33,479	BTU/hr.
Horse-power (boiler)	9.803	Kilowatts
Horse-power-hours	2547	BTU
Horse-power-hours	0.7457	Kilowatt-hours
Inches	2,540	Centimeters
Inches	25.4	Millimeters
Inches	0.0254	Meters
Inches	0.0833	Foot
Inches of mercury	0.03342	Atmospheres
Inches of mercury	1.133	Feet of water
Inches of mercury	13.57	Inches of water
Inches of mercury	70.73	Lbs./sq. ft.
Inches of mercury	0.4912	Lbs./sq. inch
Inches of water	0.002458	Atmospheres
Inches of water	0.07355	In. of mercury
Inches of water	0.5781	Ounces/sq. inch
Inches of water	5.202	Lbs./sq. foot
Inches of water	0.03613	Lbs./sq. inch
Kilowatts	56.92	BTU/min.
Kilowatts	1.341	Horse-power
Kilowatts	1000	Watts
Kilowatt-hours	3415	BTU
Liters	0.2642	Gallons
Liters	2.113	Pints (liq.)
Liters	1.057	Quarts (liq.)
Meters	100	Centimeters
Meters	3.281	Feet
Meters	39.37	Inches
Meters	1000	Millimeters
Meters	1.094	Yards
Ounces (fluid)	1.805	Cubic inches
Ounces (fluid)	0.02957	Liters
Ounces/sq. inch	0.0625	Lbs./sq. inch
Ounces/sq. inch	1.73	Inches of water
Pints	0.4732	Liter
Pounds (avoir.)	16	Ounces
Pounds of water	0.01602	Cubic feet
Pounds of water	27.68	Cubic inches
Pounds of water	0.1198	Gallons
Pounds/sq. foot	0.01602	Feet of water
Pounds/sq. foot	0.006945	Pounds/sq. inch
Pounds/sq. inch	0.06804	Atmospheres
Pounds/sq. inch	2.307	Feet of water
Pounds/sq. inch	2.036	In. of mercury
Pounds/sq. inch	27.68	Inches of water
Temp. (°C.) + 273	1	Abs. temp. (°C.)
Temp. (°C.) + 17.78	1.8	Temp. (°F.)
Temp. (°F.) + 460	1	Abs. temp. (°F.)
Temp. (°F.) − 32	5/9	Temp. (°C.)
Therm	100,000	BTU
Tons(long)	2240	Pounds
Ton, Refrigeration	12,000	BTU/hr.
Tons (short)	2000	Pounds
Watts	3.415	BTU
Watts	0.05692	BTU/min.
Watts	44.26	Foot-pounds/min.
Watts	0.7376	Foot-pounds/sec.
Watts	0.001341	Horse-power
Watts	0.001	Kilowatts
Watt-hours	3.415	BTU/hr.
Watt-hours	2655	Foot-pounds
Watt-hours	0.001341	Horse-power hrs.
Watt-hours	0.001	Kilowatt-hours

PRESSURE CONVERSION TABLES

Equivalent Inches		Pressure Per Square Inch	
Water	Mercury	Pounds	Ounces
0.10	0.007	0.0036	0.0577
0.20	0.015	0.0072	0.115
0.30	0.022	0.0108	4.173
0.40	0.029	0.0145	0.231
0.50	0.037	0.0181	0.289
0.60	0.044	0.0217	0.346
0.70	0.051	0.0253	0.404
0.80	0.059	0.0289	0.462
0.90	0.066	0.325	0.520
1.00	0.074	0.036	0.577
1.36	0.100	0.049	0.785
1.74	0.128	0.067	1.00
2.00	0.147	0.072	1.15
2.77	0.203	0.100	1.60
3.00	0.221	0.109	1.73
4.00	0.294	0.144	2.31
5.0	0.368	0.181	2.89
6.0	0.442	0.217	3.46
7.0	0.515	0.253	4.04
8.0	0.588	0.289	4.62
9.0	0.662	0.325	5.20
10.0	0.74	0.361	5.77
11.0	0.81	0.397	6.34
12.0	0.88	0.433	6.92
13.0	0.96	0.469	7.50
13.6	1.00	0.491	7.86
13.9	1.02	0.500	8.00
14.0	1.06	0.505	8.08
15.0	1.10	0.542	8.7
16.0	1.18	0.578	9.2
17.0	1.25	0.614	9.8
18.0	1.33	0.650	10.4
19.0	1.40	0.686	10.9
20.0	1.47	0.722	11.5
25.0	1.84	0.903	14.4
27.2	2.00	0.975	15.7
27.7	2.03	1.00	16.0

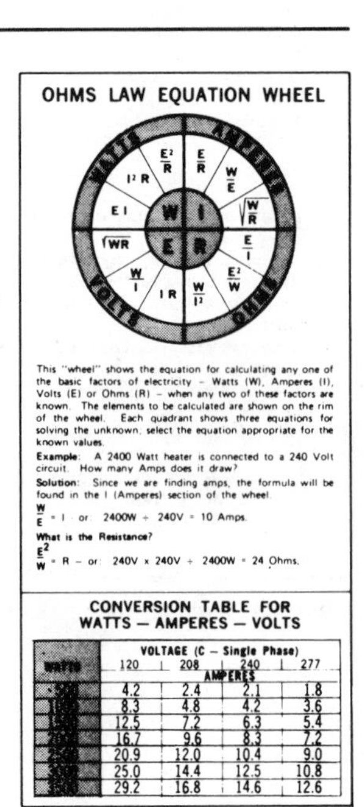

OHMS LAW EQUATION WHEEL

This "wheel" shows the equation for calculating any one of the basic factors of electricity – Watts (W), Amperes (I), Volts (E) or Ohms (R) – when any two of these factors are known. The elements to be calculated are shown on the rim of the wheel. Each quadrant shows three equations for solving the unknown; select the equation appropriate for the known values.

Example: A 2400 Watt heater is connected to a 240 Volt circuit. How many Amps does it draw?

Solution: Since we are finding amps, the formula will be found in the I (Amperes) section of the wheel.

$\frac{W}{E} = I$ or: 2400W ÷ 240V = 10 Amps.

What is the Resistance?

$\frac{E^2}{W} = R$ – or: 240V × 240V ÷ 2400W = 24 Ohms.

CONVERSION TABLE FOR WATTS — AMPERES — VOLTS

WATTS	VOLTAGE (C – Single Phase) 120	208	240	277
	AMPERES			
[illegible]	4.2	2.4	2.1	1.8
[illegible]	8.3	4.8	4.2	3.6
[illegible]	12.5	7.2	6.3	5.4
[illegible]	16.7	9.6	8.3	7.2
[illegible]	20.9	12.0	10.4	9.0
[illegible]	25.0	14.4	12.5	10.8
[illegible]	29.2	16.8	14.6	12.6

CAPACITY TABLE FOR HEATING AND COOLING

FOR CONVERTING CFM TO BTUH
AT VARIOUS REGISTER OR
DIFFUSER TEMPERATURES

(Courtesy, Hart & Cooley Manufacturing Company).

	HEATING*			COOLING**		
REGISTER TEMPERATURE	120°	140°	160°	65°	60°	55°
TEMPERATURE RISE (OR DROP)	50°	70°	90°	15°	20°	25°
CFM						
50	2,700	3,780	4,860	1,050	1,400	1,755
75	4,050	5,670	7,290	1,580	2,110	2,630
100	5,400	7,560	9,720	2,110	2,810	3,510
125	6,750	9,450	12,150	2,630	3,510	4,390
150	8,100	11,340	14,580	3,160	4,210	5,265
175	9,450	13,230	17,010	3,685	4,915	6,140
200	10,800	15,120	19,440	4,210	5,620	7,020
250	13,500	18,900	24,300	5,265	7,020	8,775
300	16,200	22,680	29,160	6,320	8,425	10,530
350	18,900	26,460	34,020	7,370	9,830	12,285
400	21,600	30,240	38,880	8,425	11,230	14,040
450	24,300	34,020	43,740	9,480	12,640	15,795
500	27,000	37,800	48,600	10,530	14,040	17,550
600	32,400	45,360	58,320	12,640	16,850	21,060
700	37,800	52,920	68,040	14,740	19,660	24,570
800	43,200	60,480	77,760	16,850	22,460	28,080
900	48,600	68,040	87,480	18,950	25,270	31,590
1000	54,000	75,600	97,200	21,060	28,080	35,100
1200	64,800	90,720	116,640	25,270	33,700	42,120
1400	75,600	105,840	136,080	29,480	39,310	49,140
1600	86,400	120,960	155,520	33,700	44,930	56,160
1800	97,200	136,080	174,960	37,910	50,540	63,180
2000	108,000	151,200	194,400	42,120	56,160	70,200
2200	118,800	166,320	213,840	46,330	61,780	77,220
2400	129,600	181,440	233,280	50,540	67,390	84,240

*Based on 70° Return Air
**Based on 80° Return Air
Total Cooling Btuh = Sensible + Latent. Latent = 30% of Total

MAIN BURNER AND PILOT BURNER ORIFICE SIZE CHART (GAS-FIRED)

	Natural gas: 1020 Btu — 0.65 SG 3 ½ in. WC manifold		*Propane: 2500 Btu — 1.5 SG 11 in. WC manifold*	
Input (Btu/h) per spud	*Drill size*	*Decimal tolerance*	*Drill size*	*Decimal tolerance*
12,000	51	0.064-0.067	60	0.038-0.040
15,000	48	0.073-0.076	58	0.040-0.042
20,000	43	0.086-0.089	55	0.050-0.052
25,000	41	0.093-0.096	53	0.056-0.059
27,500	39	0.097-0.100	53	0.060-0.063
40,000	32	0.113-0.116	49	0.070-0.073
50,000	30	0.124-0.128	46	0.078-0.081
60,000	27	0.140-0.144	43	0.086-0.089
70,000	22	0.153-0.157	42	0.090-0.093
80,000	20	0.156-0.161	40	0.095-0.098
90,000	17	0.168-0.173	38	0.098-0.101
100,000	13	0.180-0.185	35	0.107-0.110
105,000	11	0.186-0.191	34	0.108-0.111
110,000	10	0.188-0.193	33	0.109-0.113
125,000	5	0.200-0.205	1/8	0.121-0.125
135,000	3	0.208-0.213	30	0.124-0.128
140,000	7/32	0.214-0.219	30	0.124-0.128
150,000	1	0.223-0.228	29	0.132-0.136
160,000	A	0.229-0.234	28	0.136-0.140
175,000	C	0.237-0.242	27	0.140-0.144
190,000	E	0.245-0.250	25	0.145-0.149
200,000	F	0.252-0.257	23	0.150-0.154
210,000	H	0.261-0.266	21	0.154-0.159
220,000	I	0.267-0.272	20	0.156-0.161
240,000	K	0.276-0.281	18	0.164-0.169
260,000	M	0.290-0.295	16	0.172-0.177
280,000	5/16	0.307-0.312	13	0.180-0.185
300,000	0	0.311-0.316	11	0.186-0.191
310,000	P	0.318-0.323	9	0.191-0.196
320,000	21/64	0.323-0.328	7	0.196-0.201

(Courtesy, Luxaire, Inc.).

SUBJECT: NOISY BLOWER CHECK OUT PROCEDURE

Current Situation: Calls from the field requesting complete blower assemblies because of noise or vibration.

Suggested Improvement: The following procedure will enable the serviceman to pinpoint the noise source with a minimum of service time involved.

After turning off the main power to the furnace, the blower belt is removed. The belt is then inspected for cracks, loose fabric or glazing. If any are found, the belt should be replaced.

The main power should be turned back on and the motor be allowed to operate momentarily with the belt still removed. The motor should be felt for excessive vibration.

The main power should be shut off and the motor pulley observed while the motor slows down. The "V" formed by the motor pulley must run true. It is possible for the pulley to be balanced but the "V" could be out of round causing the belt to ride high and then low in the "V" groove causing a thumping noise. If this condition is found, the pulley must be replaced.

If while the motor was running there was excessive vibration, the motor pulley should be removed and the motor allowed to run again. In this way it is determined whether the motor or pulley is causing the problem.

In the event that the belt, motor and motor pulley check out properly, the problem must lie with the blower wheel, pulley or shaft. With the belt still removed, the blower pulley is rotated rapidly and observed for out of roundness and sideways motion. If the blower pulley is defective, it must be replaced.

If all the above checks prove acceptable, the blower wheel must be out of balance and must be replaced. It is virtually impossible for the blower shaft to be bent as this is a machined piece and any bend or warpage would not allow the blower wheel to slip over the shaft on our assembly line.

After any defective parts are replaced, the pulleys should be aligned to prevent undue stress on the bearings. The belt tension should be adjusted so that light finger pressure applied midway between the pulleys will depress the belt approximately 3/4 to 1 inch. The ideal tension will enable the motor to start without slipping.

(Courtesy, Luxaire, Inc.).

1.3.4 Air for Combustion and Ventilation

1.3.4.1 General Provisions

(a) The provisions of 1.3.4 apply to appliances installed in buildings and which require air for combustion, ventilation and dilution of flue gases from within the building. They do not apply to (1) direct vent appliances which are constructed and installed so that all air for combustion is obtained from the outside atmosphere and all flue gases are discharged to the outside atmosphere, or (2) enclosed furnaces which incorporate an integral total enclosure and use only outside air for combustion and dilution of flue gases.

(b) Appliances shall be installed in a location in which the facilities for ventilation permit satisfactory combustion of gas, proper venting and the maintenance of ambient temperature at safe limits under normal conditions of use. Appliances shall be located so as not to interfere with proper circulation of air within the confined space. When buildings are so tight that normal infiltration does not provide the necessary air, outside air shall be introduced.

(c) While all forms of building construction cannot be covered in detail, air for combustion, ventilation and dilution of flue gases for gas appliances vented by natural draft normally may be obtained by application of one of the methods covered in 1.3.4.2, 1.3.4.3 and 1.3.4.6.

1.3.4.2 Appliances Located in Unconfined Spaces:

(a) In unconfined spaces in buildings of conventional frame, masonry, or metal construction, infiltration normally is adequate to provide air for combustion, ventilation, and dilution of flue gases.

(b) If the unconfined space is within a building of unusually tight construction, air for combustion, ventilation, and dilution of flue gases shall be obtained from outdoors or from spaces freely communicating with the outdoors. A permanent opening or openings having a total free area of not less than one square inch per 5,000 Btu per hour of total input rating of all appliances shall be provided. Ducts may be used to convey make-up air from the outdoors and shall be of the same cross-sectional area as the free area of the openings to which they connect. The ducts may be connected to the cold air return of the heating system only if they connect directly to outside air. The minimum dimension of rectangular air ducts shall be not less than 3 inches.

1.3.4.3 Appliances Located In Confined Spaces:

(a) All Air From Inside Buildings: The confined space shall be provided with two permanent openings, one commencing within 12 inches of the top and one commencing within 12 inches of the bottom of the enclosure. Each opening shall have a minimum free area of one square inch per 1,000 Btu per hour of the total input rating of all appliances in the enclosure. These openings must freely communicate with interior areas having adequate infiltration from the outside. (*See Figure 1-5.*)

(b) All Air From Outdoors: The confined space shall be provided with two permanent openings, one commencing within 12 inches of the top and one commencing within 12 inches of the bottom of the enclosure. The openings shall communicate directly, or by ducts, with outdoors or spaces (crawl or attic) that freely communicate with outdoors.

1. When directly communicating with the outdoors, each opening shall have a minimum free area of one square inch per 4,000 Btu per hour of total input rating of all appliances in the enclosure. (*See Figure 1-6.*)

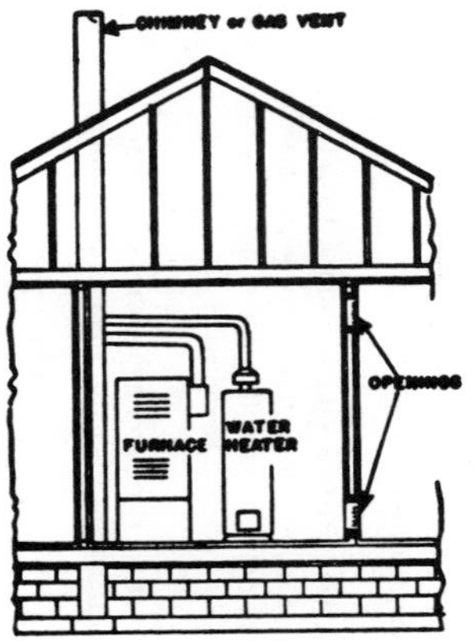

NOTE: Each opening shall have a free area of not less than one square inch per 1,000 Btu per hour of the total input rating of all appliances in the enclosure.

Fig. 1-5. Appliances Located in Confined Spaces; All Air from Inside the Building. See 1.3.4.3 (a).

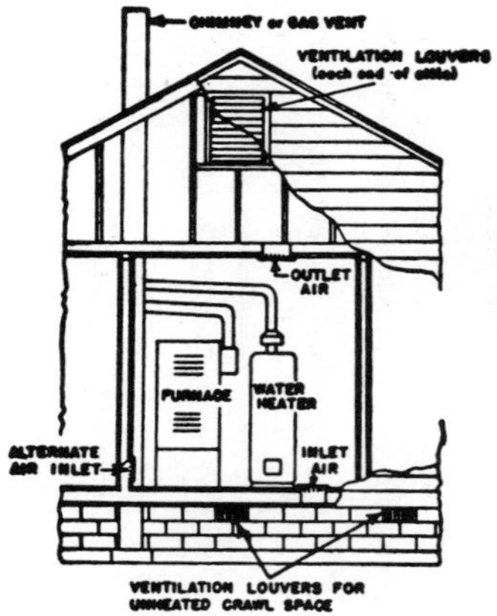

NOTE: The inlet and outlet air openings shall each have a free area of not less than one square inch per 4,000 Btu per hour of the total input rating of all appliances in the enclosure.

Fig. 1-6. Appliances Located in Confined Spaces; All Air from Outdoors — Inlet Air from Ventilated Crawl Space and Outlet Air to Ventilated Attic. See 1.3.4.3 (b).

1.3.4.3 Appliances Located in Confined Spaces (Cont'd.)

2. When communicating with the outdoors through vertical ducts, each opening shall have a minimum free area of one square inch per 4,000 Btu per hour of total input rating of all appliances in the enclosure. (*See Figure 1-7.*)

3. When communicating with the outdoors through horizontal ducts, each opening shall have a minimum free area of one square inch per 2,000 Btu per hour of total input rating of all appliances in the enclosure. (*See Figure 1-8.*)

4. When ducts are used, they shall be of the same cross-sectional area as the free area of the openings to which they connect. The minimum dimension of rectangular air ducts shall be not less than 3 inches.

(c) **Ventilation Air From Inside Building — Combustion and Draft Hood Dilution Air From Outdoors:** The confined space shall be provided with two openings located and sized as in 1.3.4.3(a). In addition there shall be one opening directly communicating with outdoors or spaces (crawl or attic) that freely communicate with outdoors. This opening shall have a minimum free area of one square inch per 5,000 Btu per hour of total input of all appliances in the enclosure. Ducts may be used to convey make-up air and shall be of the same cross-sectional area as the free area of the openings to which they connect. The ducts may be connected to the cold air return of the heating system only if they connect directly to outside air. The minimum dimension of rectangular air ducts shall not be less than 3 inches. (*See Figure 1-9.*)

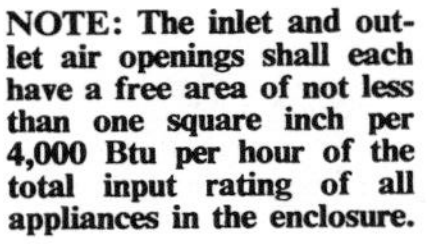
NOTE: The inlet and outlet air openings shall each have a free area of not less than one square inch per 4,000 Btu per hour of the total input rating of all appliances in the enclosure.

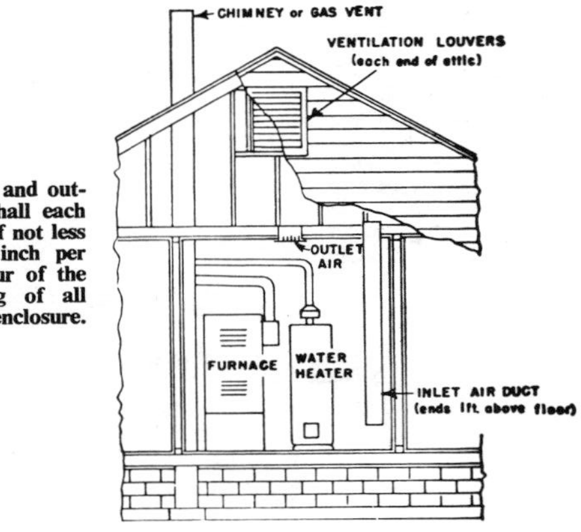

Fig. 1-7. Appliances Located in Confined Spaces; All Air from Outdoors Through Ventilated Attic. See 1.3.4.3(b)

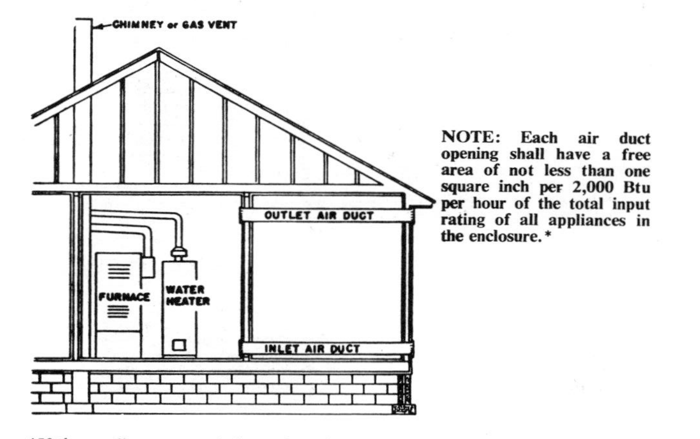

*If the appliance room is located against an outside wall and the air openings communicate directly with the outdoors, each opening shall have a free area of not less than one square inch per 4,000 Btu per hour of the total input rating of all appliances in the enclosure.

Fig. 1-8. Appliances Located in Confined Spaces; All Air from Outdoors. See 1.3.4.3(b)

1.3.4.4 Louvers and Grilles: In calculating free area in 1.3.4.2 and 1.3.4.3, consideration shall be given to the blocking effect of louvers, grilles or screens protecting openings. Screens used shall not be smaller than ¼-inch mesh. If the free area through a design of louver or grille is known, it should be used in calculating the size opening required to provide the free area specified. If the design and free area is not known, it may be assumed that wood louvers will have 20–25 percent free area and metal louvers and grilles will have 60–75 percent free area.

1.3.4.5 Special Conditions Created by Mechanical Exhausting or Fireplaces: Operation of exhaust fans, kitchen ventilation systems, clothes dryers, or fireplaces may create conditions requiring special attention to avoid unsatisfactory operation of installed gas appliances.

1.3.4.6 Specially Engineered Installations: The size of combustion air openings specified in 1.3.4.2 and 1.3.4.3 shall not necessarily govern when special engineering provides an adequate supply of air for combustion, ventilation, and dilution of flue gases.

1.3.5 Appliances on Roofs.

NATIONAL FUEL GAS CODE (Continued)
(Courtesy, American Gas Association).

1.3.4 Air for Combustion and Ventilation

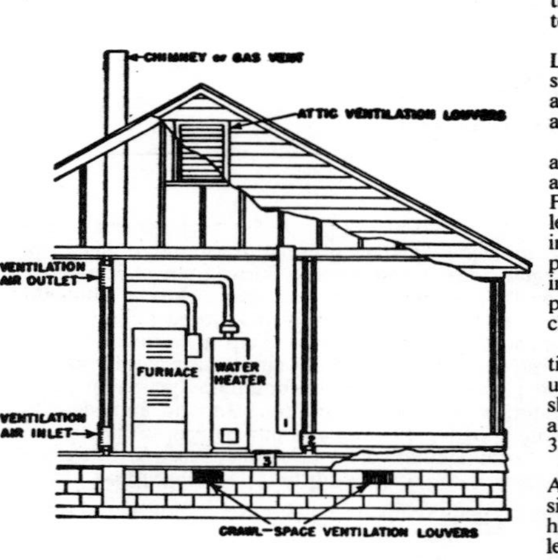

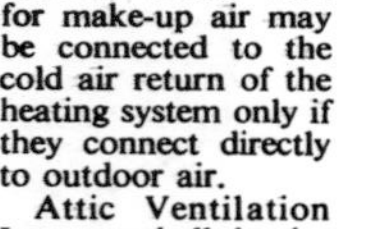

NOTE: Ducts used for make-up air may be connected to the cold air return of the heating system only if they connect directly to outdoor air.

Attic Ventilation Louvers shall be installed at each end of attic with alternate air inlet No. 1.

1, 2, and 3 mark alternate locations for air from outdoors. Free area shall be not less than 1 square inch per 5,000 Btu per hour of the total input rating of all appliances in the enclosure.

Crawl-Space Ventilation Louvers for unheated crawl space shall be installed with alternate air inlet No. 3.

Each Ventilation Air Opening from inside the building shall have a free area of not less than 1 square inch per 1,000 Btu per hour of the total input rating of all appliances in the enclosure.

Fig. 1-9. Appliances Located in Confined Spaces; Ventilation Air from Inside Building — Combustion and Draft Hood Dilution Air from Outside, Ventilated Attic or Ventilated Crawl Space. See 1.3.4.3(c)

APPENDIX 1-D

SIZING OF VENTING SYSTEMS SERVING APPLIANCES EQUIPPED WITH DRAFT HOODS AND APPLIANCES LISTED FOR USE WITH TYPE B VENTS

(This Appendix is informative and is not a part of the Code.)

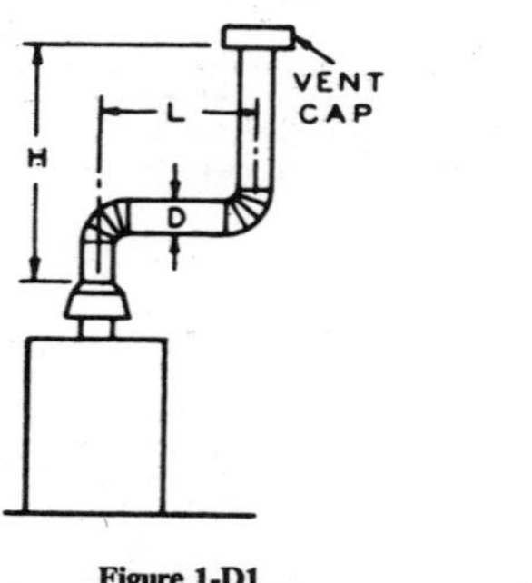

Figure 1-D1
Double Wall or Asbestos Cement Type B Vents or Single-Wall Metal Vents Serving a Single Appliance. (See Tables 1-D1 and 1-D2.)

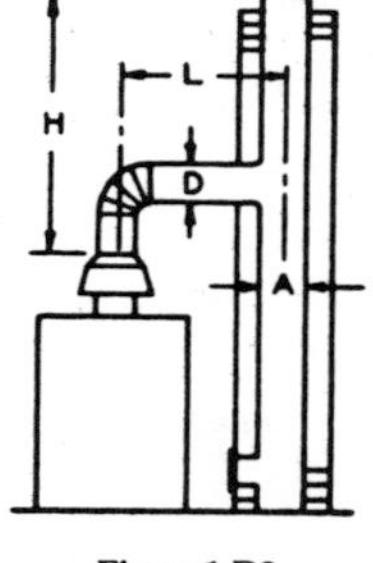

Figure 1-D2
Masonry Chimney Serving a Single Appliance. (See Table 1-D3.)

Notes for Single Appliance Vents. *(See Tables 1-D1, 1-D2 and 1-D3.)*

1. For single-wall metal pipe, use Table 1-D2.
2. If the vent size determined from the Tables is less than the size of the draft hood, the smaller sized vent may be used as long as the vent height "H" is at least 10 feet.
3. Vents for draft hoods 12 inches in diameter or less should not be reduced more than one size (12 inches to 10 inches is a one-size reduction). For larger gas-burning equipment, reductions of more than two sizes (24 inches to 20 inches is a two-size reduction) are not recommended.
4. Regardless of the vent size shown, do not connect any 4-inch draft hoods to 3-inch vents.
5. Zero (0) lateral "L" applies only to a straight vertical vent attached to a top outlet draft hood.
6. Use sea level input rating when calculating vent size for high altitude installation.
7. Designation "NR" in Tables 1-D1, 1-D2 and 1-D3 indicates not recommended.
8. Number followed by an asterisk (*) in Tables 1-D2 and 1-D3 indicate the possibility of continuous condensation, depending on locality. Consult local serving gas supplier and/or local codes.

Table 1-D1

Capacity of Type B Double-Wall Vents with Type B Double-Wall Connectors Serving a Single Appliance

(Courtesy, American Gas Association).

Height H	Lateral L	Vent Diameter — D													
		3"	4"	5"	6"	7"	8"	10"	12"	14"	16"	18"	20"	22"	24"
		Maximum Appliance Input Rating in Thousands of Btu Per Hour													
6'	0	46	86	141	205	285	370	570	850	1170	1530	1960	2430	2950	3520
	2'	36	67	105	157	217	285	455	650	890	1170	1480	1850	2220	2670
	6'	32	61	100	149	205	273	435	630	870	1150	1470	1820	2210	2650
	12'	28	55	91	137	190	255	406	610	840	1110	1430	1795	2180	2600
8'	0	50	94	155	235	320	415	660	970	1320	1740	2220	2750	3360	4010
	2'	40	75	120	180	247	322	515	745	1020	1340	1700	2110	2560	3050
	8'	35	66	109	165	227	303	490	720	1000	1320	1670	2070	2530	3030
	16'	28	58	96	148	206	281	458	685	950	1260	1600	2035	2470	2960
10'	0	53	100	166	255	345	450	720	1060	1450	1925	2450	3050	3710	4450
	2'	42	81	129	195	273	355	560	850	1130	1480	1890	2340	2840	3390
	10'	36	70	115	175	245	330	525	795	1080	1430	1840	2280	2780	3340
	20'	NR	60	100	154	217	300	486	735	1030	1360	1780	2230	2720	3250
15'	0	58	112	187	285	390	525	840	1240	1720	2270	2900	3620	4410	5300
	2'	48	93	150	225	316	414	675	985	1350	1770	2260	2800	3410	4080
	15'	37	76	128	198	275	373	610	905	1250	1675	2150	2700	3300	3980
	30'	NR	60	107	169	243	328	553	845	1180	1550	2050	2620	3210	3840
20'	0	61	119	202	307	430	575	930	1350	1900	2520	3250	4060	4980	6000
	2'	51	100	166	249	346	470	755	1100	1520	2000	2570	3200	3910	4700
	10'	44	89	150	228	321	443	710	1045	1460	1940	2500	3130	3830	4600
	20'	35	78	134	206	295	410	665	990	1390	1880	2430	3050	3760	4550
	30'	NR	68	120	186	273	380	626	945	1270	1700	2330	2980	3650	4390

NATIONAL FUEL GAS CODE

(Courtesy, American Gas Association).

Table 1-D2

Capacity of Single-Wall Metal Pipe or Type B Asbestos Cement Vents Serving a Single Appliance

Height H	Lateral L	Vent Diameter — D							
		3″	4″	5″	6″	7″	8″	10″	12″
		Maximum Appliance Input Rating in Thousands of Btu Per Hour							
6′	0	39	70	116	170	232	312	500	750
	2′	31	55	94	141	194	260	415	620
	5′	28	51	88	128	177	242	390	600
8′	0	42	76	126	185	252	340	542	815
	2′	32	61	102	154	210	284	451	680
	5′	29	56	95	141	194	264	430	648
	10′	24*	49	86	131	180	250	406	625
10′	0	45	84	138	202	279	372	606	912
	2′	35	67	111	168	233	311	505	760
	5′	32	61	104	153	215	289	480	724
	10′	27*	54	94	143	200	274	455	700
	15′	NR	46*	84	130	186	258	432	666
15′	0	49	91	151	223	312	420	684	1040
	2′	39	72	122	186	260	350	570	865
	5′	35*	67	110	170	240	325	540	825
	10′	30*	58*	103	158	223	308	514	795
	15′	NR	50*	93*	144	207	291	488	760
	20′	NR	NR	82*	132*	195	273	466	726
20′	0	53*	101	163	252	342	470	770	1190
	2′	42*	80	136	210	286	392	641	990
	5′	38*	74*	123	192	264	364	610	945
	10′	32*	65*	115*	178	246	345	571	910
	15′	NR	55*	104*	163	228	326	550	870
	20′	NR	NR	91*	149*	214*	306	525	832
30′	0	56*	108*	183	276	384	529	878	1370
	2′	44*	84*	148*	230	320	441	730	1140
	5′	NR	78*	137*	210	296	410	694	1080
	10′	NR	68*	125*	196*	274	388	656	1050
	15′	NR	NR	113*	177*	258*	366	625	1000
	20′	NR	NR	99*	163*	240*	344	596	960
	30′	NR	NR	NR	NR	192*	295*	540	890
50′	0	NR	120*	210*	310*	443*	590	980	1550
	2′	NR	95*	171*	260*	370*	492	820	1290
	5′	NR	NR	159*	234*	342*	474	780	1230
	10′	NR	NR	146*	221*	318*	456*	730	1190
	15′	NR	NR	NR	200*	292*	407*	705	1130
	20′	NR	NR	NR	185*	276*	384*	670*	1080
	30′	NR	NR	NR	NR	222*	330*	605*	1010

Table 1-D3

Capacity of Masonry Chimneys and Single-Wall Vent Connectors Serving a Single Appliance

Height H	Lateral L	Single-Wall Vent Connector Diameter — D (To be used with chimney areas not less than those at bottom)							
		3″	4″	5″	6″	7″	8″	10″	12″
		Maximum Appliance Input Rating in Thousands of Btu Per Hour							
6′	2′	28	52	86	130	180	247	400	580
	5′	25*	48	81	118	164	230	375	560
8′	2′	29	55	93	145	197	265	445	650
	5′	26*	51	87	133	182	246	422	638
	10′	22*	44*	79	123	169	233	400	598
10′	2′	31	61	102	161	220	297	490	722
	5′	28*	56	95	147	203	276	465	710
	10′	24*	49*	86	137	189	261	441	665
	15′	NR	42*	79*	125	175	246	421	634
15′	2′	35*	67	113	178	249	335	560	840
	5′	32*	61	106	163	230	312	531	825
	10′	27*	54*	96	151	214	294	504	774
	15′	NR	46*	87*	138	198	278	481	738
	20′	NR	NR	73*	128*	184	261	459	706
20′	2′	38*	73	123	200	273	374	625	950
	5′	35*	67*	115	183	252	348	594	930
	10′	NR	59*	105*	170	235	330	562	875
	15′	NR	NR	95*	156	217	311	536	835
	20′	NR	NR	80*	144*	202	292	510	800
30′	2′	41*	81*	136	215	302	420	715	1110
	5′	NR	75*	127*	196	279	391	680	1090
	10′	NR	66*	113*	182*	260	370	644	1020
	15′	NR	NR	105*	168*	240*	349	615	975
	20′	NR	NR	88*	155*	223*	327	585	932
	30′	NR	NR	NR	NR	182*	281*	544	865
50′	2′	NR	91*	160*	250*	350*	475	810	1240
	5′	NR	NR	149*	228*	321*	442	770	1220
	10′	NR	NR	136*	212*	301*	420*	728	1140
	15′	NR	NR	124*	195*	278*	395*	695	1090
	20′	NR	NR	NR	180*	258*	370*	660*	1040
	30′	NR	NR	NR	NR	NR	318*	610*	970
Minimum Internal Area of Chimney-A Square Inches		19		28	38	50	63	95	132

See Table 1-D7 for Masonry Chimney Liner Sizes.

NATIONAL FUEL GAS CODE

(Courtesy, American Gas Association).

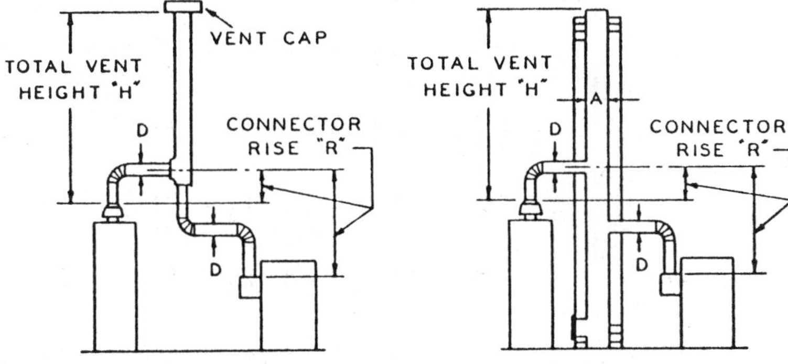

Figure 1-D3
Double-Wall or Asbestos Cement Type B Vents or Single-Wall Metal Vents Serving Two or More Appliances. (See Tables 1-D4 and 1-D5.)

Figure 1-D4
Masonry Chimney Serving Two or More Appliances. (See Table 1-D6.)

Notes for Multiple Appliance Vents. (*See Tables 1-D4, 1-D5, and 1-D6.*)

1. For single-wall metal pipe connectors, use Table 1-D5.

2. Maximum Vent Connector Length: 1½ feet for every inch of connector diameter. Greater lengths require increase in size, rise or total vent height, to obtain full capacity.

3. Each 90-degree turn in excess of the first two reduces the connector capacity by 10 percent.

4. Each 90-degree turn in the common vent reduces capacity by 10 percent.

5. Where possible, locate vent closer to or directly over smaller appliance connector.

6. Connectors must be equal to or larger than draft hood outlets.

7. If both connectors are same size, common vent must be at least one size larger, regardless of tabulated capacity.

8. Common vent must be equal to or larger than largest connector.

9. Interconnection fittings must be same size as common vent.

10. Use sea level input rating when calculating vent size for high altitude installation.

11. Designation "NR" in Tables 1-D4, 1-D5 and 1-D6 indicates not recommended.

NATIONAL FUEL GAS CODE
(Courtesy, American Gas Association).

Table 1–D4
Capacity of Type B Double-Wall Vents with Type B
Double-Wall Connectors Serving Two or More Appliances

Vent Connector Capacity

Total Vent Height "H"	Connector Rise "R"	Vent Connector Diameter — D													
		3"	4"	5"	6"	7"	8"	10"	12"	14"	16"	18"	20"	22"	24"
		Maximum Appliance Input Rating in Thousands of Btu Per Hour													
6'	1'	26	46	72	104	142	185	289	416	577	755	955	1180	1425	1700
	2'	31	55	86	124	168	220	345	496	653	853	1080	1335	1610	1920
	3'	35	62	96	139	189	248	386	556	740	967	1225	1510	1830	2180
8'	1'	27	48	76	109	148	194	303	439	601	805	1015	1255	1520	1810
	2'	32	57	90	129	175	230	358	516	696	910	1150	1420	1720	2050
	3'	36	64	101	145	198	258	402	580	790	1030	1305	1610	1950	2320
10'	1'	28	50	78	113	154	200	314	452	642	840	1060	1310	1585	1890
	2'	33	59	93	134	182	238	372	536	730	955	1205	1490	1800	2150
	3'	37	67	104	150	205	268	417	600	827	1080	1370	1690	2040	2430
15'	1'	30	53	83	120	163	214	333	480	697	910	1150	1420	1720	2050
	2'	35	63	99	142	193	253	394	568	790	1030	1305	1610	1950	2320
	3'	40	71	111	160	218	286	444	640	898	1175	1485	1835	2220	2640
20'	1'	31	56	87	125	171	224	347	500	740	965	1225	1510	1830	2190
	2'	37	66	104	149	202	265	414	596	840	1095	1385	1710	2070	2470
	3'	42	74	116	168	228	300	466	672	952	1245	1575	1945	2350	2800
30'	1'	33	59	93	134	182	238	372	536	805	1050	1330	1645	1990	2370
	2'	39	70	110	158	215	282	439	632	910	1190	1500	1855	2240	2670
	3'	44	79	124	178	242	317	494	712	1035	1350	1710	2110	2550	3040
40'	1'	35	62	97	140	190	248	389	560	850	1110	1405	1735	2100	2500
	2'	41	73	115	166	225	295	461	665	964	1260	1590	1965	2380	2830
	3'	46	83	129	187	253	331	520	748	1100	1435	1820	2240	2710	3230
60' to 100'	1'	37	66	104	150	204	266	417	600	926	1210	1530	1890	2280	2720
	2'	44	79	123	178	242	316	494	712	1050	1370	1740	2150	2590	3090
	3'	50	89	138	200	272	355	555	800	1198	1565	1980	2450	2960	3520

NATIONAL FUEL GAS CODE
(Courtesy, American Gas Association).

Table 1–D4 (Cont'd.)

Table 1-D4 (Continued)

Common Vent Capacity

Total Vent Height "H"	Common Vent Diameter													
	3"	4"	5"	6"	7"	8"	10"	12"	14"	16"	18"	20"	22"	24"
	Combined Appliance Input Rating in Thousands of Btu Per Hour													
6'	—	65	103	147	200	260	410	588	815	1065	1345	1660	1970	2390
8'	—	73	114	163	223	290	465	652	912	1190	1510	1860	2200	2680
10'	—	79	124	178	242	315	495	712	995	1300	1645	2030	2400	2920
15'	—	91	144	206	280	365	565	825	1158	1510	1910	2360	2790	3400
20'	—	102	160	229	310	405	640	916	1290	1690	2140	2640	3120	3800
30'	—	118	185	266	360	470	740	1025	1525	1990	2520	3110	3680	4480
40'	—	131	203	295	405	525	820	1180	1715	2240	2830	3500	4150	5050
60'	—	NR	224	324	440	575	900	1380	2010	2620	3320	4100	4850	5900
80'	—	NR	NR	344	468	610	955	1540	2250	2930	3710	4590	5420	6600
100'	—	NR	NR	NR	479	625	975	1670	2450	3200	4050	5000	5920	7200

See Figure 1-D3 and Notes for Multiple Appliance Vents.

(Courtesy, American Gas Association).

Table 1-D5

Capacity of A Single-Wall Metal Pipe or Type B Asbestos Cement Vent Serving Two or More Appliances

Vent Connector Capacity

Total Vent Height "H"	Connector Rise "R"	Vent Connector Diameter — D: 3"	4"	5"	6"	7"	8"
		Maximum Appliance Input Rating in Thousands of Btu Per Hour					
6'-8'	1'	21	40	68	102	146	205
	2'	28	53	86	124	178	235
	3'	34	61	98	147	204	275
15'	1'	23	44	77	117	179	240
	2'	30	56	92	134	194	265
	3'	35	64	102	155	216	298
30' and up	1'	25	49	84	129	190	270
	2'	31	58	97	145	211	295
	3'	36	68	107	164	232	321

Common Vent Capacity

Total Vent Height "H"	Common Vent Diameter: 4"	5"	6"	7"	8"	10"	12"
	Combined Appliance Input Rating in Thousands of Btu Per Hour						
6'	48	78	111	155	205	320	NR
8'	55	89	128	175	234	365	505
10'	59	95	136	190	250	395	560
15'	71	115	168	228	305	480	690
20'	80	129	186	260	340	550	790
30'	NR	147	215	300	400	650	940
50'	NR	NR	NR	360	490	810	1190

See Figure 1-D3 and Notes for Multiple Appliance Vents.

Table 1-D6

Capacity of A Masonry Chimney and Single-Wall Vent Connectors Serving Two or More Appliances

Single-Wall Vent Connector Capacity

Total Vent Height "H"	Rise Connector "R"	Vent Connector Diameter — D					
		3"	4"	5"	6"	7"	8"
		Maximum Appliance Input Rating in Thousands of Btu Per Hour					
6'-8'	1'	21	39	66	100	140	200
	2'	28	52	84	123	172	231
	3'	34	61	97	142	202	269
15'	1'	23	43	73	112	171	225
	2'	30	54	88	132	189	256
	3'	34	63	101	151	213	289
30' and up	1'	24	47	80	124	183	250
	2'	31	57	93	142	205	282
	3'	35	65	105	160	229	312

Common Chimney Capacity

Total Vent Height "H"	Minimum Internal Area of Chimney — "A" Square Inches					
	19	28	38	50	78	113
	Combined Appliance Input Rating in Thousands of Btu Per Hour					
6'	45	71	102	142	245	NR
8'	52	81	118	162	277	405
10'	56	89	129	175	300	450
15'	66	105	150	210	360	540
20'	74	120	170	240	415	640
30'	NR	135	195	275	490	740
50'	NR	NR	NR	325	600	910

See Table 1-D7 for Masonry Chimney Liner Sizes.

See Figure 1-D4 and Notes for Multiple Appliance Vents.

Example of Multiple Vent Design Using Table 1-D4
Double Wall Type B Vent

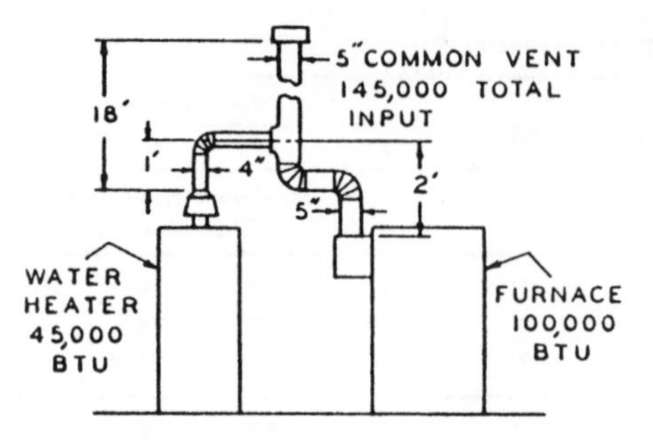

Figure 1-D5 Example: Connect a 45,000 Btu water heater with a 1 foot connector rise "R" and a 100,000 Btu furnace with a 2 foot connector rise "R" to a common vent with a minimum total vent height "H" of 18 feet.

1. WATER HEATER VENT CONNECTOR SIZE. Using Table 1-D4, read down Total Vent Height "H" column to 15 feet and read across 1 foot connector rise "R" line to Btu rating equal to or higher than water heater input rating. This figure shows 53,000 Btu and is in the column for 4-inch connector. Since this is in excess of the water heater input it is not necessary to find the maximum input for an 18 foot minimum total vent height. Use a 4-inch connector.

2. FURNACE VENT CONNECTOR SIZE. Under Vent Connector Tables read down Total Vent Height "H" column to 15 foot and read across 2 foot Connector Rise "R" line. Note 5-inch vent size shows 99,000 Btu per hour or less than furnace input. However, with 20 foot Total Height read across 2-foot connector rise line. Note 5-inch vent size shows 104,000 Btu per hour. Since 18-foot height is 3/5th of difference between 15- and 20-foot heights, take difference between 99,000 and 104,000 or 5,000 and add 3/5 of this to 15 foot figure of 99,000, 99,000 + 3,000 = 102,000 which is maximum input for 18-foot Total Vent Height. Therefore a 5-inch connector would be the correct size for the furnace, providing the furnace had a 5-inch or smaller draft hood outlet.

3. COMMON VENT SIZE. Total input to Common Vent is 145,000 Btu. Note that for 15-foot Total Vent Height "H" maximum Btu for 5-inch vent is 144,000. For 20-foot Total Vent Height "H" maximum Btu for 5-inch vent is 160,000.

Therefore for 18-foot Total Vent Height maximum allowable input would be 3/5 of difference between 144,000 and 160,000 = 3/5 x 16,000 or 9,600; 144,000 +9,600 = 153,600 which is greater than total input to common vent. Therefore common vent can be 5-inch-diameter pipe.

NATIONAL FUEL GAS CODE
(Courtesy, American Gas Association).

Table 1-D7

Masonry Chimney Liner Dimensions with Circular Equivalents

Nominal Liner Size Inches	Inside Dimensions of Liner in Inches	Inside Diameter or Equivalent Diameter Inches	Equivalent Area Square Inches
4 x 8	$2\frac{1}{2} \times 6\frac{1}{2}$	4	12.2
		5	19.6
		6	28.3
		7	38.3
8 x 8	$6\frac{3}{4} \times 6\frac{3}{4}$	7.4	42.7
		8	50.3
8 x 12	$6\frac{1}{2} \times 10\frac{1}{2}$	9	63.6
		10	78.5
12 x 12	$9\frac{3}{4} \times 9\frac{3}{4}$	10.4	83.3
		11	95
12 x 16	$9\frac{1}{2} \times 13\frac{1}{2}$	11.8	107.5
		12	113.0
		14	153.9
16 x 16	$13\frac{1}{4} \times 13\frac{1}{4}$	14.5	162.9
		15	176.7
16 x 20	13 x 17	16.2	206.1
		18	254.4
20 x 20	$16\frac{3}{4} \times 16\frac{3}{4}$	18.2	260.2
		20	314.1
20 x 24	$16\frac{1}{2} \times 20\frac{1}{2}$	20.1	314.2
		22	380.1
24 x 24	$20\frac{1}{4} \times 20\frac{1}{4}$	22.1	380.1
		24	452.3
24 x 28	$20\frac{1}{4} \times 24\frac{1}{4}$	24.1	456.2
28 x 28	$24\frac{1}{4} \times 24\frac{1}{4}$	26.4	543.3
		27	572.5
30 x 30	$25\frac{1}{2} \times 25\frac{1}{2}$	27.9	607
		30	706.8
30 x 36	$25\frac{1}{2} \times 31\frac{1}{2}$	30.9	749.9
		33	855.3
36 x 36	$31\frac{1}{2} \times 31\frac{1}{2}$	34.4	929.4
		36	1017.9

When liner sizes differ dimensionally from those shown in Table 1-D7, equivalent diameters may be determined from published tables for square and rectangular ducts of equivalent carrying capacity or by other engineering methods.

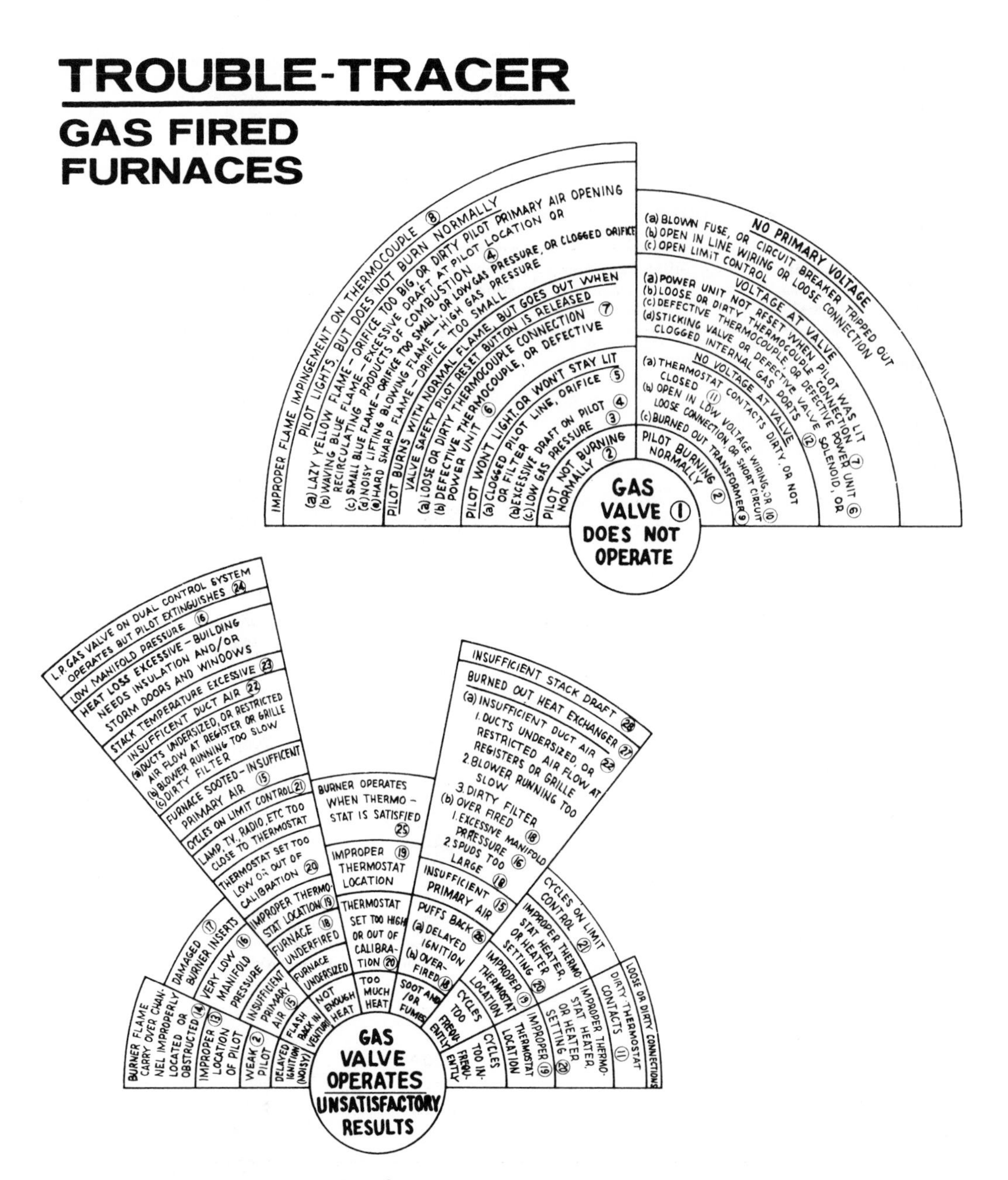

INSTRUCTIONS

Center of chart states basic trouble. Begin there.

Adjacent partial circles state specific troubles that could cause basic trouble. Determine which of these specific troubles exists before proceeding further.

(Courtesy, SJC Corp. [formerly Tappan Air Conditioning]).

Remaining partial circles state various contributing troubles to specific trouble. Determine contributing trouble or troubles before attempting correction.

Numbers in chart are for explanation and guidance. Where no numbers are shown, the statement is considered to be self-explanatory.

(1) There are many types of automatic gas valves used on gas fired residential and small commercial furnaces, suspended unit heaters, and duct heaters. Some have built-in gas pressure regulators, and automatic safety pilot mechanism; while others have external gas pressure regulators, and automatic safety pilot mechanism. Some valves provide for 100% pilot gas shut off when the pilot flame is extinguished, while others provide for only burner gas shut off when the pilot flame is extinguished.

The function of the gas pressure regulator is to regulate line gas pressure to a suitable pressure in the gas burner manifold.

The function of the automatic safety pilot mechanism is to provide gas shut off in case the pilot is extinguished.

The most commonly used automatic safety pilot mechanism is the thermoelectric power unit type. A constant burning pilot performs the dual function of igniting the burner gas, and supplying heat to the "hot junction" (tip) of a thermocouple. The thermocouple generates electricity (millivolts) which energizes a power unit. The mechanism also contains a reset button. The reset button must be depressed on 100% pilot shut-off valves (valves in which the gas to the pilot is shut off if the pilot flame is extinguished), in order to get gas at the pilot, and must be held in after the pilot lights approximately three minutes to allow sufficient time for the thermocouple to heat to operating temperature.

Caution: For safety when ever the pilot is not burning shut off the main gas supply and make certain combustion chamber is free of unburned gas before lighting pilot.

(2) Pilot Not Burning Normally: Pilot should burn with a steady blue flame, and envelop 3/8 to 1/2 inch of the thermocouple tip. The pilot flame or the main burner flame must not impinge on the thermocouple "cold junction" (base).

(3) Low Gas Pressure: Check with gas company, or check liquid in L.P. tank.

(4) Excessive Draft on Pilot, or Waving Blue Flame: Excessive draft conditions are often responsible for improper pilot operation by floating the flame on the thermocouple causing the flame to heat both the "hot junction" and "cold junction". If the draft condition cannot be corrected, it will be necessary to shield or baffle the pilot flame.

(5) Clogged Pilot Line, Orifice, or Filter: Clean all openings and lines through which pilot gas passes.

(6) Defective Thermocouple, or Defective Power Unit: Check thermocouple with millivoltmeter, or use a test thermocouple known to be good. Thermocouples should generate not less than 16 millivolts with normal pilot flame on "open circuit", that is, when the thermocouple is disconnected from the power unit. However, the exact millivolts of the thermocouple should conform to the recommendation of the manufacturer of the thermocouple.

Checking with Millivoltmeter (open circuit)

a. Make sure that the pilot flame is normal.
b. Disconnect the thermocouple from the power unit.
c. Attach one lead of the meter to the end of the thermocouple that makes contact with the power unit. Attach the other lead of the meter to the thermocouple immediately beyond the insulation.

Checking with Test Thermocouple:

a. Remove the old thermocouple.
b. Install test thermocouple.
c. Depress the reset button and hold it in about three minutes (In 100% pilot shut-off installations, the reset button must be depressed while the pilot is being lit and held in until the the thermocouple heats to operating temperature).
d. If the power unit holds in when the reset button is released, gas should flow through the valve.
e. If the power unit holds in with the test thermocouple but fails to hold in with the old thermocouple, the old thermocouple must be replaced, as no field service can be performed on a defective thermocouple.
f. Make certain there is a clean, tight connection of the thermocouple with the power unit. (Don't tighten more than 1/4 turn past finger tight).

(7) Loose, or Dirty Thermocouple Connection: Disconnect thermocouple from the power unit and wipe cone and socket with a lint free cloth or chamois. Replace thermocouple and tighten 1/4 turn beyond finger tight.

(8) Improper Flame Impingement on Thermocouple: See (2) and (4)

(9) Burned Out Transformer: Check primary to transformer with test lamp or voltmeter. If no light or voltage, check power source and check for improper connections, broken wiring, or loose connections.

(10) Open in Low Voltage Wiring, or Loose Connection, or Short Circuit: Remove wires from low side of transformer. Check low side with 24 volt test lamp, or voltmeter. If there is voltage, replace wires and check at valve connection. If no voltage at valve connection, there is an open or short in the wiring.

(11) Thermostat Contacts Dirty or Not Closed: Make sure thermostat contacts close. Clean contact with business card or other unprinted hard surface paper.

(12) Sticking Valve or Defective Valve Solenoid, or Clogged Internal Gas Ports:

a. Before concluding that this condition exists, check for power at the valve with 24 volt test lamp, or voltmeter, and make sure the pilot flame is normal, and the power unit of the safety pilot mechanism is holding in when the reset button is released.
b. On 100% shut off pilot valves, the pilot flame will extinguish when the reset button is released if the power unit is not holding in.
c. On Non 100% shut off valves, the pilot flame will continue to burn when the reset button is released. To check if the power unit is holding in, back off on the nut that connects the thermocouple to the power unit. If the power unit was holding in, there will be a "click" when the nut is backed off.
d. If the power unit is holding in, and there is power to the valve, and the valve does not open, it is either sticking or clogged, or the solenoid is defective. To check if solenoid is defective: (a) Make sure there is 24 volt power at the valve. (b) Remove one wire at the valve terminal and then touch it to the terminal, and listen for a "click". If no "click", the

(Courtesy, SJC Corp. [formerly Tappan Air Conditioning]).

solenoid is defective. Replace the entire valve or parts as required by the manufacturer of the valve.

e. If the power unit is holding in, and the solenoid "clicks", and the valve does not open, the valve is sticking or its internal ports are clogged. Cut off the line gas valve and the pilot valve, and disconnect wiring to valve. Disassemble valve and carefully clean it, or replace it. If you disassemble valve; make sure new gaskets are used when reassembling it.

(13) Improper Location of Pilot: The pilot must be located so as to immediately light the burner, or burners adjacent to it, and also provide flame to the tip of the thermocouple. The burner flame carry over channel provides ignition to the burners that are not adjacent to the pilot.

(14) Burner Flame Carry Over Channel Improperly Located or Obstructed: Since the burner flame carry over channel provides ignition to the burner, or burners not adjacent to the pilot, it must be in close proximity to them, and must be clean, otherwise delayed ignition of the outer burners will occur.

(15) Insufficient Primary Air: The proper burner flame is one in which no yellow flame appears. To obtain this, first close primary air shutter until you get a yellow flame, then re-open until all traces of the yellow flame have disappeared. After adjustment, tighten screw to lock shutter in position. The burner flame should be checked at least at the beginning of each heating season to determine if it is burning properly. Any dust, or other foreign matter, lodged in the primary air shutter will cause a yellow flame, thus sooting and fumes.

(16) Manifold Pressure: Manifold pressures for all gases except L.P. gas should be 3" to 3-1/2" W.C. (Water Column); for L.P. approximately 11" W.C. Always check manifold pressure with a manometer and adjust gas pressure regulator to provide correct pressure. A removable 1/8" pipe plug is located on the center of the manifold for ease of attaching a manometer. Some utilities require that appliances equipped with pressure regulators be adjusted for manifold pressure of 2-1/2" on Manufactured Gas, and 3-1/2" on Natural Gas, plus or minus 0.3" W.C. Very low manifold pressure may cause flash back in the burner venturi, underfiring, and probably insufficient heat.

Excessive manifold pressure will cause overfiring, and may result in a burned out heat exchanger.

(17) Damaged Burner Inserts: If the burner inserts are damaged they may not retain the flame on top of the burner, thus the flame may flash back and burn in the burner venturi. Replace damaged inserts.

(18) Furnace Firing: The B.T.U. input to the burners must never exceed that shown on the rating plate. The burners are equipped with fixed orifices intended to give the correct input with a manifold pressure as stated in (16) To check input at the meter for gases other than L.P. "clock" the meter. That is, check the number of seconds required for 1 cu. ft. of gas to pass through the meter, then determine the input by the following formula:

$$\text{BTU input} = \frac{3600 \times \text{Gas Heating Valve (BTU/cu. ft.)}}{\text{No. seconds per cu. ft. of gas}}$$

If you do not know the B.T.U. value of the gas in your area, check with the gas company. Since meters are not generally provided on L.P. installations, accurate B.T.U. input cannot be "clocked". However, with reference to charts (obtain from L.P. supplier) on orifice size, manifold pressure, and B.T.U. heating valve per cu. ft., the input may be accurately determined. Overfiring is due to spuds too large or excessive manifold pressure (See (16)). Underfiring is due to spuds too small, or manifold pressure too low (See (16)). Spuds should be sized in accordance with the recommendations for the manufacturer of the furnace.

(19) Thermostat Location: The thermostat should normally be located in the living room (or dining room) about five feet from the floor, on an inside wall between wall studs, in an area of natural air circulation. It should not be located behind door, furniture or drapes, in cold drafts from an open door, next to concealed pipes or ducts, in direct rays of sun or fireplace, near discharge air outlets, lamp, radio, TV sets, etc. The wall hole behind the thermostat should be sealed with putty or similar substance to prevent wall drafts from affecting its operation.

(20) Thermostat Setting, Heater, and Calibration: Thermostat is generally calibrated to maintain temperature within one degree of its setting in the area where it is located. If the thermostat does not maintain close temperature control, it may be out of calibration, and should be recalibrated in accordance with recommendations of the manufacturer of the thermostat; or the heater may be improper, or improperly adjusted. The amp rating of the heater should be the same as the amp rating of the gas valve.

(21) Cycles on Limit Control: The limit control is a safety switch operating by temperature. Its purpose is to prevent over heating. Limit setting depends on model of furnace and limit control used. No attempt should be made to increase manufacturer's setting. If the furnace is cycling on limit control see (16), (18), (22). It is highly important that corrections be made to prevent cycling on limit control to prevent possible burn out of the heat exchanger.

(22) Insufficient Duct Air: This is generally the most common cause of unsatisfactory operation of a ducted heating system. It can cause (1) not enough heat (2) cycling on limit control (3) burned out heat exchangers, (4) excessive temperature rise through furnace (it must not exceed 105°). It is caused by (a) ducts undersized, or restricted air flow at register or grille. (b) Blower running too slow. If the blower (belt driven type) is running too slow the belt may be slipping, or the motor variable speed pulley is too many turns open. Too increase blower speed, loosen Allen set screw on outer side of variable speed pulley and screw the pulley together. Whenever the speed of the blower is increased, the motor current draw must be checked to determine if it is excessive. The current draw should never exceed 115% of the amp rating on the motor name plate, or 105% of the S.F.A. (Service Factor Amps), which ever is less. Make sure belt tensión is carefully adjusted so that when the belt is gripped by hand there is approximately a one-inch total deflection when "finger" pressure is applied. If the belt is too tight, the motor will draw excessive current, and the blower bearings will burn out. (c) Dirty filters. They are a very common cause of insufficient duct air. Filters should be replaced or cleaned at least once a year. If the furnace is used with summer air conditioning, this should be done at least twice a year. (d) If

the furnace is used with air conditioning, the evaporator (cooling coil) must be clean. Filter must be installed and clean.

(23) Stack Temperature Excessive: Stack temperature (down stream from back draft diverter) should not exceed 480° plus room temperature. Stack temperature in excess of this temperature may be caused by (a) insufficient duct air (See (22)). (b) Over firing (See (18).) (c) Misplaced flue gas baffles.

(24) L.P. Gas Valve on Dual Control System Operates but Pilot Extinguishes: The pilot on dual control system remains on natural gas when the burner operation changes to L.P. gas. If the pilot should extinguish when the L.P. gas valve opens, it is most probably due to improper L.P. gas manifold pressure. L.P. gas manifold pressure should be approximately 11" W.C. (See (16)). If the L.P. Gas Manifold pressure is too high or too low it may cause delayed ignition, thus an excessive "puff" which may extinguish the pilot. If the pilot does not become relit, the pilot thermocouple will not have sufficient heat to generate sufficient electricity to hold the safety switch closed, thus, the safety pilot switch will open the electrical circuit to the L.P. gas valve and the valve will close.

(25) Burner Operates When Thermostat is Satisfied (Contacts Open): This condition is due to (a) short in wiring to thermostat. To check this, remove one wire from the valve. If the valve closes, there is a short in the wires to the thermostat. (b) Valve sticking open. To check this, remove one wire from the valve. If the valve does not close, it is sticking open. Replace it.

(26) Puffs Back (Flame puffing Out into Vestibule): Puffs back may be caused by (a) Delayed ignition (See (2), (13), (14)). (b) Over firing (See (18)). (c) Sooting (See (15)). (d) Burned out heat exchanger (See (27)).

(27) Burned Out Heat Exchanger: To check for burned out heat exchanger (a) Operate the burners (b) Start the blower, and watch closely to see if the burner flame pattern changes when the blower starts. If the flame waves when the blower starts, it is an indication of a burned out heat exchanger. For causes of heat exchanger burn out (See (16), (18), (22)).

(28) Insufficient Stack Draft: Although the stack draft on a properly designed gas furnace does not effect the burner operation, the draft must be sufficient to vent the products of combustion for safety reasons, in case of improper combustion due to abnormal conditions such as over firing, burning with yellow flame, etc. and prevent "spillage" (products of combustion spilling into the furnace area rather than venting up the stack). To check for spillage, hold a match, or lighted cigarette at the edge of the back draft diverter. If the match goes out, or the smoke does not vent up the stack, the draft is insufficient.

(29). When ordering repair parts, please furnish to the distributor the model and series and serial number of the furnace. Also if you have a repair parts catalog, furnish the part number and description of the part. If you cannot furnish the part number and description of the part, please advise if the part is for a natural gas or L.P. gas unit.

FLUE GAS DAMPER: SERVICE ANALYSIS CHART

Symptom	*Check Procedure*	*Remedy*
Damper will not open	1. Remove cover from damper operator. With power to furnace "on" and thermostat calling for heat, check voltage (24 V nominal) across terminals 1 and 4.	1a. If voltage is present, go to step 2. 1b. If no voltage is seen: (b1) the white lead is open or loose at its terminations. (b2) the problem is not in the damper.
	2. Remove white lead from terminal 4 and blue lead from terminal 1. Check continuity between terminals 1 and 4.	2a. If there is continuity with some resistance, go to step 3. 2b. If there is no continuity, replace the motor operator assembly.
	3. Is return spring broken?	3. Replace spring.
	4. Is damper housing or plate distorted?	4. Repair or replace housing assembly.
Damper will not close	1. Is 115-V power connected to furnace?	1. Damper will remain open if there is no power supply.
	2. Is thermostat calling for heat?	2. Thermostat must be "off" for damper to close.
	3. Is damper connected to printed circuit board in furnace electrical box?	3. Make sure the edge connector is positioned so that the terminals contact the printed circuit.
	4. Remove cover from damper operator. Check voltage (24 V nominal) across terminals 1 and 2.	4a. If voltage is present: (a1) but motor does not operate, replace the motor operator assembly. (a2) and the motor operates, the coupling or cotter pin retainer for the damper shaft is broken. Replace as necessary. (a3) and the motor stalls, the damper shaft is binding. Repair or replace the housing assembly. 4b. If no voltage is seen: (b1) there is an open circuit in the blue or red lead leading to terminals 1 and 2. (b2) the problem is not in the damper.

Symptom	*Check Procedure*	*Remedy*
Damper closes and opens, but gas valve or ignition system does not operate.	1. Remove cover from damper operator. With power to furnace "on," thermostat calling for heat, and damper in open position, check voltage (24 V nominal) across terminals 1 and 3 on damper terminal board.	1a. If voltage is seen, problem is not in the damper. 1b. If no voltage is seen: (b1) the black lead in the edge connector is loose. (b2) replace the motor operator assembly.

PRESEASON PREVENTATIVE MAINTENANCE FOR GAS-FIRED HEATING SYSTEMS

(Courtesy, Luxaire, Inc.).

1. **Thermostat:** (Check while hands and shoes are clean!)
   (a) Mercury bulb type
      1. Check level of mounting base.
      2. Blow dust from sensing bimetal.
      3. Check heat anticipator setting (if adjustable) to match the amp draw of gas valve.

   (b) Open-contact (snap action) type
      1. Blow dust from working parts.
      2. Clean oxidation and dust from contacts. (warning!) Do not clean silver contacts with an abrasive, use a calling card or paper currency.
      3. Check heat anticipator setting (if adjustable) to match the amp draw of gas valve.

2. Before you start for the basement or utility area, check the registers and grilles to be sure they have not been closed off at the valves or covered by furniture, rugs, or other obstructions to air flow.

3. **Power:**
   (a) Check fuses in 115-V power supply to the unit, and for power in the unit at the blower control with Tattle-lite or Amprobe.

   (b) Check across the transformer secondary terminals with a low-voltage Tattle-lite or voltmeter. (warning!) Do not short across these terminals with a wire or screwdriver. This will shorten the life of, or burn up, the secondary windings of the transformer. Some transformers are internally fused and shorting across the terminals would blow this fuse ruining the transformer.

4. **Filters:** (Be sure the homeowner can find them in unit.)

   (a) Throwaway type should be replaced at the start of each season and replaced after two cleanings.

   (b) Washable permanent type—Remove, wash, drain, and spray with filter oil.

   (c) Plastic permanent type—Wash, drain, and reinstall.

Note: Standard filters rated at 300 fpm per square foot of filter area should not be confused with high velocity filters rated at 600 fpm per square foot of filter area. High velocity filters are most often used on heating-cooling systems. Be sure to replace with the same type.

5. **Blower Assembly:**

   (a) Check squirrel cage for dust buildup on vanes that would cause cage to handle less air; clean if necessary.

   (b) If belt-drive blower, adjust shaft retainer collar on shaft to give about 3/8 in. of end play between collar and bearing for pulley alignment.

   (c) Inspect bearings for wear and oil with a few drops of SAE No. 30 motor oil. This does not apply to "packed" bearings.

   (d) Inspect belt for cracks indicating "breakdown" and replace if any are found.

   (e) Clean fuzz and lint from air openings in motor housing and oil each bearing with 3 or 4 drops of SAE No. 30 motor oil.

   (f) Replace and adjust blower "V" belt. One inch total depression of belt on pulleys is proper tension.

   (g) Put fan control in "summer" position so blower will run and with blower compartment closed, check the running amp draw of motor with an Amprobe to check for motor overload. Correct amps are on the rating plate of motor.

   (h) Inspect vent pipe for soundness.

6. **Pilot and Thermocouple Lead:**

   (a) Remove the pilot tubing and thermocouple lead from the gas valve.

1. Remove burner to which the pilot is attached.
2. Remove pilot assembly from the bracket and blow or brush lint and dirt from same.
3. Clean the thermocouple head and contact (opposite end) with fine emery cloth or steel wool and replace in pilot frame.

7. **Main Gas Burners:**

(a) Remove from unit and clear of lint or dust by blowing a high velocity stream of air or water down through the stainless steel top of the burners and flame runners. Replace burners in unit.

(b) Attach pilot burner to mounting bracket and gas valve connections.

(c) Check source and remedy—unvented driers or workshop dust.

8. Light pilot and adjust flame to a soft fire enveloping the upper $5/8$ in. flame extending from pilot hood 1-$1/4$ in. The needle valve controlling the pilot gas is located in the swivel head on the gas valve.

9. With gas valve still in pilot position, attach manifold pressure gauge to the main burner manifold, turn gas valve to the full "on" position, also check to see that any manual shut off valve in gas line is "full" open, turn power on to energize gas valve and adjust regulator to maintain proper manifold pressure. L.P. gases should have two stage regulation.

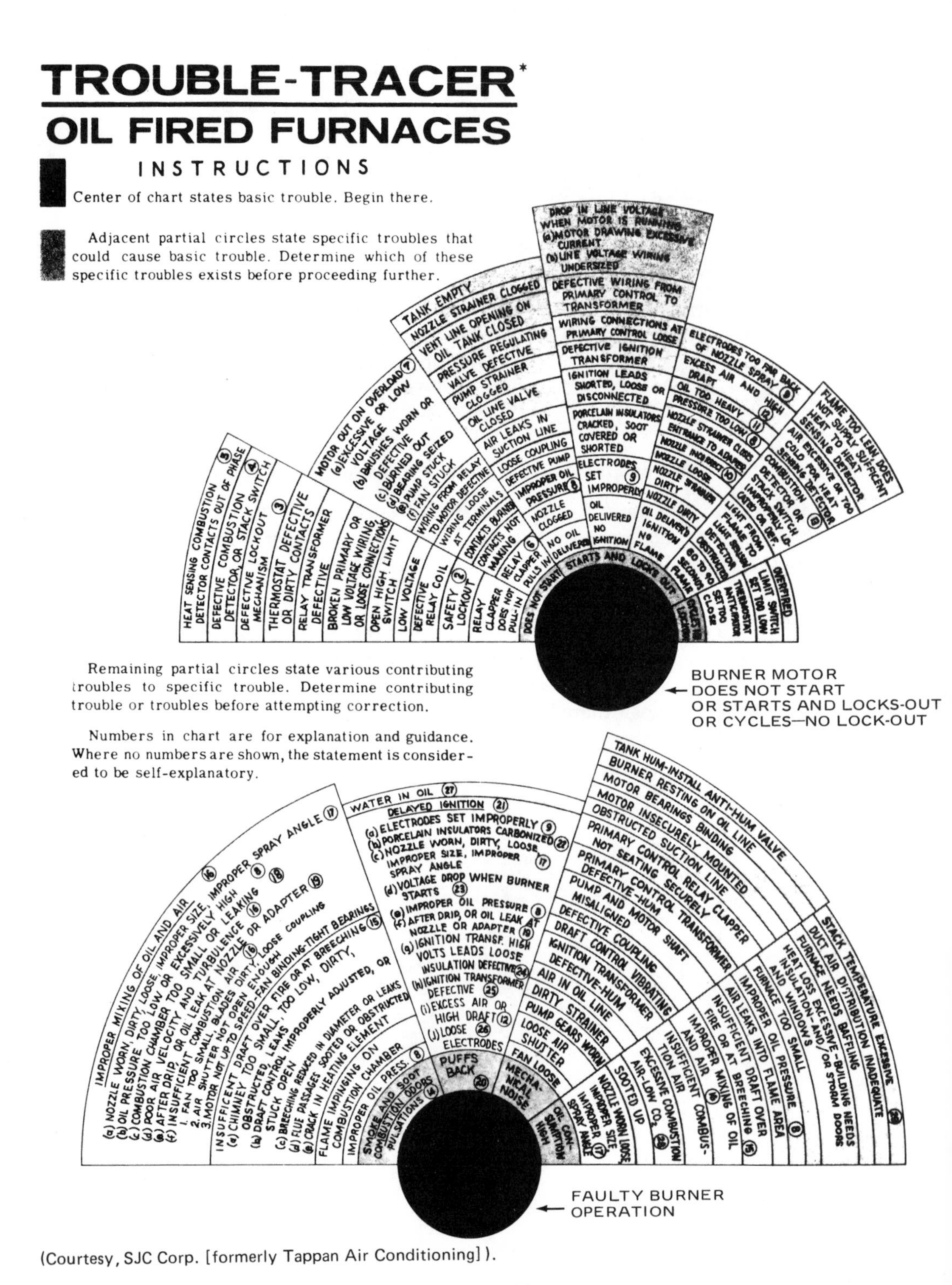

(Courtesy, SJC Corp. [formerly Tappan Air Conditioning]).

Oil Heating Terminology

A. One-Stage Fuel Unit: Unit with one set of gears (pressure gears). This type is generally used where the fuel supply tank is above the burner, so that gravity provides a constant flow of oil to the fuel-unit. It may also be used with a two-pipe system if the tank is not more than 10' below the burner.

B. Two-Stage Fuel Unit: Unit with two sets of gears, a pressure set same as in one-stage unit, and a vacuum set to create a vacuum within the fuel-unit which will draw oil from a submerged tank, or a tank located below the oil burner, and discharge it into an intermediate reservoir which supplies the pressure gears.

C. One-Pipe System: One line from tank to fuel-unit.

D. Two-Pipe System: Two lines from tank to fuel-unit. This type should always be used whenever the oil supply level is below the fuel-unit, whether it be a one-stage or two-stage unit. A two-pipe system eliminates necessity of bleeding air from the lines as the air is then returned to the tank.

E. By-Pass Plug: 1/8" pipe plug used to close internal port and convert unit from one-pipe to two-pipe system.

F. Bleed: To remove air from the system.

G. Lift: Oil under vacuum.

H. Head of Oil: Column of oil above fuel-unit.

I. Inlet Port: Port which receives oil from tank.

J. By-Pass Port: Port which discharges surplus oil back to tank in two-pipe system.

K. Return Line Port: Same as By-Pass Port.

L. Valve Differential: Number of pounds per square inch pump pressure must drop to close nozzle cut-off valve.

M. Delivery: Gallons Per Hour (GPH) pumped from the nozzle outlet assembly of the fuel-unit.

N. Primary Control: Primary controls provide for the safe start and operating procedure for the burner. They consist of a transformer, a relay, a lockout mechanism with manual reset button; and terminals for wiring to line, burner motor, ignition transformer, thermostat, and safety lockout initiating means (Heat Sensing Combustion Detector, Light Sensing Combustion Detector). When Stack Switch is employed with Primary Control, the lockout mechanism is internally wired.

O. Hollow Cone Nozzles: Hollow cone nozzles spray oil in a cone pattern with only the edges of the cone pattern containing major portion of oil.

P. Solid Cone Nozzles: Solid cone nozzles spray oil in a cone pattern with the entire cone containing oil.

Test and Service Equipment

The service man should be equipped with the following:

1. Draft gauge (.0" to .12")
2. Pressure gauge (0 lbs to 200 lbs)
3. Vacuum gauge (0" to 30")
4. Smoke Tester
5. CO_2 Indicator
6. Box of nozzles
7. Nozzle changer
8. Ohmmeter
9. Voltmeter
10. Ammeter
11. Flame Inspection Mirror (3-7/8" x 1-15/16")
12. 6" Pocket Rule
13. Stack Thermometer (200° to 1200°)
14. Hand Tools.

(1) Caution: Whenever unsatisfactory operation of an oil burner has occurred, and the burner motor is not operating, be sure to open burner power switch and check the furnace to see if it is not full of oil vapor by opening the furnace inspection door. If excessive oil vapor is present, do not attempt to start the unit until the vapor has been dispersed as there may be danger of ignition and serious puff-back. Also check the combustion chamber for liquid oil. If it is present, and excessive vapors have dispersed, ignite it with a burning newspaper and let it burn out with the burner not operating. Then check (a) high limit reset (if counterflow furnace) before closing burner power switch (b) check main fuse and power supply (c) check oil burner fuse (d) check oil tank and make sure it contains oil (e) check oil valves (f) close burner power switch (g) reset lockout (h) jump thermostat terminal.

(2) Safety Lockout: The purpose of safety lockout is to prevent burner operation in event of ignition or flame failure within a predetermined time (generally 60 to 90 seconds) after the burner motor starts; or also to cut the burner motor off if the flame should decrease appreciably or become extinguished while the burner is operating. There are several methods of initiating safety lockout. The more commonly used are; (a) Stack switch. It has a helix that is sensitive to temperature change. It is mounted in the stack near the furnace where the temperature rise is approximately 100° in one minute from normal start, and at least 18" ahead of the automatic draft adjuster. (b) Heat sensing combustion detector. It has a hermetically sealed switch that opens and closes its contacts in response to temperature changes produced by radiant heat of the flame. It must be mounted so the flame sensing "face" is in direct line with a stable part of the flame. (c) Light sensing combustion detector. It has a hermetically sealed light actuated cell that has low electrical resistance in the presence of light rays from the flame, permitting current flow; and high electrical resistance in the absence of light rays from the flame, prohibiting current flow. It is generally mounted in the blower end of the burner blast tube, and must be so located that the light sensing "face" is in direct line with a stable part of the flame. Whenever safety lockout occurs, the lockout mechanism must be reset manually before burner operation may be resumed.

(3) Defective Lockout Mechanism: Occasionally the lockout mechanism may become "fatigued" and will not hold when the reset button is pushed. To check for this, make sure there are no oil vapors in the furnace by opening the inspection door. If vapors are present, follow instructions in Item (1) If no vapors, press reset button and release it immediately. If the burner fails to start when the button is released, the lockout mechanism is defective. Remove the primary control and inspect the lockout mechanism.

(4) Defective Combustion Detector or Stack Switch: The purpose of the combustion detector or stack switch is to initiate safety lockout in event of ignition or flame failure within a predetermined time (generally 60 to 90 seconds) after the burner motor starts; or also to cut the burner motor off if the flame should decrease appreciably or become extinguished while the burner is operating. To check for defective combustion detector, momentarily jump the detector terminals on the primary control. If the burner starts, the detector, or wiring to it, is defective; except that in the case of the heat sensing detector its contacts

may be out of phase (See Item(5)). To check for defective stack switch, remove its cover and examine "cold" contacts. They should be closed. If they are not closed, the helix is not properly returning to "cold" position, thus it is defective.

(5) Heat sensing combustion detector contacts out of phase: These contacts may become out of phase (open) due to ambient temperature rise while burner is off (at least 20°), or due to removal of burner drawer assembly for servicing. To put contacts in phase, momentarily short across detector terminals of primary control to cause relay to pull in and initiate burner cycle.

(6) Relay Clapper Pulls in: Occasionally the relay clapper may pull in but the burner motor does not start due to relay contacts not making or burned. To check this, remove primary control and inspect contacts.

(7) Motor Out on Overload: Burner motors are equipped with manual overload reset. Push button to reset overload.

(8) Improper Oil Pressure: Two of the most valuable tools in serving any fuel-unit are the pressure gauge and the vacuum gauge. It is a well known fact that if these gauges were used consistently, 75% of the fuel-units sent in for repair would never need to be removed from the burner. The pressure gauge is used to determine whether the gears are pumping and building up sufficient pressure. Oil pressure should be 100 p.s.i. at the nozzle with the pump operating. When the pump cuts off the pressure should hold at approximately 80 p.s.i. If the pressure is not 100 p.s.i. turn adjusting screw clockwise to increase pressure, counterclockwise to reduce pressure. If the pressure drops back to zero p.s.i., it indicates the cut-off valve is leaking and should be cleaned or replaced. The vacuum gauge is used to determine vacuum, lift, air leaks or obstruction in suction line. If the tank is below the burner level and the oil passes through a line filter, the vacuum gauge must show a reading. If it does not, an air leak is present. If the tank is above the level of the burner, the vacuum gauge should read zero. If it shows a reading, it is an indication of obstruction in the line. The vacuum should never exceed 17". If it does, the light portion of the oil will separate from the heavy portion, resulting in a milky appearance of the oil, and lower pump pressure. This indicates lift too high, run too long, line too small, or obstruction in line.

(9) Electrodes set improperly: Follow unit manufacturer's specification. However, if it is not available, the electrode gap should generally be 1/8" to 5/32", 5/8" above center of nozzle, and 1/8" in front of nozzle. The electrode tips must not be closer than 1/4" to the nozzle or any other metal part, and the oil spray must not strike the electrode tips.

(10) Nozzle Incorrect: Nozzle size and spray pattern should conform to unit manufacturer's recommendations. If unavailable on name plate or otherwise, it is generally wise to select the smallest firing rate that will provide sufficient heat on the coldest day. If the unit is not undersized, the GPH rating may reasonably be determined by dividing the BTUH Bonnet rating by 112,000. The spray pattern should be hollow cone nozzle below 1 GPH, and may be hollow cone or solid cone from 1 GPH to 2 GPH. Use solid cone above 2 GPH. We use a solid cone on 1.25 GPH and above.

(11) Oil too heavy: Oil should not be heavier than #2 as designated by CS-12-48, or #3 as designated by CS-12-40.

(12) Excess air and high draft: Excess air and high draft may move the oil spray so quickly that it does not ignite. A dense white cloud of vapor on starting shows this to be occurring. Reduce the air shutter and/or the draft in this case so that normal ignition may occur. The draft over the fire should be .02" W.C. to .04" W.C., and .04" W.C. to .06" W.C. at the flue connection to the unit.

(13) Combustion detector or stack switch improperly located, sooted, or defective: For proper location see Item(2). Excessive soot or dirt will prevent the heat sensing detector from absorbing sufficient heat to open its contacts, and will prevent the light sensing detector from absorbing sufficient light to reduce resistance in the detector sufficiently to permit current flow, thus lockout occurs. Defective detector also causes lockout. To check for either of these conditions, in the case of the heat sensing detector, (a) disconnect one detector lead. (b) jump thermostat terminal. (c) reset lockout. If burner operates, the detector circuitry is shorted. If burner does not operate (d) momentarily jump detector terminals. If burner operates beyond 60 to 90 seconds, the detector is improperly located, sooted, or defective. In the case of the light sensing detector (a) disconnect one detector lead. (b) jump thermostat terminal. (c) reset lockout. If burner operates, detector circuitry is shorted. (d) reconnect lead (e) remove detector from holder. If burner operates, the detector is defective or its leads shorted. In the case of the Stack switch (a) jump thermostat terminals (b) reset lockout (c) close "cold" contacts manually. If burner operates, the stack switch is defective.

(14) Smoke and Soot — Combustion Odors — Pulsations: These are the three most common results of Faulty Burner Operation. The four essentials for eliminating faulty burner operation are (a) proper oil pressure (b) proper draft (c) proper mixing of oil and air (d) proper ignition.

(15) Insufficient Draft Over Fire or at Breeching: Drafts should be checked with an accurate draft gauge. Draft over the fire should be .02" W.C. to .04" W.C. Draft at the breeching between the flue connection to the unit and the automatic draft control should be .04" W.C. to .06" W.C. and it should be at least .02" W.C. higher than over the fire. If it is more than .02" W.C. at the breeching than over the fire, it is an indication that the unit flue passages are sooted, dirty or obstructed; or excessive combustion air. Leaks into fire box, or crack in heating element will cause low draft over the fire. To check for crack in heating element (a) check draft over fire without furnace blower operating (b) check draft over fire with blower operating. If the draft is less with blower operating than when not operating, it is an indication of a crack in the heating element.

(16) Improper mixing of oil and air: Proper mixing of oil and air is essential for proper burner operation. The delivery of air by the fan is not enough. The turbulence must be such that the oil and air will mix properly in the combustion chamber. Without this turbulence, the oil and air cannot mix properly and a smoking fire will result. The less turbulence you have, the more the air will escape the oil completely and the greater will be the excess amount that will have to be furnished to achieve the proper oil-air ratio. This will lower the CO_2 reading and operating efficiency. The CO_2 reading should generally not be lower than 7% or higher than 12%. A reading within this range should generally give a good smoke reading (#2). A higher smoke reading is an indication of insufficient combustion air or insufficient draft over the fire. A lower smoke reading is an indication of excessive combustion air or excessive draft over the fire. The turbulator and static disc must be clean and conform to the firing rate. They are stamped with the firing rate range for which they should be used.

(17) Nozzle worn, dirty, loose, improper size, improper

spray angle: Nozzles become worn after long operation. In general, they should be changed every two years or oftener depending on the number of hours of operation per heating season. If the orifice of a nozzle becomes dirty, do not attempt to clean it as cleaning may alter its size. Replace the nozzle. A loose noozle will permit oil leak into the combustion chamber. It will burn in an oxygen starved atmosphere and cause smoke and soot. It may sometimes be necessary to reduce nozzle size in order to eliminate smoke; or change the spray pattern and/or spray angle.

(18) Combustion Chamber too Small or Leaking: If the combustion chamber is too small, or the nozzle too large, impingement of the fire or oil on the sides of the chamber will occur. A leaking combustion chamber will upset the proper oil-air mix. Both conditions will cause soot.

(19) After drip, or oil leak at nozzle or adapter: An after drip is caused by the pump pressure regulating valve not cutting off properly when the pump stops. Proper pressure and proper cut-off must be maintained. These can be determined only by the use of gauges. See Item (8) for use of gauges. Oil after drip, or leak at nozzle or adapter permits oil leaking into the combustion chamber. It will burn in an oxygen starved atmosphere and cause smoke and soot.

(20) Puffs Back: Puff back results from the oil vapor not igniting instantaneously as it leaves the nozzle when the burner starts. Once the pressure regulating valve has opened, the oil issues in a fine, highly inflammable spray from the nozzle in considerable volume. The ignition, of course, should be on at this time, and the spark, but not the electrode, should be directly in the path of the spray. Any delay for any reason whatsoever in the igniting of this oil vapor results in a considerable volume of it collecting in the chamber and flue passages. This is highly dangerous. The greater the time delay in igniting the oil, the greater the amount of oil in the chamber and flue passages, thus, the greater puff-back. Delayed ignition, therefore, is the basic cause of puff-back.

(21) Delayed Ignition: Delayed ignition is the basic cause of puffs-back, and the longer the delay, the greater the puff.

(22) Porcelain Insulators Carbonized: The porcelain insulators that hold the electrodes in place and insulate them from the metal holder are porous. With dirty, sooty operation, they become coated with soft carbon and are changed into conductors. When this occurs, the ignition spark may occur back in the blast tube, and since it takes a few seconds for the oil vapors to work back into the blast tube, an accumulation of vapors will be in the chamber, thus puff-back occurs. If there is any indication that the insulators are not performing their job, replace them.

(23) Voltage Drop When Burner Starts: Occasionally the burner motor may draw excessive current but will not lockout on its overload. Check the motor current draw and compare it with its name plate rating. If the current draw is excessive, it may leave the ignition transformer with a weak spark, or practically none at all. To check on ignition spark (a) close oil line valve, (b) insert mirror in inspection door, (c) jump thermostat terminals, (d) inspect spark (ignition should lockout in 60 to 90 seconds), (e) disconnect leads to burner motor, (f) reset lockout, (g) recheck spark. If a good bluish white spark is not visible, transformer is weak or defective.

(24) Ignition Transformer High Voltage Leads Loose, Insulation Defective: Anything that causes reduction in ignition voltage will cause weak spark, or none at all, at igniters. To check for spark at igniters follow instructions in Item (23).

(25) Ignition Transformer Defective: To check it, follow instructions in Item (23). If the spark is too weak, or-none at all, replace the transformer.

(26) Loose Electrodes: The electrodes must be firmly clamped in their holder so that their setting cannot be changed by any vibration resulting from operation of the burner. For electrode setting see Item (9).

(27) Water in Oil: Any quantity of water in the oil will cause puff-back. The sudden passage of water through the nozzle extinguishes the flame, but the chamber remains red hot. As the oil replaces the water, the vapor ignites off the sides of the hot chamber. To remove the water, bleed the line.

(28) Excessive Combustion Air-Low CO_2: CO_2 reading should generally not be lower than 7% or higher than 12%. A reading within this range should generally give good smoke reading (#2). A higher smoke reading is an indication of insufficient combustion air, or insufficient draft over the fire. A lower smoke reading is an indication of excessive combustion air or excessive draft over the fire. The draft over the fire should be .02" W. C. to .04" W. C. Excessive combustion air also causes excessive stack temperature. Stack temperature should generally be 500° to 600°.

(29) Stack Temperature excessive: The most common cause of high stack temperature, which means inefficient operation and high oil bills, is excessive air. Seal up all air leaks, and adjust combustion air and draft stabalizer to obtain stack temperature of 500° to 600^{0} with smoke reading #2.

(30) When Ordering Repair Parts, please furnish to the distributor the model and series and serial numbers of the furnace. Also if you have a repair parts catalog, furnish the part number and description of the part.

(Courtesy, Sundstrand Hydraulics, Inc.).

ROUTINE PERFORMANCE CHECKS

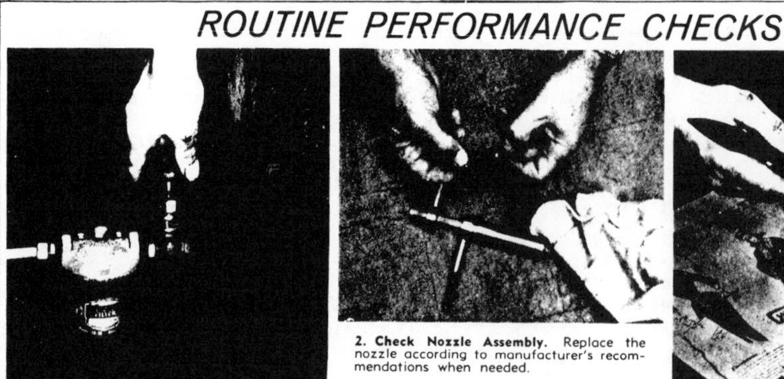

1. **Check Shut-Off Valve and Line Filter.** Replace or clean cartridge in line filter if dirty. Be sure to open shut-off valve.

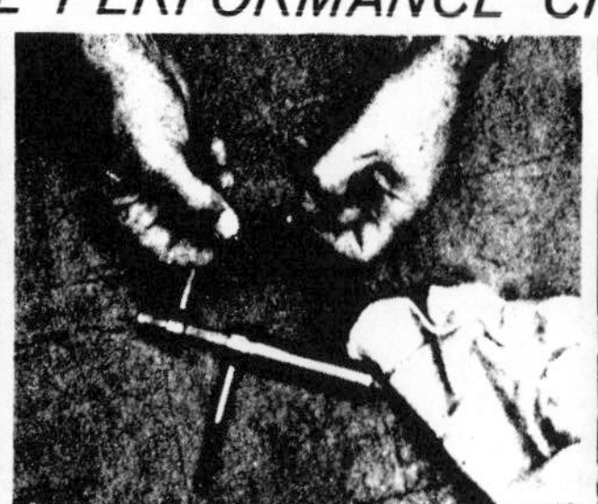

2. **Check Nozzle Assembly.** Replace the nozzle according to manufacturer's recommendations when needed.

Important:
Use proper designed tools for removal of nozzle from firing head.

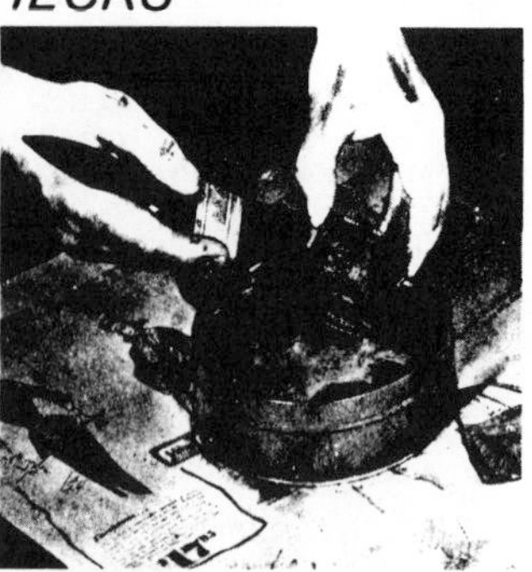

3. **Check Strainer.** Clean strainer using clean fuel oil or kerosene. Install new cover gasket. Replace strainer if necessary.

4. **Check Connections.** Tighten all connections and fittings in the intake line and unused intake port plugs.

5. **Pressure Setting.** Insert pressure gage in gage port. Normal pressure setting should be at 100 PSI. Check manufacturer's pressure setting recommendation on each installation being serviced.

6. **Insert Vacuum gage in unused intake port.** Check for abnormally high intake vacuum.

TROUBLE SHOOTING

	cause	remedy
NO OIL FLOW AT NOZZLE	Oil level below intake line in supply tank	*Fill tank with oil.*
	Clogged strainer or filter	*Remove and clean strainer. Repack filter element.*
	Clogged nozzle	*Replace nozzle.*
	Air leak in intake line	*Tighten all fittings in intake line. Tighten unused intake port plug. Tighten in-line valve stem packing gland. Check filter cover and gasket.*
	Restricted intake line (High vacuum reading)	*Replace any kinked tubing and check any valves in intake line.*
	A two pipe system that becomes airbound	*Check and insert by-pass plug.*
	A single-pipe system that becomes airbound (Model J unit only)	*Loosen gage port plug or easy flow valve and drain oil until foam is gone in bleed hose.*
	Slipping or broken coupling	*Tighten or replace coupling.*
	Rotation of motor and fuel unit is not the same as indicated by arrow on pad at top of unit	*Install fuel unit with correct rotation.*
	Frozen pump shaft	*Return unit to approved service station or Sundstrand factory for repair. Check for water and dirt in tank.*

	cause	remedy
OIL LEAK	Loose plugs or fittings	*Dope with good quality thread sealer.*
	Leak at pressure adjusting end cap nut	*Fibre washer may have been left out after adjustment of valve spring. Replace the washer.*
	Blown seal (single pipe system)	*Check to see if by-pass plug has been left in unit. Replace fuel unit.*
	Blown seal (two pipe system)	*Check for kinked tubing or other obstructions in return line. Replace fuel unit.*
	Seal Leaking	*Replace fuel unit.*
NOISY OPERATION	Bad coupling alignment	*Loosen fuel unit mounting screws slightly and shift fuel unit in different positions until noise is eliminated. Retighten mounting screws.*
	Air in inlet line	*Check all connections.*
	Tank hum on two-pipe system and inside tank	*Install return line hum eliminator, in return line.*
PULSATING PRESSURE	Partially clogged strainer or filter	*Remove and clean strainer. Replace filter element.*
	Air leak in intake line	*Tighten all fittings and valve packing in intake line.*
	Air leaking around cover	*Be sure strainer cover screws are tightened securely.*
LOW OIL PRESSURE	Defective gage	*Check gage against master gage, or other gage.*
	Nozzle capacity is greater than fuel unit capacity	*Replace fuel unit with unit of correct capacity.*

IMPROPER NOZZLE CUT-OFF

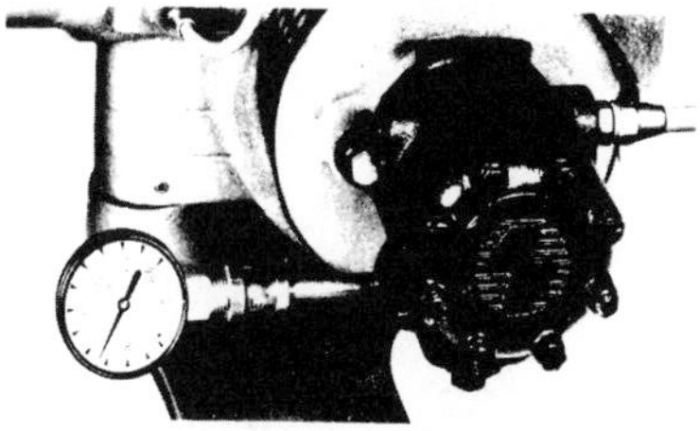

To determine the cause of improper cut-off, insert a pressure gage in the nozzle port of the fuel unit. After a minute of operation shut burner down. If the pressure drops and stabilizes above 0 P.S.I., the fuel unit is operating properly and air is the cause of improper cut-off. If, however, the pressure drops to 0 P.S.I., fuel unit should be replaced.

cause	remedy
Filter leaks	*Check face of cover and gasket for damage.*
Strainer cover loose	*Tighten 8 screws on cover.*
Air pocket between cut-off valve and nozzle	*Run burner, stopping and starting unit, until smoke and after-fire disappears.*
Air leak in intake line	*Tighten intake fittings and packing nut on shut-off valve. Tighten unused intake port plug.*
Partially clogged nozzle strainer	*Clean strainer or change nozzle.*

OIL BURNER NOZZLE
INTERCHANGE CHART

STEINEN 30° — 90°	**DELAVAN** 30° — 90°
H/PH	A
S (.5 — 2.0) or Q	A or W
S/SS (2.25 +)	B

HAGO/ [illegible] 30° — 90°	**DELAVAN** 30° — 90°
SS	A
H	A or W
ES/P	B

MONARCH 30° — 90°	**DELAVAN** 30° — 90°
R/NS/PL	A
AR	A or W
PLP	B

EFFECTS OF PRESSURE ON NOZZLE FLOW RATE

NOZZLE RATING AT 100 PSI	NOZZLE FLOW RATES IN GALLONS PER HOUR (Approx.)					
	80 PSI	120 PSI	140 PSI	160 PSI	200 PSI	300 PSI
.50	0.45	0.55	0.59	0.63	0.70	0.86
.65	0.58	0.71	0.77	0.82	0.92	1.12
.75	0.67	0.82	0.89	0.95	1.05	1.30
.85	0.76	0.93	1.00	1.08	1.20	1.47
.90	0.81	0.99	1.07	1.14	1.27	1.56
1.00	0.89	1.10	1.18	1.27	1.41	1.73
1.10	0.99	1.21	1.30	1.39	1.55	1.90
1.20	1.07	1.31	1.41	1.51	1.70	2.08
1.25	1.12	1.37	1.48	1.58	1.76	2.16
1.35	1.21	1.48	1.60	1.71	1.91	2.34
1.50	1.34	1.64	1.78	1.90	2.12	2.60
1.65	1.48	1.81	1.95	2.09	2.33	2.86
1.75	1.57	1.92	2.07	2.22	2.48	3.03
2.00	1.79	2.19	2.37	2.53	2.82	3.48
2.25	2.01	2.47	2.66	2.85	3.18	3.90
2.50	2.24	2.74	2.96	3.16	3.54	4.33
2.75	2.44	3.00	3.24	3.48	3.90	4.75
3.00	2.69	3.29	3.55	3.80	4.25	5.20
3.25	2.90	3.56	3.83	4.10	4.60	5.63
3.50	3.10	3.82	4.13	4.42	4.95	6.06
4.00	3.55	4.37	4.70	5.05	5.65	6.92
4.50	4.00	4.92	5.30	5.70	6.35	7.80
5.00	4.45	5.46	5.90	6.30	7.05	8.65
5.50	4.90	6.00	6.50	6.95	7.75	9.52
6.00	5.35	6.56	7.10	7.60	8.50	10.4
6.50	5.80	7.10	7.65	8.20	9.20	11.2
7.00	6.22	7.65	8.25	8.85	9.90	12.1
7.50	6.65	8.20	8.85	9.50	10.6	13.0
8.00	7.10	8.75	9.43	10.1	11.3	13.8
8.50	7.55	9.30	10.0	10.7	12.0	14.7
9.00	8.00	9.85	10.6	11.4	12.7	15.6
9.50	8.45	10.4	11.2	12.0	13.4	16.4
10.00	8.90	10.9	11.8	12.6	14.1	17.3
11.00	9.80	12.0	13.0	13.9	15.5	19.0
12.00	10.7	13.1	14.1	15.1	17.0	20.8
13.00	11.6	14.2	15.3	16.4	18.4	22.5
14.00	12.4	15.3	16.5	17.7	19.8	24.2
15.00	13.3	16.4	17.7	19.0	21.2	26.0
16.00	14.2	17.5	18.9	20.2	22.6	27.7
17.00	15.1	18.6	20.0	21.5	24.0	29.4
18.00	16.0	19.7	21.2	22.8	25.4	31.2
19.00	16.9	20.8	22.4	24.0	26.8	33.0
20.00	17.8	21.9	23.6	25.3	28.3	34.6
22.00	19.6	24.0	26.0	27.8	31.0	38.0
24.00	21.4	26.2	28.3	30.3	34.0	41.5
26.00	23.2	28.4	30.6	32.8	36.8	45.0
28.00	25.0	30.6	33.0	35.4	39.6	48.5
30.00	26.7	32.8	35.4	38.0	42.4	52.0
32.00	28.4	35.0	37.8	40.5	45.2	55.5
35.00	31.2	38.2	41.3	44.0	49.5	60.5
40.00	35.6	43.8	47.0	50.5	56.5	69.0
45.00	40.0	49.0	53.0	57.0	63.5	78.0
50.00	44.5	54.5	59.0	63.0	70.5	86.5

(Courtesy, Delavan Corporation).

PROPER FLOW RATES

Oil burner nozzles are available in a wide selection of flow rates, all but eliminating the need for specially calibrated nozzles. For example, between 1.00 GPH and 2.00 GPH inclusive, seven different flow rates are available. The following guidelines may be used for determining the proper flow rates:

The proper size nozzle for a given burner unit is sometimes stamped on the name plate of the unit.

If the unit rating is given in BTU per hour input, the nozzle size may be determined by . . .

$$GPH = \frac{\text{BTU Input}}{140{,}000}$$

If the unit rating is given in BTU output . .

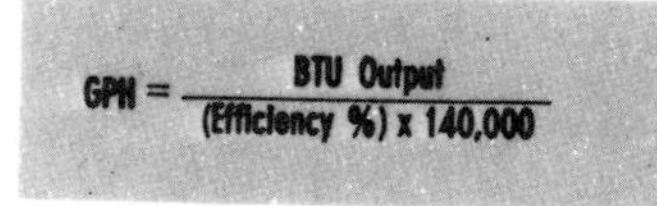

$$GPH = \frac{\text{BTU Output}}{(\text{Efficiency }\%) \times 140{,}000}$$

On a steam job, if the total square feet of steam radiation, including piping, is known.

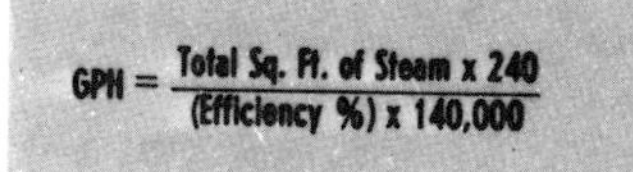

$$GPH = \frac{\text{Total Sq. Ft. of Steam} \times 240}{(\text{Efficiency }\%) \times 140{,}000}$$

If the system is hot water operating at 180° and the total square feet of radiation, including piping, is known . . .

$$GPH = \frac{\text{Total Sq. Ft. of Hot Water} \times 165}{(\text{Efficiency }\%) \times 140{,}000}$$

Generally, with hot water and warm air heat, the smallest flow rate that will adequately heat the house on the coldest day is the most economical in operation.

RECOMMENDED COMBUSTION CHAMBER DIMENSIONS

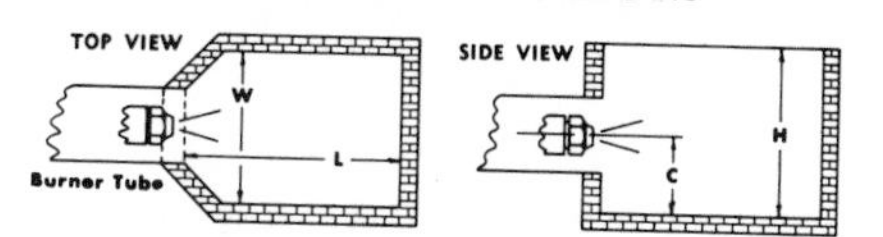

Nozzle Size or Rating (GPH)	Spray Angle	Square or Rectangular Combustion Chamber				Round Chamber (Diameter in Inches)
		L Length (In.)	W Width (In.)	H Height (In.)	C Nozzle Height (In.)	
0.50 – 0.65	80°	8	8	11	4	9
0.75 – 0.85	60°	10	8	12	4	*
	80°	9	9	13	5	10
1.00 – 1.10	45°	14	7	12	4	*
	60°	11	9	13	5	*
	80°	10	10	14	6	11
1.25 – 1.35	45°	15	8	11	5	*
	60°	12	10	14	6	*
	80°	11	11	15	7	12
1.50 – 1.65	45°	16	10	12	6	*
	60°	13	11	14	7	*
	80°	12	12	15	7	13
1.75 – 2.00	45°	18	11	14	6	*
	60°	15	12	15	7	*
	80°	14	13	16	8	15
2.25 – 2.50	45°	18	12	14	7	*
	60°	17	13	15	8	*
	80°	15	14	16	8	16
3.00	45°	20	13	15	7	*
	60°	19	14	17	8	*
	80°	18	16	18	9	17

* **Recommend oblong chamber for narrow sprays.**

NOTES: These dimensions are for average conversion burners. Burners with special firing heads may require special chambers.

Higher backwall, flame baffle or corbelled backwall increase efficiency on many jobs.

Combustion chamber floor should be insulated on conversion jobs.

For larger nozzle sizes, use the same approximate proportions and 90 sq. in. of floor area per 1 gph.

For FlameCone installations, use approximately 65 sq. in. of floor area per 1 gph.

(Courtesy, Delavan Corporation).

TROUBLE-TRACER*

ELECTRIC FURNACES

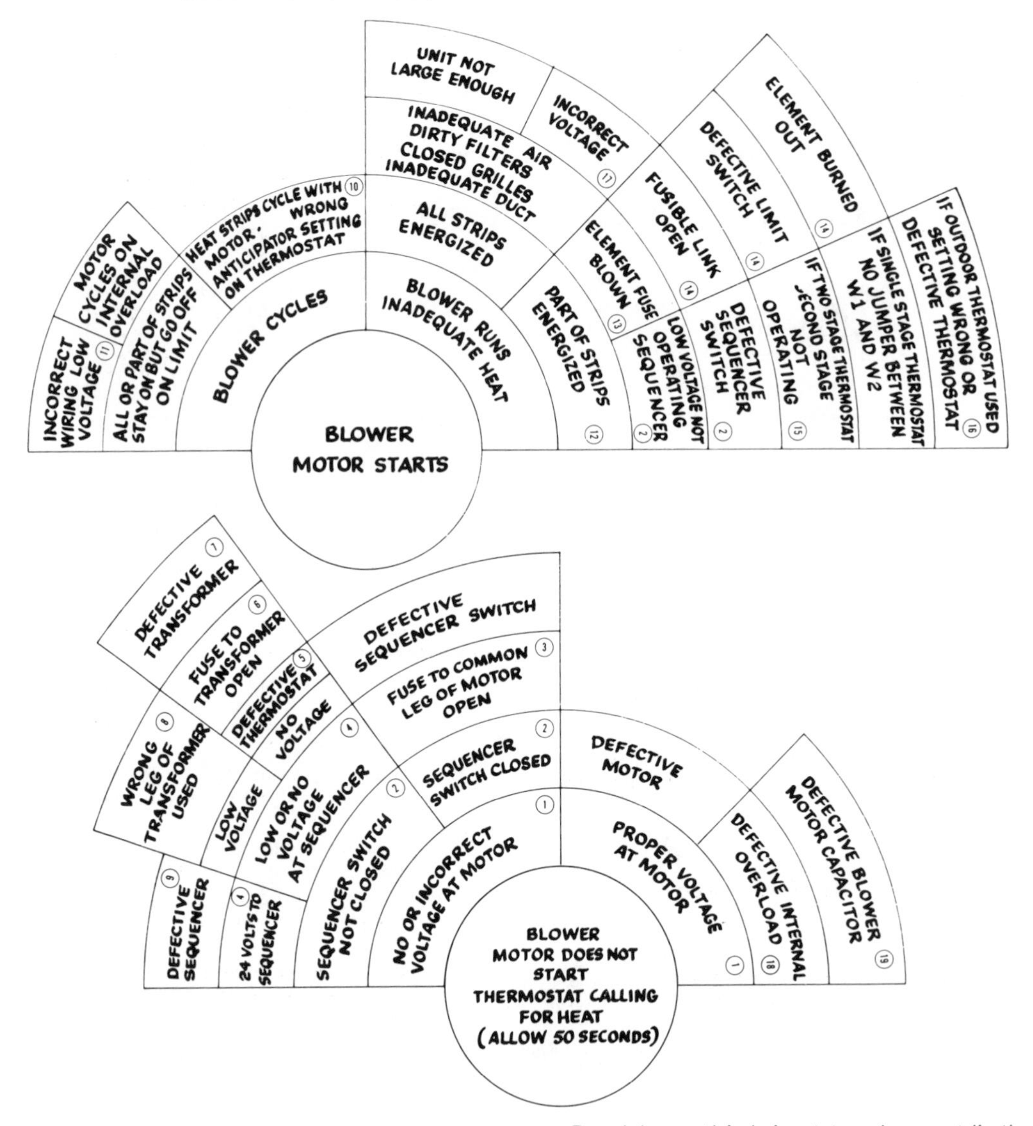

Center of chart states basic trouble. Begin there.

Adjacent partial circles state specific troubles that could cause basic trouble. Determine which of these specific troubles exist before proceeding further.

Remaining partial circles state various contributing troubles to specific trouble. Determine contributing trouble or troubles before attempting correction.

Numbers in chart are for explanation and guidance. Where no numbers are shown, the statement is considered to be self-explanatory.

(Courtesy, SJC Corp. [formerly Tappan Air Conditioning]).

1. Check voltage at motor leads. Be sure to use common (white) and lead going to sequencer switch. Black, if high speed, blue, if medium speed and red, if low speed. Voltage should be 230 plus or minus 10%.

2. Remove wires from line voltage slave switch on number 1 sequencer. Check with continuity tester to determine that switch is closed and making contact. When attaching wires, make sure they are connected to top and bottom terminals. The center post is not used.

3. Check fuse on fuse block that feeds common leg (white) to blower motor.

4. Check to see that 24 volts is present at heater terminals on first sequencer.

5. Jumper R to W_1 on terminal board. If unit starts to operate after normal delay (not more than 50 seconds), trouble is in stat or stat wiring. Next, remove stat and jumper R and W terminals. If unit starts problem is in thermostat, not in wiring.

6. Check fuses on block feeding primary side of transformer.

7. If 230 or 208 volts is being supplied to primary side, check for 24 volts on the secondary. If 24 volts is not present on secondary, the transformer is defective.

8. Check line voltage to transformer for proper primary lead connections. Yellow lead for 230V; red lead for 208V.

9. If there is 24 volts to the heater and switch does not operate, the switch is defective.

10. Check model and size of furnace. Refer to the installation instructions to determine if thermostat is set for proper heat anticipation.

11. Check low voltage wiring at terminal block.

12. Check amp draw on each strip to determine if they are heating.

13. Check fuses feeding each leg of elements not heating.

14. With a voltmeter, ohmmeter or continuity light, check fusible link, limit switch and element. Be sure to shut off power and disconnect voltage leads if using ohmmeter or continuity light.

15. Place jumper across W_1 and W_2 terminals — first on terminal board, then on stat to determine if stat is operative. If this energizes the second stage, the stat is defective.

16. Place jumper across terminals of outdoor stat. If strip energizes, stat is defective.

17. If unit is operating on 208 volts instead of 240 volts, the heating capacity is 75% of rating.

18. Shut off power, remove cap and pull out internal overload. Check terminals 1 and 3 with a continuity tester. (See figure 1).

19. Check capacitor. If capacitor is defective, shorted or burned out, motor will not run.

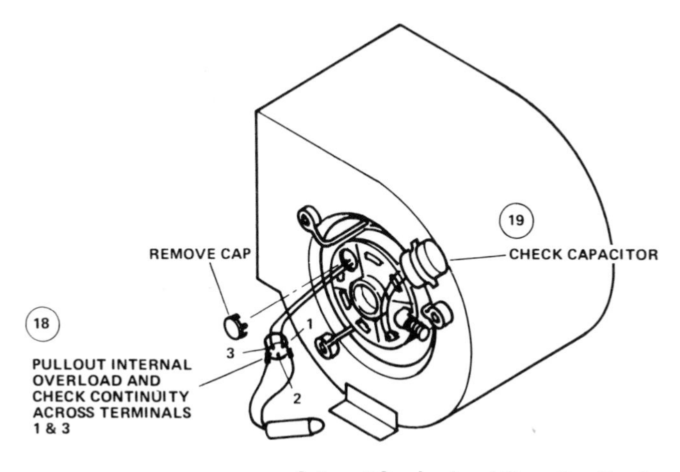

Internal Overload and Capacitor Check

HOW TO DETERMINE C. F. M. USING AMPERES AND TEMPERATURE DIFFERENCE
240 VOLT-SINGLE PHASE

CORRECTION FACTOR FOR VOLTAGE

$$\frac{\text{Actual Voltage}}{\text{Established Voltage}} = \frac{235 \text{ Volts}}{240 \text{ Volts}} \quad 0.98 \times \text{CFM}$$

235 Volts	0.98	230 Volts	0.96
225 Volts	0.94	220 Volts	0.92

AMP	Btuh	CFM 10° ΔT	CFM 15° ΔT	CFM 20° ΔT	CFM 25° ΔT	CFM 30° ΔT	CFM 35° ΔT	CFM 40° ΔT	CFM 45° ΔT	CFM 50° ΔT	CFM 55° ΔT	CFM 60° ΔT
1	819	75	50	37	30	25	21	19	17	15	14	12
2	1638	151	101	75	60	50	43	37	33	30	27	25
3	2457	227	151	113	91	75	65	56	50	45	41	38
4	3276	303	202	151	121	100	86	75	67	60	55	51
5	4095	379	252	189	151	126	108	94	84	75	68	63
6	4914	455	303	227	182	151	130	113	101	91	82	76
7	5733	530	353	265	212	176	155	132	117	106	96	88
8	6552	606	404	303	242	202	173	151	134	121	110	101
9	7372	682	455	341	273	227	195	170	151	136	124	114
10	8191	758	505	379	303	252	216	189	168	151	137	126
15	12286	1137	758	568	455	379	325	284	252	227	206	190
20	16382	1516	1011	758	606	505	433	379	337	303	275	253
25	20478	1896	1264	948	758	632	541	474	421	379	344	316
30	24573	2275	1516	1137	910	757	650	568	505	455	413	379
35	28669	2654	1769	1327	1061	884	758	663	589	530	482	442
40	32764	3033	2022	1516	1213	1011	866	758	674	606	551	506
45	36860	3412	2275	1706	1365	1137	975	853	758	682	620	569
50	40956	3792	2528	1896	1516	1264	1083	948	842	758	689	632
55	45051	4171	2780	2085	1668	1390	1191	1042	926	834	758	695
60	49147	4558	3033	2275	1820	1516	1300	1137	1011	910	827	758
65	53242	4929	3286	2464	1971	1643	1408	1232	1095	985	896	822
70	57338	5309	3539	2654	2123	1769	1516	1327	1179	1061	965	885
75	61434	5688	3792	2844	2275	1896	1625	1422	1264	1137	1034	948
80	65529	6067	4045	3033	2427	2022	1733	1516	1348	1213	1103	1011
85	69625	6446	4297	3223	2578	2148	1824	1611	1432	1289	1172	1074
90	73720	6826	4550	3413	2730	2275	1950	1706	1516	1365	1241	1138
95	77816	7205	4803	3603	2882	2401	2058	1801	1601	1441	1310	1201
100	81912	7584	5056	3792	3033	2528	2166	1896	1685	1516	1378	1264

A. Volts x amps = Watts

B. Watts x 3.413 = BTUH

C. $\text{Heating CFM} = \frac{\text{BTUH Capacity}}{1.08 \times (T2\text{-}T1)}$

D. BTUH Capacity = CFM x 1.08 x (T2-T1)

E. $\Delta T = \frac{\text{BTUH Capacity}}{1.08 \times \text{CFM}}$

ΔT = Temperature rise across unit

Note: The above values are additive in the table.

(Courtesy, Luxaire, Inc.).

GUIDE FOR SIZING FUSES

(Courtesy, Bussman Manufacturing, Div. of McGraw-Edison Company).

Dual-Element Time-Delay Fuse

● **Main Service**—Size fuse according to method in ●

● **Feeder Circuit With No Motor Loads.** The fuse size must be at least 125% of the continuous load† plus 100% of the non continuous load. Do not size the fuse larger than the ampacity of the conductor*.

● **Feeder Circuit With All Motor Loads.** Size the fuse at 150%♦ of the full load current of the largest motor plus the full load current of all other motors.

● **Feeder Circuit With Mixed Loads.** Size fuse at sum of
a. 150%♦ of the full load current of the largest motor.
b. 100% of the full load current of all other motors.
c. 125% of the continuous non-motor load†.
d. 100% of the non-continuous non-motor load.

● **Branch Circuit With No Motor Load.** The fuse size must be at least 125% of the continuous load† plus 100% of the non continuous load. Do not size the fuse larger than the ampacity of the conductor*.

● **Motor Branch Circuit With Overload Relays.** Where overload relays are sized for motor running overload protection, the following fuses provide back-up protection and short-circuit protection:
a. Motor 1.15 service factor or 40°C rise: size fuse at 125% of motor full load current or next higher standard size.
b. Motor less than 1.15 service factor or over 40°C rise: size the fuse at 115% of the motor full load current or the next higher standard fuse size.

● **Motor Branch Circuit With Fuse Protection Only.** Where the fuse is the only motor protection, the following fuses provide motor running overload protection and short-circuit protection:
a. Motor 1.15 service factor or 40°C rise: size the fuse at 110% to 125% of the motor full load current.
b. Motor less than 1.15 service factor or over 40°C rise: size fuse at 100% to 115% of motor full load current.

● **Large Motor Branch Circuit—**Fuse larger than 600 amps. For large motors, size KRP-C HI-CAP time-delay Fuse at 150% to 225% of the motor full load current, depending on the starting method; i.e. part-winding starting, reduced voltage starting, etc.

Non-Time-Delay Fuse

● **Main Service**—Size fuse according to method in ●.

● **Feeder Circuit With No Motor Loads.** The fuse size must be at least 125% of the continuous load† plus 100% of the non continuous load. Do not size the fuse larger than the ampacity of the wire*.

● **Feeder Circuit With All Motor Loads.** Size the fuse at 300% of the full load current of the largest motor plus the full load current of all other motors.

● **Feeder Circuit With Mixed Loads.** Size fuse at sum of
a. 300% of the full load current of the largest motor.
b. 100% of the full load current of all other motors.
c. 125% of the continuous non-motor load†.
d. 100% of the non-continuous non-motor load.

● **Branch Circuit With No Motor Load.** The fuse size must be at least 125% of the continuous load† plus 100% of the non continuous load. Do not size the fuse larger than the ampacity of the conductor*.

● **Motor Branch Circuit With Overload Relays.** Size the fuse as close to but not exceeding 300% of the motor running full load current. This fuse size provides short-circuit protection only.

● **Motor Branch Circuit With Fuse Protection Only.** Non-time-delay fuses **cannot** be sized close enough to provide motor running overload protection. If sized for motor overload protection, non-time-delay fuses would open due to motor starting current. Use dual-element fuses.

Conductor Ampacity Selection

● **Feeder Circuit And Main Circuit With Mixed Loads.** Conductor ampacity at least sum of:
a. 125% of continuous non-motor load†.
b. 100% of non-continuous non-motor load.
c. 125% of the largest motor full load current.
d. 100% of all other motors' full load current.

● **Feeder Circuit With No Motor Load.** Conductor ampacity at least 125% of the continuous load† plus 100% of the non-continuous load.

● **Feeder Circuit With Motor Loads.** Conductor ampacity at least 125% of the largest motor full load amperes plus 100% of all other motors' full load amperes.

● **Feeder Circuit And Main Circuit With Mixed Loads.** Conductor ampacity at least sum of:
a. 125% of continuous non-motor load†.
b. 100% of non-continuous non-motor load.
c. 125% of the largest motor full load current.
d. 100% of all other motors' full load current.

● **Branch Circuit With No Motor Load.** Conductor ampacity at least 125% of the continuous load† plus 100% of the non-continuous load.

●, ●, & ● **Motor Branch Circuits.** Conductor ampacity at least 125% of the motor full load current.

†100% of the continuous load can be used rather than 125% when the switch and fuse are listed for continuous operation at 100% of rating. Most bolted pressure switches and high pressure contact switches 800A to 6000A with Class L fuses are listed for 100% continuous operation.

*Where conductor ampacity does not correspond to a standard fuse rating, next higher rating fuse is permitted when 800 amperes or less (240-3, Exc. 1).

♦In many motor feeder applications, dual-element fuses can be sized at ampacity of feeder conductors.

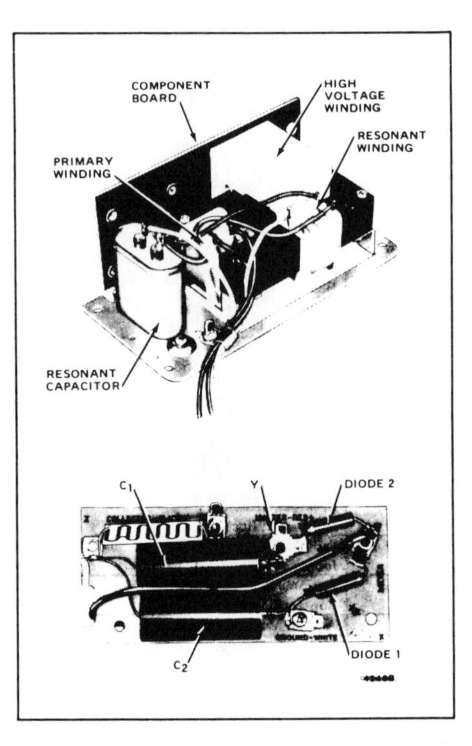

GENERAL DESCRIPTION

The F50 Electronic Air Cleaner replaced the F45 and F46 and was introduced in 1972 for use in central systems with capacities up to 2000 cfm. It uses the FC37A Electronic Cells.

INTERNAL SCHEMATIC DIAGRAM

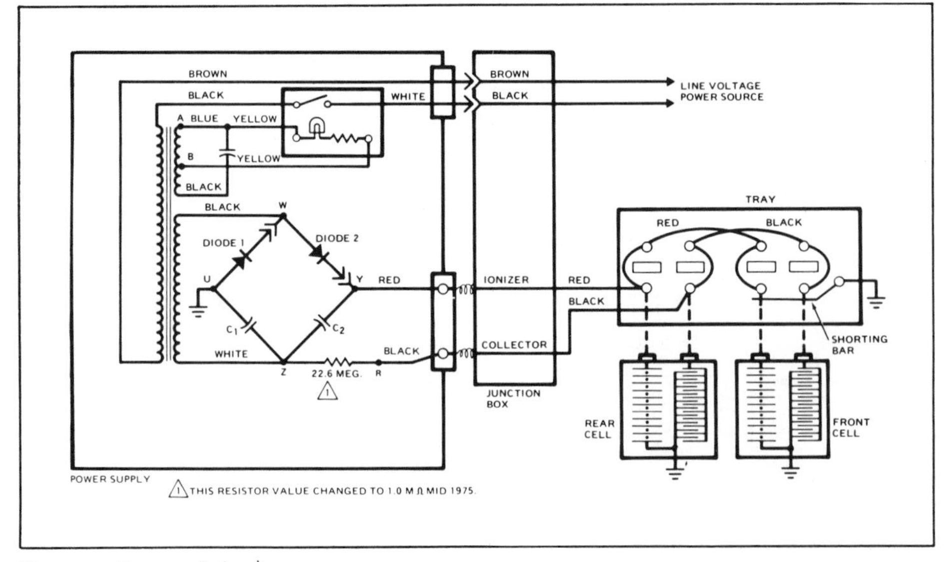

(Courtesy, Honeywell, Inc.).

CHECKOUT AND TROUBLESHOOTING

SERVICE AIDS

☐ High voltage test meter–ac and dc.
☐ Spare diode (137073A) with alligator clip.

PREPARATION

1. Check to see that the electronic cells are clean, dry, and properly installed in the air cleaner cabinet.

2. To energize the air cleaner, turn on the system fan and turn air cleaner switch ON.

NORMAL VOLTAGES

	WITH CELLS	WITHOUT CELLS
Ionizer	7,500–8,500V dc	8,500–9,600V dc
Collector	3,000V dc minimum	3,500–4,800V dc
Transformer Secondary	–	3,200–4,100V ac

ELECTRICAL TROUBLESHOOTING

Check out the electrical components of the F50 by observing the indicator light under various operating conditions. Follow the flow chart and refer to the instructions for checking components when indicated.

CHECK ELECTRONIC CELLS

When diagnostic checks indicate a possible problem in the electronic cells, inspect them carefully for any sign of mechanical damage. Check for short circuits from contacts to ground.

CHECK VOLTAGE DOUBLER CIRCUIT

1. Check voltage across each capacitor with opposite diode unplugged. If both capacitor and diode are good, the voltage will be over 3,500V dc.

2. Check diode by substitution and if the voltage still isn't right, replace the capacitor.

CHECK HIGH VOLTAGE TRANSFORMER

1. Disconnect black wire from the resonant capacitor.

2. Energize the air cleaner and measure *resonant winding* voltage.

   a. If this voltage is over 150V ac, the resonant capacitor is defective or the transformer secondary winding is open.

      (1) De-energize the power supply and check for continuity of the secondary winding.

   b. If this voltage is under 150V ac, the transformer is defective and must be replaced.

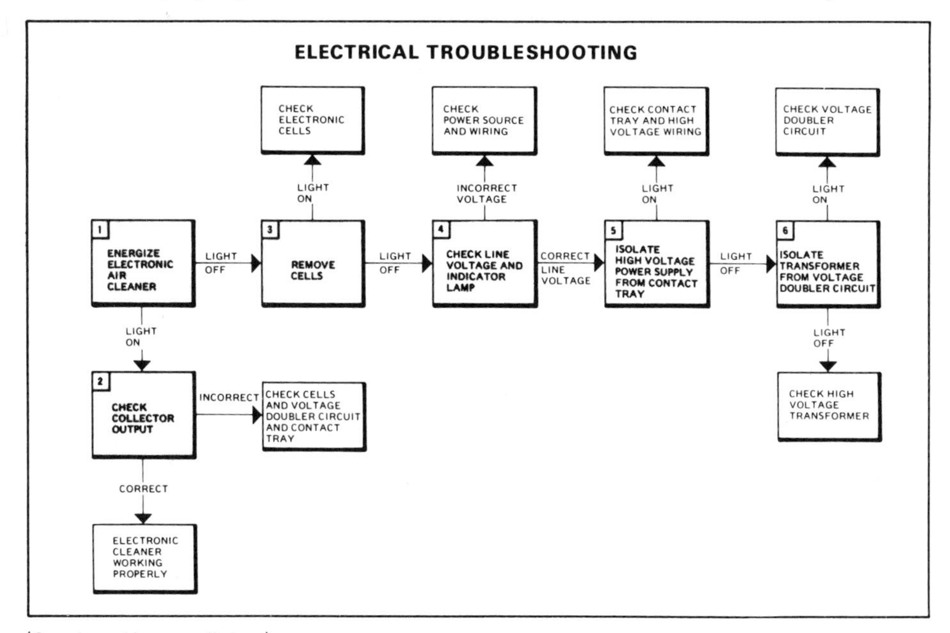

(Courtesy, Honeywell, Inc.).

GLOSSARY

AC:	Auxiliary Contacts
AMP:	Amperes
BK:	Black—wire color
BL:	Blue—wire color
BTU:	British Thermal Unit
BTUH:	British Thermal Units per hour
C:	24V Common connection
°C:	Degrees Celsius
CAD:	Cad cell
CAP:	Capacitor
CC:	Compressor Contractor
CCH:	Compressor Crankcase Heater
CFH:	Cubic feet per hour
CFM:	Cubic feet per minute
CO:	Carbon Monoxide
CO_2:	Carbon Dioxide
CR:	Control—Relay
CU:	Cubic
DPST:	Double-pole single-throwswitch
DPDT:	Double-pole double-throwswitch
E:	Electromotive force—volt
EAC:	Electrostatic air cleaner
EP:	Electric Pilot
°F:	Fahrenheit
FC:	Fan Control
FD:	Fuse Disconnect
FL:	Fuse Link
FM:	Fan Motor
FR:	Fan Relay
FT:	Foot
FU:	Fuse
G:	Green—wire color
G:	Neutral—Ground
Gal:	Gallon
GFCI:	Ground fault circuit interrupter
GV:	Gas value

H:	Humidistat
HI:	High speed
H_2O:	Water
HR:	Heat relay
HS:	Humidification system
HU:	Humidifier
I:	Ampere
IFR:	Inside fan relay
L:	Limit
LA:	Limit auxiliary
LO:	Low speed
L_1 L_2:	Power supply
N:	Nitrogen
N:	Neutral
NC:	Normally closed
NO:	Normally open
Ω:	Ohms
O:	Oxygen
OBM:	Oil burner motor
OL:	Overload
OT:	Outdoor thermostat
PPC:	Pilot power control
PSC:	Permanent split phase capacitor
R:	Red—wire color
R:	Resistance—Ohms
RH:	Resistance heater
S:	Cad cell relay terminal
SOL:	Solenoid valve
SPDT:	Single-pole double-throwswitch
SPST:	Single-pole single-throwswitch
SQ:	Sequence
T:	Thermostat terminal
TFS:	Time fan start
THR:	Thermocouple
TR:	Transformer
V:	Volt—electromotive force
W:	White—wire color
W1:	First stage heat—thermostat
W2:	Second stage heat—thermostat
wc:	Water column in inches
Y:	Yellow—wire color
Y,R,G,W:	Thermostat terminals

ANSWERS TO REVIEW QUESTIONS

CHAPTER IN BOOK

QUESTION	1	2	3	4	5	6	7	8	9	10	11	12	13	14	15	16	17	18	19	20	21	22
1	A	A		B	C	B	C	A	D	C	E1	E1		C	C	D	B	B	B	C	C	C
2	C	A		D	C	C	A	C	D	B	J2	H2		C	B	C	B	D	D	A	B	C
3	B	B	HEAT LOAD COMPUTATION	C	B	A	D	B	B	A	K4	A3	WIRING DIAGRAMS	C	D	B	A	C	C	D	C	E
4	C	D		A D	A	D	A	A	B	B	B6	I4		A	B	A	B	B	C	A	B	C
5	C	A		A	D	B	C	D	A	B	A7	B5		B	A	C	A	B	C	C	A	A
6	B	D		C D	B	C	B	C	B	A	H8	K6		A	C	B	B	A	A	B	C	A
7	C	B		D	D	D	A	A	C	A	G9	J7		D	A	B	D	B	A	B	D	B
8	A	C		A	A	B	D	B	A	D	I10	C8		A	D	B	B	A	D	D	D	A
9	B	C		D	B	A	D	D	A	C	D12	D10		B	B	C	B	C	B	C	C	D
10	B	A		A	A	A	A	C	B	B	L13	G14		B	D	D	A	B	D	D	A	A
11	•	•	•	C	•	•	•	•	•	•	F14	L15	•	•	•	B	•	•	•	•	•	•
12	•	•	•	D	•	•	•	•	•	•	C16	F16	•	•	•	B	•	•	•	•	•	•

INDEX

C

D

E

F

G

H

I

K

L

M

N

O

P

R

S

T

U

V

W